S. Dillon Ripley, Secretary of the Smithsonian Institution,
with René Dubos (right) at a symposium session.

How Humans Adapt
A Biocultural Odyssey

DONALD J. ORTNER, *Editor*

Foreword by **S. Dillon Ripley**

Epilog by **Wilton S. Dillon**

Smithsonian Institution Press
Washington, D.C.

Library of Congress Cataloging in Publication Data.

Main entry under title:
How humans adapt.
   (Smithsonian international symposia series)
   Includes bibliographical references and
index.
   Supt. of Docs. no.: SI 1.2:H88
   1. Sociobiology—Congresses.   2. Adaptation
(Biology)—Congresses.   3. Human evolution
—Congresses.   4. Man—Influence of
environment—Congresses.   I. Ortner, Donald J.
II. Series.
GN365.9.H68 1982   304.5   82-600233
ISBN 0-87474-726-0
ISBN 0-87474-725-2 (pbk.)

Other books in the Smithsonian International Symposia
Series

Simon and Schuster:
Knowledge Among Men (1966)

Smithsonian Institution Press:
The Fitness of Man's Environment (1968)
Man and Beast: Comparative Social Behavior (1971)
The Cultural Drama: Modern Identities and Social Ferment
(1974)
The Nature of Scientific Discovery (1975)
Kin and Communities: Families in America (1979)

The Smithsonian Institution gratefully acknowledges financial support of the seventh international symposium received from the following benefactors, without whose generous assistance this program could not have taken place:

Charles F. Kettering Foundation
Ellis L. Phillips Foundation
Exxon Education Foundation
Fogarty International Center, NIH
Government Employees Insurance Company
Joseph H. Hazen Foundation, Inc.
National Institute of Child Health and Human Development
Rockefeller Foundation

Contents

Epilog

Foreword

S. DILLON RIPLEY
Secretary, Smithsonian Institution

In 1848 Joseph Henry, the first Secretary of the Smithsonian, launched a publishing enterprise, "Smithsonian Contributions to Knowledge," which continued up through its thirty-fifth volume in 1916 and still inspires us today.

It is not only for nostalgia for the greater wholeness of scholarly outlook in the nineteenth century that we began, in 1966, the Smithsonian's International Symposia Series with *Knowledge Among Men.*[1] The Office of Smithsonian Symposia and Seminars, organizer of this international symposia series, was invented as our version of a College of Liberal Arts. Through it, we wanted to restore communications among specialists who had begun to lose a sense of community. We also wanted to help restore the art of writing essays. And we are pledged to keeping our trusteeship of Mr. Smithson's ecumenical legacy that, I think, includes an enduring effort to help integrate data and insights from both the sciences and the humanities.

Knowledge, the *raison d'être* and prime commodity of the Smithsonian, is increased and diffused by various means. The vital advent of the electronic media should not atrophy the older modes of face-to-face meetings of scholars, their patrons, and the users of their knowledge. As a research institute and center of learning, the Smithsonian attracts to its museums, laboratories, and lecture halls scholars from around the world to meet with our own. We pool our intellectual resources to take inventories of old questions and new questions. We pose these inquiries before scholars from diverse fields of research who otherwise, because of the extreme specialization today, might not meet. The Smithsonian serves as an intellectual free-trade zone. We strive to be a theater of ideas.

St. Thomas Aquinas's *Summa Theologica* would be a too-ambitious prototype for what we do in the international symposia series. Our prose has little to do with syllogisms. Yet, we attempt to find some rudimentary forms of synthesis or generalization through shifting a concept or an artifact from one context to another, turning them around so that they might attract light from various angles. We dare to show curiosity outside our fields of expertise and even enjoy the audacity to touch upon questions of ethics and the links between scholarship and public policy. The republic of learning includes both scholars and public servants. The mixture of "hard data," "hard choices," and speculation about the past and future continues to spark some exciting discourse manifest in the essays and commentaries of this particular volume on "How Humans Adapt." How do we adapt—to disease, malnutrition, population pressures, and depleted resources? These are not casual questions for those of us who love and celebrate life. What are the limits of adaptation? The questions need to be asked often.

As a zoologist, it is particularly interesting for me to see so well articulated the nature of human adaptations—physiological, biological, cultural—and to note how these differ from adaptations among animals that lead to the same results— population control, resource utilization, social organization. In spite of the fascinating work done in the last few years by sociobiologists,[2] it remains clear that the distinction between man and the other animals remains the possession of free will—our ability to plan, discover, and invent the mechanisms that, in turn, affect the course of our evolution. Many contributors to this symposium point out that our evolution has not kept pace with the speed at which we have altered our culture and the demands we place on our environment to maintain this culture. Adaptations for population growth and resource utilization that will allow mankind to live in a stable environment are now at least several hundred years behind the adaptations that enable us to exploit the environment in a manner that leaves fewer and fewer satisfactory options for long-term survival. The more light symposia such as this one can shed on how our forebears, including *Homo erectus* and *Homo habilis* as well as Paleolithic, Pleistocene, and recent *Homo sapiens*, adapted to their environments, the better equipped we will be to face our future.

Adaptations of hunter-gatherers, such as late adolescence, teenage sterility, low fertility, and wide birthspacing, are, one might say, just what we need today to dampen some of our own most pressing, seemingly nonadaptive, social problems. We see that some of these adaptations survive in contemporary hunter-gatherers, such as the Yanomama of Venezuela and Brazil, where the number of offspring that survive to maturity is, to a degree, correlated inversely with fertility.

The advent of agriculture reversed the adaptive value of many critical aspects of hunter-gatherer culture. The demand for labor, not the increase in available food, best explains the growth in populations at this time, according to Fekri Hassan. Cavalli-Sforza notes that, even today, African Pygmies, who love children as much as do agriculturalists, do not ascribe any economic value to them and continue to have fewer than do their agricultural neighbors.

Cavalli-Sforza also points out that agricultural systems allow for an increased rate of cultural evolution in that they encourage one-to-many transmissions of information. This contrasts with the many-to-one and one-to-one transmissions typical of hunter-gatherers, which, just as in genetic evolution, are less likely to lead to change.

Today, when one-to-many transmissions have achieved unprecedented dimensions, we find that they are too little used to encourage the evolution of human adaptations in directions that may enhance the chances of our long-term survival. Ayensu, for example, notes that today nearly half of all human food energy and proteins is derived from only three plants—wheat, rice, and corn. Not only are population and increasing energy demands destroying the areas where the wild ancestors of these plants may offer us the genetic diversity future crops will require, but cultural prejudices, adaptive in the past but no longer so, are making it difficult for peoples surrounded by untapped sources of nutrition to learn to exploit them.

I believe the overriding conclusion of this symposium is that modern culture cannot sustain the reproductive behavioral patterns that were suited to the first agricultural societies. Not only must new food and energy sources be found (or perhaps be more effectively utilized), but mankind must accelerate the rate of adaptive evolution to live harmoniously on earth. This will require not only the technological adaptations at which our species has become so facile, but also, as Midgley points out, a reexamination of our concepts of individualism and of the rights and obligations of every member of society.

In *Man and Beast*[3] I wrote, without reference to nuclear war as a technique of reducing our numbers, that "The greatest problem today is human population growth. The next greatest problem is our lack of innate or learned constraints in regard to our biosphere—that thin envelope of water, land, and air within which we live. If religion fails us in imposing constraint, if our culture encourages waste and destruction in the name of an expanding economy, then we need a new morality. Let there be a new print-out in the programmer before it is too late, and let it start somewhere in the mother's womb or just thereafter—and let it say, "Nature is my friend." The biocultural odysseys outlined by René Dubos and James V. Neel, among others, in this volume reinforce my views, moreover, that if we fear nature and have no innate constraints, then we will destroy nature. We must not pit the optimists against the pessimists but try to mobilize that "free will" that sets us apart from other creatures and provides a basis for seeing ourselves alive in our children and their grandchildren.

Notes

1. Paul H. Oehser, ed. 1966 (New York: Simon and Schuster).
2. E. O. Wilson's *Sociobiology* and *On Human Nature* reflect his participation in the Smithsonian's 1969 symposium, "Man and Beast: Comparative Social Behavior," which he has described as "a major intellectual event in recent decades." The public and scholarly interest in his provocative propositions is reflected, moreover, in publishing records having been broken by the Harvard University Press in the sale of his work on the as yet unfinished debate about nature versus nurture, ie., what is inherited and what is learned.
3. Eisenberg, J. F., and W. S. Dillon, eds. 1971. *Man and Beast: Comparative Social Behavior* (Washington, D.C.: Smithsonian Institution Press).

S. DILLON RIPLEY

Preface

One of the major doctrines of science and scholarship is that the new builds on the old—that the process of learning is cumulative. The fact that each generation of human society seems only to rediscover the wisdom of Plato and repeat many of the mistakes of human history should not obscure the fundamental truth of this observation. The work of a scholar is to make his ideas accessible to others so the ideas may be tested, evaluated, added to, accepted, modified or rejected, as appropriate. The genesis of the Seventh International Symposium and this book is a microcosm of this process.

In 1970 I read an essay in *Science* (vol. 170, pp. 815-22) titled *Lessons from a "Primitive" People,* written by James V. Neel. This essay impressed me for two reasons. First, it represented a succinct, clearly stated report of what was obviously an enormous amount of careful research by many scholars over a long period of time. But beyond this, the essay was an attempt to transcend the usual, restricted bounds of scientific reports and offer some modest speculations on the broader implications of the research conducted by Neel and his colleagues. On the basis of his research on the genetic relationships among the Yanomama Indians of South America, Neel offered some suggestions on problems and strategies which human societies should consider for the future.

From Neel's essay came the germ of the idea for a conference in which a group of scholars would take a broader look at the implications of their research on various aspects of the human experience. In the spring of 1977 I mentioned this idea to Wilton S. Dillon, director of the Office of Smithsonian Symposia and Seminars, who was searching for a theme to develop into the Institution's seventh international symposium. He was at once receptive and enthusiastic about taking a broad multidisciplinary look at the process of human biocultural adaptation and the implications of what we know of this process for contemporary and future human society. The concept was approved by the Advisory Committee appointed by the Secretary of the Smithsonian Institution to oversee the general planning of the office, and from the beginning it had the keen interest and support of the Secretary, S. Dillon Ripley.

The process of refining this idea into a workable topic and a series of essays for a symposium was, by its nature, more than could reasonably be expected of a single scholar, certainly more than my own knowledge would permit. We thus called on several distinguished colleagues for advice both in fairly formal planning conferences and informal discussions. The first of these conferences was held on October 25, 1978, and included the following participants: J. Lawrence Angel, Geoffrey Best, Roy Branson, J. Grahame D. Clark, Wilton S. Dillon, John F. Eisenberg, William H. McNeill, Stuart Marks, James V. Neel, and Donald J. Ortner. Dorothy Richardson, Carla Borden, and Barrick Groom of the Office of Smithsonian Symposia and Seminars also attended. The result of this conference was a further refinement of the symposium objectives and some tentative suggestions regarding essay topics and essayists. Following the conference we established a working group which included James V. Neel, Symposium Chairman; Stephen Toulmin, Symposium Rapporteur; Donald J. Ortner, Symposium Editor; and Wilton S. Dillon and Dorothy Richardson of the Office of Smithsonian Symposia and Seminars. The symposium was held at the Smithsonian Institution November 8 through 12, 1981.

From the early stages of our planning we wanted to keep the number of participants under fifty in order to enhance discussion and debate. Our hope was to have essays circulated prior to the symposium, with a brief summary by the essayist during the symposium followed by critical reviews by commentators and general discussion by all participants. The essays, as well as the edited commentaries and discussions, were to be published as the proceedings of the symposium. We encouraged a writing style for the essays which would be appropriate for the general, serious reading public as well as for the academic community.

Even with the relatively modest scale proposed for the symposium, the projected cost was great. I am greatly indebted to the Herculean efforts of Wilton S. Dillon and the staff of his office, who raised the necessary funds for the symposium. I specifically wish to acknowledge the efforts of Dorothy Richardson, Carla Borden, Barrick Groom, and Helen Leavitt, all of the Office of Smithsonian Symposia and Seminars, for their efforts in logistic support. Without experiencing it as closely as I have done, it is difficult to imagine the planning and time it takes to coordinate all the details necessary for a successful international symposium. To my editorial and administrative assistant, Mary D. Bancroft, I owe a great debt. Her sensitivity to the subtleties of English grammar and usage was a major help, as was her administrative experience and skill.

Many other staff members associated with the Smithsonian Institution have contributed to the success of the symposium and the published proceedings. They include: Marcia Bakry, Marguerite Brigida, Bruno Frohlich, Robert Haile, James Hobbins, Katherine Holland, Richard Howland, George R. Lewis, Barbara Russell, Agnes Stix, and David Yong. Recognition is due Felix Lowe, Maureen Jacoby, Lawrence Long, Stephen Kraft, and Kathleen Brown of the Smithsonian Press for their excellent cooperation and assistance before the symposium and especially for expediting prompt publication of the proceedings. To Hope Pantell goes special thanks for her excellent editorial work.

In editing the proceedings of the symposium I have encountered several problems and have had to live with what might be considered unsatisfactory solutions. All participants emphasized the need for timely publication. All essays were in hand by the time of the symposium; subsequently some authors made changes and I suggested a few changes which were given appropriate consideration by the authors. Our stenographic service produced a typewritten transcript of the commentary and discussion within twenty-four hours, so we were able to have some of the material reviewed by the participants. While the transcription was remarkably good there were some problems which are understandable given the conditions of varied speech patterns and background noise levels. This is particularly true with regard to the transcription of the discussions. The requirement of prompt publication has not permitted as much review by commentators and discussants as would be necessary for maximum accuracy. I have, however, compared the audio tape with the typewritten transcript of the discussion and corrected some of the problems. I have also been granted liberal editorial license by the participants to edit the commentaries and discussions. Despite my efforts, I fear that I may have, in some cases, done an injustice to the meaning and intent of the commentator or discussant. For this, of course, I must take responsibility.

Varying word meanings also posed a hazard in a group as diverse as the symposium participants. Even the words "adapt" and "adaptation" meant different things to different participants. As a human biologist I use the term to describe a process by which a group of people achieve an optimum interaction with their biocultural environment. Used in this sense the term is descriptive and lacking in any connotation of value. Some of our participants used or perceived the term in the more conventional usage, in which to adapt implies accommodation, plasticity, perhaps even acquiescence to conditions which might be unfavorable. With this connotation, the meaning of the terms takes on attributes which can be viewed as good or bad, and indeed at least one of our participants expressed strong feelings against any process involving adaptation used in the sense of acquiescence.

Despite these problems, the vigor of the ideas, the contrasting and clarifying opinions, the occasional humor, as well as the stimulating dialogue, are apparent in the proceedings. I have not managed to make the proceedings as accessible to the general reader as I would have liked. Some essays and discussion will be easier to follow than others. As an aid to the reader I have prepared brief editorial summaries which highlight the major ideas presented in each essay. I hope that some, perhaps even most, of the excitement and stimulation of the symposium will be evident in the edited proceedings and that the reader will be more aware of the challenges and problems that confront human society as we explore, litter, learn, fail to learn, experiment, and otherwise make our way toward the twenty-first century.

I wish to note with deep regret the deaths of René Dubos and Theodore Wertime shortly after the symposium.

Donald J. Ortner
Editor, *Symposium Proceedings*
May 1982

List of Participants

Essayists

Edward S. Ayensu, Director
Office of Biological Conservation
Smithsonian Institution

Richard J. Barnet, Senior Fellow
Institute for Policy Studies

Kenneth E. Boulding
Professor Emeritus of Economics
University of Colorado at Boulder

Lord Asa Briggs, The Provost
Worcester College
Oxford University

George Morrison Carstairs
Woodrow Wilson International
    Center for Scholars
Smithsonian Institution

L. L. Cavalli-Sforza
Professor of Genetics
Stanford University Medical Center

James M. Gustafson
Divinity School
University of Chicago

Fekri A. Hassan
Department of Anthropology
Washington State University

Jane B. Lancaster
Department of Anthropology
University of Oklahoma

Peter Laslett
Fellow of Trinity College
Cambridge University

Geoffrey McNicoll
Deputy Director & Senior Associate
Center for Policy Studies, The
    Population Council

Betty Meggers
Department of Anthropology
Smithsonian Institution

Mary Midgley
Lecturer in Philosophy
University of Newcastle upon Tyne

Moni Nag
Senior Associate
Center for Policy Studies, The
    Population Council

James V. Neel
Lee R. Dice University Professor of
    Human Genetics
University of Michigan Medical
    School

Donald J. Ortner
Department of Anthropology
Smithsonian Institution

John A. Passmore
The Research School of Social
Sciences
Australian National University

Nevin S. Scrimshaw, Director
International Food & Nutrition
Program
Massachusetts Institute of
Technology

Stephen Toulmin
Committee on Social Thought
The University of Chicago

Commentators

J. Lawrence Angel
Department of Anthropology
Smithsonian Institution

Allen B. Bassing
Department of Education
National Museum of American Art
Smithsonian Institution

Mary Catherine Bateson
Dean of Faculty
Amherst College

Wilfred Beckerman
Balliol College
Oxford University

Anna K. Behrensmeyer
Department of Paleobiology
Smithsonian Institution

Roy Branson
Senior Research Scholar
Kennedy Institute
Georgetown University

Claire Cassidy
Department of Anthropology
University of Maryland

Napoleon A. Chagnon
Department of Anthropology
Northwestern University

Mark Nathan Cohen
Department of Anthropology
State University of New York

James Frederic Danielli
Danielli Associates
Worcester, Massachusetts

Audrey B. Davis
Department of the History of
Science and Technology
Smithsonian Institution

Wilton Dillon, Director
Office of Smithsonian Symposia and
Seminars
Smithsonian Institution

Ido de Groot
Faculty of Medicine
University of Cincinnati

John F. Eisenberg
Department of Zoological Research
National Zoological Park
Smithsonian Institution

John C. Ewers
Ethnologist Emeritus
Department of Anthropology
Smithsonian Institution

Stuart Marks
Department of Anthropology
Saint Andrews Presbyterian College

Roger Masters
Department of Government
Dartmouth College

Nathan Reingold
Joseph Henry Papers
Smithsonian Institution

Bruce D. Smith
Department of Anthropology
Smithsonian Institution

Paula J. Thompson
Director of Education
Brooklyn Botanic Garden

Elizabeth Torre
School of Social Work
Tulane University

Marx Wartofsky
Department of Philosophy
Boston University

Andrzey Wiercinski
Department of Historical
  Anthropology
University of Warsaw

Theodore Wertime
Department of Anthropology
Smithsonian Institution

Session Chairmen

James V. Neel
Lee R. Dice University Professor of
  Human Genetics
University of Michigan Medical
  School

Donald S. Fredrickson
Scholar in Residence
National Academy of Sciences
Former Director
National Institutes of Health

Caryl P. Haskins
Former President
Carnegie Institution of Washington
Regent Emeritus
Smithsonian Institution

Philander P. Claxton, Jr.
President
World Population Society

Harry Woolf
Director
The Institute for Advanced Study

Ernest L. Boyer
President
Carnegie Foundation for the
  Advancement of Teaching

Prolog:

Technological and Social Adaptations to the Future

RENÉ DUBOS
Professor Emeritus, The Rockefeller University

I shall devote most of my remarks to the creative effects of adaptation, but I must first acknowledge that many objectionable aspects of human life are the consequences of the ease with which we usually can adapt to undesirable conditions. The dangers of adaptation can be illustrated by many different medical examples: we adapt to air pollution by physiological mechanisms that eventually result in chronic pulmonary disease; we adapt to noisy environments by anatomical changes that result in impairment of hearing; we adapt to crowded environments by developing behavioral patterns that result in an impoverishment of the human encounter; most urban dwellers readily become adapted to starless skies, treeless avenues, shapeless buildings, tasteless bread, joyless celebrations. In other words, we adapt to conditions that lower the quality of life.

I could go on and on to illustrate that much of what we call adaptation is in reality a tolerance of bad conditions, achieved at the cost of some biological, psychological, or social loss. The prevalence of such mutilations is evidence that the wisdom of the body and the mind is all too often a short-sighted wisdom.

There are many examples of dangerous adaptations. I shall mention two which reveal that societies, like individual human beings, commonly respond to threatening situations by attitudes or measures which appear adaptive at first but are destructive in the long run.

Every sensible person knows that the inevitable result of nuclear warfare would be not only immense suffering for humans and immeasurable damage to every living and inanimate thing on earth, but also the virtual collapse of Western civilization. In an official document just published concerning the effects of the atom bombs on Hiroshima and Nagasaki, it is shown that the number of deaths up to 1950 was about 200,000 in Hiroshima alone. The postwar nuclear arms race has resulted in the accumulation of weapons thousands of times more powerful than those used in Japan and stockpiles with an explosive power millions of times greater. Even while peace lasts, the cost of the nuclear and antinuclear arsenals is so great that it prevents the development of much-needed social programs. Yet, except for a few manifestations here and there and now and then, most of us behave as if we were not much concerned about the possibility of nuclear warfare, certainly less than we are, for example, about the questionable role of saccharin in the causation of cancer. We have become adapted to the astronomical cost and the potential threats of nuclear warfare simply by not letting these dangers be conscious preoccupations of our minds. The only reason to retain some hope is the increase in the intensity of antinuclear manifestations in Europe and Japan during recent months.

Another situation to which many people display little concern is massive unemployment, especially of youth. Yet youth unemployment is, in my opinion, the greatest social tragedy of peacetime. I am as ignorant concerning social techniques that could solve the unemployment problem as I am about solutions to nuclear disarmament, but I can at least express opinions that differ somewhat from those that govern present attitudes and policies.

It is generally assumed that the plight of unemployed young people can be made bearable by providing them with some form of welfare help, including entertainment, but this seems to me an almost irrelevant approach to the problem. Human beings always function as parts of structured social groups. If they are not given the opportunity to function in normal society by being meaningfully employed, they will organize themselves in social groups of their own—as young unemployed people are doing now—a situation that will inevitably lead to destructive social conflicts. Despite this enormous potential threat, we behave as if we had achieved a kind of social adaptation to youth unemployment by providing the unemployed with food and shelter— a shameful way of avoiding the responsibility of dealing with one of the deepest needs of human nature, namely, socialization. Yet the success of several "youth conservation programs," either publicly or privately financed, leaves no doubt that even delinquent youth can be reintegrated into normal society by being given the opportunity for meaningful work.

As a last example of our frightening ability to adapt to bad conditions,

I shall merely mention that the increase in terrorism has been readily absorbed by the social order and even by the stock market—a chilling reminder of the desensitization of peoples and their institutions. There is nothing like repetition, no matter how outrageous the events, to dull the senses.

I shall now turn to creative aspects of adaptation and shall particularly emphasize technological and social adaptations to the future.

At first sight, the phrase "adaptation to the future" appears nonsensical. In order to adapt to any physicochemical or social change, it would seem necessary that we be actually exposed to the changed conditions—for example, to cold, heat, crowding, solitude, poverty, or any stressful experience that requires some form of adjustment. How then can we possibly adapt to the future since we have not experienced the conditions that it will bring about?

The fact is, however, that we can and often do adapt *biologically* to future situations, because our bodily and mental mechanisms can be activated by the processes of imagination. The mere thought that we shall have to run in order to meet an appointment or catch a train causes our heart to beat faster and modifies some of our metabolic mechanisms. Similarly, various physiological and mental processes undergo profound changes when we know, or simply fear, that we shall have to deal with difficult emotional situations at the office or at some social gathering. Consciously or more often unconsciously, we thus adapt biologically and mentally to the future almost every moment of our life by *anticipating* the challenges likely to result from an expected situation and by mobilizing in advance mechanisms that may help us to cope successfully with these anticipated challenges.

I shall not deal further with the purely biological and mental aspects of *individual* adaptation to the future. Instead, I shall show that modern societies are learning to anticipate some of the likely consequences of actions they are contemplating and of the natural and social conditions to which they are likely to be exposed in the future. I must first emphasize, however, that anticipating is not the same as predicting the future. Prediction is impossible because it would require complete knowledge of past and present circumstances, as well as the ability to achieve complete control over all human beings and the rest of nature. By contrast, we can increasingly anticipate the long-range consequences of the courses of action we envisage.

Since the ability to anticipate consequences often enables modern societies to deal adaptively with the problems they will face in the future, I am inclined to believe that the "future shocks" that have been predicted as an almost inevitable consequence of technological and social innovations will not be as painful as commonly feared—

precisely because we will have anticipated and even experienced many of them in our minds and also because we can take in advance measures that will prevent undesirable consequences from occurring, at least in the form in which they have been predicted.

I shall now mention a few examples selected to illustrate that advanced industrial societies have actually begun to undergo changes that make them better adapted, not only to present situations, but also to the anticipated consequences of situations that have not yet occurred.

There are great tragedies in the world today. Paradoxically, however, much of contemporary gloom comes not from actual tragic situations but from the prospect of social and technological difficulties that have not yet occurred and may never materialize. We are collectively worried because we anticipate that, if demographic and technologic growth continues at the present rates, the earth will soon be overpopulated and its resources depleted. There will be catastrophic food shortages; pollution will alter the climate, spoil the environment, poison us, rot our lungs, and dim our vision. I believe, as do many others, that industrial civilization will eventually collapse *if* we do not change our ways—but what a big *if* this is.

Human beings are rarely passive witnesses of threatening situations. Their responses to threats may be unwise, but they inevitably alter the course of events and make mockery of any attempt to predict the future from extrapolation of existing trends. In human affairs, the *logical* future, determined by past and present conditions, is less important than the *willed* future, which is largely brought about by deliberate choices—made by the human free will. In my opinion, our societies have a good chance of remaining prosperous because they are learning to anticipate, long in advance, the shortages and dangers they *might* experience in the future if they did not take adequate preventative measures.

For example, the North American continent is far from overpopulated. Its population density is much lower than that of Europe and Asia. Yet, the fear of overpopulation in the *next century* has been one of the factors responsible for the decrease in average family size. Birthrates are below replacement rates in most American social groups. They have also fallen in many other countries, and the one-child family is being advocated in China. Although—barring ecological disasters—the world population will certainly continue to increase for a few decades, the increase will not be as rapid as was believed only twenty years ago. Indeed, the world population is likely to reach a plateau in the course of the next century.

Environmental degradation became a widely recognized danger only during the 1960s, and most antipollution programs are barely ten years old. Yet, many are the places in which environmental quality has already been vastly improved. Air pollution has markedly decreased in several large European, American, and Asian cities, for example, in London and even in Tokyo. Several streams and lakes that were so grossly polluted as to be qualified "dead" a decade ago have been brought back to a level of purity compatible with a rich and desirable aquatic life. A fish as ecologically exacting as the salmon has returned to the Thames in London and to the Seine in Paris. Various species of finfish and shellfish, including oysters, can once more be harvested in the Jamaica Bay of New York City.

Forests that had been devastated are being allowed to recover spontaneously and massive reforestation programs are being carried out in several parts of Africa and Asia, especially in China. Areas that had become desertic are reacquiring a diversified flora and fauna as a result of being protected against browsing, for example, in Israel and even in some parts of North Africa and the Sahel. While environmental degradation is still increasing in many parts of the earth—and is particularly alarming in the tropical rain forest as well as in areas undergoing desertification or being affected by acid rains—there are signs that modern societies are learning to work with nature so as to minimize the loss of agricultural soil and of biological species.

The most interesting aspect of the environmental improvements that have occurred during the past ten years is that many of them resulted from measures taken long before environmental degradation had become extreme. The air of most cities was not really poisonous at the time when steps were taken to decrease urban air pollution. Most lakes and rivers were far from dead at the time control of water pollution was begun. Environmental policies were therefore responses, not to *actual* emergencies but rather to the *anticipation* of emergencies.

Furthermore, adaptation to the future is beginning to be conceived as going beyond preventing or correcting dangerous situations. The phrase "good environment" is no longer taken to mean only freedom from noxious influences. It implies also surroundings that provide emotional, aesthetic, and social satisfactions. City planners used to be almost exclusively concerned with problems of public health and with greater mobility and efficiency in the various aspects of urban economic life. They are now beginning to emphasize, in addition, factors that contribute in other ways to the quality of life, such as the improvement of parks and waterfronts and the role of "city centers" in the furtherance of social and cultural activities.

Adaptation to the future is also apparent in our concern for natural resources, as illustrated by the case of copper. Rich copper ores have become scarce and, even though methods have been worked out to use ores of low grade, a shortage of this metal appeared likely some ten years ago. Long before there was any real shortage, however, technologists developed substitutes for copper and established that aluminum could serve as such in many types of uses, including telephone wires, with, of course, appropriate modifications. Furthermore, it was recently discovered that glass optical fibers, and even synthetic fibers, can be used for the transmission of messages and are indeed superior to copper in some respects. Cables of glass optical fibers are now used in several urban telephone systems. Copper is still as important as in the past in many technologies, but a shortage of this metal is no longer an immediate danger.

There are many other examples of possible replacement of a particular metal by another one or by some synthetic product. We have entered the "age of substitutability." Whether we like it or not, plastics are likely to become more and more common in many different aspects of our lives. Admittedly, the use of substitutes often entails greater expenditure of energy but, as we shall now see, adaptation to the future is also taking place in the production and use of energy.

The evolution of policies for the production of energy constitutes a spectacular example of the rapid adaptive changes that commonly occur now in technological societies. When it became apparent that the supplies of certain fossil fuels were being depleted, especially with regard to petroleum and natural gas, industrial countries turned to nuclear reactors as sources of energy. It soon became apparent, however, that the natural supplies of uranium also are limited. In view of this fact, attention was focused on the breeder reactor, which produces nuclear fuel while generating energy. In several countries, the highest priority was therefore given to theoretical and practical research on the breeder reactor technology.

During the 1960s and 1970s, however, several groups of scientists and citizens took a stand against the breeder program, not necessarily because they doubted its technical or economic feasibility, but because they were alarmed by the dangers inherent in its operation—in particular by the inevitable accumulation of plutonium. Whatever its merits and dangers, the breeder program thus provides a striking illustration of the rapid effect of social forces on the evolution of energy policies.

During the early 1970s, more than 90 percent of the funds for energy research were earmarked for the nuclear program. There has not yet been any catastrophic accident resulting from this program (not even at Three Mile Island). From many points of view, in fact, nuclear technologies have been *so far* the safest method for the production of energy. Yet, public concern for the dangers inherent in them has

RENÉ DUBOS

profoundly affected many aspects of energy research, production, and use. In the budgets of all industrialized countries, an ever- increasing percentage of the funds for research is now allocated to energy sources that used to be neglected: solar, wind, waves, tides, geothermal, biomass, and so forth.

It can truly be said that the recent evolution of research policies for the production of energy has been an adaptive process, determined not by the existence of shortages or by actual disasters but by the *anticipation* of shortages and of disasters that have not yet materialized.

As a last example, which can serve as a caricature illustrating how our societies are trying to adapt to the future, I shall mention a biological problem that has recently caused a great deal of public alarm. Scientific techniques now make it possible to modify the hereditary genetic constitution of microbes. From this limited scientific fact, many persons have concluded that it will eventually be possible to modify also the genetic constitution of human beings. Such genetic engineering of humans is not possible now, and many are the scientists who share my view that it may never be possible on any significant scale. Yet, several institutes of bioethics have been created in which physicians, biologists, sociologists, jurists, and theologians discuss the medical, ethical, legal, and theological aspects of any change that might eventually be brought about in some physical or mental aspect of human nature—by genetic manipulations yet to be imagined. Thus, we try to adapt not only to the future by anticipating the changes that might be brought about by scientific technology, but also to changes imagined by the authors of science fiction.

I realize, of course, that the ability to anticipate long-range consequences does not mean that modern societies will always act early and vigorously enough to prevent the disastrous effects of certain courses of action. Pessimists have good reasons to believe that some day, somewhere, an innovation will be carried so far that it will cause irreversible damage to the human species or to global ecology. I find hope against the likelihood of a catastrophe—following an overshoot— in the phenomenal resiliency of natural systems and of human beings. By cultivating the anticipation of consequences and thus adapting to the future, our societies will be able to overcome the myth of inevitability.

I also believe that adaptation to the future can have highly *creative* effects. For example, the fear that our societies will soon become unmanageable because of their excessive complexity is beginning to encourage the design of smaller communities and enterprises with dimensions and values that are more humane and easier to apprehend.

Creative adaptations to the future will increasingly become the mechanisms by which human beings can invent new social structures and ways of life.

There is a widespread tendency, and I have been guilty of it in this essay, to emphasize chiefly the technical aspects of the future, but the aspects bearing on ideologies, philosophies, religions, and arts are at least as important as those derived from science and technology, because they are the ones that give a favorable direction to human activities and to the growth of knowledge. The word "favorable," however, inevitably implies parascientific questions of value.

Modern societies know how to solve, or can learn to solve, most of the technical problems that preoccupy them today and that they assume to be the determinants of their future. But the really important problems of our times have their origin in our uncertainties or poor judgment concerning values—a fact that makes us accept the possibility of nuclear war for reasons of national prestige and the tragedy of youth unemployment for the sake of economic considerations. Social adaptations to the future should therefore be focused on the formulation of value systems suitable to new ways of life, yet compatible with the unchangeable needs of human nature and with human aspirations.

Parascientific values make human life qualitatively different from animal life. Admirably adapted as animals are to the natural habitats in which they have evolved, they are prisoners of Darwinian evolution, which is irreversible and which determines where and how they must live. Human beings, by contrast, are blessed with the freedom and flexibility of social evolution, which is almost always reversible. While retaining the biological characteristics of the species *Homo sapiens*, we human beings have been able to adapt, socially, to many different ways of life. We have been members of roving bands, of sedentary villages, of towns, and of cities. We have lived in cloistered religious or scholarly communities. Today, as in the past, some of us function as hunter-gatherers, as pastoralists, as farmers, as sailors, as artists, as factory workers, as industrial managers, or as reclused scholars.

Through history, and also prehistory, we have had the freedom to choose our course. We have often been able to change direction and even reverse our steps in order to reach the goals we had first selected or new ones we select upon further reflection. In the words of Charles Kettering, "Nothing ever built arose to touch the skies unless some man dreamt that it should, some man believed that it could, and some man willed that it must." The *deterministic* future operates in human life as it does in other forms of life, but we have continuously and increasingly supplemented it by a *willed* future that we invent to fit our values and aspirations.

8   RENÉ DUBOS

We are still on the way, renewing and enriching ourselves by moving on to new places and experiences. Wherever human beings are involved, social adaptations and evolution make it certain that *trend is not destiny*, because life starts anew, for us, with each sunrise. In the words of one of my favorite French writers: "Demain, tout recommence."

The Symposium Essays:

1. The Natural Past and the Human Future: An Introductory Essay

STEPHEN TOULMIN
Member, Committee on Social Thought, University of Chicago

Editor's Summary.   Toulmin's essay places the issues of the symposium in the broader context of philosophy, asking whether modern human society has gotten out of phase with the biocultural checks and balances of the past—to the point where the future of humanity as well as other life forms is threatened. He discusses various modes of biocultural adaptation, ranging from laissez faire to highly interventionist practices of high technology, and emphasizes the need for greater social responsibility.

The task of this introduction is to pose the general questions for this Smithsonian symposium as cogently and precisely as possible, without trespassing too grossly onto other essayists' topics. We shall be facing issues of many distinct but interconnected kinds, scientific and human issues, historical and anthropological ones, theoretical and practical, ethical and political, philosophical and even theological issues; and our discussion will move between two historical poles or termini.

The terminus *a quo* is the situation in which human beings existed during the million and more years of nomadic life that ended ten or fifteen thousand years ago. The terminus *ad quem* is the situation of humanity today, in which we are compelled to recognize the limits to further growth and, in the light of that recognition, find ways of taking intelligent command over our own future, as well as that of the other living creatures with whom we share the terrestrial habitat.

In mapping the biological and cultural journey by which, in the course of the intervening millennia, humanity passed from the first of these conditions to the second, we shall have to consider the following questions, among others:

(1) How far did the original nomadic condition of humanity embody a genuine balance or harmony between the human mode of life and its habitat? And how far, by contrast, did it already involve minor imbalances whose effects on the overall human population would in due course necessitate the transitions from nomadism to sedentism, urbanism, industry, and scientific technology?

(2) What distinctive changes have characterized each of the major transitions in subsequent human history: from the great self-sustaining bureaucratic structures of the ancient Middle Eastern and Oriental empires, by way of the medieval Agricultural Revolution and the subsequent transformations from the sixteenth century on, down to the widespread industrial exploitation of scientific technology that began only during the present century?

(3) Have these changes carried human beings away from a stable and harmonious state of nature into a state of disharmony with—if not actual threat to—the processes of the natural world? Or have they, rather, required humans to create quite new kinds of social and cultural equilibria, where previously there had been no such stability or harmony?

(4) In what respects, in particular, are the activities of existing human populations truly "out of harmony" with the natural habitat in ways that may threaten the survival of the species? And in what respects, by contrast, have human beings already begun to control their population size, and other modes of adaptation, in ways that were never effective during the previous history of nature?

(5) Finally, what further "adaptive procedures" can be devised, through which humans can improve the relationship between themselves and their habitat? And what new risks will they run, if they fail to devise such novel procedures of adaptation?

By way of preface, let me concentrate here on two preliminary sets of issues —one historical, the other analytical. Historically speaking, our question, "How do humans adapt?" is only a present day restatement of a much older question; and the problems created by the impact of human agriculture, industry, and population growth on the natural habitat have their own proper need to be seen in their relation to several long-established intellectual traditions. To begin with, they are one contemporary expression of a long-standing debate about humanity's place in, and responsibility for, the world of nature.

STEPHEN TOULMIN

Current disputes about exploitation and conservation, for instance, echo older themes of stewardship and domination, which were already familiar in the Middle Ages. The same contrast also reflects alternative cosmological views about the basic character of the created world itself.

Some of our contemporary problems have been in dispute ever since Stoics and Epicureans disagreed about them in late classical antiquity. Is it our duty as humans to keep our lives in harmony with the processes of nature? And do we offend against our own proper status as human beings if we allow our modes of life and conduct to run counter to, and even to disrupt, those natural processes? (So argued the Stoics some 2,000 years ago and their voice is familiar to anyone who follows the contemporary discussion of environmental issues.) Or, alternatively, do the activities of human beings differ in crucial respects from the processes of nature? Do social and cultural activities take place on a level, and involve concerns, quite foreign to the phenomena and mechanisms of the nonhuman or prehuman world? (To the Epicureans, the forces of nature were, for good or for ill, wholly indifferent to the interests of humanity; and this Epicurean voice, too, is familiar in the contemporary debate about economic development and natural resources.)

Analytically speaking, meanwhile, the questions posed for discussion in this symposium rest on the idea that human beings can arrive at their proper place in the natural world, and solve their resulting problems, by "adapting" to that world. We should take care to ask, in response, just how the term "adapting" is to be understood in this context.

- What is involved in "adapting?"
- How do we know when something is "well adapted?"
- What makes a mechanism or relationship an "adaptive" one?
- What criteria of success or failure are relevant to human "adaptation?"

These questions need to be elucidated in any cogent and coherent account of human adaptation, whether biological or cultural or (as here) "biocultural." Further, we must here disentangle the chief ambiguities embodied in the terms "adapt," "adapted," "adaptation," and so on. Some four or five distinct modes of adaptation are, in fact, embodied in human life and history, so that any comprehensive account of human adaptation must be *multi-modal*.

Finally, throughout the area that we are exploring in this symposium, considerations of human nature and human culture overlap so substantively that we can no longer effectively separate them. How far, then, is the human species "well adapted" to its terrestrial habitat? How we answer that question depends, to a great extent, on how we choose to understand the term "adapted"; and recognizing how that term can best be interpreted, as it applies to human beings and human societies, turns out to be a task of some delicacy.

Has Humanity Fallen from Natural Grace? Nothing in the world (as Bertrand Russell liked to remark) has the same power to convince human beings of God's existence, omniscience, and perfect benevolence as the afflictions of war, pestilence, and natural disaster. By some quirk of thought, we recognize blessings withheld better than blessings forthcoming. The same quirk works also in reverse: if failures highlight our full human potential more effectively than successes, the successes alert us to the limits of human power more effectively than failures. So, though in our own century we have mastered smallpox and controlled many other infectious diseases, have made striking improvements in agricultural productivity and infant survival rates, and have brought the convenience of electric power to all parts of the world, we are only now fully ready to listen to the prophets of ecological doom.

A symposium on human adaptation must, therefore, have two faces. The more obvious and fashionable aspect is the *negative* one. A sense that we live in a threatened world, whose resources are limited, fragile, and at immediate risk, links many groups today: the industrialists and economists in the Club of Rome, the demographers and anthropologists who appreciate the drastic effects of reduced death rates on the patterns of traditional life through the growth of vast and unmanageable new cities, and the environmental biologists who understand what over-cultivation can do to regions of marginal agriculture and how poisonous chemicals are contaminating our food and drinking water.

Yet the same coin has another, less gloomy face. Alongside all these legitimate fears, there has grown up a novel recognition that our destiny is largely in our own hands. Unthinking expansion and uncontrolled appetites may have put at risk both human life and the biosphere itself, but intelligent analysis and conscious human restraint may undo that damage. Indeed, the damage can perhaps be undone only in that way.

This second thought rests on one *positive*, and less often noticed, conviction. Until very recently, even in the most developed societies, people regarded the traditional human scourges—wars and depressions, starvation and infant mortality, leprosy and venereal disease—as natural and inescapable afflictions, no more open to human intervention than storms and droughts. Fluctuations in the business climate were suffered without despair or even complaint, like hurricanes; while loathsome diseases were even welcomed, Job-like, as the interest payable to the Divine Bookkeeper on account of our moral debts. For many of us today, this attitude is no longer possible. Some may view the undiscriminating use of new medical and technical skills with mixed feelings, as a threat to the natural conditions of human survival, but those powers have also had a positive value, by undercutting the earlier apathy or resignation, and strengthening our hope of rising to new levels of collective intelligence and personal self-command.

I myself would go further. We not only possess that ability, we also have the basic will—or at least the basic desire—to use it. In this direction, the full resources of human charity and sympathy have barely been touched, far less stretched. The more we learn how to do things for the good of our fellow humans, the less content are we that these good things should be left undone. In point of theory, no doubt, we might hold the demographic threat at arm's length by denying modern obstetrical and gynecological skills to a sufficient part of mankind as a deliberate act of policy; but we all know quite surely that, in practical terms, any such suggestion is unthinkable. Nor can we in the more developed countries tolerate mass starvation in any part of humanity, without reflecting on the obligations that it creates for us. By consciously refusing to demand more of ourselves, we would make ourselves less human. In the long run, therefore, the realistic question for policy analysts is not *whether* we are to do better, but only *how* we are to do so.

Callousness is the stepchild of despair. Once we have acknowledged how far the future destiny of our planet and species depends on our intelligence and will in the present, we can no longer leave that destiny to Thomas Malthus's grim reapers: war, starvation, and epidemic disease. On the other hand, the practical exercise of human will and intelligence takes time, and that fact itself has consequences. The greater our appetite for improvement, the greater is our potential for frustration, impatience, and ill-considered decisions. Having lifted our eyes to the hills, we can scarcely wait to get there. So, what we have yet to do can breed guilt about what we have not yet done; and this, in turn, exposes us to the danger of premature despondency, cargo cults, and conspiracy theories. The United States of America is "the Great Satan"; the moral fiber of the richer nations is "weakened by affluence and permissiveness"; the poorer nations "are being robbed blind by predatory multinationals"; and so on. The traditional resources of pathological guilt and collective anxiety bear their poisoned fruit of resentment and self-doubt, and we can hardly keep our attentions focused on the central problems of the human agenda.

It should be no surprise, therefore, if the problems of human destiny and direction are couched and discussed nowadays in more comprehensive, cosmological, and even eschatological terms than has been customary for some time. Viewed on a worldwide scale, problems of agricultural economics cease to be problems only of agricultural economics; rather, they become one aspect of our global ecological destiny. Population planning and control may call, in practice, for a mixture of fiscal and social incentives, but in their larger significance for humanity they are no longer matters of fiscal or social policy alone. Energy production, similarly, creates major tasks for technology and finance, but it involves also some crucial issues about patterns of human settlement, environmental damage, and international—not to say, interspecies—equity. None of these problems can be solved by technical specialists alone; in this area, interdisciplinary analysis is no longer an intellectual luxury, but has become a practical necessity. That is why this symposium has brought together such a varied group of speakers.

At the heart of the whole symposium lies, in fact, a cosmological question. A century ago, human thinking in the developed countries was dominated by the Idea of Progress; today, we are more inclined to believe in a Human Fall from Natural Grace. For many centuries, human beings saw themselves either as the rulers, or as the fulfillment, of the Natural Order; either as having Dominion over the Lower Creation or, more recently, as Victors in the Evolutionary Contest. Now, some members of the human species are ready, almost for the first time, to see our species not as the ruler or savior of nature, but as its curse. Where nineteenth-century thinkers were confident about the human role in the world, we have moved close to despair. Formerly, human beings never imposed insupportable demands on the natural world, but now we seem to be monopolizing its products, disrupting its processes, and exhausting its resources. Somehow (it is argued) human policies and practices have fallen out of step with natural processes. The old harmony between Humanity and Nature has been lost.

This belief, in the destructive consequences of human life on the natural world, is one twentieth-century counterpart to the belief in the Fall. In some preceding Golden Age, the human species kept within its natural niche and took its tithe from the rest of the living creation, but never threatened to engulf it. At some point in history, however, everything began to change. Through greed, human beings overreached their habitats, and so upset the balances of nature. The goats introduced into North Africa in late antiquity ringbarked the trees, so that the Sahara overwhelmed the former granary of the Roman Empire; and

STEPHEN TOULMIN

the legendary Gardens of the Hesperides inland from where Benghazi now stands quickly became brackish lagoons. By now, indeed, we ourselves are endangering not only our own habitats, but also those of all our fellow species. In short, the human species is no longer properly "adapted to" the situation in which it lives.

Our forebears exploited the resources of Earth, but never to the point of destruction. Above all, the physical traces and chemical residues they produced were strictly localized, and barely affected their fellow species. In all these respects, therefore, human life was biologically "adapted to" its terrestrial situation. More recently, human activities, human technologies, and even human cultures seem radically to have changed the demands that Humanity places on Nature. As a result, we are overpowering the "adaptive mechanisms" which used to keep those effects in check. So, the central issue for our human agenda, on both a theoretical and a practical level, comes to be formulated in the question, "How can human life be brought back into mutual adaptation with the resources of the terrestrial habitat?"

The foundations of this argument are by no means beyond dispute. On the contrary, there are some real grounds for doubt about them and one of our foremost tasks must be to consider how far the supposed contrast between an earlier, natural condition of "adaptive equilibrium" and the newer, humanly created state of "biospheric catastrophe" is a valid contrast. Are the earlier and the later phases in Humanity's relations with Nature truly so different in kind? Or does the current situation only continue (while perhaps accelerating) processes that were present all along? Whichever way we answer, this argument remains a useful starting point. Valid or fallacious, clear or confused, for good or for ill, we must come to terms with this picture of the natural past and the human present. We need to frame our policies for the terrestrial future in terms that allow both Humanity and Nature—both Humanity and *the rest of* Nature—their own proper places and prospects.

In passing, let me touch briefly on some general issues. Our present questions rise above all academic dichotomies between the human sciences and the natural sciences. A century ago it was natural for the founders of the human sciences to welcome Wilhelm Dilthey's distinction between the explanatory sciences of nature (*Naturwissenschaften*) and the interpretative sciences of human conduct and culture (*Geisteswissenschaften*). For some sixty to eighty years, the new fields established themselves alongside, but independently of, the older and more self-confident physical sciences. Still, it was one thing to *distinguish* the human from the natural sciences, or to claim a proper methodological independence for them; it was quite another to *separate* them, or to imply that human activities can be substantively independent of their natural contexts. So, during the last twenty years, there has been a fresh move to reunify the human and natural sciences, spurred by a new appreciation of the biological significance of social behavior and the ecological impact of human activities. The topics of this symposium thus fall squarely across the accepted boundaries between *Naturwissenschaft* and *Geisteswissenschaft*: they are concerned equally with the natural significance of culture and with the cultural relevance of nature.

The current picture of Humanity's Fall from Nature embodies certain ambiguities. For instance, when exactly is that Fall supposed to have taken place? Different people would place the crucial transition at different times. Some see the most damaging encroachments by humanity on nature as very recent. For them, all went reasonably well until scientific technology led to the bulk release of toxic substances into the atmosphere and waterways, and until modern scientific medicine generated an explosive growth in human population.

Others place the transition a full century or more back. So long as human beings disposed only of the physical strength of domestic animals, they could do no irreversible damage, and the trouble became significant only when they harnessed first steam power, and later, electrical and nuclear power, to their industrial processes. Still others locate the essential change nearer to 1,000 years ago. The use of iron axes and plowshares to clear and cultivate the fertile plains of Northern Europe led to the agricultural surplus that made possible the growth of cities, the development of mercantilism and capitalism, the emergence of a "middle class" and the secular intellectual renaissance responsible for modern science.

Finally, there are those for whom the crucial transition took place very much earlier. For more than a million years, our forebears were adapted to a nomadic life of hunting and gathering, but some 10,000 years ago the available hunting grounds could no longer support a continually growing population and some humans were forced to adopt a new, settled way of life based on domestication and agriculture. This enabled them to produce, laboriously, the food that was no longer there for the taking—and from that moment the pressure was on. The move toward disharmony had begun.

This disagreement leaves it quite unclear, also, what the supposed "disharmony" between Humanity and Nature consists in. Was this dislocation associated with the excesses of modern industrial technology or with our great- great-grandfathers' mechanical modes of production, with the late medieval turn to modernization, or with the prehistoric change from nomadism to sedentism? Alternative diagnoses

STEPHEN TOULMIN

imply different prescriptions. Here the antitechnological radicals of the New Left clearly differ from the machine-breaking Arcadians; Jacques Ellul from William Morris; the Ayatollah Khomeini from the author of *Genesis*; and the advocates of nuclear fusion or planetary colonization from the apostles of Zero Population Growth.

The ambiguities about the contrast between a Golden Age of natural harmonies, and a subsequent Leaden Age of human disruption, may lead us to wonder whether this contrast is really as sharp as it seems. Can we assume that for millions of years the early nomadic hominids lived in balance with the other species that shared their natural habitats, so that the original human beings were well adapted to those habitats? In that case, our present task would appear to be a Stoic one: to modify human modes of life in ways that will *restore* a proper "adaptive balance" between the demands of Humanity and the resources of Nature.

Or should we, alternatively, assume that the populations of early humans increased, from the very beginning, more or less continuously (though at first, no doubt, rather slowly) so that, all along, human beings were too successful at reproducing themselves and, thus, were *too well adapted* to their habitats? In that second case, nomadism was doomed from the start. The moment the growth of human population began to fill the available hunting grounds, some new mode of life was bound to appear and the subsequent transitions to urbanism, industrialism, and scientific technology were merely further responses to the same basic pressure of growing population. Correspondingly, our present task would then appear not a Stoic but an Epicurean one; i.e., to *create* a just "adaptive balance" between the demands of human beings and the needs of all other natural things for the first time.

Which of these is the truer perspective? Shall we join Hegel in contrasting Nature with Culture—the natural, repetitive, value-free processes characteristic of prehistory with the cultural, progressive, value-creating characteristics of human history? Or do the cultural activities of human beings, as Engels implies in *The Dialectics of Nature*, merely carry forward the processes of nature, but by other means?

During the last 10,000-15,000 years human culture has differentiated itself from prehistoric nature, but that transition did not take place all at once, in a saltatory manner, nor did human modes of life take on their distinctive aspects at a single stroke. Instead, this transition has involved a whole series of stages. Sedentism, urbanism, industrialism, and scientific technology have carried the life of humanity progressively further away from the life of our evolutionary precursors, but this does not mean that human life is, in itself, "unnatural" or inimical to the lives of other species. Instead, it remains to be seen how far, in point of fact, the history of human culture represents a break with the earlier history of organic species, how far it simply carries farther natural developments already in process before the appearance of the human species.

The Multiple Modes of Adaptation. The central question of this symposium—How Do Humans Adapt?—is both a nodal question and a thorny one. It is a nodal question because our approach to problems of practical policy depends on our attitude toward a whole range of theoretical issues, e.g., whether the evolutionary adaptation of an organic species normally involves a stable population level. It is also a thorny question because the task of presenting all these issues in clear and consistent proportion remains to be completed. So long as that is still undone, the ambiguities about human "adaptation" are liable to confuse and entangle our ideas just as much as to illuminate them.

To move from historical to analytical issues, then, the central family of terms with which this symposium deals—"adapt," "adapted," "adaptive," "adaptation," and the rest—involve hidden complexities. In their primary, everyday usage, all of these refer to processes and relationships by which one person, thing, or way of acting becomes *apt* or *fit* to another, in some respects and for some purposes. An electric razor manufactured to operate on 115 volts is used with a 230-volt supply by interposing an "adaptor" between the razor and the electrical outlet. A craftsman who loses his job after twenty years shows how "adaptable" he is by the speed with which he develops the novel skills he needs to take up another line of work. We come into a shaded room from the bright sunshine outside, and it takes a few seconds for our vision to "adapt" to the new conditions. In each case, we see human beings "adapting" to novel situations, even though none of them involves "adaptation" in its specifically evolutionary sense.

Darwinian "adapting" and "adaptedness," which result from variation and selective perpetuation of favored subgroups within natural populations of animals or plants, define *one particular mode* of adaptation. By contrast, human adaptation is a *multi-modal* activity, and a good "fit" between human life and its biocultural habitats depends on a variety of different processes. Nor are these different kinds of "adapting" alternative routes to the same goals. Different human purposes are served in different ways: electric transformers, industrial retraining, and the physiology of the visual system are

STEPHEN TOULMIN

different kinds of "adaptors," "adaptability," and "adaptive mechanisms," and they exemplify quite different modes of adaptation.

To go further: by asking how human beings are to adapt to the terrestrial environment *as a whole*, we are raising some fresh theoretical problems, even for the purposes of evolutionary biology. The Darwinian concepts of "adaptation" and "adaptedness" account for the relative survival values of competing subpopulations in a particular habitat, by considering what factors affect their relative rates of reproduction. One subpopulation is *relatively* "better adapted" than a competing subpopulation, to the extent that it can outbreed its competitors *in a particular shared habitat*. And, since the breeding test is only a relative (or comparative) test, its use brackets out all questions about the overall "carrying capacity" of the habitat. So the Darwinian concept of "adaptedness" tells us, in itself, nothing whatever about a population's merits on any absolute scale, let alone about its merits relative to the entire terrestrial biosphere.

Once we begin discussing the "adaptation" of organic species to the entire biosphere—that is, to the *totality* of available biological niches—we cannot automatically reapply terms originally defined with references to *particular* niches. If we are to judge the "adaptedness" of different species or populations— including the human species—to the terrestrial environment *as a whole*, some new theoretical decisions and definitions will be required. This takes us beyond the scope of the original Darwinian concept. The totality of available biological niches is not merely one more niche among others; in this case, we are obliged to reintroduce explicitly all those considerations—above all, those questions about "carrying capacity"—which we had previously bracketed out. Indeed, those considerations themselves will now become the heart of our theoretical problem.

We may distinguish at least four distinct forms or modes of human adaptation:

(1) In cases of *calculative, conscious, or rational adaptation*, human beings explicitly recognize alternative courses of action open to them and make deliberate choices between them. A man loses his job/his wife/his sight/his heating oil supply and he adapts to his new situation by retraining for a new career/moving in with friends/ taking a crash course in Braille/converting to natural gas. In this first sense, *adapting* means choosing, on the basis of a conscious calculation, to act differently for a purpose: to maximize the gains and minimize the losses resulting from the changed situation.

(2) In cases of *homeostatic, autonomic, or feedback adaptation*, physiological, psychological, economic, or social mechanisms al-

ready in place counteract the disequilibrium produced by external changes so that the *status quo ante* is restored. The weather heats up or the supply of beef goes down; in response, our pores enlarge so that more perspiration evaporates, or the price of beef goes up so that people consume less. In this second sense, *adapting* means responding functionally so as to maintain some "normal" (equilibrium/health/desired) condition in the face of external changes.

(3) In cases of *developmental, maturational, or progressive adaptation,* individual human beings become increasingly able to deal with the problems of life in their given habitat as they grow up. With the development of mobility, speech, and deliberation, children become progressively less dependent on adults and so better adapted to their conditions of life. In this third mode, *adapting* means developing the ability to cope effectively and independently with the problems of living—whether by "adapting" in modes (1) and (2) above, or in any other way.

(4) In cases of *evolutionary, populational, or selective adaptation,* certain physiological, behavioral, or cultural variants establish themselves at the expense of other variants, as equipped to meet current ecological, social, or intellectual demands. The deserts enlarge, social mobility increases, computers become generally available to scientists; in response, new forms of plants evolve with thicker seed pods capable of withstanding the drought, new dialects and modes of speech come to serve as social differentiators, and the procedures of science are increasingly slanted toward computer programming. In this final sense, *adapting* means evolving novel features, which first appear without conscious foresight and subsequently establish themselves selectively, as being fitter or apter to the novel conditions than their predecessors and rivals.

Notice that (4) the evolutionary mode of adapatation—like (2) the homeostatic mode—is here defined in terms that cover both organic and behavioral changes. It includes (4a), *organic adaptations* of the classic Darwinian type, in which the variants are encoded genetically and the selective demands are ecological, but also (4b), *behavioral adaptations* that are transmitted educationally rather than genetically, in which the variants are encoded in language, techiques, and cultural norms, and the selective demands are correspondingly social, intellectual, or cultural.

These two submodes (4a and 4b) are associated here, not out of perversity, but to underline an important point. Modern evolutionary biology teaches us that selection always operates on the *phenotype*,

STEPHEN TOULMIN

i.e., on the pool of mature adults, which can alone demonstrate their "well-adapted" characteristics by contributing offspring to the next generation—and this fact helps in two ways to blur the line between Nature and Culture.

How, then, are we to frame the central problems of human adaptation, i.e., the means for insuring a better *match* or *fit* between human populations and their terrestrial habitat? In analyzing these problems, we cannot confine ourselves to any single mode of adapting, because human individuals, cultures, and societies demonstrably maintain themselves and respond to novel conditions of life in ways that illustrate all these different modes of adaptation.

(1) In the political and managerial realms, the role of conscious evaluation and choice (*rational adaptation*) is clear enough. If anything, we are liable to exaggerate this role: human agents rarely refuse to take credit for favorable turns of events, even when these were not in fact outcomes of their deliberate, conscious decision.

(2) In the social realm, similarly, self-sustaining institutions, economic equilibria, and standard-operating procedures play an equally familiar role, and all of these exemplify *homeostatic adaptation*. Since the work of Emile Durkheim and Talcott Parsons, indeed, sociologists have relied extensively on parallels between the feedback loops characteristic of physiological systems and the interlinked constraints operative in "social systems" in their theoretical analyses.

(3) In the realm of child development and education, the whole process of growing up brings with it an increase in mastery over, and therefore in "adaptedness to," the individual's conditions of life. To the extent that educational reform makes this process itself open to change and refinement, *developmental adaptation* is an additional channel for the improvement of human modes of life.

(4) Finally, variation and selective perpetuation (*selective adaptation*) help to maintain the match between the technical repertoires of our arts and sciences and the demands that they currently face. The conditions for improving that match chiefly concern either protecting the opportunities for innovation and experimentation or insuring that the resulting variants are tested and appraised in a sufficiently critical manner.

Each mode of adaptation, however, also has a limited scope, and when pressed beyond its proper limits can display corresponding shortcomings. The conditions on which conscious, calculative adaptation (1)—the explicit, rational appraisal of alternative courses of action—is practicable at all are rather stringent; the alternatives between which we choose must be clearly definable and close at hand, and we must have sufficient command to make the resulting choice effective. Where the proposed alternatives are too remote, the practical limitations on (say) social forecasting will limit our power to choose between them as realistic goals. Unless effective "handles" exist for controlling the factors involved, we can too easily deceive ourselves about our own power of decision. It is still highly uncertain, for instance, whether effective levers exist in a nontotalitarian society for directly controlling the level of the money supply. Mapping the limits of social diagnosis, forecasting, and control—and learning to respect those limits in practice—are thus preconditions for judging the practicability of directive planning (dirigisme) within a democratic state.

(2) The homeostatic operation of "adaptive mechanisms" (2) is also subject to inherent limits. In physiology as much as in social practice, the primary role of feedback loops is conservative: they serve to maintain existing adaptive relations, but are little suited to modifying those relations in response to external changes, still less to creating entirely new adaptive relationships. As Claude Bernard himself understood very well, such systems can never explain the physiological processes of morphogenesis and developmental adaptation; and sociological theories based on direct analogies with Bernard's physiology have been essentially conservative in their force.

To the extent that homeostatic social systems operate "healthily" or "functionally," changes in social relationships become, by definition, "dysfunctional" or "pathological"; so human adaptation cannot rely exclusively on such homeostatic "systems." As Max Weber warned us, bureaucracies and other self-maintaining social organizations may begin by operating effectively, even "adaptively," relative to their original situations, but, in the face of changing needs and conditions, they too easily respond by protecting their own powers and structures, instead of modifying them as occasion requires. (I recall a graduate student from the Caribbean, who complained that American sociological ideas were inherently conservative. "In my island," he said, "it is no use saying that change is dysfunctional. We know our institutions have to change; the only question is how to change them for the best." It was no great surprise to me when that same student reemerged last year in the new revolutionary government of Grenada. After all, whatever their other weaknesses may be, Marxist social critics have never disregarded questions about the proper directions of social change!)

Developmental adaptation (3), too, has a definite but limited part to

STEPHEN TOULMIN

play in the overall processes of human adaptation. All our hopes and plans for "doing better" as a species, of course, depend on what human beings are capable of "growing up into being," with or without conscious educational shaping. But once again, the directions in which we can best guide an individual's development are chosen by criteria external to the process of maturation itself. Likewise, the value of free variation and critical selection—i.e., selective "trial and error" adaptation (4)—is well established in many technical fields of work, where it serves to improve the detailed content of our arts, technologies, and sciences. But it is not yet fully clear on just what conditions, and in just what kinds of cases, the virtues of selective (mode 4) adaptation outweigh the advantages of direct, consciously planned (mode 1) adaptation. From the International Monetary Fund down, many leading national and international agencies today need to learn how to strike this crucial balance between central (mode 1) direction and pragmatic (mode 4) experimentation, with greater discrimination and effectiveness.

Humans thus adapt in half a dozen interlinked ways: through genetic survival, functional morphogenesis, physiological responsiveness, social flexibility, cultural cohesion, intellectual variability, and technological refinement, in addition to the deliberate exercise of conscious observation, judgment, and choice. During the millions of years before the emergence of *Homo sapiens*, our hominid precursors lived the communal life of nomadic hunters and gatherers, and became biologically adapted to their natural habitats. To some extent, no doubt, we still carry that heritage in our genes.

Meanwhile—since biological evolution did not stop abruptly two million years ago—historical development has modified the social and cultural expression of our specific characteristics. As a result, human modes of life have become more refined, complex, and elaborate. Cultural, social, and technical adaptation took its place in human life alongside biological adaptation and the selective adaptation of social, cultural, and technical forms still remains one of our most powerful means of "adaptability."

The resulting dialogue between Nature and Culture plays itself out in the lives of human individuals both through the processes of morphogenesis and through the varied patterns of education, socialization, and enculturation, by which each human child becomes developmentally adapted to its native culture or chosen habitat. But it plays itself out also in the arena of collective life, through the existence of shared bodily systems and through the development of local cultures, techniques, and institutions which respond adaptively, and so conservatively, to short-term changes in the local situation. Finally, as fully *human* beings, we have an exceptional capacity, in suitable cases, to master and change our own situations on the basis of observation, calculation, and choice, thus adding the reflective power of conscious, rational adaptation to the other, older modes.

None of these modes of adaptation operates in isolation. At one extreme, rational, calculative adaptation is limited by the physiological character of our central nervous systems and by the scope of our preexisting arts and techniques, so that the best we can do is to "do the best we can"—given where we are and what resources we have to work with biologically and culturally. At the other extreme, human society, culture, and technology are continually modifying the selective pressures acting on the "gene pools" of all of the species occupying the terrestrial habitat. So, increasingly, the character of the human phenotype will reflect not just *how we were made*, but also *what we have made of ourselves*—as well as what we have made of our fellow species and of the world that we share with them.

Looking to the Future. Going beyond all these general theoretical issues, this symposium considers more specifically what instruments are available for handling our present difficulties most intelligently, and so for "bettering ourselves." Are the practical problems of human adaptation more typically, today, matters of population policy or economic development? Are they best dealt with by influencing people's personal self-interests, by improving their educations, by fashioning new institutions, or by developing new economic resources? How much real power do we have, in any case, to predict and control social, cultural, and technical changes, whether on a national or an international scale and what are the natural limits to that capacity? And, when developing novel institutions for dealing with our problems, how can we avoid creating on a worldwide scale the self-defensive bureaucracy whose rigidities Max Weber rightly deplored on the level of the nation-state? How, that is, can we develop styles of institutional administration and organization that are conservationist without being purely conservative; institutions which will pursue their own proper concerns and functions, without being too protective of their own narrower interests?

Finally, larger cosmological presuppositions underlie our answers to all these questions. Can we hope to agree on these answers, regardless of our other intellectual and religious commitments? Or do our positions depend essentially on what vision of Humanity's Place in the World of Nature we are already committed to, for quite other reasons? Can these questions be answered in the light of "natural reason" aside from all doctrinal differences or is our assessment of

STEPHEN TOULMIN

humanity's future and its agenda condemned to sectarian division and theological misunderstanding? Specifically, how far are the policies that we need, in order to escape from our current difficulties, wished on us by the requirements of nature itself; and how far do we still have significant elbow room to shape the world to our own hearts' desires? As human beings living social and cultural lives in natural habitats, how far are we still free to decide for ourselves what kinds of beings and lives, cultures and habitats we prefer to fashion, i.e., what we should seek to make of ourselves, and for ourselves?

There remain some people today in whose eyes the very belief in a Fall from Nature, and in the Limits to Growth, is itself the most direct cause of despondency about our current situation; others regard this belief as being, rather, one expression of a despondency whose true roots lie elsewhere, e.g., in the social conflicts internal to modern industrial-technological society. A third group is not convinced that anything in our situation really warrants despondency, rather that we have the means of salvation in our own hands and all we need is the intelligence and courage to put them to work. So, with that divergence of views in mind, let me close by putting the final question: "*Which way are we then to go?*"

Are we to take the conservative, or conservationist, route? Should we minimize the impact of Humanity on Nature by reducing the scale of human activities, rather than by changing their basic character?

Are we to follow, rather, the radical road to the same goals? Should we concentrate on creating new social and economic organizations which would give industry, for example, positive incentives for building into its operations safeguards to protect the natural habitat?

Or, would we do better to set aside our current anxieties and proceed full steam ahead, with renewed confidence in our human capacities? Indeed, have we not, perhaps, already embarked on a fifth historical transition, as significant as the earlier four, with the switch from energy-hungry industrial manufacture to information-intensive activities and enterprises? And will not this change do much to counter and correct the current damaging effects of human activities on the resources of the natural world?

These questions may well appear highly abstract and theoretical, yet few general questions have more concrete and practical implications for the governments, public agencies, and private corporations of the world. So, as we seek to clarify and answer those questions in this symposium, we are taking on a task of applied science and scholarship that is as complex as it is urgent. We have been given a large canvas, yet even the smallest details of our picture will be none the less significant; and only a sense of their practical importance can give us the courage we need to attack them with the clear-headedness they demand.

Commentary by Mary Catherine Bateson
Dean of Faculty, Amherst College

One of the very important changes that lies behind our gathering here and that Stephen Toulmin has taken notice of in his paper is the fact that when we think of adaptation now, we must think of the problem of adapting to the totality of available biological niches rather than any given niche. I believe that as we think about the problems of human adaptation, we are newly in the situation where we must think of the species and of the environment as a unity.

This change in the problem of adaptation has to be seen in the light of the discussion of the similarities and differences between true biological evolution and cultural evolution. Biological evolution—and all of these discussions of socio-biology that are developed in many of the papers—supposes transmission through a line of descent.

Of course, much of the discussion of the analogy of biological and cultural evolution developed as a result of the realization that culture does the same thing to a different degree, and we can speak of a cultural inheritance just as we can speak of a biological inheritance. A great deal of cultural knowledge is transmitted by older members of the group to younger members of the group (vertical transmission). Increasingly, however, we deal with the fact that people are learning from their contemporaries (horizontal transmission). This also creates a kind of unity that I think is new.

It has been so important to emphasize cultural diversity that we have perhaps underemphasized the situation that exists today, in which, as we try as a single species to adapt to our single environment, new ideas anticipating the future can be transmitted horizontally and are being transmitted in that way. Our basic technological ideas are moving across cultural boundaries, national boundaries, and lines of descent, and are being passed from place to place both for good and for ill.

These are factors that very much affect the question of what is the subject in a sentence that has the verb "adapt" in it. Especially when we think of that kind of adaptation that involves anticipation, that involves feed-forward, it seems very important to know who is doing the adapting.

Commentary by Wilton S. Dillon
Director, Office of Smithsonian Symposia and Seminars, Smithsonian Institution

In addition to my administrative activities, I am engaged constantly in the practice of anthropology in the Smithsonian Institution. One of the great pleasures of that serious process had to do with the celebration in 1979 of the Centennial of the birth of Albert Einstein. That was also the year in which we were celebrating the International Year of the Child, and we combined those two observances and tried to ask the question that seems pertinent once again today in the light of Stephen Toulmin's essay, namely, "how to keep alive a sense of wonder and curiosity?" I would like to wonder out loud with you about several points of departure in Stephen's paper.

One of the questions that he did properly get into almost immediately has to do with which behavioral adaptations are transmitted educationally. Part of my wondering is: how does one learn to adapt? What are one's parental or political models? If not political in the sense of following political leaders, whose behavior do we follow as we begin to try to cope with solving a whole series of problems on a daily basis?

What do we remember from our parents' capacity to adapt? The differentiation between a father and a mother gives us two different styles of approaching problem-solving, because both parents are not identical. How does socialization actually occur? One of the aspects of adaptation which I raise as a question here, to be followed up in future research and investigations that could spring out of the Toulmin essay, has to do with the question of play and playfulness. Alfred Kroeber, the late great general anthropologist at Berkeley, Columbia, and other places, had a proposition that still needs to be tested—and I think this is something for philosophers as well as physiologists to look at—the proposition that art, science, and technology are the products of the play impulse in mammals.

Mammals seem to have a greater playfulness than birds, who are constantly seeking food and have not enough leisure time left from obtaining their protein supply to engage in apparently useless activity called play. I suggest that playfulness is an extraordinary impetus to the solution of problems via the temporary removal of literalism and rigidity.

There is something else that has to do with the biological factors affecting the interplay of culture and biology. And that has to do with the human life cycle and the ceremonies marking off our movement from birth to death. Each ceremonial as we move from birth through early childhood into adolescence and adulthood, including the mating and dating process and finally the glorious entry into the hereafter, gives us an opportunity to look at how we have adapted to these new stages of life.

I wonder, when does one's own notion of being an adult—mature and dignified—rule out the willingness to express oneself playfully and to know that one's dignity is not impaired? There is, therefore, in these ceremonial looks at the life cycle and rites of passage an analytical mode for approaching the links between biology and culture.

One other whole area of adaptation that I would hope philosophers could look at is how we deal, in any human group, with the question of reciprocity and the obligations which go not only with the exchange of literal gifts but symbolic gifts; how we deal with the giving and repaying, receiving and repaying process, because much of the hostility in the world, it seems to me, in interpersonal relationships and going on up from diads to triads to nation-states and alliances, has to do with the failure to find in these systems adaptive devices for the discharge of obligations, which means that dependent persons, groups, or nations feel a great hostility toward their protectors and their benefactors when they cannot find commodities of exchange to which both parties attach value. This particular problem relates to the question of how we manage and control violence in the world.

One book I would very much urge you to look at again in the light of this symposium is one of Margaret Mead's classics, *New Lives for Old* (1956), written about the transformation of the people in New Guinea, in the Admiralty Islands. There is a wonderful man there named John Kilipak, whose picture you will find in the book, who was age fourteen when Margaret Mead went to study the group and who continued to be her friend and informant until her death.

John Kilipak came out of a chiefly tradition. His adaptations from the Stone Age to the Atomic Age and the world of the United Nations are well documented in Mead's book. In no way in his adaptive processes did he ever lose his preindustrial, prescientific, pre-world-organization roots. There was a great self-assurance about him, which I think some followers need in times of great change, some kind of rootedness with the past, so that they can better adapt to changes when they have in view one leader, one visible leader who seems to be able to cope.

Discussion

DR. TOULMIN: I enjoyed Catherine Bateson's intervention particularly because she is absolutely right. In posing the questions that I did, I spoke too much about what we "adapt to" and how we "adapt," and not at all about who the "we" are (or is) that do/does the adapting. In retrospect, it seems that there are some important questions here and that the current debate in this area embodies certain ambiguities about who this "we" are/is.

There are strands in what I call the "public" philosophy that I would describe as kind of Neo-Epicureanism (e.g., the psychotherapy movement). This represents an attempt to perpetuate a kind of individualism, in which salvation comes to each individual independently and that the individual can be saved without anything necessarily happening to anybody else. There are other strands that I would describe as Stoic or Neo-Stoic elements, which link back to traditions of thought that see humankind as having a moral unity, so that the idea of an individual being saved in a community that is damned, or vice versa, becomes inconceivable. So there is a problem of seeing who we are, who now have to come to terms with the styles of our own adaptation, both to our biological environment and to our socio-cultural situation.

The fact is that we ourselves oscillate between speaking of ourselves as "the species" and realizing it is quite artificial to talk about ourselves in this way. To some extent, therefore, the problem I raised in my paper, in abstract terms, about how the different modes of adaptation fit together will be resolved only when we better understand the different roles that we take on when we think about what this "adapting" involves.

DR. MIDGLEY: I want to say a very simple and obvious thing. I think several other speakers as well as Dr. Toulmin want to resist the Rousseauish extreme to which one so readily goes when one is horrified by the present state of affairs and, in general, have had difficulty in knowing how far to go. I think his essay was very helpful about this.

May I suggest some very simple mapping concepts here? We do have some reason for saying that our own civilization represents a good adaptation of our species towards the environment (I am being terribly generalist here, like a biologist), as we might with any other species that we found was living in such a way that it was using the gifts it had and that it was not so adapted if it was failing to use those gifts.

There is another very crude and obvious sense of adaptation, namely, stability. Long ago I read with fascination a splendid book called *Voles, Mice and Lemmings*, which followed the cycles that these creatures got into whereby their population would build up and then crash. I am sure the author used the term "imperfect adaptation" or the like about this instability, which he suggested could not have been the original state and wasn't the original state of many species, but probably had been produced by people killing predators. When predators had been there, a more stable population curve had been maintained and that had now been lost. Now, while I take the point of the discussants who wanted, as it were, to get nearer and say, "but what is man, what particular detailed point of view are we taking?", it does seem to me that the task of adapting is to stand back and take this rather big evolutionary perspective. I want to say even from that big evolutionary perspective one can see this as a clash, because it is quite true that our civilization has succeeded in mobilizing and using capacities which were previously unused.

So we ought not just to stand back and stop it all, but of course then when we look at what we call more primitive civilizations, we see some human capacities in each of them which were also getting very well used. We need to look at things like Indian and Chinese civilization very seriously and ask if we are really right in saying that human gifts are being more fully used here.

I just want to finish up by saying that I think we probably don't fully enough control the tendency to say that the gifts we are using, notably mathematics and science, are the central human gifts. So that there can really be no alternative, you see, to the kind of bargain that we have got. I think we are all here because we are fairly worried and may be prepared to look at that problem a little more open-mindedly than many contemporary people.

DR. TOULMIN: I was struck by one thing that René Dubos said, the sense I often have that the problems people are really talking about intensively are problems that very often are on their way to being solved. I remember the pea soup fogs of our childhood in London before they started cleaning up the air, and what Dubos says is right. When everybody is talking about air pollution, in fact a lot begins to get done about it. I wonder whether in a hundred years' time people may not look back at us and wonder whether we too were not getting preoccupied about the population problem just at the stage at which actually it was beginning to be effectively dealt with. Then what will appear to be the central issue for us? If we are able to achieve some kind of stability in terms of population, where will the tension then appear to have shifted most urgently?

STEPHEN TOULMIN

2. Parental Investment: The Hominid Adaptation

JANE B. LANCASTER
Associate Professor of Anthropology, University of Oklahoma
CHET S. LANCASTER
Adjunct Associate Professor of Anthropology, University of Oklahoma

Editor's Summary. Jane and Chet Lancaster provide an evolutionary perspective on the family. Many culturally patterned practices have their roots in a survival strategy that places great emphasis on parental care for children who depend, long after weaning, on both mother and father for protection and food. The Lancasters suggest that the secret estrous cycle of the human female minimizes male competition and promotes male social involvement with the female and her children. They also note that the most common marriage pattern is monogamous, which enhances male parental investment in the welfare of their offspring.

The major features of the evolutionary history of the human skeleton, dentition, and brain have been known for decades now. "Missing links" have been found, and hundreds of fossils of protohominids and early humans are known from many parts of Africa and Eurasia. Recent discoveries from the rift valleys of Africa are important, not only for the human fossils they reveal, but also for the careful attention to the physical and ecological context in which the fossils belong and because of the radiometric dating which places sites in temporal order (Campbell 1979; Isaac 1976). The new ordering, dating, and establishment of context for these fossils forces a different interpretation of the fossil record, emphasizing not just how early hominids looked but also how they lived.

The fossil record clearly shows us that small (around four feet tall), fully bipedal protohominids lived in the rift valley areas of Tanzania and Ethiopia during the Plio-Pleistocene (Johanson and Edey 1981). Their skulls reveal small, ape-size brains and only slightly projecting canine teeth. We have their footprints, 3.7 million years old, and a skeleton, 40 percent complete, dating back 3.5 million years. They reveal that the unique pattern of human bipedal striding was established nearly four million years ago. What is surprising is that this freeing of the hands is not directly associated with two other human hallmarks: the manufacture of stone tools and the enlargement of the brain, evidence for which appears at least a million years later—the earliest stone tools coming from Ethiopia about 2.5-2.7 million years ago and the enlarged brain of *Homo habilis* appearing around 2 million years ago in Kenya (Lewin 1981a).

These early bipedal hominids, the Australopithecines, though human in their mode of locomotion, now appear much less human in their behavior in terms of what can be inferred about their diet and social organization. We begin to see them now more as bipedal "apes," foraging together in bisexual groups in the woodland-savannahs of Africa—probably using informal tools of wood, stone, and fiber like the modern chimpanzee and pursuing a basically frugivorous diet (McGrew 1981; Tanner 1981; Zihlman 1981). The use of such informal tools would permit relatively small-bodied apes to extract embedded food items so common to woodland-savannah environments, such as nuts and hard-shelled fruits or termites and rhizomes hidden under the ground (Parker and Gibson 1979). In short, these early protohominids are the perfect "missing links": man-apes exploiting a new, more open environment, their use of informal tools, and perhaps the carrying of food, expedited by a new locomotor pattern, bipedalism. The record of their behavior, though very incomplete, has not revealed the significant hallmarks of a truly human pattern, one based on hunting and gathering, the division of labor, and stone tools.

The Division of Labor. The division of labor refers to a uniquely human dietary/ecological shift which forms the fundamental platform upon which all human behavior and elaborations of culture are built. Isaac (1978) describes this platform as composed of a few simple behavioral elements which combined into a coherent pattern fundamentally changing the relationship of our species to its environment. The division of labor is a pattern of feeding in which the two sexes specialize in obtaining food from different levels of the food chain (Lancaster 1978).

Males hunt for meat—energy from high on the food chain, concentrated in a large "package." Their endeavors are risky, unpredictable, and often dangerous. Females, by contrast, concentrate on gathering sources of energy from lower on the food chain: high-grade plant food or an occasional small protein package, such as insects or small vertebrates. Their work is less dangerous, success is more predictable,

JANE B. LANCASTER AND CHET S. LANCASTER

and the food provides the social group with a nutritional baseline adequate to sustain adults and young, even if hunting is poor for weeks on end. Among contemporary tropical hunter-gatherers, women's gathering contributes about 65 to 70 percent of the caloric intake of the group's diet (Gaulin and Konner 1977). Men's contribution of meat represents fewer calories, but provides essential amino acids.

The effectiveness of the human pattern rests on a behavioral complex which includes bipedalism and the ability to carry food, water, equipment, and infants; a psychology of postponed consumption of food and a willingness to share, even when food is scarce; and, finally, a home base or camp where group members can return to share food after their dispersed food quests.

In contrast to humans, who regularly share food, nonhuman primates are individual foragers. Their diet contrasts with the human, not only in what is eaten but in terms of how it is obtained (Gaulin and Konner 1977; Teleki 1975). Although most monkeys and apes spend their lives in social groups, aiding each other in mutual defense (protection of the group's territory) and socializing daily by grooming and resting together, they do not regularly share food. A sick or injured monkey can die of hunger or thirst. Even though its close relatives may show protective concern about its condition, none will comprehend its need for nourishment. Small, weak, young, and subordinate animals suffer when food is scarce because of the disadvantage in finding, competing for, and processing food (Dittus 1980). The earliest humans evolved a simple, yet unique solution to this problem. Food-sharing and the division of labor is a kind of food insurance policy which smooths out fluctuations in individual success in the food quest at the same time that it permits regular feeding at two levels of the food chain. When game is scarce, human omnivores do not face starvation as do carnivores. In short, early humans discovered a way to regularly eat high-quality proteins without taking the risks of carnivore specialists.

The archeological record tells us something about the history of the division of labor. Isaac (1978) describes excavations at Koobi Fora on the banks of Lake Turkana in Kenya. These excavations show clear indications of postponed consumption of food. Some sites were used for the killing and butchering of a single large animal like a hippopotamus, whereas others include a variety of stone tools and the bones of many different kinds of animals. The latter sites must have been home bases where humans camped, brought back parts of large animals as well as small animals to share, and manufactured and used stone tools. The oldest of these camps dates to just a little under two million years ago.

These sites provide the first evidence of the regular eating of meat taken from animals larger in body size than that of the individual primate hunter. Although both chimpanzees and baboons are known to capture small mammals and sometimes even share bits by tearing the carcass apart with bare hands (Teleki 1975), such behavior is clearly not functionally comparable to the regular butchering and division of large mammals. These sharp flakes and broken bits of bone give tangible evidence of the unique human pattern—the division and sharing of meat and the postponed consumption of food at home base.

The division of labor between human males and females and the regular sharing of food represents the true watershed for differentiating ape from human life ways. Bipedalism was a necessary precursor to human behavior because it facilitated the carrying of food and implements, but it antedated the division of labor, stone tools, and expansion of the brain by between 1.5 and 2 million years.

The Division of Labor and the Feeding of Juveniles. The division of labor does not represent simply a sharing of resources collected at two different levels of the food chain and shared between males and females. Although such behavior represents a kind of feeding insurance policy for the two sexes, the division of labor is associated with a much more adaptive pattern of behavior. Unlike the young of other mammals, human juveniles do not have to feed themselves once they are weaned. The division of labor takes on added significance because its feeding insurance covers not only adults but also the much more vulnerable juvenile population.

It is generally agreed that the human species is characterized by prolonged juvenile dependency (Jolly 1972; Lancaster 1975). However, figure 1 shows clearly that even the great apes have prolonged social dependency for offspring, lasting to the ages of ten or more years. Mother monkeys and apes wean their infants after one to five years of lactation, depending on the species. The years of infancy and nursing are followed by an equally long period during which a juvenile enjoys the social and psychological protection of its mother and close relatives and the knowledge of its social group about food resources and environmental dangers. In contrast to humans, however, the juvenile monkey or ape feeds itself. The basic diet may be typical of that of many primates living in the tropics: fruits, nuts, leaves, and some animal protein. The important contrast between human and nonhuman primates is not so much in what is eaten but in whether each individual must forage for itself, and in whether there is a collective responsibility for gathering and sharing food between adults for their young.

Altmann's (1980) account of mother-infant relations among free-ranging baboons in Kenya demonstrates how pressing the problem of

JANE B. LANCASTER AND CHET S. LANCASTER

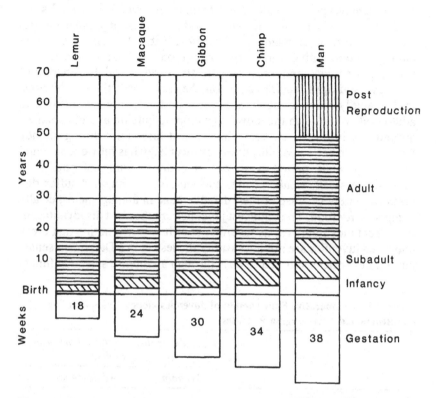

Figure 1. Progressive prolongation of life phases and gestation in primates. Note the proportionality of the four indicated phases. The postreproductive phase is restricted to humans, and is probably a recent development (after Lovejoy 1981).

feeding juveniles can be. Mother baboons spend 81 to 89 percent of their waking hours foraging during the later weeks of lactation. Time-energy budgets of lactating females have little margin for error, and their death rate doubles. The exact timing of weaning coincides with the maximal availability of weaning foods during the annual cycles, which suggests that the events of mating and birth nearly two years earlier are timed by this need. Even so, many juvenile baboons are unable to find adequate forage, and the improvement of their mothers' life expectancy at weaning coincides with a drop in their own. Altmann's data strongly suggest that mother baboons are at the limits of their ability to provide nourishment for their young and exemplify the fact that female monkeys and apes are only able to provide for a single, nutritionally dependent young at a time, although they are able to protect and foster the development of two or more.

The demographic effects of the feeding of juveniles by adults are evident when comparing the survivorship curves of human hunter-gatherers with other mammalian species. Table 1 demonstrates that among group-hunting carnivores, about one in four of those born survives weaning, and only one in six survives adolescence. For monkeys and apes feeding lower on the food chain than carnivores, in the rapidly growing toque macaque, about one in eight survives to adulthood, whereas in the slower developing baboon and chimpanzee around one in three survives. The greatest loss is not during infancy but during the early juvenile period. Dittus (1980) has fully documented why the juvenile period is so perilous: during times of poor food supply, the small, weak, young, and subordinate animals suffer disproportionately because of their disadvantage in finding and especially competing for food. The youngest juveniles are the most disadvantaged; life expectancy improves as juveniles age. It is worth noting that these imperiled juveniles are not necessarily the less fit in a Darwinian sense, but are nevertheless forced to pass through an intense selection funnel

TABLE 1 Comparative Survivorship of Juveniles among Group-hunting Carnivores and Free-ranging Primates

	Percentage of those born who survive	
	Weaning	Adolescence
Group-hunting Carnivores		
Lion (Schaller 1972)	26.0	15.8
Lion (Bertram 1975)	29.0	14.0
Wolf (Pimlott et al. 1969)	23.7	18.5
Free-ranging Primates (stable populations)		
Yellow baboon (Altmann et al. 1977)	45.0	33.0 (second year)
Toque macaque (Dittus 1977, 1979)	42.0	12.5
Gombe chimpanzee (Teleki et al. 1976)	48.0	38.0
Provisioned Primates (expanding 13–35% per year)		
Japanese macaques (Masui et al. 1975)	80.6 (female)	39.8 (female)
	82.4 (female)	67.1 (female)
Rhesus macaque (Sade et al. 1977; Drickamer 1974)	82.0	58.0
Barbary macaque (Burton and Sawchuck 1974)	71.1	54.3

JANE B. LANCASTER AND CHET S. LANCASTER

at a tender age. The relationship of food to the survival of juveniles is clarified when a comparison is made between wild, self-feeding groups and those which are regularly provisioned by humans. When food is a free good or nearly so, juvenile primate survival is nearly double the rate of wild groups living in stable populations. For monkeys, such a survival rate of juveniles gives an annual population growth rate of 13 to 35 percent.

Table 2 reveals that human hunter-gatherers and horticulturalists, even without benefit of modern medicine, successfully raise to adulthood about one out of every two children born to them. Human survivorship to five years is between 62 to 71 percent, nearly the same as survivorship to that age for primates which are freely provisioned. In other words, for humans living in simple economies, survivorship to adulthood is comparable to nonhuman primate groups with no food limitations on population growth.

A comparison of the reproductive parameters of the life histories of African apes and human hunter-gatherers indicates no major contrasts (table 3). All three species have a four-to-five-year birthspacing which is locally variable, but always long. The shortest spacing is found in the gorilla, where juvenile weaning foods are ubiquitous and require less extractive skill. The longest spacing belongs to the chimpanzee, where adult feeding patterns are more complex and require experience. Similarly, average completed fertility (the number of offspring surviving infancy per female) is nearly the same for all three species as is the total reproductive span in a female's life (about twenty-five years).

TABLE 2 Comparative Survivorship of Juveniles among Human Hunter-Gatherers and Horticulturalists

| | Percentage of those born who survive | | |
	1 year	5 years	15 years
!Kung San (Howell 1979)	79.8	65.8	58.3
Libben Paleo-Indian (Lovejoy et al. 1977)	82.5	69.3	52.9
Yąnomamö (Neel and Weiss 1975)	73.0	62.0	50.0
Paleo-Indian (Blakeley 1971) 3 populations combined	86.3	71.0	46.4
Etowah Mound (Blakeley 1977)	90.0	68.0 (9 years)	54.0 (19 years)

TABLE 3 Reproductive Life Histories from Populations of African Apes and Human Hunter-Gatherers

	Birthspacing of surviving offspring	Average completed fertility	Reproductive span
Chimpanzee (Tutin 1980)	5.75 years	3–5 offspring	age: 13–14 to 40–45 years years: 26–32
Gorilla (Harcourt et al. 1980)	3.83 years	2–5 offspring	age: 10–11 to 35–36 years years: 25–26
!Kung San (Howell 1979)	4.12 years	4.7 offspring	age: 18.8–45 years years: 26.2

The feeding of juveniles may explain why the variety and success of ape species contracted sharply in the Pliocene and Pleistocene, whereas humans had constant population expansion. There were no changes in birthspacing or the number of young produced in a human female's lifetime, but a major shift may have occurred in the numbers of children who survived the juvenile period.

Comparative studies of mammalian life history strategies clearly indicate that increases in both total life span and the length of the prereproductive period are correlated with stable, predictable food resources. Clearly, a delay in the onset of reproduction by the expansion of the more vulnerable juvenile period will not evolve unless there is a greatly improved adult product, i.e., a more successful reproducer as a result of the investment of those juvenile years.

Parker and Gibson (1979) argue that the developed extractive foraging techniques associated with human food-sharing and tool-using require a long period of juvenile dependency. Beck (1980) in a recent review of animal tool-using behavior gives some insight into what such a protracted, protected period of juvenile dependency might mean to evolving hominids. He catalogs an impressive array of tool-using behaviors by captive species which are never known to use tools in the wild. These species, ranging from many varieties of birds and monkeys—even to ungulates—had only one thing in common: lots of leisure time in captivity, especially during the juvenile period of development. Beck argues that their seemingly more intelligent behavior compared to wild relatives is based on many opportunities to explore the properties of objects through random manipulation. Even-

JANE B. LANCASTER AND CHET S. LANCASTER

tually, a pool of responses develops which can be fortuitously reinforced by the unexpected attainment of a food reward.

For early hominids, the evolution of a protected period of juvenile dependency spent at home base could have led to major improvements in tool-using and tool-making techniques even before any significant change in brain size or organization, because of the creation of leisure time to be spent in play, object manipulation, and the development of skilled performances without a need to participate in the food quest. Perhaps this explains why the earliest stone tools currently predate any documented expansion of the hominid brain by nearly half a million years.

The Evolution of "Father." Humans, then, follow a very special pattern in terms of juvenile dependency. Unlike other animals, human juveniles do not have to feed themselves. The feeding dependency of humans is long-term, lasting among hunter-gatherers for an additional twelve years beyond the first four years of infancy and lactation. Although the mechanism by which the remarkable human success rate in rearing young is attained involves important changes in the behavior of both females and males, a crucial step lies in the evolution of the role of "father."

Although common in birds, male parental investment is relatively rare in mammals, probably because the evolution of lactation placed the major burden of parenting squarely on the shoulders of the female. However, male paternal care is surprisingly frequent in primates compared to other mammals. Of the more than 200 species of primates living today, 18 percent are monogamous compared to less than 4 percent for other mammals (Hrdy 1981; Kleiman 1977). In nearly all of these primate cases, males invest heavily in their offspring by defending them and their food resources or even by assuming the burden of carrying the infant for most of the day. Even in promiscuously mating monkeys such as baboons and macaques, males form special protective relationships with the offspring of females with whom they have mated (Altmann 1980; Busse and Hamilton 1981; Taub 1980).

Male baboons, for example, permit selected infants to feed next to them and scrounge scraps of hard-to-process foods. These same infants may be protected in dominance encounters or snatched up during a fight or when a predator threatens the group. Certainly male monkeys and apes, even those from promiscuous and polygynous mating systems, contribute to the survival of their young in a variety of ways: defense of the group against predators, protection of the group's resources from competitors, and sometimes in the location of abundant food supplies. The one thing they do not do is to regularly bring food to their mates and offspring.

Lovejoy (1981) points to the evolution of paternal provisioning of females and young as a crucial step in the improvement of human reproductive efficiency. It is certainly clear that the evolution of the role husband/father is unique to the human species and represents major ecological and social specializations. However, it should also be emphasized that human extractive efficiency is much greater than that of nonhuman primates. Hunting and gathering with tools is about twice as effective as primate foraging (Gaulin and Konner 1977). Where female baboons spend between 66 percent (nonlactating) and 81 to 89 percent (lactating) of each day foraging for food, a woman gatherer spends only a total of about two and a half days a week in the food quest. According to Lee (1980), !Kung women provide about 70 percent of the calories for the group and each gathers for an average of five individuals. Clearly, human female gathering is much more efficient than baboon female foraging, which barely supports one adult and one infant. Lee also points out that gathering by women is about 67 percent more productive than male hunting, with its high element of risk and energy expenditure. In fact, male hunters can only afford to take these risks because of the great extractive efficiency of women.

The efficiency of the human system is underscored when it is remembered that hunter-gatherers are able to support a social group with a weekly work effort of about two and a half days for healthy adult men and women (Gaulin and Konner 1977). Infants, juveniles, the sick, and the aged are not expected to contribute to the group's food supply. The division of labor and the feeding of juveniles does not simply represent male provisioning of females and young but rather a joint effort of both sexes, using much more efficient extractive foraging techniques based on skilled tool-using and intelligent food searches.

It is interesting to speculate on the effect of the economic division of labor on the social organization of the earliest humans. Since studies of Old World monkeys and apes indicate that attachment based on links through females is the prime organizer of most primate societies (Hrdy 1981; Lancaster 1975; Zihlman 1981), it is reasonable to assume that the first social network through which food was shared was one joined by uterine links. Much of this food probably was gathered by females. The division of labor and the specialization of male hunting must have added a new element to the equation. For the first time, males and females shared responsibility for feeding their offspring. Eventually, this mutual economic and reproductive interest probably gave rise to the formulation of a second set of emotional attachments—

JANE B. LANCASTER AND CHET S. LANCASTER

ones linking specific males to specific females and young. The uniqueness of the human family does not rest on the division of labor alone. Human males do not relate to females and young simply as members of an age-sex class (West and Konner 1976). The human pattern is one in which specific males relate to specific females and their children, taking on special responsibility for their welfare. This relationship, with its reciprocal economic and social obligations, is summed in the role of "father-husband"—a role with no true counterpart among nonhuman primates.

Parenting and the Life Cycle of the Human Female. When the life cycle of the human female is viewed from the perspective of parenting behavior, its major features clearly support a pattern of high levels of long-term investment and the dependency of multiple young of differing ages. For a variety of reasons, most modern cultures present poor models for the female life cycle during most of human history because of their wide variation in women's activities, health, nutrition, numbers of children born and reared, and the use of artificial techniques to alter fertility. A real understanding of women's reproductive biology must start with a reconstruction of how it unfolds in the hunting-gathering lifestyle in which it evolved. The most striking features of women's biological legacy in hunter-gatherer society all support the human pattern of parental investment. They include a long period of adolescent sterility following menarche, late age for first birth, a pattern of nearly continuous nursing, long lactation, low natural fertility, a birthspacing of four years between siblings, a low frequency of menstrual cycling during the life course, and early menopause.

A striking feature of development in hunter-gathering societies is the relatively late age at which women bear their first child, even though sexual activity begins early and may even precede puberty. Montagu (1979), reviewing both cross-cultural and historic data, demonstrates a general trend for menarche between thirteen to sixteen years, followed by several years of adolescent sterility so that fertility is unlikely to occur before age sixteen or seventeen. Earlier ages of menarche are associated with sedentary agricultural societies and later ages for the onset of reproduction with mobile hunter-gatherer groups. Montagu's data agree with Howell's (1979) careful work on !Kung hunter-gatherer demography, in which girls reach menarche between sixteen and seventeen, and birth of the first infant is at nineteen. These girls, then, are fully mature both physically and mentally before they become parents, and their investment of time and energy in their first infant does not suffer from competing interests in terms of their own biological, social, or cognitive development. This figure of menarche at sixteen contrasts sharply with records from modern developed societies in which menarche now occurs at 12.5 years (Malina 1978; Roche 1979; Tanner and Eveleth 1976).

Although the literature is now full of debate about details of the secular trend for lowering the age of puberty at various historical periods, there seems to be no real argument about the fact that both men and women complete their physical growth much earlier now than in times past and that the change is linked to diet. Stini (1971, 1979) shows that growth and diet are so closely linked that in societies under nutritional stress, the development of both sexes follows a slow growth trajectory and male growth may extend into the middle twenties, omitting the adolescent growth spurt typical of modern societies in which growth ceases in the late teens. A long period of adolescent sterility is not unique to humans of times past but is, in fact, typical of mammals (Montagu 1979), probably because undertaking reproduction before full adulthood is counterproductive. For a female mammal it allows time for full growth, the storage of energy for lactation, the establishment of a home territory, and experience in mating with a wide variety of males before undertaking the costs and risks of pregnancy, birth, and lactation.

A second major feature of women's life history, but one that was only made obvious in recent years from studies of hunter-gatherers, is a pattern of long, nearly constant lactation, which demands high levels of body contact and suckling during the first years of life. Blurton-Jones (1972) argues that the very composition of human breast milk indicates a species' adaptation for nearly continuous suckling, and Konner (1976a) documents that among !Kung hunter-gatherers women nurse their infants two to three times per hour during the day and several times during the night. The mother keeps her infant next to her body both day and night, a feature universal among hunter-gatherers (Konner and Super, ms.). Short (1976) argues that for most of human history, women spent their reproductive years in lactation rather than in menstrual cycling as they do today. He estimates that hunter-gatherer women experience nearly fifteen years of lactational amenorrhea and just under four each of pregnancy and menstrual cycles. By contrast, in the reproductive life history of the modern woman with fewer pregnancies and a longer reproductive period, the average women will spend only two years pregnant and lactating and nearly thirty-five years in menstrual cycles (see fig. 2). Given the recent linkage of estrogen (at high levels during the follicular phase of the menstrual cycle) with cancer, these figures are particularly alarming.

The importance of lactation to successful reproduction is further underscored by the fact that, if enough fat has not been stored prior

JANE B. LANCASTER AND CHET S. LANCASTER

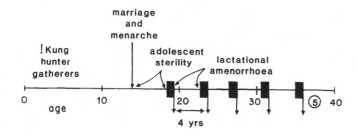

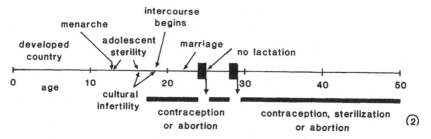

Figure 2. Changing patterns of human fertility. Nomadic hunter-gatherers have a relatively late menarche and adolescent sterility defers the birth of the first child until nineteen. Lactational amenorrhea keeps births four years apart. Early menopause results in a completed family size of about five—two to three of whom would survive into the reproductive years. In a typical developed country, menarche occurs at twelve to thirteen years of age, cultural infertility breaks down in the late teens, and intercourse before marriage requires the use of contraception or abortion. Lactation is so short that birthspacing is dependent on contraception. If the desired family size is two, contraception, sterilization, or abortion are necessary for a further twenty years. The inevitable result of this imposed sterility is an enormous increase in the number of menstrual cycles (after Short 1976:16).

to pregnancy, fetal growth will be sacrificed to maternal fat deposition (Winick 1981). It is especially interesting in this connection to note that, while human females do not advertise cyclic fertility by estrous swelling as do chimpanzees, they do continuously advertise their ability to lactate. Short's (1979) diagram (fig. 3) of a male's view of a female primate underscores the unique development of permanent breasts in pubescent women, made conspicuous and stable with deposits of fat in contrast to the other primates, whose breasts experience an increase in glandular tissue late in each pregnancy which resorbs again after

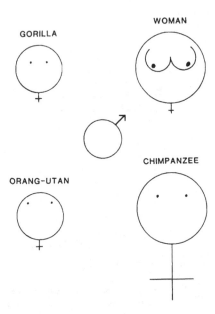

Figure 3. The male's view of the female, comparing the great apes with humans. The most significant contrasts are seen in the estrous swelling of the chimpanzee, advertising fertility, and the permanent breasts of women, advertising systemic fitness and the storage of fat for lactation (after Short 1979:17).

each weaning. The deposition of fat during human adolescence on the breasts and buttocks is a unique feature of human sexual dimorphism, which advertises reproductive fitness and healthy systemic function (Huss-Ashmore 1980), indicating not fertility so much as a woman's ability to give parental investment.

Maternal care in hunter-gatherer societies is characterized by high levels of indulgence and parental investment (Berndt 1981; Konner 1976a). The mother's breasts are bare, and the infant is permitted to nurse whenever it wants, day or night. This arrangement provides the infant with nutrition on demand, continuous body contact, an upright position during the day, and constant visual, auditory, tactile, and social stimulation. No wonder such babies are described as alert, social, content, and even precocious in development!

This extremely permissive mother-infant relationship is embedded in a highly social context. Although the mother is continuously with

JANE B. LANCASTER AND CHET S. LANCASTER

her infant, she is rarely alone with it. Her daily work and leisure activities occur within the context of a group of close friends and relatives, frequently of all ages and both sexes. As the infant matures, and begins to toddle on its own, its attention is captured by the activities of this social group, particularly its youngest members. By the time it is four years old or so and its mother directs her attention to a new infant, the attraction of a play group helps to compensate for the partial loss of her care.

Such high levels of maternal care are only possible because human hunter-gatherers experience wide birthspacing and small family size (Short 1976). Howell (1979) reports a birthspacing of nearly four years among the !Kung and a completed family size (numbers of children surviving infancy) of five. Birdsell (1980) confirms such figures for Australian aboriginals. Such a wide birthspacing is not typical of modern noncontracepting societies, but it is virtually the same as that of the great apes. It is interesting to note in this connection that Maccoby and her associates (1979) found, in measuring concentrations of sex hormones in umbilical-cord blood of Western newborns, that the levels of sex hormones in second-born children only reach first-born values when there was a spacing of four years or more between maternal pregnancies.

The mechanism underlying a natural birthspacing of four years for hunter-gatherers is currently debated, but most observers relate it one way or another to a long, four-year period of lactation, diet, and maternal activity patterns. Lee (1980) fully discusses the need of the mobile hunter-gatherer woman to balance her role as mother with that of gatherer. Although her two-and-a-half-day work week is not demanding, her gathering trips are long, and require carrying the combined burdens of her nursing infant and gathered food. Lee (1980:324) calculates that !Kung women walk 1,500 miles a year carrying loads of fifteen to thirty-three pounds. As a child matures, the mother carries it less, but over the four-year period of dependency, she will carry her child a total distance of around 4,900 miles. The hunter-gatherer woman thus finds herself at the intersection of two critical systems; the productive system, in which she supplies nearly two-thirds of the food, and the reproductive system, which demands high levels of maternal investment.

Probably several different mechanisms contribute to maintaining a wide birthspacing for hunter-gatherers (Cohen 1980; Hassan 1980; Lee 1980 for recent discussions). Birdsell (1979) argues for high levels of infanticide among hunter-gatherers. Whiting (quoted in Cohen 1980:286–87) notes that hunter-gatherers, in contrast to shifting horticulturalists, show a marked preference for infanticide as opposed to postpartum abstinence as a mechanism of maintaining birthspacing. He argues that a postpartum sex taboo more generally operates as a systematic spacer of births; infanticide more often functions in a remedial fashion to correct "mistakes." The preference for infanticide suggests, then, a low level of natural fertility mantained by a biological rather than cultural mechanism.

The most favored current hypothesis explaining wide birthspacing is based on the unusual frequency of suckling by hunter-gatherer infants who nurse on demand. When permitted, human infants naturally nurse briefly and frequently; two to three times per hour during the day and several times during the night (Konner and Worthman 1980; Konner and Super, ms.), the same high frequency as reported for gorilla (Short 1980) and chimpanzee (Clark 1977) infants. Maternal gonadal function is apparently suppressed by such frequent nipple stimulation, which leads to regular prolactin surges and suppression of estrogen and progesterone in the blood serum (McNeilly 1979). Since prolactin clears rapidly from the bloodstream, the high frequency of nipple stimulation appears to be the critical factor. As Lee puts it (1980:343), it is rather like carrying your contraceptive around on your hip.

A second factor in the naturally low fertility of hunter-gatherer women relates to critical levels of fat storage, already recognized as highly significant in human reproduction. First proposed by Frisch (1978) and recently reviewed by Cohen (1980) and Huss-Ashmore (1980), the critical fatness hypothesis suggests that human women will not ovulate unless adequate stores of fat have been deposited. These fat deposits represent enough stored energy (around 150,000 calories) to permit a woman to lactate for a year or more without having to increase her caloric intake. Frisch (1978) believes that not only birthspacing but also the timing of the onsets of menarche and menopause rest on the storage of energy in fat. Cohen (1980) notes the appeal of this hypothesis for explaining the secular trend in both the earlier onset of menarche and also the loss of a long period of adolescent sterility in modern society. Sedentism combined with high levels of caloric intake lead to early deposition of body fat in young girls and "fool" the body into early biological maturation long before cognitive and social maturity are reached.

One unexpected effect of wide birthspacing in combination with the small group size typical of hunter-gatherer society is a characteristic sibling relationship and a special composition of juvenile play groups. These features are important because they represent the context of child development for most of human history. According to Draper (1976) and Konner (1976b), the play group is continuously available,

JANE B. LANCASTER AND CHET S. LANCASTER

composed of perhaps ten to fifteen children of all ages and both sexes. Because of the wide range of ages of its members, from about two to fifteen, its activities tend to stress noncompetitive games, fantasy playing of adult roles, the practice of adult skills, and cooperative projects. It also includes both boys and girls, so that the development and practice of adult skills through play are shared by both sexes and tend to be less sexually dimorphic than in the sex-and-age-segregated play groups typical of larger societies. Older members of the play group, as well as adults in camp, supervise the play of younger ones so that no single child is burdened with the continuous care of a young sibling, as is typical of many sedentary societies where mothers must work away from home (Weisner and Gallimore 1977).

Growing up in a hunter-gatherer society, then, is very different from growing up in ours. Mothers and other adults are highly indulgent. Small children have less need to compete intensely for their parents' attention because siblings are widely spaced. The social context of life is rich with continuous opportunities for socializing with different but familiar adults and a multi-age, bisexual playgroup. This is exactly the kind of context which Beck (1980) would describe as optimal for the learning of complex skills and interaction patterns.

A final feature of the human female life-cycle is menopause. Menopause is possibly unique to human females and can best be understood from the perspective of parental investment. The cessation of fertility at a point when the organism is still physically healthy and vigorous makes no sense if natural selection favors fertility alone. Researchers in human biology and evolution have only recently looked at menopause (Frisch 1978; Gaulin 1980), and virtually no field research has been published except to show that in tribal societies women assume wider social roles once they are free of child-care responsibilities (reviewed by Fushs 1977).

Howell (1979:129–31) notes that among the !Kung, menopause follows a very different pattern than in modern society. Many !Kung women experience the birth of their last child in their late thirties or early forties. This birth is often followed by a long period of lactation lasting from four to as many as nine years. Often, the menstrual cycling never resumes after this. The long lactation period for the last child of a hunter-gatherer mother may have important effects in regulating or masking some of the more unpleasant symptoms of menopause reported by modern women. Certainly the superimposition of lactation over the hormonal fluctuations and reductions associated with menopause must give a very different hormonal, emotional, and behavioral profile. The life expectancy of the hunter-gatherer woman may be quite good once she has reached adulthood. Howell (1979) reports that among the !Kung about a third of the girls who reach age fifteen survive into the sixty plus age category. The relatively early cessation of fertility in the life cycle of hunter-gatherer women is adaptive when the long-term dependency of human juveniles and adolescents is fully appreciated. A human woman is, in fact, maximizing her reproductive potential by ceasing fertility early enough so that her last child can remain dependent for fifteen-sixteen years.

The major features of the human female's life cycle, when played out in the context of hunter-gatherer life, maximize her ability to give high levels of parental investment to a small number of dependent children spaced at roughly four-year intervals, but dependent for as much as sixteen years each. The normative biological cycles and patterns of the hunting-gathering past with late adolescence, teenage sterility, long lactation, low fertility, wide birthspacing, and early menopause look very different from those of the present (see fig. 2). The implications of these differences for the interpretation of rising frequencies of early teenage pregnancy and parenting, child abuse, sudden-infant-death syndrome, the division of labor, and changing male and female roles, birth-control techniques, human learning of parenting behavior, the timing of reproduction in the life cycle, and the experience of menopause must be enormously far-reaching and important to us if we are ever to find effective solutions to these modern problems.

Discussion. Humans, along with their great ape relatives, enjoyed a long evolutionary history of restrained fertility, late parenting, wide birthspacing, and the production of a few valued offspring needing high levels of parental investment. The principal difference between the great ape and human pattern of parental investment was an increased success in the rearing of juveniles to adulthood as a result of direct input from both parents into the feeding of offspring.

Sedentism (Hassan 1980), the development of civilized society (Goody 1976; Lancaster and Lancaster, n.d.), colonial domination by European and Asian centers (Etienne and Leacock 1981), and modern development of the Third World (Page and Lesthraeghe 1981; Rogers 1981) have all had profound and cumulative impact on the lives of women. Whereas the hunter-gatherer woman stood at the intersection of the crucial systems of production and reproduction, the combined effects of historic change since sedentism have been to increase the role and value of women as producers of children and decrease their significance in the economic and politicial systems of their societies.

Many women in the world today have reproductive histories almost diametrically opposed to those of the hunter-gatherer past. In the

JANE B. LANCASTER AND CHET S. LANCASTER

Developing World, women begin their reproductive careers early, before the completion of full growth, and they last well into the late forties. Eight or more children are born in closely spaced succession, lactation shortens, and both child and maternal morbidity and mortality are high. The burdens of parental investment in so many children become too great for each child to receive the high levels of physical and emotional care it needs. In the Developed World the high cost of raising children has led to a seeming return to the pattern of the hunter-gatherer past in which only a few children are reared. This lowered fertility, however, has been obtained through artificial means with no thought as to the evolutionary history behind women's reproductive biology. Extant birth-control techniques are all based on the assumption that menstrual cycling and not lactation amenorrhea is the normal condition of the human female.

The problem facing the world today is the relentless expansion of population, especially the high fertility of countries which have not witnessed the demographic transition. In spite of this, little real attention has been directed to the lives of Third World women, whose future behavior will undoubtedly determine the course of world population growth. History and evolution both indicate that a reduction in fertility is not correlated with improved survivorship of children alone, but rather with significant roles played by women in both the productive and reproductive systems. Yet virtually no international development programs are concerned with creating opportunities for women in economic or political life or in evaluating the impact of improved standards of living on fertility. A return to the age-old species' pattern of parents producing only a few, highly invested young can only occur if major effort is devoted to reshaping the life histories of women.

Conclusion. Evolutionary theorists argue that differences between animal forms reflect an evolutionary history of species' adaptations to particular strategies of both feeding and reproduction. For human beings, the fundamental platform of behavior for the genus *Homo* was the division of labor between male hunting and female gathering, which focused on a unique human pattern of parental investment—the feeding of juveniles. The importance of contributions by both the male and female parent toward juvenile survival reduced the significance of sexual selection in human reproductive strategies and emphasized parental investment and parental partnerships for the rearing of children. This uniquely human pattern of parental investment was so significant in human evolutionary history that it shaped the essential features of human reproductive biology and the forms and details of the human life cycle. This ancient pattern of restrained reproduction appears optimal for the production of healthy, intelligent young. A return to such a pattern after nearly 10,000 years of high fertility is not in opposition to human nature or reproductive biology. It would, in fact, be most compatible with the patterns of parental investment and reproductive biology shaped by millions of years of evolutionary history.

Acknowledgments. The stimulus for writing this article came from several years of planning and discussion with members of the Social Science Research Council's Committee for Biosocial Perspectives on Parents and Offspring. The support and encouragement of the Social Science Research Council is gratefully acknowledged. Partial funding for the activities of the committee was granted by the National Institute for Child Health and Human Development (Grant no. 5R13–HD11777–02). Data on the survivorship of juveniles was developed during a Summer Research Fellowship awarded by the Faculty Research Council of the University of Oklahoma. Special thanks are due to Barbara King and Wayne McGuire for their help in the collection of data and library references.

References

Altmann, Jeanne 1980. *Baboon Mothers and Infants*. Cambridge, Mass.: Harvard University Press.

Altmann, J., S. A. Altmann, G. Hausfater, and S. McClusky 1977. "Life Cycles of Yellow Baboons: Infant Mortality, Physical Development, and Reproductive Parameters," *Primates* 18:315–30.

Beck, B. B. 1980. *Animal Tool Behavior*. New York: Garland STPM Press.

Berndt, C. H. 1981. "Interpretation and 'Facts' in Aboriginal Australia." Pages 153–204 in F. Dahlberg, ed., *Woman the Gatherer*. New Haven, Conn.: Yale University Press.

Bertram, B. C. R. 1975 "Social Factors Influencing Reproduction in Wild Lions." *Journal of Zoology* (London) 177:463–82.

Birdsell, J. B. 1979. "Ecological Influences on Australian Aboriginal Social Organization." Pages 177–52 in I. S. Bernstein and E. O. Smiths, eds., *Primate Ecology and Human Origins*. New York: Garland STPM Press.

Blakely, R. L. 1971. "Comparison of the Mortality Profiles of Archaic, Middle Woodland, and Middle Mississippian Skeletal Populations." *American Journal of Physical Anthropology* 34:43–54.

———1977. "Sociocultural Implications of Demographic Data from Etowah, Georgia." Pages 45–66 in R. L. Blakely, ed., *Biocultural Adaptation in Prehistoric America*. Athens: University of Georgia Press.

Blurton-Jones, N. 1972. "Comparative Aspects of Mother-Child Contact." Pages 305–29 in N. Blurton-Jones, ed., *Ethological Studies of Child Behavior*. Cambridge: Cambridge University Press.

Burton, F. D., and L. A. Sawchuck 1974. "Demography of *Macaca sylvanus* of Gilbralter." *Primates* 15:271-78.

Busse, C., and W. J. Hamilton III 1981. "Infant Carrying by Male Chacma Baboons." *Science* 212:1281-282.

Campbell, B. 1979. *Humankind Emerging*. Boston: Little Brown.

Clark, C. B. 1977. "A Preliminary Report of Weaning among Chimpanzees of the Gombe National Park, Tanzania." Pages 235-60 in S. Chevalier-Sknolnikoff and F. E. Poirer, eds., *Primate Bio-Social Development*. New York: Garland Publishing.

Cohen, M. N. 1980. "Speculations on the Evolution of Density Measurement and Population Regulation in *Homo sapiens*." Pages 275-304 in M. N. Cohen, R. S. Malpass, and H. G. Klein, eds., *Biosocial Mechanisms of Population Regulation*. New Haven, Conn.: Yale University Press.

Dittus, W. P. J. 1977. "The Social Regulation of Population Density and Age-Sex Distribution in the Toque Monkey." *Behaviour* 63:281-322.

——1980. "The Social Regulation of Primate Populations: A Synthesis." Pages 263-86 in D. Lindburg, ed., *The Macaques: Studies in Ecology, Behavior, and Evolution*. New York: Van Nostrand Reinhold.

Draper, Patricia 1976. "Social and Economic Constraints on Child life among the !Kung." Pages 199-217 in R. B. Lee and I. DeVore, eds., *Kalahari Hunter-Gatherers*. Cambridge, Mass.: Harvard University Press.

Drickamer, L. C. 1974. "A Ten-Year Summary of Reproductive Data for Free-Ranging *Macaca mulatta*." *Folio Primatologica* 21:61-80.

Etienne, M., and E. Leacok 1980. *Women and Colonization: Anthropological Perspectives*. South Hadley, Mass.: Bergen.

Frisch, R. E. 1978. "Population, Food Intake, and Fertility." *Science* 199:22-30.

Fuchs, E. 1977. *The Second Season*. Garden City, N.Y.: Anchor Press/Doubleday.

Gaulin, S. J. 1980. "Sexual Dimorphism in the Human Post-Reproductive Life-Span: Possible Causes." *Journal of Human Evolution* 9:227-32.

Gaulin, S. J., and M. J. Konner 1977. "On the Natural Diet of Primates, Including Humans." Pages 2-86 in R. and J. Wurtman, eds., *Nutrition and the Brain*. Vol. I. New York: Raven Press.

Goody, Jack 1976. *Production and Reproduction: A Comparative Study of the Domestic Domain*. London: Cambridge University Press.

Harcourt, A. H., Dian Fossey, K. J. Stewart, and D. P. Watts 1980. "Reproduction in Wild Gorillas and Some Comparisons with Chimpanzees." *Journal of Reproduction and Fertility*, Supplement 28:59-70.

Hassan, Fekri 1980. "The Growth and Regulation of Human Population in Prehistoric Times." Pages 305-20 in M. N. Cohen, R. S. Malpass, and H. G. Klein, eds., *Biosocial Mechanisms of Population Regulation*. New Haven, Conn.: Yale University Press.

Howell, N. 1979. *Demography of the Dobe !Kung*. New York: Academic Press.

Hrdy, S. 1981. *The Woman That Never Evolved*. Cambridge, Mass.: Harvard University Press.

Huss-Ashmore, R. 1980. "Fat and Fertility: Demographic Implications of Differential Fat Storage." *Yearbook of Physical Anthropology* 23:65–91.

Isaac, G. L. 1976. "The Activities of Early African Hominids." Pages 483–514 in G. Isaac and L. McCown, eds., *Human Origins*. Menlo Park, Calif.: W. A. Benjamin, Inc.

Johanson, D., and Maitland Edy 1981. *Lucy: The Beginnings of Humankind*. New York: Simon and Schuster.

Jolly, Alison 1972. *The Evolution of Primate Behavior*. New York: Macmillan.

Kleiman, D. G. 1977. "Monogamy in Mammals." *The Quarterly Review of Biology* 52:39–69.

Konner, M. J. 1976a. "Maternal Care, Infant Behavior, and Development among the !Kung." Pages 218–45 in R. B. Lee and I. DeVore, eds., *Kalahari Hunter-Gatherers*. Cambridge, Mass.: Harvard University Press.

———1976b. "Relations among Infants and Juveniles in Comparative Perspective." Pages 99–129 in M. Lewis and L. Rosenblum, eds., *Friendship and Peer Relations*. New York: Wiley.

Konner, M., and C. Super, ms. "Sudden Infant Death: An Anthropological Hypothesis."

Konner, M., and C. Worthman 1980. "Nursing Frequency, Gonadal Function, and Birth Spacing among !Kung Hunter-Gatherers." *Science* 207:788–91.

Lancaster, C. S. and J. B. ms. "Eve and Adam: Making a Living and the Origins of Sex and the Family."

Lancaster, J. B. 1975. *Primate Behavior and the Emergence of Human Culture*. New York: Holt, Rinehart and Winston.

———1978. "Carrying and Sharing in Human Evolution." *Human Nature* 1:82–89.

Lee, R. B. 1980. "Lactation, Ovulation, Infanticide, and Women's Work: A Study of Hunter-Gatherer Population Regulation." Pages 321–48 in M. N. Cohen, R. S. Malpass, and R. G. Klein, eds., *Biosocial Mechanisms of Population Regulation*. New Haven, Conn.: Yale University Press.

Lewin, R. 1981. "Ethiopian Stone Tools Are World's Oldest." *Science* 211:806–07.

Lovejoy, C. O. 1981. "The Origin of Man." *Science* 211:341–50.

Lovejoy, C. O., et al. 1977. "Paleodemography of the Libben Site, Ottawa County, Ohio." *Science* 198:291–93.

Maccoby, E. E., C. H. Doering, C. N. Jacklin, and H. Kraemer 1979. "Concentrations of Sex Hormones in Umbilical-Cord Blood: Their Relation to Sex and Birth Order of Infants." *Child Development* 50:632–42.

Malina, R. M. 1978. "Adolescent Growth and Maturation: Selected Aspects of Current Research." *Yearbook of Physical Anthropology* 21:63–94.

Masui, K., Y. Sugiyama, A. Nishimura, and H. Ohsawa 1975. "The Life Table of Japanese Monkeys at Takasakiyama." Pages 401–06 in *Contemporary Primatology: Fifth International Congress of Primatologists, Nagoya 1974*. Basel: Karger.

McGrew, W. C. 1981. "The Female Chimpanzee as a Human Evolutionary Prototype." Pages 35–74 in F. Dahlberg, ed., *Woman the Gatherer*. New Haven, Conn.: Yale University Press.

McNeilly, A. S. 1979. "Effects of Lactation on Fertility." *British Medical Bulletin* 35:151–54.

Montagu, M. F. Ashley 1979. *The Reproductive Development of the Female: A Study in the Comparative Physiology of the Adolescent Organism.* Littleton, Mass.: PSG Publishing Company.

Neel, J.V., and K. Weiss 1975. "The Genetic Structure of a Tribal Population, the Yąnomamö Indians." Biodemographic Studies XII. *American Journal of Physical Anthropology* 42:25–52.

Page, H. J., and R. Lesthaeghe 1981. *Child-spacing in Tropical Africa: Traditions and Change.* New York: Academic Press.

Parker, S. T., and K. R. Gibson 1979. "A Developmental Model for the Evolution of Language and Intelligence in Early Hominids." *The Behavioral Sciences* 2:367–408.

Pimlott, D. H., J. A. Shannon, and G. B. Kolenosky 1969. *The Ecology of the Timber Wolf in Algonquin Park.* Ontario Department of Lands and Forests.

Roche, A. F. 1979. "Secular Trends in Human Growth, Maturation, and Development." *Monograph in Social Research and Child Development* 44, nos. 3–4.

Rogers, Barbara 1981. *The Domestication of Women: Discrimination in Developing Countries.* London: Tavistock.

Sade, D. S., et al. 1977. "Population Dynamics in Relation to Social Structure on Cayo Santiago." *Yearbook of Physical Anthropology* 20:253–62.

Schaller, G. B. 1972. *The Serengeti Lion: A Study of Predator-Prey Relations.* Chicago: University of Chicago Press.

Short, R. V. 1976. "The Evolution of Human Reproduction." *Proceedings, Royal Society,* Series B, Vol. 195:3–24.

———1979. "Sexual Selection and Its Component Parts, Somatic, and Genital Selection, as Illustrated by Man and the Great Apes." *Advances in the Study of Behavior* 9:131–58.

———1980. "The Great Apes of Africa." *Journal of Reproduction and Fertility,* Supplement 28:3–11.

Stini, W. A. 1971. "Evolutionary Implications of Changing Nutrition Patterns in Human Populations." *American Anthropologist* 73:1019–30.

———1979. "Adaptive Strategies of Human Populations under Nutritional Stress." Pages 387–407 in W. A. Stini, ed., *Physiological and Morphological Adaptation and Evolution.* The Hague: Mouton.

Tanner, J. M., and P. B. Eveleth 1976. "Urbanization and Growth." Pages 144–66 in G. A. Harrison and J. B. Gibson, eds., *Man in Urban Environments.* New York: Oxford University Press.

Tanner, N.M. 1981. *On Becoming Human.* New York: Cambridge University Press.

Taub, D. M. 1980. "Female Choice and Mating Strategies among Wild Barbary Macaques (*Macaca Sylvanus L.*)." Pages 287–345 in D. Lindburg, ed., *The Macaques: Studies in Ecology, Behavior and Evolution.* New York: Van Nostrand Reinhold.

Teleki, G. 1975. "Primate Subsistence Patterns: Collector-Predators and Gatherer-Hunters." *Journal of Human Evolution* 4:125–84.

Teleki, G., et al. 1976. "Demographic Observations (1963–1973) on the Chimpanzees of the Gombe National Park, Tanzania." *Journal of Human Evolution* 5:559–98.

Trivers, R. L. 1972. "Parental Investment and Sexual Selection." Pages 136–79 in B. H. Campbell, ed., *Sexual Selection and the Descent of Man.* Chicago: Aldine.

Tutin, C. E. G. 1979. "Mating Patterns and Reproductive Strategies in a Community of Wild Chimpanzees (*Pan troglodytes schwinfurthii*)." *Behavioral Ecology and Sociobiology* 6:29–38.

———1980. "Reproductive Behaviour of Wild Chimpanzees in the Gombe National Park, Tanzania." *Journal of Reproduction and Fertility,* Supplement 28:43–57.

Weisner, Thomas S., and Ronald Gallimore 1977. "My Brother's Keeper: Child and Sibling Caretaking." *Current Anthropology* 18:169–90.

West, Mary M., and Melvin Konner 1976. "The Role of the Father: An Anthropological Perspective." Pages 185–218 in M. Lamb, ed., *The Role of the Father in Child Development.* New York: Wiley.

Winick, M. 1981. "Food and the Fetus." *Natural History* 90:76–81.

Zihlman, A. L. 1981. "Women as Shapers of the Human Adaptation." Pages 75–120 in F. Dahlberg, ed., *Woman the Gatherer.* New Haven, Conn.: Yale University Press.

Commentary by Anna K. Behrensmeyer

Associate Curator of Paleobiology, Smithsonian Institution

My own view of human adaptation involves a long time perspective, over millions of years, and some direct experience with the fossil and archeological materials in Africa. As a geologist and paleontologist. I am not really qualified to discuss the fascinating physiological and behavioral aspects of human reproduction, but instead will present some thoughts based upon what the fossil and paleontological record has to say about early hominoid adaptation and behavior. I am particularly concerned with separating hypotheses based upon studies of ourselves and modern primates from those that can be based at least in part upon the fossil record.

The Lancasters' essay is well reasoned and convincing as an argument for the development of the family unit. It is consistent with the fossil evidence, such that it is. Behavior is difficult to fossilize, but many forms of evidence regarding fossil behavior actually do exist. What is lacking at present is an ability among archeologists and paleontologists to interpret the fossil behavioral evidence. This is improving by leaps and bounds, but we still have a long way to go.

I would like to go briefly over some of the evidence we do have that bears on early social structure. We have sites containing bone and stone, or just bone clusters, that point to the provisioning hypothesis and to the formation of groups of hominoids that may have scavenged or killed animals and brought them back to a central place, or campsite. We cannot, however, say how many hominoids were influential in bringing these "provisions" back or whether the male or the female of the species was doing this. We do not know how long these campsites may have existed, whether for months, years, or even hundreds of years. The stone circle at Olduvai, 1.7 million years old, which is believed to represent some sort of dwelling, suggests group activity. Again, it does not give us any information about the numbers, the timing, or the relationships of the individuals.

The "first family" that has been mentioned in the press for the Ethiopian site of Hadar, dated at about 3.5 million years ago, may in fact be simply the accumulated remains of many carnivore meals at a spot on a floodplain rather than the catastrophic demise of a whole family of early hominid individuals. In fact, there is no evidence for a flood or other catastrophe, such as a volcanic eruption, a Mount St. Helens, that would have preserved a family group intact. I do not think that this particular piece of evidence should be called "the first family."

Paleodemographic studies of some of these South African cave sites indicate delayed maturation of the young. I think this is some of the most interesting evidence in support of the idea of early parental investment.

And finally, the footprints at Laetoli, 3.7 millions years old, which according to Mary Leakey show two adults and one juvenile striding across the dusty volcanic slopes, are the most provocative of all, I think. There we have a small group, most likely a family, walking bipedally, 3.7 million years ago. There is just no doubt about their being bipedal. If you see the footprints, you can readily interpret them, much more readily than isolated scraps of bones and stone. This discovery of early bipedality might indicate that other adaptations were also earlier: gathering, hunting, scavenging, and a considerable degree of technical ability.

I think too much importance is placed on the advent of stone tools between 2.5 and 2 million years ago. Actually, tool use may go back much further into the past. There is the potential for finding sites that do not have stone tools—clusters of bones that could only have been put together by hominid activity. We are beginning to understand the processes that result in accumulations of bones. This is the subject matter of taphonomy, and in the future it may be possible to say that there are earlier campsites than there are stone tools. What this means is simply that the hunter-gatherer or scavenger-gatherer adaptation may be much older, throwing the recent changes, say, of the past thousands or tens of thousands of years into much sharper relief. And then you might ask, what was happening during all of that time after bipedality evolved—three to five million years ago or whatever it turns out to be?

I am concerned particularly with the paleoenvironments of East Africa and South Africa, where the hominid fossils and artifacts have been found. These can give us some idea of what the early adaptive strategies and pressures might have been. Something that stands out above all factors in my mind is that fluctuations of resources became more extreme during the course of the late Pliocene and Pleistocene, in time as well as in space. With the advent of drought cycles and increased seasonality there was much greater patchiness in terms of where the food and water was, the shade and salt, and the other things that were important to hominids and to humans as well.

If the early hominids were provisioning their young and raising fewer with more care, they would have needed to know very specifically where the provisions were to be found. So I think that one of the points in the Lancasters' paper that should be emphasized is the point about mental maps. That is, maps in the minds of the hominids about where and when resources would be available. I think this may actually be one of the crucial adaptations, the ability to predict over seasons and over periods of years, aided by a longer lifespan, so that knowledge was gradually accumulated about different kinds of cycles and resource fluctuations.

What made this an even more wonderful adaptation is the ability to communicate these maps in time and space within a group and to the young. Other animals can make and follow resource maps; some can even communicate these to each other. But no other animal that we know of can do it as well as humans today, and as I suspect hominids did it in the past. In the savanna habitat of the Pleistocene, being able to maximize resources through maps and communication—one might even call it planning—would have given hominids a great adaptative advantage, particularly in times of stress.

Perhaps the long period of time when little seems to have been happening was the time when this ability developed. Group resource knowledge would have been selected for or against, and the transmission to the young would have been all-important.

Discussion

DR. TOULMIN: Could I quickly raise a point arising out of Jane Lancaster's presentation that is terribly important. Right at the end of her remarks, she stated that high investment in a small number of children was the pattern that evolved in traditional hunter-gatherer societies. She suggested that the low investment in a very large number of children, which was characteristic of a lot of industrializing cultures, with a high degree of social mobility and so on, was an exception to what had been the natural state of affairs, and that what one wanted to look for was ways in which the developing countries could create the conditions which would make it possible for women, after having had a small number of children, to go back to the idea of being satisfied with that and investing in them, rather than simply going on to have more children.

I remember at a meeting in Stockholm there was a paper putting forward very much the same argument from the standpoint of economics, pointing out that in many countries, having a large number of children is the best old-age pension mechanism, the best social security mechanism one could have, and that it is the absence of communally accepted responsibility for the old which forces people, for economic reasons, into multiplying the number of children for fear of what will otherwise happen to them in old age.

I think it is a pity that we weren't able to persuade Dr. Lee Kuan Yew to come here and talk to us about Singapore, a culture which, as a matter of political policy, has adopted precisely the kind of measures that Jane Lancaster, I think, was rightly talking about.

I pick on this particular thing to underline, just because it is a very nice example of the sort of policy which makes sense both in biological and in socioeconomic terms. And what I think is very interesting about this whole debate is how often, when we do really get down to these crucial issues, you find that these are issues which optimize our situation both in biological and in socioeconomic terms, so that it no longer matters which theoretical interpretation you put on them. It becomes purely a lot of polemics, insisting that you have to see this as a biological matter rather than from a socioeconomic angle, or vice versa.

DR. HASKINS: I think that is an extremely important point, and it is a part, of course, of the thing that Betty Meggers discusses [chap. 6]—

that evolutionary forces have been operating, the selection kind of forces, from the earliest cell right straight through.

DR. TOULMIN: And operating phenotypically on the behavior and on the structures.

DR. LASLETT: But every one of those points can be refuted. The notion that people procreate children for old age is a pretty shaky one.

DR. HASSAN: The interplay between socioeconomic conditions, I think, is very well illustrated by what Dr. Lancaster said, and can be seen all through the agricultural continuum and in industrial societies. It may in fact underlie some of the problems the Yanomama Indians are going through [see Neel, chap. 3]. It may also have implications for economic conditions in general and the relationship between males and females. If, in the Pleistocene we had a division of labor that lasted two million years or more and which seems to have continued to the present, beginning to break down only in industrial times, one would wonder, to what extent do we have any built-in biological mechanisms that have to be overcome, or is this strictly cultural? One of the things that has changed, obviously, is longevity. That accelerated very much in the seventeenth and eighteenth centuries, reaching now a stage where women reach seventy years of age, twenty or more years beyond menopause, when, in the Pleistocene, they used to die.

Also, because now we have medical facilities, we can save a lot of the children and therefore we do not have the 40 or 50 percent child mortality. This means that women do not have to carry as many children to reach the two or three needed for replacement. Because we also have contraceptives that are more effective than in the past and changes in the requirements for labor, the conditions for women, whether socioeconomic or biological, that existed in the Pleistocene times, have visibly vanished and created a totally different environment for women in the world. I think the point to be raised is, how are we going to adapt to these changes and how are we adapting to them? Are we considering these long-term changes and implications, and are there any genetic underpinnings that we have to wrestle with in the process?

DR. CAVALLI-SFORZA: There seems to be a great deal of faith in the idea that lactational amenorrhea is responsible for the suspension of menstruation for three years. But it seems that there is really no evidence for that. The evidence, if anything, is that after one and a half or even less years of lactation menstruation will begin again, on average. There is some indirect evidence that lactation alone is not sufficient to induce amenorrhea as indicated by the fact that there are many cultures that have very long sex taboos. In fact, among the African pygmies sex taboos last three years.

JANE B. LANCASTER AND CHET S. LANCASTER

There has been a recent analysis of sex taboos in Africa in a book called *Child Spacing in Tropical Africa*, which shows that this taboo is extremely common and has disappeared in only a few places there, in some cases as a result of religious influence. But there would be no sex taboo, which apparently is respected, unless there were a need for it to prevent further pregnancies. The motivation for the taboo is a fairly reasonable one. It is that the last born child would suffer if there were a newborn before the former was three years old. So I think that the lactation amenorrhea idea has to be perhaps restricted. It is certainly a component, but not a sufficient one.

There are other things that I would like to discuss. I think there should be a voice of dissent in response to sociobiology. I think that sociobiology has to be considered as a dangerous prejudice for many reasons if it is not used with caution. One reason is that it is a very spotty theory, which has recently been shown to need correction—inclusive fitness, for instance [discussed briefly in chap. 5].

DR. CHAGNON: Would you repeat that?

DR. CAVALLI-SFORZA: Inclusive fitness is not a useful idea. It does not lead to the theory—well, I can give you references to that.

I have another thing to say in response to Dr. Masters. I think he said that one of the statements in my essay apparently indicated that culture has the function of giving rigidity rather than flexibility. I think that if culture gives rigidity it is just because we have sensitive periods for learning.

That is a very important concept that is not very widely accepted. I think that we are ready to learn certain things in certain times, and later we become less likely to learn. I think I can still learn something at my age.

DR. LANCASTER: I just wanted to make two points on the lactation amenorrhea. One is that it appears to be totally dependent upon the frequency of nursing, so that women can nurse for years, but if they are going to do it three or four times a day they don't get the same effect. So many African horticulturalist women can go out and work in the field for hours with their children on their hip and could, as a result, have a very different hormonal profile.

The other thing is that the return to menstrual function does not indicate a return to ovulation. This is something that has got to be more fully worked out, but there are current studies being done now on American women who are following the !Kung pattern of continuous breast feeding essentially, and they are getting birthspacing three to four years on that basis, and their nutrition is optimal, so you can't. . .

DR. CAVALLI-SFORZA: Isn't it correct that the idea of the frequency of nursing is more a hypothesis that was published about a year ago?

DR. LANCASTER: There are a number of different publications that have come out in the last year and a half on the question of monitoring the levels of prolactin and progesterone and their relation to nursing frequencies.

PARTICIPANT: There is something wrong in this discussion, because in Bangladesh the period of postpartum afertility, which is not quite the same as postpartum amenorrhea, is much longer than in some of the hunter groups, for example, which are supposed to represent the maximum reproductive possibilities.

It has been postulated there that it is the malnutrition of these women that is synergistic with the prolonged lactation. They do get a longer period of postpartum afertility than would be predicted. Certainly the Guatemalan women who lactate for two years, and are reasonably continuously breast feeding, become pregnant while they are still breast feeding—often during the second year.

DR. LANCASTER: The thing is that data have to be gotten on those women as to their frequency of breast feeding. Breast feeding alone means nothing in terms of amenorrhea.

PARTICIPANT: But these are women carrying their children essentially continuously and nursing them on demand.

DR. LASLETT: A statistical study of the European peasant population would show that there is a correlation with the present character, that women do have a strongly entrenched breast feeding custom in society. You get intervals of about two and a half years between children, but of course with very considerable variance.

That doesn't mean to say that you get no conceptions. You can get areas of Europe, in certain parts of France for example, where the breast feeding is very brief, and attempts are made, successfully in a few cases or a number of cases, to feed children from cow's milk or sheep's milk or goat's milk. And there, of course, the birth interval is very much shorter. So you do get, even in historic times, statistics to authenticate the importance of the breast-feeding hypothesis. But I don't think that anybody who holds that hypothesis believes it is the whole explanation.

I just would like to put the point to Dr. Lancaster about menarche. The observation is made among the peasant populations that menarche is certainly later than in hunter-gatherer groups. But what is interesting about this is the variance effect on the child. You get records of sixteenth- and seventeenth-century Italy, where there obviously is some sexual maturity very early—ten, eleven, and twelve years of age—and some who are not sexually mature until eighteen or nineteen. What seems to happen over time is a contraction of the variation

JANE B. LANCASTER AND CHET S. LANCASTER

around the mean, and of course in very recent times a reduction in the variance.

DR. LANCASTER: You get exactly the same variance in great apes.

PARTICIPANT: I have a question that has nothing to do with breast feeding. In our modern world, are we going to be redefining the role of parental investment? As we become increasingly more interdependent are we going to be needing to do things differently in order to ensure the survival of our offspring or are we going to be moving from a short-term focus on the nuclear family, a defense of the competitive way of parenting to a more long-term look at the family of humanity and toward a more cooperative way of parenting than now?

DR. PASSMORE: In response to the question—how far are we in any position to make social changes—I would note that we still have a tendency to think of society as something completely maleable, even though we are more and more aware that nature isn't completely maleable. So we have a vision of ourselves as deciding tomorrow to have a completely new way of bringing up children, and this is, of course, totally absurd.

We have powerful traditions, customs, habits, beliefs running through the society and exerting a profound influence, even on those who regard themselves as maximum dissidents in relation to the social structure. We had anthropologists and sociologists not so very long ago who produced a kind of determinism that would have made it impossible for us to ever make any such changes. We were so much the product of society that social change became a very mysterious thing.

Clearly, it isn't like that either. If a seed for social change already exists, perhaps independently, then it is quite astonishing how rapidly people can be swung and changed. I think this has happened, for example, in respect to pollution. People had a considerable awareness of the discomfort of pollution, of the London fogs and the pneumonia and what-have-you. As soon as people began thinking that something really had to be done about that, it was done with extraordinary rapidity.

On the other hand, if you're dealing with something like the preservation of an insect in some remote area, there is very little in Western culture which makes anybody think this is desirable or necessary. You have an awful job convincing anybody to do something because the whole tradition has been against this.

PARTICIPANT: You are bringing up some important points. If we are talking about adaptation, do we need to challenge our assumptions about our roles in social change?

DR. PASSMORE: I think everybody has been doing that in one way or another. One tends to get rather absolutist statements here, but I think we need a far more detailed look. I don't think anyone can solve the problem of how to direct social change, it does occur, but we are very uncertain about the limits of persuasion in those circumstances in which you are able to change your direction.

DR. HASSAN: Doesn't that process start with the voice of dissent? How does it start?

DR. PASSMORE: It may be a voice of dissent or a discomfort which dissent takes up.

DR. TOULMIN: It makes an awful difference if the voices of dissent actually echo around the place.

DR. MIDGLEY: The real central idea in Jane Lancaster's paper is what is the specific point at which people became human, and you have got to say pretty quickly that there are several points. But the one she fixes on, and it really does come across, is that it is when males began to share responsibilities which had previously been confined to females.

The discussion, it seems to me, missed this point in saying that the bones don't have the mental maps of where to find the food. They *do* have that; they just haven't got the time. It seems to me that the point at which primate evolution was pushing was that the females were doing what was needed for cooperation, the males were not doing it; at a certain point they began to do it.

Now, that is a motivational change of enormous force. It is not an intellectual change. The intellectual both precedes and follows it, but that is the corner. And if this is the corner, then somebody who might sympathize, who says, damn it, let's get a more interesting life for women, let's dissipate this very strong motivation to care for one's own children and turn it into a care for the whole universe, that seems to me unrealistic.

We have here an enormous emotional tendency, which has very deplorable and sinister sides, namely, that you do not care for other people's children half so much as you readily hate the people across the river. But this motivation toward one's own children is a tremendous resource, and when one talks of getting rid of it, I think one needs to think twice.

DR. HASSAN: It's interesting, because in certain segments of the population in industrial nations we are returning to a stage of primate ecology in which the male leaves the family and you end up with the female and a child. So you are breaking the unit along the weak line that has been cemented through evolution. That is a key problem we have to address.

DR. HASKINS: I think we have time for just two more comments, in view of our time constraints.

DR. ORTNER: Just a very brief anecdote. I have been interested in the Near East for several years now, although I am not intimately acquainted with its culture. But, as many of you know, there is an impending crisis developing in Egypt and other Near Eastern countries where the population size is skyrocketing. The Egyptian Government is very much aware of this, and indeed Mrs. Sadat was very much in the forefront of trying to convey a sense of immediacy to the Egyptian people about this problem. She was quite successful in doing so, at least until the tragedy of her husband's assassination. One of the things that I have asked some of my Arab friends who are women is how the Arab family will react to this attempt. I think the response of these friends is perhaps important in the context of the questions we have been raising. The response that I have been getting is that Arab women are eager to reduce the number of children they have; it is their husbands who are causing the problems. They seem to be more entrenched in the past cultural traditions. Here, if my anecdote is relevant and generally applicable, we have a situation where the women of a society seem to be much more sensitive to the biological realities than the men.

PARTICIPANT: Might I supplement that, because although it does tend to be true for Muslim society and more true generally in Egypt, there is another factor and that is the high infant and preschool mortality. When the bank came in with the first population project, so-called, and also the U.S. AID and the Population Council and so on, they made relatively little progress and came to the conclusion that they were going to continue to make poor progress until this frightfully high infant and preschool mortality was brought down. As a result, the second bank population project in Egypt, which is supposed to cover about one-third of the country and be a pioneering project, says very little about family planning per se and is concentrating almost entirely on bringing down the infant and preschool mortality.

DR. DANIELLI: How good was the evidence that mortality really was the cause?

PARTICIPANT: Well, of course, this gets us into another area. It is easy to make the statement that women don't want to accept family planning because they have an idea of a certain number of living children, but mortality has to be brought down first.

Of course, it is much harder to prove. But there are a few individual studies, which have been based on exhaustive interviews with women, that seem to confirm this. And then there is the other side, the empirical evidence that in societies where infant and preschool mortality has been brought down relatively quickly, family planning has been acceptable. But it has been associated with a whole lot of other economic and social changes.

3. Some Base Lines for Human Evolution and the Genetic Implications of Recent Cultural Developments

JAMES V. NEEL
Lee R. Dice University Professor of Human Genetics
University of Michigan Medical School

Editor's Summary. In his essay James Neel warns that human beings are confronting conditions and challenges which are quite different from those which influenced the earlier course of human evolution. He foresees potential long-range problems for the species arising from culturally induced changes in the biocultural patterns associated with our hunter-gatherer ancestors. These problems include a loss of genetic diversity, preservation of deleterious genes resulting from improved medical care, and heightened vulnerability to new epidemic diseases.

The human animal (whose genetic kinship with the higher primates is, on the basis of current work, being shown to be even closer than we had thought) now finds itself in a world quite alien to that in which it evolved. Outside of conditions of incipient domestication—and the ruthless anthropocentric selection that captivity usually entails for animals which "measure up"—no species has ever before, in its entirety, been called upon for such rapid biological adjustments.

The uniqueness of the human dilemma is that the need is self-inflicted, arising from the culture we ourselves have created. In this presentation I would like first to direct your attention to the conditions of human evolution as seen through the eyes of the geneticist. Then I would like to address the question of the extent to which we have altered those conditions. Finally, I shall discuss the time scale on which these altered conditions may be affecting our genetic endowment, and try to place in perspective our long-range genetic problems as compared with our short-range nongenetic problems.

Our Kinship with the Primates. Before proceeding, I feel obliged to document that opening statement of closeness to the other higher primates. The old criteria of comparative anatomy are being supplemented by two powerful sets of new criteria. The first of these is based on the precise composition of the chains of amino acids, termed polypeptides, found in the corresponding proteins of different primates. Since the exact sequence of the amino acids in a polypeptide can now be readily determined, we can ask by how many amino acids does a given protein—hemoglobin, for example—differ in man, the chimpanzee, and the gorilla? Table 4 presents these data with respect to twelve

TABLE 4 Difference between Humans and Chimpanzees and Gorillas with Respect to Amino-acid Composition of Selected, Sequenced Polypeptides

	Difference from Man	
Polypeptide	*Gorilla*	*Chimpanzee*
Myoglobin (153)	1*	1*
Hemoglobin α (141)	1	0
Hemoglobin β (146)	1	0
Hemoglobin δ (146)	1**	1**
Hemoglobin Gγ (146)	. . .	0
Hemoglobin Aγ (146)	. . .	0
Fibrinopeptide A (16)	0	0
Fibrinopeptide B (14)	0	0
Cytochrome c (104)	. . .	0
Carbonic Anhydrase I (260)	. . .	3****
Carbonic Anhydrase II (85)	2***	. . .
Myelin basic protein (170)	. . .	1

Note—Figures in parentheses indicate number of amino acids in the polypeptide.
*   Different substitution
**   Same substitution
*** R. E. Tashian (unpublished data)
**** D. Hewett-Emmett (unpublished data; partial sequences)

JAMES V. NEEL

of the polypeptide chains found in various important proteins, but especially hemoglobin. Normal adult hemoglobin is composed of 2 α and 2 β polypeptide chains. There is also a minor component composed of 2 α and 2 δ polypeptides. During most of fetal life the principal hemoglobin present is composed of 2 α and 2 γ chains, the γ chains being of two different kinds.

Finally, in early embryonic life there is a quite different type of hemoglobin, composed of 2 ξ and 2 ε polypeptides. The convenience with which hemoglobin can be sampled has made these chains favorite objects of research, but six other kinds of protein have also been sequenced. Man and the chimpanzee, as indicated in table 4, differ in only six of the 1,442 amino acids present in those eleven of the proteins for which data are available, whereas man and the gorilla differ in six of the 701 amino acids in the seven polypeptides for which common data are available.

Each amino acid is specified by a DNA-based genetic code in which a sequence of three so-called nucleotides in the code specifies which amino acid occurs at each position in the polypeptide. Thus a polypeptide of 100 amino acids corresponds to the coding instructions contained in 300 nucleotides. Each of the amino acid differences just mentioned can be explained by a single nucleotide difference between the two species in the coding instructions for that polypeptide. By the criterion of DNA finding expression in polypeptides, man and the chimpanzee differ in only six among the 4,326 nucleotides which specify the eleven polypeptides for which a detailed comparison is possible, whereas man and the gorilla differ in six among the 2,103 specifying the seven polypeptides for which we have data. The true differences at the DNA level are probably somewhat greater because some code changes are not expressed in protein structure, but the principle stands. These similarities are so striking that I find it tempting to consider our symposium not as the Human but as the Primate Odyssey.

A set of even more powerful techniques for comparing our species with the higher primates is now becoming available. It is recently possible to determine *directly*, and then compare, the precise composition of the genetic code (i.e., the sequence of nucleotides) in corresponding regions of the chromosomes in different species, rather than reasoning indirectly from protein structure. This enormously extends the range of possible comparisons.

Two regions under particular scrutiny in man at the present time are those coding for β-δ-γ and for the α hemoglobin polypeptides mentioned in table 4. Figure 4 is a simplified map of the β "gene." (The term "gene" is in a state of transition; I use it as synonymous

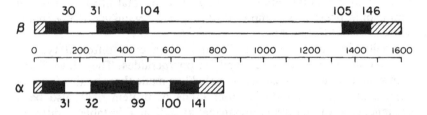

Figure 4. Fine structure of the gene coding for the β-polypeptide of human hemoglobin. Although the total length of the genetic code involved is, as shown on the scale, some 1,600 nucleotide pairs, only the portion shown in solid black finds expression in the hemoglobin molecule. The diagonally hatched sequences shown at the beginning and end of the gene are transcribed, from DNA to RNA (and that to the left apparently plays a role in initiating transcription), but do not find the functional hemoglobin molecule. Reproduced by permission of Dr. T. Maniatis and Annual Reviews, Inc.

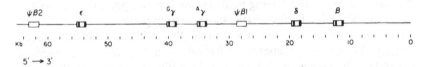

Figure 5. Fine structure of the region of chromosome immediately adjacent to the β-globin gene pictured in figure 4. Within a sequence of some 60,000 nucleotides (60 kilobases) are found five genes expressed in hemoglobin at various stages in human development, plus two pseudogenes (ψβ₁, ψβ₂). These latter are sequences of DNA with unmistakable homologies to the DNA encoding for the β-polypeptide, but do not find expression at any stage in development. All five of the genes that do find expression have the complexity of the gene shown in figure 4. Reproduced by permission of Dr. T. Maniatis and Annual Reviews, Inc.

with a sequence of DNA that finds some phenotypic expression at some organismic level.) One of the amazing discoveries of recent years is that interspersed within the DNA that gets translated into polypeptides are stretches of DNA, termed introns, that are never translated.

Figure 5 is a map of the chromosomal region immediately adjacent to the locus for the β polypeptide. Within a stretch of about 50,000 of the units (nucleotides) that comprise DNA, we find the genes coding for five of the different hemoglobin polypeptides just mentioned, these separated by stretches of DNA that are not translated into amino acids. Within the latter there are what are called "pseudo genes," stretches of DNA very similar to functional genes but not actually translated. The precise nature of each of the nucleotides of the DNA that finds expression as a polypeptide, of the nucleotides within the introns, and of those within the intervening sequences, are all rapidly being worked out.

The sole purpose of burdening you with this level of complexity is to make the point that these (and other details) now becoming available for man are also being studied in several laboratories with respect to the other higher primates. We will very shortly be able to make the same comparisons at the DNA level which I just made for the protein level, not only for these genes but for many other portions of our genetic material and perhaps, some day, for all of it. The ultimate in precise comparison between the primates will then be at hand.

Conditions of Human Evolution. Let us now consider the conditions under which we diverged from the other primates. While we cannot recreate, with the desired accuracy, the human population structure of two to three million years ago, a reasonable facsimile of more recent events is provided by some of the surviving primitive human populations. The term "primitive" is employed in the anthropological tradition as synonymous with "preliterate; technologically simple; subsistence based on hunting, gathering, and elementary agricultural practices; a social structure in which concepts of kinship play a dominant organization role; relatively untouched by civilization."

My own experience with such populations is largely limited to the Amerindians of Central and South America, especially the Yanomama of southern Venezuela and the northern half of Brazil. In drawing on that experience, I recognize that even the most remote of the surviving Amerindian primitive groups undoubtedly have been substantially influenced by post-Columbian developments. Furthermore, all of the tribes we have been among rely heavily on slash-and-burn agriculture for their subsistence, a practice that cannot be much more than 10,000 years old. On the other hand, some of the populations we have studied still practice hunting-gathering as an integral part of their economy, and must retain the mobility of even more primitive cultures. Finally, let me recognize the occupational risk inherent in generalizing from any one experience; I can only invite those whom I offend to attempt their own biocultural synthesis, at the same time reminding them that my goal has been to develop a framework for the basic biology rather than to elaborate details of the kinship or ceremonial systems.

We can conveniently divide the conditions to be treated into three subsets, namely, genetic, infectious, and dietary, which will be considered in turn.

Genetic Structure.   Manlike creatures (hominids) split off from the other primates some five million years ago. Up until perhaps 10,000 years ago— 99.8 per cent of hominid existence—the basic social unit had been the band, bands with similar speech and customs being loosely organized into tribes. In almost all the relatively unacculturated tribal societies studied thus far, the band—now usually dignified by the term village—has an acknowledged leader. He has achieved this position through a combination of well-demonstrated attributes that inspire trust and confidence, and sometimes fear. His position is usually best described as first among equals. He certainly does not have the power we have delegated to our elected representatives. His constituents are much better informed than we are concerning the essential elements of the society in which they function, and so in a much better position to engage in informal criticism. He usually has several wives, whereas the average member of the village has one. This is one of the few ways a nonmaterial society can reward achievement.

Most hunter-gatherers had to be mobile much of the time. As the ability of the species to protect its young increased, as the human adolescence lengthened, the potential for reproduction of the human female was such that the need to regulate the entry of new life into the community was recognized. By a poorly understood combination of prolonged lactation, intercourse taboos, crudely induced abortion, and infanticide, a Yanomama woman limits her introduction of a new child to this culture to about once every three years. There is a stereotype of primitive man, like other animals, reproducing at near capacity just to stay even. This is incorrect. In the Yanomama, among liveborn children not killed at birth, the probability of death prior to maturity is about 20 percent. Were a woman to reproduce on a more rapid schedule, the death rate among children would undoubtedly be higher, since this would imply both a shortening of the nursing period and loss of mobility in foraging.

If the band was not decimated by some disaster, its natural rate of increase was perhaps 0.5 to 1 percent per year. In time the band-village grew to an unwieldy size in terms of its technology. In the Yanomama this is 150 to 200 persons. But as the village expanded there arose the kind of contention for leadership among strong men documented in many tribal societies. Tensions accumulated to the point where the band or village split along lineal lines. We may surmise—although it has never been documented and now never will be—that occasionally after an especially bitter intravillage dispute, one

JAMES V. NEEL

of these fission products wandered so far afield that it lost contact with the rest of the tribe and evolved into a new tribe.

The primary concern of our fieldwork has been to bring some genetic substance, some quantification, to bear on the preceding rather hackneyed description of primitive man (Neel and Salzano 1964; Neel 1971, 1975, 1976, 1978a, 1978b, 1980; Chagnon et al. 1979; Smouse in press). Three points stand out:

(1) *Differential fertility*:   As noted earlier, village headmen (and other prominent villagers, such as shamans) are more apt to be polygynous than their fellow villagers. Among the Yanomama, the average number of living children *claimed* by a headman aged thirty-four or older was 8.6 ± 4.6 and by non-headman, 4.2 ± 3.4 (Chagnon, Flinn, and Melancon 1979). Our blood-typing studies reveal that about 9 percent of children are not the biological children of the parents who claim them, but exclusions are no more common for headmen than for other Indians (Neel and Weiss 1975). Reproductive differentials among males are even more striking if we consider expectation at birth. Because of the shallow time depth in the Yanomama, resulting from the lack of a written language but compounded by cultural factors as well, we have been forced to resort for some demographic questions to a computer simulation based on our field observations of Yanomama demography in four villages (MacCluer, Neel, and Chagnon 1971; Li, Neel, and Rothman 1978).

Figure 6 depicts the number of grandchildren born to eighty-six males who were in the cohort aged up to nine at the time the simulation was initiated. The striking spread in the number of grandchildren born to the members of this cohort is undoubtedly enhanced by the fact that a successful son of a headman has the opportunity to manipulate the marriage of his sisters in such a way that he himself has an increased opportunity of acquiring multiple wives (see next section). To the extent that the man who survives and becomes a headman has a superior genetic endowment as compared with the average Indian, the stage is set for rigorous natural selection.

Differential fertility among the women is less striking (Neel and Weiss 1975). As mentioned above, our studies suggest that those who survive to the age of reproduction have a child about every three years. Infertility is very rare among Yanomama women. The picture that emerges is of the men in fierce and lethal competition for positions of leadership, which ensure

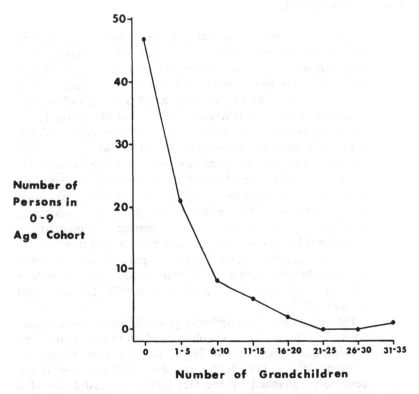

Figure 6. The number of grandchildren produced in a computer-simulation program by a cohort of 86 Yanomama males aged 0–9 years (after Neel, 1981).

enhanced reproduction, but of women throughout reproductive life striving for a culturally imposed goal of a child about every three years. Although obviously the more able mother will raise more children, the opportunity for natural selection is greater via the male. The Yanomama may present an extreme picture of male aggression, but the generality seems sound.

(2) *Inbreeding*: One is entitled to suspect that in the small and isolated bands of early hominoids, inbreeding was by contem-

porary standards very high. Such high levels persisted in many tribal societies. We have examined this subject in detail in the Yanomama. The organizing principle in most primitive cultures is kinship rather than religious or political beliefs, and the Yanomama are no exception. For instance, within each village all Yanomama categorize themselves as members of a defined lineage, each village of 100 to 150 persons containing representatives of 2 to 6 lineages, membership in a lineage being patrilineal.

Marriage is lineage exogenous; a highly preferred form of marriage is for men of two lineages to exchange younger sisters. In the following generation, the female offspring of such an exchange must again marry outside the lineage. Again, a highly preferred form of arranged marriage, which meets this requirement, is for a man to marry the daughter of his father's sister or of his mother's brother, technically a prescriptive bilateral cross-cousin marriage system. Even when a cousin marriage is not feasible, there is a preference for village endogamy.

The absence of written records for the Yanomama and a reluctance to discuss the dead result in very poor genealogies. However, for thirty-seven marriages in which all four grandparents could be identified in a complex of four villages (out of a total of 124 marriages), thirteen involved first cousins (MacCluer, Neel, and Chagnon 1971).

In the absence of good genealogies, we have resorted to a computer simulation program designed to determine how inbreeding builds up with time in the Yanomama (MacCluer, Neel, and Chagnon 1971; Spielman, Neel, and Li 1977). A first-cousin marriage corresponds to a coefficient of inbreeding of 0.06. We find that each 100 years (four generations), the coefficient of inbreeding of the average marriage accumulates by 0.01. This process is greatly abetted by the differential fertility discussed earlier. A person often has no choice but to marry a cousin. Let us assume that the ancestors of the present Amerindians arrived in the Americas some 20,000 years ago (conservative), and further assume that from the moment of that arrival inbreeding has been accumulating more or less along the lines indicated by our simulation. Then the coefficient of inbreeding in the child of the usual Amerindian marriage (the probability that the pair of alleles with which each gene is represented is identical by virtue of descent) could be as high as 0.30 to 0.50.

(3) *Genetic consequences of village fissions*: Earlier I referred to the fissioning of villages as they reached a size incompatible with Indian lifestyle and as tensions for leadership accumulated. These fissions are not such that each product represents a random sampling of the village gene pool, but rather are along family lines. We have had the opportunity to document the genetic results of three such fissionings among the Yanomama. One was a friendly event, with frequent exchanges between the two villages following the separation, but the other two resulted in the two villages "maintaining their distance" subsequent to the event. The first fission resulted in villages which, when studied by ourselves, differed genetically no more than might be expected from random sampling with no reference to bloodlines, but the other two fission events resulted in differences between the two daughter villages, which were very unlikely under a random scheme of dividing the villages (Smouse, Vitzthun, and Neel, 1981).

We believe that repetitions of this fissioning of small villages, in spite of limited migration between villages and occasional fusions of two villages which have come upon hard times, accounts for the marked genetic microdifferentiation we have observed in our studies of thirty-seven different genetic marker systems in the villages of thirteen different tribes. These structured social events create a genetic model in which groups, the genetic differences between which have been near maximized by the manner of their origin, are in vigorous competition with one another (a process sometimes termed group selection).

At the risk of gross oversimplification, let me say that Mother Nature is repeatedly trying out new gene combinations. If one of these new villages wanders so far from the other villages that it can become the nucleus for a new tribe—which then, expanding, engages in a lethal competition with other tribes— again there will be relatively large genetic differences between the competitors. This pattern of population propagation encourages human diversity; the interdeme [between group] competition that seems to have characterized early man (and down to the present) should ensure a statistical edge to the more adaptive gene combinations and, in consequence, rapid evolution.

Before leaving this discussion of the lifestyle of our tribal ancestors, I would like to depart for a moment from character as a scientist. As one driven by a mixture of youthful fantasy and scientific logic to the study of some of the most nearly intact primitive cultures in our hemisphere, I must confess how quickly, once there, it all seemed perfectly natural, how quickly the players sorted themselves out into all the same types I knew and dealt with back home—even in academia.

Disease Pressure. The impact of disease on primitive man was quite different than on ourselves. This discussion will of necessity be illustrative rather than exhaustive. We and others have so far done no more than collect a few vignettes relating to a very complex situation. For these purposes, let us distinguish between the infectious/contagious diseases and the constitutional. The infectious/contagious diseases of primitive man were, as far as we know, primarily endemic, whereas those of civilization are recently primarily epidemic. There is a medical world of difference between these two. The small, subdivided bands in which primitive man lived cannot permanently support such epidemic diseases as small-pox and measles, i.e., there is not the necessary continuous renewal of susceptibles in adequate numbers.

On the other hand, in our own studies, such conditions as tropical pyoderma (Lawrence et al. 1979a), filariasis (Beaver et al. 1976; Lawrence et al. 1979b), and multiple intestinal parasitism (Lawrence et al. 1980) were common. As best we can judge, the degree of infestation as regards the latter was not heavy (as contrasted, for instance, with what is encountered in many civilized populations). As regards viral disease, we find evidence from serological studies for the strong representation of the enteric and arboviruses (Neel et al. 1968). Malaria pressure is now heavy, but this may be a post-Columbian development. In consequence of such disease pressures as I have mentioned, immunoglobulin levels are about twice as high as our own.

Elsewhere I have argued that the transplacental transmission of these immunoglobulins to the child during pregnancy, plus universal and prolonged nursing of the child, should result in a high degree of passive immunity during the early weeks of life and a relatively smooth transition to active immunity to the various endemic diseases (Neel et al. 1964; Neel and Salzano 1967). In short, the matrix within which disease operated in primitive man simply was much less supportive of decimating epidemics than became the case with the population densities and practices of civilization.

We have had the unusual and unsought experience of very active involvement in what was probably the first epidemic of measles among the Yanomama. Much has been written about the allegedly great sensitivity of primitive man to the epidemic diseases of civilization. At the time the epidemic struck, we had just come into the field, with 2,000 doses of measles vaccine for routine immunization, plus a well-equipped medical chest. Our observations cover both the response to measles and to measles vaccine, with follow-up studies a year later with respect to antibody titer. The attack rate among unvaccinated Indians approached 100 percent. Virtually all the inhabitants of entire villages were ill simultaneously. Mortality in villages where there was neither a vaccination program nor access to antibiotics for the treatment of post-measles complications was at least 20 percent.

On the basis of observations documented elsewhere (Neel et al. 1970), it is my strong impression that much of the mortality was a reflection of the epidemiology of an infectious disease in a so-called virgin soil population. When all of the villagers are ill simultaneously, including mothers and the infants they are nursing, with no elementary care, and food and water in short supply because it must be provided daily, when the psychological response is a retreat to the hammock to await death, there are many factors militating against survival in addition to the primary response. While—in the face of the literature on experimental infections as well as the evidence for disease susceptibility related to tissue histocompatibility types—I do not doubt there are some genetically determined susceptibilities to infectious diseases, I believe they have been over-emphasized. This has an important practical implication: we cannot shrug off the poor health of acculturating peoples, continuing down to the present on Amerindian reservations in the United States, with vague allusions to innate susceptibilities. The root cause is unequal access to, or poor motivation to profit from, medical care.

To a virgin soil population overtaken by an epidemic, it is strictly academic whether their decimation is for primary or secondary reasons. Estimates of the size of the native populations of the Americas at the time of first European contact are notoriously controversial. If we accept Denevan's (1976) figure for America north of Mexico of 4,400,000, then the U.S. census figure of 1910 of 210,000, plus allowance for Canadian Indians, reveals about 95 percent decimation over a 400-year period. To quote Dorris (1981:47): "One cannot begin to fathom the trauma that must have been experienced by those few Native people who, by genetic chance, survived the onslaught of previously unknown diseases, only to watch most of their families and friends perish." I suggest that the mind frame engendered by that historical experience is one factor reflected in all recent health statistics on Amerindians.

We have on several occasions attempted to collect "base line" data among the Indians which would serve as a point of contrast for some of the undesirable consequences of life in an industrialized society. One such effort involved an attempt to derive a base line for chromosomal damage (Bloom et al. 1970). When cultured lymphocytes from human adults living in industrialized societies are examined for chromosomal damage, about one or two cells per 100 will exhibit a significant finding, in the form of chromatid breaks, free chromatid fragments, or more complex damage.

It is currently believed that a significant fraction of the cells exhibiting chromosomal damage result from the concomitants of civilization, notably, exposure to radiation and certain chemicals. Accordingly, we undertook cytological studies of the Yanomama. To our surprise, not only was the percentage of cells with evidence of chromosomal damage higher in the Yanomama than in the controls (4.1 vs. 1.0 percent) but among a total of 4,875 cells examined, we encountered in twenty-one cells an extreme form of damage not previously reported for normal subjects. When we attempted to follow this observation up two years later, we did not encounter these cells (Bloom et al. 1973). Knowing that both in cell culture and in individuals viral infections may produce chromosomal damage (rev. in Nichols 1970), our tentative explanation has been that shortly before our observations some virus had infected numerous members of the village. These strange cells were so abnormal that they could not possibly pass through a mitotic division.

Our explanation of our failure to observe them two years later was that in the interval, at cell division, they had been unable to complete the mitotic cycle because of the extensive abnormalities. I'm afraid that most of our associates had a less charitable interpretation, attributing the original findings to some type of technical error on our part. Very recently, however, Dr. Awa, who has been responsible for the very extensive cytogenetic studies undertaken on the survivors of the atomic bombings and their children, has reported fourteen such cells among a total of 60,000 examined *from persons not exposed to radiation* (Awa in press). We feel vindicated as to the validity of the finding, but unable to explain its apparently higher frequency in this particular group of Amerindians. Be this as it may, on the basis of this one probe we have no reason to believe the kinds of chromosomal damage we see in cells from urban dwellers is a recent phenomenon.

Dietary Patterns. Hunters and gatherers are extremely broad-minded as to what they eat. The Yanomama—who now probably derive some 60 to 70 percent of their caloric intake from the cooking banana, cultivated under slash-and-burn agricultural techniques—retain that broad-mindedness. Caterpillars, beetle larvae, snakes and lizards—as well, of course, as fish and larger game—are relished. We believe that primitive man, although undoubtedly occasionally hard-pressed for food at unfavorable times of the year, was in general adequately nourished, albeit seldom obese.

Lee (1968) points out that the Australian aboriginese usually achieve an adequate level of nutrition with a twenty-hour work week. When carefully examined, the reports of grossly malnourished primitive groups which surface from time to time usually demonstrably involve unfortunates pushed off onto marginal land by representatives of Western culture intruding into their traditional territory. In this connection. I do not regard the rather marginal nutritional circumstance of the new Guinea natives as typical of primitive man, but as a "special case" with complex (and debatable) origins.

Mineral balance is, of course, one aspect of diet. In examinations of some 506 Yanomama of all ages, we encountered no hypertension. This in turn led to a study of salt metabolism. The Yanomama add no salt to their diet. Our first detailed studies of the physiological adjustments to this involved males, from whom it is much easier to obtain the necessary urine samples than from females (Oliver et al. 1975). The urinary excretion of sodium per twenty-four hours for twenty-six Yanomama males averaged 1.02 ± 1.51 mEq Na+, about 1 percent of the excretion of the average salt-using inhabitant of the United States.

Salt retention is regulated primarily by hormones known as renin and aldosterone. The average plasma level of renin for eleven males was some three times higher than commonly seen in U.S. Caucasians, but the amount of aldosterone excreted in a twenty-four-hour urine collection was twenty to twenty-five times above the average level in the United States, a level encountered in U.S. males with the usual salt intake only in the presence of certain endocrine tumors.

We then began to wonder how a woman could make a baby under these circumstances, since a normal new-born infant and its placenta contains about 500–700 mEq of sodium (MacGillivnay and Buchanan 1958; Gray and Plentl 1954). Although the series of pregnant women studied on a later expedition is small, only four, the findings do not require statistics. Their plasma renin values were again clearly elevated, but the urinary aldosterone levels exceeded anything previously reported except in a few individuals with aldosterone-secreting tumors (Oliver et al. 1981). Since no other people have ever been observed under such conditions of extreme, chronic salt deprivation, there are no other data for comparison. It seems clear, however, that what we found was probably the norm during human evolution and serves as an example of the remarkable physiological resourcefulness of our species.

Earlier I mentioned one of our occasional efforts to establish base lines from which to estimate certain impacts of civilization. Another of these involved the estimation of trace metal levels in the body (Hecker et al. 1974). In a comparison of four such metals in the hair, blood, and urine of 100 Red Cross blood donors in Ann Arbor, Michigan, and 90 Yanomama, copper levels were found to be the same in both groups, but lead and cadmium markedly lower in the Yanomama. The surprise was provided by mercury. The Yanomama blood levels were significantly higher than those of the controls, most of the difference the result of higher levels in males. However, urinary mercury levels in the Yanomama were lower than those observed in U.S. controls. This discrepancy was thought to imply an exposure to alkyl rather than inorganic or aryl mercury. The most likely source of the mercury is dietary, but we have no concrete suggestions regarding the male-female difference, or, for that matter, the health impact (if any) of this finding.

Studies comparable to our own have been performed on a variety of other primitive groups, but the sum total of our knowledge of the biomedical world of primitive man remains disjointed and fragmented. The general conclusion that, viewed from the perspective of primitive man, civilization is requiring a remarkable array of physiological adjustments has been obvious for many years. We are, however, light years away from a comprehensive picture of the full scope of these adjustments (and their possible genetic implications) and, with the rapid disruption of the remaining primitive groups, now can never hope for a detailed understanding.

Our Genetic Kinship with Primitive Man. How different are we from the men and women who evolved and until very recently in the history of our species lived under the conditions just described? Obviously we have no "before" and "after" specimens for comparisons. But for two of the three principal ethnic groups which have developed major civilizations—Negroids and Mongoloids—we can still find remnant groups who did not get caught up in the culture spiral. Only the Caucasoids have thoroughly eliminated/assimilated all traces of their primitive predecessors.

At this point it is important to emphasize that the conditions of living which obtain in Europe, the United States, Japan, China, and so forth, are really very, very recent. On a comparative scale, until a few thousand years ago, most of the world has lived far more like hunter-gatherers and primitive agriculturalists than the way we find ourselves living today. Genetic change, in contrast to cultural, can occur only slowly in a long-lived animal of low fecundity such as man. The very longevity that permits the transmission of a complex culture dampens our rate of biological adjustment. Thus, we would not expect much genetic change in a few millennia in a relatively constant environment, even for a quite advantageous trait introduced through mutation.

On the other hand, it can be argued that among the highly heterogeneous human populations of 3,000 years ago, only those capable of assimilating what we are pleased to call the advances of civilization survived, so that the average man, whatever his ethnicity today, does differ appreciably from the average of his forebearers. Likewise, the possibility cannot be discounted that the great plagues of the Middle Ages exerted some selective influence, although after the experience with measles described earlier, I do not put the emphasis on this possibility that some do.

It is, then, a reasonable working hypothesis that during the relatively brief span of civilization, selection can at most have produced only minor differences between ourselves and primitive man. This hypothesis can be tested in a limited way today, and much more accurately in the near future. On the basis of the study of approximately 100 genetic traits whose precise mode of transmission is understood, we can say that no absolute differences between populations of primitive and civilized humans are known, although there are, of course, differences in the frequency with which certain traits are present in various groups. This finding is in contrast to our statement concerning chimpanzees, gorillas, and people: the amino acid differences which were described are the result of absolute genetic differences.

In the future, the ability to characterize DNA precisely, as discussed in the introductory section of this paper, will permit a much more exact treatment of the question of genetic differences between primitive and civilized man. That additional differences in the frequency of defined genetic traits will come to light is certain, including some traits that relate to intellectual functioning, just as there are already differences in anthropological characteristics. But the existence of differences in themselves, of course, has no implication that we are genetically better prepared than they for the civilization that has evolved during the past several thousand years. The differences might simply reflect the chance of which ethnic groups first got caught up in the culture spiral and then decimated their neighbors.

Potential Genetic Consequences of Recent Cultural Changes. Others in this book speak to the array of immediate problems that essentially (if not absolutely) the mind which evolved under the conditions we have been discussing now must address. I would like to consider briefly some of the genetic implications of recent cultural changes, the time scale on which these changes might occur, and their relevance to the immediate challenges confronting our species. Although, as argued above, I believe that developments since the advent of civilization have had relatively little genetic impact on our species, a continuation of these changes, especially those initiated by the Industrial Revolution, does have clear *long-range* implications: the human odyssey is entering a new phase. The genetic implications of these changes can be grouped under five categories:

1. Loss of Human Diversity. Any agricultural scientist, looking to our uncertain future, would like at his disposal just as much genetic diversity as possible in the plant and animal strains with which he deals. I cannot avoid the conclusion, given how poorly we still understand the biological basis of our own gene pool, that the same applies to ourselves. But we are losing human diversity rapidly, by virtue of two great historical trends. The first is the continuing physical disappearance of primitive groups, the decimations continuing right down to the present. The other is an accelerating rate of interethnic marriage. The result of this latter trend will, of course, be greater variation between the members of the resultant population than between the members of the contributory populations. In this sense, human diversity is not lost, but in the sense of circumscribed groups of people who on statistical balance clearly stand apart from other groups, it is.

Obviously it will be a long, long time before the lines between the major ethnic groups disappear, but if one aspect of natural selection has been to produce combinations of favorable genes in defined ethnic groups—a phenomenon the geneticist terms coadaptation (Dobzhansky 1951)—then our species is in the process of dissolving some of these combinations.

There is waiting in the wings another major source of loss of human diversity: localized decimations because of loss of commitment to population regulation. Primitive groups all over the world have been observed attempting to regulate the rate of entry of new life into their community (cf. esp. Firth 1957). The origins of that commitment and then of its apparent loss (which has characterized recent human history), and the consequences thereof, could by themselves be the subject of several conferences. I take it as a given that even with a superior system of food distribution, our species has achieved numbers perilously close to exceeding the dependable food supply of the earth. The genetic consequences of the tragedy implied by widespread famine are again of two types.

On the one hand, given the certainty that famine, though widespread, will be spottily distributed, there will be a still further impoverishment of the human gene pool. This has happened in the past, but now could be on a wider scale. On the other hand, given that the relative shortage of food implies *chronic* malnourishment for large numbers of people (a situation I presume occurred only rarely in primitive man), such people will effectively be disenfranchised from reaching their genetic potential. The success of the human odyssey is not to be measured by the numbers in which man overruns the earth, but by an equation which must include the quality of existence as well.

2. Changes in the Nature of Natural Selection. Biological (sometimes termed natural) selection has two functions. One we may term the housekeeping function; the other, the evolutionary. Although the latter gets much more attention, the former is in fact what most biological selection is all about. Two erosive forces tend to disrupt genetic adaptations. One is mutation. Considerable uncertainty still exists as to the exact rate of mutation in man, but that its effects are much more often unfavorable then favorable, there is little doubt. The other disruptive force is chance.

The principal vectors thus far identified which can offset these deleterious influences—and possibly leave a little room for positive selection—are the opportunities for selection implicit in high mortality rates and the kind of differential selection based on ability provided by the institution of headmanship (cf. Neel 1970, 1981). It is by no means clear what proportion of the prereproductive mortality of primitive people—which, exempting infanticide, we have placed at roughly 20 percent—had selective implications, but we can certainly suspect that with prereproductive mortality among live-born infants now reduced to a few percent in most civilized populations, more individuals with diseases in which genetic factors are either the chief or a contributory cause are now surviving and reproducing than in the past.

Furthermore, there is certainly less differential fertility based on leadership, and the genetic consequences of the meek inheriting the earth are by no means clear. There are some other changes in our population structure whose genetic consequences are equally obscure. For instance, whereas sterility among women in primitive cultures appears to be quite uncommon by contrast, as noted earlier, almost 10 percent of couples in the United States and Western Europe are either unable to conceive or are highly infertile. Although this infertility sometimes has an anatomical basis, more often the reason is unclear. To the extent that the basis is a physiological response to aspects of our rapidly changing culture, the potential for genetic selection is obvious.

3. *Our New Epidemiological Vulnerability.* The crawling continuum of humanity over the earth's surface has created a dangerous epidemiological setting. We now have vaccines against almost all the known agents of epidemics and are exulting over having eradicated smallpox. The fact is, we do not understand the origin of any of the previous epidemic scourges. With the growing evidence that significant evolutionary events may as often rest on poorly understood reorganizations of the genetic material as on the slow accumulation of minute genetic changes, our ability to anticipate the emergence of new agents of disease is limited. Clearly, should a significant new agent emerge, the present combination of population density and mobility is an epidemiologist's nightmare.

A similar nightmare is, of course, the potentiality for changes in the gene pool inherent in nuclear warfare. By virtue of a long association with the genetic follow-up studies in Hiroshima and Nagasaki, I have on several occasions had the dubious privilege of participating in Congressional hearings on the impact of nuclear war on this and other countries. The very well-documented consequences of a nuclear exchange—which I assume very unlikely to be "limited"—are such that those who speak of any kind of nuclear war as a tenable aspect of foreign policy are clearly out of touch with reality.

I mention the possibility of new epidemics and nuclear warfare in the context of the human odyssey for a very simple reason. Either development, to the extent it is somewhat localized, as well as circumscribed areas of starvation, has the potentiality for causing enormous changes in the human gene pool, through mortality schedules not uniformly distributed among the people of the world. Nuclear warfare also carries the possibility of an increased mutation rate. Since, however, the average amount of whole body, acute radiation which is lethal is approximately 450 r units, and since by our recent calculations the amount of radiation which will double the human mutation rate is of the order of 150 r (Schull, Otake, and Neel 1981), the absolute worst genetic impact of nuclear warfare would be to triple the mutation rate for one generation. Undesirable as this is, the net impact upon the qualities of the total human gene pool of increased mutation rates from nuclear warfare is almost certainly much less than the patchily distributed deaths of millions of people, not only from nuclear warfare and nuclear fallout but from the resultant social disorganization.

4. *More Inertia in the Gene Pool(s).* Because of the breakdown of isolates and the development of modern transportation, the sizes of human interbreeding populations will increase. This development, plus

changes in the mortality structure, implies that the role of chance in eliminating "good" genes will be diminished. Otherwise stated, large populations are less prone to major fluctuations in the frequency of the genes represented in them than are small populations. There is more inertia in the larger human gene pool(s). On the one hand, this can be regarded as a stabilizing genetic development. On the other hand, it means that, for defined population groups, genetic change in response to altered selective pressures will be slower than when these groups were numerically much smaller, but the process will be much less subject to disruption by chance.

5. Accumulation of Deleterious Genes Due to Relaxation of Inbreeding. The final consequence to be mentioned is a minor one, but it serves to illustrate some of the nuances of our changing genetic structure. Several hundred very serious, recessively inherited diseases are now recognized, whose presence in the population is almost certainly maintained by simple mutation pressure. In a large population at genetic equilibrium with no or very little inbreeding, the frequency of a recessive disease which is incompatible with reproduction, but has no effect when heterozygous, will be equal to the mutation rate. Thus at a mutation rate of 1×10^{-5}/locus/generation, the disease frequency will be 1×10^{-5}. The frequency of the responsible gene, (q), will be the square root of the disease frequency, in this case: about 0.003. If p = frequency of the normal gene, in this case, 0.997, then by the Hardy-Weinberg expression the frequency of carriers is 2 pq, or about 6/1000 persons.

In the presence of inbreeding, the frequency of carriers is less, since the marriage of cousins, who have a greater probability of carrying similar genes, is more apt to result in a defective child than the marriage of noncousins. Thus in a community practicing inbreeding, defective genes are eliminated more rapidly. For instance, if the cumulative coefficient of inbreeding for the Yanomama were really as high as we postulate, then carrier frequency should be less than one tenth of the figure mentioned above. Since some of the recessive genes associated with severe disease may have slightly deleterious effects even in single dose, the population as a whole is better off with the higher level of inbreeding. If, in addition to relaxation of inbreeding, civilization and its medical miracles also make it possible for these defective children to reproduce (as is the case now for phenylketonuria) and mutation pressure continues the same, the frequency in the population of genes with deleterious effects will increase over the expectation with random breeding given above, but on a *very* slow time scale.

Perspective. Most of the genetic impacts of cultural change, as just enumerated, proceed on a rather leisurely time scale. The exceptions are, of course, famine and nuclear war. Given the apparent inability of any of our political systems to plan more than a few years ahead, the genetic time scale is so deliberate that I doubt genetic problems will attract much attention in the near future. In my opinion, this is as it should be. There are more urgent questions which, if unmet, will dwarf concerns over any short-range genetic changes. In fact, in view of the many difficult straits our species must navigate in the near future, it may even seem a bit of self-indulgence to contemplate our evolutionary trends. I do so in almost (but not quite) the same philosophical vein that permits the astronomer to remind us that on a far more deliberate schedule, the sun is dissipating its energy.

To me, as a biomedical scientist, the human odyssey has carried us into truly strange and uncharted seas. Like Odysseus, whose ordeals returning from the siege of Troy will forever be associated with the title of our symposium, we can anticipate enormous challenges ahead, but—unlike Odysseus—there is no return to familiar shores once we have met our challenges. The estimate is that by the year 2000, unless checked by unanticipated developments, there will be 6.4 billion people on Spaceship Earth, an increase of 55 percent since 1975. By that time—although this projection is also subject to considerable uncertainty—erosion, urbanization, desertification, and saline buildup will have claimed 20 to 30 percent of the present arable land. In an effort to offset this and provide for population increases, an area of humid tropical forest the size of the State of Delaware is being converted every week to other uses. On an annual basis, this amounts to an area the size of Great Britain. Much of this conversion appears to be under circumstances which will rapidly exhaust the agricultural potential of this land. Finally, you are all familiar with the litany of disappearing mineral resources. There are limits to the ability of technology to offset the implications of those trends.

The enormous series of interlocking challenges posed by these developments will have to be met by minds essentially those of our primitive ancestors. At this point I part company with Ohno (1976) in his thesis of Promethean evolution, i.e., that "human intelligence and certain other refined functions of higher vertebrates have evolved by rising above the Epimethean law of evolution by natural selection." There is no magic genetic endowment waiting to be capitalized on as times get tougher. For the immediate future, we will make do with primate brains which have undergone some poorly understood modifications. The challenges mentioned above will be upon us well before the long-range genetic trends just described take effect. This is perhaps just as well, since so many of those long-range trends, as now understood, will either result in an erosion of our basic mental endowment or a nutritional setting in which that endowment can't be realized.

Now, I apparently have much more respect for the quality of the mind of primitive man than many. A. E. Wallace wondered, "how could selection have established an organ with so much unused potential if the brain of 'savages' is equal in innate endowment to the brain of civilized Europeans?" I respectfully suggest that the calculations that in primitive times entered into obtaining a second wife were no less demanding than aspects of the calculus, if less formalized, but with the additional flavor that the primitive man might lose his life if he calculated incorrectly.

In addition, in those times the consequences of a miscalculation were almost immediately apparent: evolution did not prepare us to deal with technologies so strong that the full implications of their application would not be apparent for 50 to 100 years. We are perilously close to being unable to cope with the complexity we have created.

I would suggest—with little hope of finding an influential audience— that the first step in moving toward meeting the short-range challenges to be treated in this symposium would be a clear recognition and statement of this magnitude by political leaders. This requires setting aside a long accumulation of petty national differences. It is shocking that Russia and the United States should be putting such effort and money into armaments when the governments of these countries could shortly (in the historical sense) be overwhelmed by the ecological and population trends bearing down upon their own and other national governments. We might far better be providing examples of how responsible nations work to restore a precarious ecological balance, a balance which could only be further disturbed by nuclear war. Somehow we must find time to catch our intellectual breaths. The rate of change we now take for granted is really at great variance with all our past evolutionary experience. My favorite statistic concerning the rate of technological change to which our species is more accustomed is that, in Western Europe during the Pleistocene, it required some 75,000 years for the rough Chellean hand-axe to be replaced by the more even symmetric Acheulian axe. Contrast that with the recent pace of events!

The view of the immediate prospects for the human odyssey which our experience with primitive cultures, combined with an analysis of current trends, thrusts upon me is gloomy. But it is not my intention to convey unmitigated pessimism or to appear to hint even obliquely to engineering a return to our more primitive days, although through inadvertence we may be heading in that direction. If in fact we can master our immediate problems and return to population numbers and a system of planetary management more reasonable with reference to current technology (expanding only when the additional technology is clearly in view), the potentialities for our continuing odyssey approach the sublime.

But as the human odyssey continues, humankind faces a biological dilemma of great magnitude. In the twin interests of humanitarianism and progress, we seem bent on changing the ground rules under which we evolved. I hope I have made it abundantly clear that we currently understand only the broad outlines of these rules. Even so, that they are being changed, I have no doubt. Gene frequencies change slowly, but advantageous combinations of genes can be relatively quickly disrupted. No responsible geneticist foresees short-term genetic problems in any way comparable to the obvious nongenetic perils in our immediate future. But, assuming we somehow deal with the latter and move to that higher plateau with respect to the quality of life which we can now imperfectly visualize, then some day we will face that moment when we do understand our genetic structure, and how it arrived at its present state, well enough to foresee the ultimate impact of the changes we have wrought. That will be a very interesting moment in human history. We will surely not attempt to offset the slow erosion of our genetic potential by breaking up into small demes, each with a genetically superior headman with multiple wives—but by what alternative will we meet this day?

In any current discussion of the problems of our transition, the question of the role to be played by the new recombinant DNA technologies is sure to arise. This issue is easily dealt with. Each day seems to bring new discoveries of the complexity of organization of our genetic material. While one can scarcely argue with the thrust of engineering bacteria to produce human insulin or growth hormone or to attack the oil slicks with which we periodically bespatter our planet, it is for the foreseeable future premature—and certainly arrogant even for us—to imagine that we might begin to remake something as poorly understood as ourselves.

Recently, Freeman Dyson (1979)—under the phrase, "The Greening of the Galaxie"—has written with dash and imagination on the subject of space colonization. While I agree with his plea that we maintain genetic diversity to ensure maximum probability of successful adaptation by some people to these new environments, that is about the extent of our agreement. Until man has adapted better to managing the only spaceship of which he will be certain—our own planet—how presumptuous to consider carrying our present maladaptations to other planets which, from all indication, are even more fragile than our own! There is challenge in abundance on every side of us, today, here!

And what of primitive man in the meantime? By the end of this century as a result of the pressure of those 6.4 billion people, there will be no more intact primitive cultures, with the exception of a few small relics. Some genetically intact, recently primitive people, yes, but no longer functioning within a matrix approximately similar to that within which we evolved. Now that we finally have the intellectual and laboratory tools to analyze ourselves evolving, the opportunity is gone. This final loss is a very small event in the perspective of the magnitude of the human tragedy, which is the history of the Western world's contact with its primitive contemporaries.

References

Awa, A. A. "Present Status of Cytogenetic Studies on the Children of Atomic Bomb Survivors." *Japanese Journal of Human Genetics,* in press.

Beaver, P. C., J. V. Neel, and T. C. Orihel 1976. "*Dipetalonema Perstans* and *Mansonella Ozzardi* in Indians of Southern Venezuela." *American Journal of Tropical Medicine and Hygiene* 25:263–65.

Bloom, A. D., J. V. Neel, K. W. Choi, S. Iida, and N. A. Chagnon 1970. "Chromosome Aberrations among the Yanomama Indians." *Proceedings of the National Academy of Sciences, U.S.A.* 66:920–27.

Bloom, A. D., J. V. Neel, T. Tsuchimoto, and K. Meilinger 1973. "Chromosomal Breakage in Leukocytes of South American Indians." *Cytogenetic and Cell Genetics* 12:175–86.

Chagnon, N. A., M. V. Flinn, and T. F. Melancon 1979. "Sex-Ration Variation among Yanomama Indians." Pages 290–320 in N. A. Chagnon and W. Irons, eds., *Evolutionary Biology and Human Social Behavior: An Anthropological Perspective.* North Scituate, R. I.: Duxbury Press.

Chagnon, N. A., and colleagues 1979. Various chapters in N. A. Chagnon and W. Irons, eds., *Evolutionary Biology and Human Social Behavior: An Anthropological Perspective.* North Scituate, R. I.: Duxbury Press.

Denevan, W. M. 1976. Chapter in W. M. Denevan, ed., *The Native Population of the Americas in 1492.* Madison: University of Wisconsin.

Dobzhansky, Th. 1951. *Genetics and the Origin of Species.* 3d ed. New York: Columbia University Press.

Dorris, M. A. 1981. "The Grass Still Grows, the Rivers Still Flow: Contemporary Native Americans." *Daedulus* 110, vol. 2:43–69.

Dyson, F. 1979. *Disturbing the Universe.* New York: Harper and Row.

Firth, R. 1957. *We, the Tikopia.* 2d ed. London: George Allan and Unwin, Ltd.

Gray, M. J., and A. A. Plentl 1954. "The Variations of the Sodium Space and the Total Exchangeable Sodium during Pregnancy." *Journal of Clinical Investigation* 33:347.

Hecker, L. H., H. E. Allen, B. D. Dinman, and J. V. Neel 1974. "Heavy

Metal Levels in Acculturated and Unacculturated Populations." *Archives of Environmental Health* 29:181–85.

Lawrence, D. N., R. R. Facklam, F. O. Sottnek, G. A. Hancock, J. V. Neel, and F. M. Salzano 1979a. "Epidemiologic Studies among Amerindian Populations of Amazonia. I. Pyoderma: Prevalence and Associated Pathogens." *American Journal of Tropical Medicine and Hygiene* 28:548–58.

Lawrence, D. N., B. Erdtmann, J. W. Peet, J. A. Nunes de Mello, G. R. Healy, J. V. Neel, and F. M. Salzano 1979b. "Epidemiologic Studies among Amerindian Populations of Amazonia. II. Prevalence of *Mansonella Ozzardi.*" *American Journal of Tropical Medicine and Hygiene* 28:991–96.

Lawrence, D. N., J. V. Neel, S. H. Abadie, L. L. Moore, L. J. Adams, G. R. Healy, and I. G. Kagan 1980. "Epidemiologic Studies among Amerindians of Amazonia. III. Intestinal Parasitoses in Newly Contacted and Acculturating Villages." *American Journal of Tropical Medicine and Hygiene* 29:530–37.

Lee, R. B. 1968. "What Hunters Do for a Living, or How to Make Out on Scarce Resources." Pages 30–48 in R. B. Lee and I. DeVore, eds., *Man the Hunter.* Chicago: Aldine Publishing Corporation.

Li, F. H. F., J. V. Neel, and E. D. Rothman 1978. "A Second Study of the Survival of a Neutral Mutant in a Simulated Amerindian Population." *American Naturalist* 112:83–96.

MacCluer, J. W., J. V. Neel, and N. A. Chagnon 1971. "Demographic Structure of a Primitive Population: A Simulation." *American Journal of Physical Anthropology* 35:193–207.

MacGillivray, I., and T. J. Buchanan 1958. "Total Exchangeable Sodium and Potassium in Non-pregnant Women and in Normal and Preeclamptic Pregnancy." *Lancet* 2:1090–1093.

Mariatis, T., E. F. Fritsch, J. Lauer, and R. M. Lawn 1980. "The Molecular Genetics of Human Hemoglobins." *Annual Review of Genetics* 14:145–78.

Neel, J. V. 1967. "The Genetic Structure of Primitive Human Populations." *Japanese Journal of Human Genetics* 12:1–16.

———1970. "Lessons from a 'Primitive' People." *Science* 170:815–22.

———1971. "Genetic Aspects of the Ecology of Disease in the American Indian." Pages 561–90 in F. M. Salzano, ed., *The Ongoing Evolution of Latin American Populations.* Springfield, Ill.: C. C. Thomas.

———1975. "The Study of 'Natural' Selection in Man: Last Chance." Pages 355–68 in F. M. Salzano, ed., *The Role of Natural Selection in Human Evolution.* Amsterdam: North-Holland Publishing Co.

———1976. "The Circumstances of Human Evolution." *The Johns Hopkins Medical Journal* 138:233–44.

———1978a. "The Population Structure of an Amerindian Tribe, the Yanomama." *Annual Review of Genetics* 12:365–413.

———1978b. "Rare Variants, Private Polymorphisms, and Locus Heterozygosity in Amerindian Populations." *American Journal of Human Genetics* 30:465–90.

———1980. "Isolates and Private Polymorphisms." Pages 175–93 in A.

Eriksson, ed., *Population Structure and Genetic Disorders*. London: Academic Press.

————1981. "The Major Ethnic Groups: Diversity in the Midst of Similarity." *American Naturalist* 117:83–87.

————1981. "On Being Headman." *Perspectives in Biology and Medicine* 24:277–94.

Neel, J. V., and F. M. Salzano 1964. "A Prospectus for Genetic Studies of the American Indian." *Cold Spring Harbor Symposium on Quantitative Biology* 29:85–98.

Neel, J. V., F. M. Salzano, P. C. Janqueira, F. Keiter, and D. Maybury-Lewis 1964. "Studies on the Xavante Indians of the Brazilian Mato Grosso." *American Journal of Human Genetics* 16:520–40.

Neel, J.V., and F. M. Salzano 1967. "Further Studies on the Xavante Indians. X. Some Hypotheses-Generalizations Resulting from these Studies." *American Journal of Human Genetics* 19:554–74.

Neel, J. V., A. H. P. Andrade, G. E. Brown, W. E. Eveland, J. Goobar, W. A. Sodeman, G. H. Stollerman, E. D. Weinstein, and A. H. Wheeler 1968. "Further Studies on the Xavante Indians. IX. Immunologic Status with Respect to Various Diseases and Organisms." *American Journal of Tropical Medicine and Hygiene* 17:486–98.

Neel, J. V., W. R. Centerwall, N. A. Chagnon, and H. L. Casey 1970. "Notes on the Effect of Measles and Measles Vaccine in a Virgin Soil Population of South American Indians." *American Journal of Epidemiology* 91:418–29.

Neel, J. V., and K. M. Weiss 1975. "The Genetic Structure of a Tribal Population, the Yanomama Indians. XII. Biodemographic Studies." *American Journal of Physical Anthropology* 42:25–51.

Neel, J. V., M. Layrisse, and F. M. Salzano 1977. "Man in the Tropics: The Yanomama Indians." Pages 109–41 in G. A. Harrison, ed., *Population Structure and Human Variation*. Cambridge: Cambridge University Press.

Nichols, W. W. 1970. "Virus-Induced Chromosome Abnormalities." *Annual Review of Microbiology* 24:479–500.

Ohno, S. 1976. "Promethean Evolution as the Biological Basis of Human Freedom and Equality." *Perspective in Biology and Medicine* 19:527–32.

Oliver, W. J., E. L. Cohen, and J. V. Neel 1975. "Blood Pressure, Sodium Intake, and Sodium Related Hormones in the Yanomama Indians, a 'No-Salt' Culture." *Circulation* 52:146–51.

Oliver, W. J., J. V. Neel, R. J. Grekin and E. L. Cohen 1981. "Hormonal Adaptation to the Stresses Imposed upon Sodium Balance by Pregnancy and Lactation in the Yanomama Indians, a Culture Without Salt." *Circulation* 63:110–16.

Schull, W. J., M. Otake, and J. V. Neel 1981. "The Genetic Effects of the Atomic Bombs: A Reappraisal. *Science* 213:1220–227.

Smouse, P. E. "Genetic Architecture of Swidden Agricultural Tribes from the Lowland Rain Forests of South America." In M. Crawford and J. Mielke, eds., *Current Developments in Anthropological Genetics*. New York: Plenum Press, in press.

Smouse, P. E., V. J. Vitzthun, and J. V. Neel 1981. "The Impact of Random and Lineal Fission on the Genetic Divergence of Small Human Groups: A Case Study among the Yanomama." *Genetics* 98:179–97.

Spielman, R. S., J. V. Neel, and F. H. F. Li 1977. "Inbreeding Estimation from Population Data: Models, Procedures, and Implications." *Genetics* 85:355–71.

Wallace, A. E. 1975. Quoted in S. J. Gould, "Charles Darwin's Natural Selection." *Science* 188:824–26.

Commentary by Napoleon A. Chagnon
Department of Anthropology, Northwestern University

When Jim Neel and I began our work together among the Yạnomamö in 1964, the anthropological study of human kinship was largely structural in its overall scope and largely taken to be symbolic cultural guidelines that defined social rules about proper economic, ritual, political, and matrimonial relationships.

Whatever kinship is about, according to one view, it is certainly not about biology. Now, that is a fairly extreme position in social anthropology, but many social anthropologists in fact lean in that direction and traditionally have tended to repudiate any suggestion that kinship can have certain biological dimensions. Another view—and one that is being rapidly eroded by animal behavior studies—is that human beings, because they possess symbolic communication, possess kinship.

About the time that Jim and I began the field work among the Yạnomamö, the English entomologist William D. Hamilton published his two now-classic papers in 1964, the theoretical consequences of which are only now beginning to be felt in anthropology, particularly in the study of social behavior, which largely—among tribesmen—means kinship behavior.

Hamilton's solution to the problem of genetic altruism, i.e., compromising your own individual reproductive success in order to enhance the survival or reproductive success of a neighbor, focused sharply on the enormous significance of the social acts—and I want to emphasize the word "social"—of favoring and disfavoring neighbors, particularly genetic relatives, and argued that the social behavior of the species would largely evolve in ways that were affected by the environment of neighbors. I think it is quite appropriate to draw attention to one's neighbors as a significant dimension of the environment.

Rather than go into any details on the technical dimensions of Hamilton's arguments, I might say that they were probably anticipated in a very puckish way by the distinguished, now deceased, geneticist J.B.S. Haldane when he said: "I would lay down my life for eight cousins." Some of us in anthropology are only now exploring the implications of Hamilton's theoretical argument.

I want to add that there is another solution to the problem of genetic altruism that time will prevent me from discussing. It is Robert Trivers' ingenious solution in 1971 to the problem of genetic altruism via reciprocal altruism.

Some of us are now beginning to explore the implications of Hamilton's argument, looking at man as a social strategist who interacts with his neighbors in a way that benefits or affects individual reproductive interests and by extension inclusive fitness interests. This is all new, and much that has been said on human sociobiology to this very day, in particular some of the recent books that have been published, has been based either on existing data or, to a large extent, speculations about what yet-to-be-collected data might look like.

It is true that very little anthropological research has been done in the field that is in fact inspired by and designed around attempts to verify or reject hypotheses that emanate specifically from these new theoretical ideas in biology. Some of the data that I collected with Jim Neel among the Yąnomamö were obtained to get geneological and sociological information as close to a biological reality as possible—I probably spent more time attempting to do that than is generally true of social anthropologists traditionally.

Some of that Yąnomamö data are in fact germane to several of the issues that have been raised by the work of Hamilton and Trivers, and those issues essentially have to do with social behavior as adaptation. Kinship can be looked at in behavioral terms. One of the things that has arisen in the study of kinship and in response to the development of sociobiological theory, is that a colleague, Professor Marshall Sahlins—in fact, he is also one of my former teachers—has made the argument that no human society on earth organizes the calculus of its social life in terms that are consistent with predictions from kin-selection theory. Now that is a fairly profound and challenging statement, and one of the things that I have done is reexamine my Yąnomamö data, which show that for at least one society this proposition can be tested. I will give you some of the results in a moment.

For some reason "axes" got involved in this conference. I have observed an axe fight in the Yąnomamö village. If Professor Sahlins's claim is true, we would expect that (in a fight in a village comprised of individuals who had varying numbers of kin and of individuals whose kin were of varying degrees of relatedness) one would recruit oneself in this fight pretty much independently of coefficients of relationship and kinship. It turns out that the coefficient of relatedness calculated from geneology among the members of that team, was 0.2124, which is roughly equivalent to a half-sibling relationship among seventeen people (males and females). Their coefficients of relatedness, on average, to the members of the other team was 0.0633, which is about a half-cousin. And their average coefficient of relatedness to the members of the village at large (268 people) was slightly higher than that.

It is not likely that this fighting team was a random sample from the village at large, and I think this is the kind of evidence that will have to be collected before champions or opponents of the approach that kinship might be an active kind of behavioral strategy can make either critical or supportive statements about the ultimate biological meaning of kinship in not only our own species but other species as well.

Jim mentioned briefly the significance of studying village fissioning. As two villages begin moving away from each other, but have not yet separated completely, the result would be what Jim referred to as a relatively friendly fission. One of the things that happens in such a fission is the formation of villages whose individual members have varying degrees of relatedness to any other person in the village. Figure 7 shows six different villages and the variation that exists in these villages, all of which are from the same population block, in the number of relatives individuals have.

In the figure you see four quartiles. What this says is that about 1 percent of the people in that village are related to 25 percent or fewer of the members of the village. The next quartile shows that about half of a percent are related to 26 to 50 percent. And about 1 percent are related to 51 to 75 percent of the other members of the village. And finally, the last quartile really reveals the consanguinity of these villages. It shows that nearly 90 percent of the members of that village are related to 76 percent or more of the rest of the members of the village.

To test propositions about human kinship in an adaptive sense, the new ideas in biology as they affect the study of human behavior would

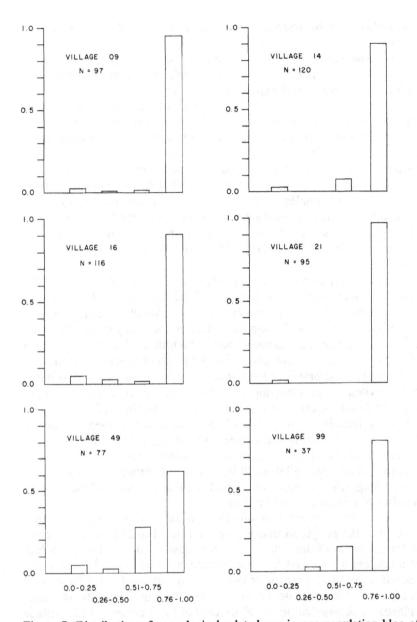

Figure 7. Distribution of genealogical relatedness in one population bloc of Yąnomamö Indians. The values are scaled as a proportion reflecting what fraction of the village residents is related to what percentage of all residents. In four of the six villages, 90 percent of the members are related in some demonstrable way to at least 76 percent of all residents in that village (from Chagnon and Irons 1979: figure 4-5).

JAMES V. NEEL

stimulate research projects in which a researcher would pick individuals who had numerous vs. others who had few kin in the village, and compare the quality of their social interactions in such things as giving mutual aid or taking sides in fights, to see if in fact this has any long-term net effect on their reproductive success. Only when that kind of research is done will it be possible to make programmatic statements on the degree of veracity that exists on either side of the argument about the degree to which sociobiological theory applies to humans—in the past or in contemporary cultures of differing social, technological, and environmental circumstances.

One of the fundamental arguments always made in anthropology textbooks is that the "nuclear" family is universal. I started worrying about this and reexamined some of my data. It is quite significant in a sense that people have to interact as social beings with other members around them. The "family" is fundamental to the whole idea of where you get your resources as you go through the life cycle. My studies show how rapidly, because of mortality, divorce, and migration, the nuclear family deteriorates as an individual goes through life.

One of the most important dimensions of human social behavior, at least historically, has been the extent to which humans must rely on aid from others, and to what extent we have argued that that aid comes from parents. The data just presented give a rough idea of how rare parental aid really is after individuals reach the age of puberty, at least in this kind of population. In effect, by the third age (eleven-fifteen years) category, they (particularly young men just entering their reproductive careers, in a society where there is some competition over females) can really expect to get very little social aid from both of their biological parents, but must instead rely on neighbors. The extent to which they do rely on these neighbors is, I think, going to be the name of the sociological and anthropological game for the future, and I think this is predicted by, and makes sense in terms of, Hamiltonian arguments in the study of behavior.

To add a somewhat lighter note to an otherwise grim kind of conference, I thought I would end my comments with a few slides showing what field research is like and what Jim Neel *really* did down there. He did some fishing. Sometimes he caught little ones. But sometimes he caught bigger ones. That fish, when we got it out on the sand, was actually much bigger than either Jim Neel or me (fig. 8). There were giants on the Orinoco in those days. Here is the intrepid evolutionary biologist or biologist-geneticist being a tourist on a jungle trail on the upper Orinoco. But this is probably the view that most of us should try to keep in our mind of Jim Neel, a view that shows what

Figure 8. Symposium chairman James V. Neel and Napoleon Chagnon with giant catfish caught in the upper Orinoco river in Venezuela during field studies of the Yąnomamö Indians.

he has been after most of his career in his field research. And I would like to pay him a personal and special tribute.

I would like to end with this comment. I was struck that many of the emergent grim developments that Jim described, and that we must adapt to, really have to be met with essentially the same kind of mind that our primitive ancestors had, the kind of mind that we probably witness today in primitive society. And I think that that is a very important point. I don't think our mind has changed that much.

The new developments in behavioral biology view man as a social strategist, making critical decisions. Social life requires a lot of complex decisions, and I think that it is important that we keep this in mind, because there are many misconceptions about what the study of evolutionary biology is. I really believe that those of us who are doing the work in terms of field research pay a great deal of attention to the role of the mind, the environment, and behavior and how the mind synthesizes data as we go about making our social choices.

In any event, I think there is much to be learned as we look toward the future and adapt to the coming environment. There is much to learn yet about the kinds of things that we acquired in the past, and much to learn about the mind that we are using, by the study of some of the disappearing people whose loss Jim and I both lament.

One final point: We may be entirely wrong, at least those of us who are pursuing these new kinds of ideas in the study of human social behavior. I will finish with a puckish comment that is alleged to have been in an exchange between Richard D. Alexander and one of his skeptical biology students, who said: "All right, I understand now what you sociobiologists are up to. We used to ask the question 'why did the chicken walk across the road?' and the answer used to be 'to get to the other side.' Now, since Hamilton and Trivers, the correct answer is 'to increase its inclusive fitness.'"

References

Chagnon, Napoleon A. and William Irons, eds., 1979. *Evolutionary Biology and Human Social Behavior*. North Scituate, R. I.: Duxbury Press.

Discussion

DR. ANGEL: The question I want to ask concerns the care of children among the Yanomama as compared to, well, any primitive agricultural group or modern industrial groups. I had the impression that the care was perhaps better, more intense. I will use the word "love," which hasn't occurred so far and which I think is rather important, between the parent and child. And I would ask whether this does apply or whether they think this applies more to a group like the Yanomama than to the same group after it has been acculturated and, as in the case of modern bushmen, forced to adopt agricultural practices? Is there a real difference and does this affect the survival of the children? My strong impression is that in comparison with the statistics I have from prehistoric farming societies, where a great many children die and die very young, that there is a slightly bigger survival in these hunting groups. Is this true?

DR. CHAGNON: The quantity and quality of love shown to children by the Yanomama Indians is of a very high quality and exceedingly abundant. They are no different than parents of other societies in that regard.

DR. NEEL: I have never followed cultures through the transition, so I can't speak to the before and after of this. But I would certainly subscribe to what Dr. Chagnon says about the quality of affection for their children in this very harsh world in which they function.

One could get into a long discussion about the extent to which that motivates them in the control of reproduction they exercise. I happen to think that one of the reasons for child spacing as practiced in many primitive people is the realization that if a child comes too quickly again it forces a nursing child off the breast before it is yet ready to cope, and I view that as one manifestation of the concern for offspring.

DR. FREDERICKSON: Might I ask one question about the degree of affection shown children by adults, although they are not related to them? Is it remarkable?

DR. NEEL: Well, as Nap brought out, in those villages practically everybody is related to everybody else, and I believe it is fair to say that if a mother is occupied, the child will nurse at the mother's sister's breast or at the grandmother's breast not infrequently. Children are a shared asset.

DR. CHAGNON: This is actually a statistical question and one can only answer this by going out in the field and collecting these kinds of data. You recall, Jim, the time we went to the village where the young man's feet were infected with insects. He got that way because he was an orphan. He was literally dying from blood poisoning, as I recall, and Jim spent two days picking little bugs out of his feet. Since there is variation in an orphan's experience, the question to be asked is how the variation correlates with the coefficiency of relatedness in terms of who is giving aid to the orphan children. That is an empirical question.

DR. TOULMIN: That is why I think it would be very interesting to look into the role of adoptive children in societies of this kind. Jim Neel made the perfectly legitimate point that a lot of the things the sociogeneticists discussed as though they were entirely the product of genetic factors can be regarded as having a mixed background, and it is not clear to me that the results you have been presenting enable us to differentiate between those which could be called sociogenetic and those which are sociocultural.

DR. CHAGNON: I happen to think that distinction is absurd.

DR. TOULMIN: Fine, but let it be made clear why it is absurd, if we are going to avoid irrelevant prejudices against the methodological approach. I remember Gary Becker of the economics department at

Chicago giving a very interesting paper about group loyalty and about the generalized economic payoff from group loyalty. And he was able to show clearly enough how great this was, and how the existence of this payoff would, in the long run, be of such a kind as to make group loyalty clearly advantageous to anybody who lived in society for any length of time.

It seems to me all of these results are explicable on a sociobiological basis if one takes this phrase "sociobiology" in a very broad sense, which embraces Gary Becker's generalized economics. And I think the reason why Don Frederickson's question cuts deep is because it has to be made clear that the basic fact is that blood relations tend to show altruism, whereas it seems to me that the basic fact is that we tend to show altruism toward blood relatives.

That is a different fact which could be the product of a whole lot of different considerations, as I think Jim rightly showed. I don't yet know of any evidence which enables us to differentiate between the respective contributions of the different factors. And I'm not prejudiced in either direction. I just want to know how much we know.

DR. CHAGNON: Well, I would go back to raise an even more general kind of issue: the question really asked by evolutionary biology is "why be social in the first place?" Most social scientists assume that living in social groups is somehow or other "natural." Well, it is not natural. There are a lot of animal species that are solitary.

Once you start thinking about that question, then it becomes very important to understand what made it possible for organisms to live together in a relatively amicable way. To that question the social sciences, including economics and Gary Becker, really do not have an answer. That is a biological question.

DR. TOULMIN: Except that Gary Becker's answer is that if you start doing it, then all kinds of things are going to result in your continuing to do it. And that is a kind of answer.

DR. CHAGNON: I would accept that. But let me also make one final remark. It is by no means inconsistent with evolutionary biological theory that the needs of the individual and the needs of the group are totally incompatible. There are lots of situations where it would pay the individual to live in a group, and that is the essential dimension of sociobiology. So where the individual and group interests are consistent, you would expect cooperative groups to develop.

DR. WIERCINSKI: I would like to raise three questions. The first is how fast micro-evolutionary change can take place? I have shown that in skeletal material from one small original population in Poland, very rapid micro-evolutionary changes may occur in the polygenic traits, in this case involving the shape of the neurocranium. At the same time, some cultural traits did not change as much: for example, the Polish language. We should always be very careful in expressing general statements such as that genetic change usually occurs slowly while cultural change may occur rapidly. It may be very different.

Second, has there been a reduction in the impact of natural selection in recent times? I would like to cite the value of the so-called index for the opportunity of selection, published on the basis of data by the anthropologists from the Poznine Center in Poland. These calculations were based on skeletal data from Poland, beginning with the Neolithic period. And since Poland is very well investigated both archeologically and anthropometrically, it is possible to have good data for such an investigation.

So the so-called index for the opportunity of the natural selection runs as follows in Neolithic times: the greater the figure, the less the impact of the selection. In the Neolithic it was 56; in the Bronze Age, 61; in the Late Medieval Period, 74; beginning the first half of the nineteenth century, 73. In 1966 in Poland, the index runs to 99. It means that there is almost no natural selection, better to say no opportunity for natural selection.

Third, what is the relationship between intrapopulational variation and interpopulational variation? My own investigations dealing with the craniological data for the evolution of hominids—for which, of course, I can't calculate standard deviations in the proper way—the figures behave very regularly. I took into consideration only simple intraindividual differences, and I have averaged them for particular chrono-territorial groups of hominids, at big intergroup differences.

There was a very gradual decrease of intergroup differences against intragroup differences. Then, if we rely upon the materials of Ilse Szwidecki, for the last 5,000 years based upon the multivariate distances, the same process continued. I have received exactly the same result for the Polish material, starting from the Neolithic and ending in recent times.

So I am afraid that Professor Neel is right when he speaks in his essay about all of the dangers because of genetic erosion of the genetic makeup of men, and I am afraid that it must occur if the situation proceeds. That is to say, this erosion must occur if natural selection is reduced to zero.

4. The Transition to Agriculture and Some of Its Consequences

L. L. CAVALLI-SFORZA

Professor of Genetics, Stanford University Medical Center

Editor's Summary. Luca Cavalli-Sforza emphasizes the transcendent importance of culture in human adaptation. Recent studies suggest the genetic component of an individual's potential intelligence is less important and the environment more important than had been thought previously. Culture allows great flexibility, partly because cultural information can be spread more quickly and extensively than genetic traits. Contemporary problems in human society may be due, in part, to the fact that cultural change, even with its great flexibility, is not always able to absorb rapid technological change.

Culture in its most general sense—the use of tools and the means of communication—is not unique to man, but is unquestionably more developed in man than in his nearest cousins by orders of magnitude. The development of the human brain is a direct witness to the importance of language and of tool production and use. Our brain is some four times larger than that of our nearest cousins, and much of the increase is in the brain structures that have to do with the production of language and with toolmaking.

Culture is one other method of adaptation that makes us more flexible, more versatile, and more ready to occupy successfully new niches—so much so that we are now close to having saturated our old planet. In fact, we are seriously worrying about how many more years we can keep multiplying at the mad rates of growth which have become prevalent in much of the world. Shall we be able to stop it by appropriate cultural measures, or is it too late? We have certainly made big mistakes in letting this crazy population explosion happen, and we now seem to be unable to regulate it. Does equilibrium have to come in the form of the old, painful remedies by which overpopulation was checked in historical times: pestilence, famine, and wars?

My interest will be devoted to considering some of the regulatory mechanisms of society—with some emphasis on population growth—that may have played a role in the past, to see if we can learn anything from them, and to inquire generally into our mechanisms of regulation, both cultural and biological.

Culture versus Nature. It may be good to replace the word "nurture" with culture in the old dichotomy of nature and nurture. Culture is nurture of the intellect and, at least in the affluent society, this kind of nurture is more important on average than food alone. Moreover, the production of food is part of culture.

Scientific thinking has shown a cyclic tendency to magnify the relative importance of either nature or culture. From 1915 to 1930 there was a strong hereditarian tendency among the human geneticists of the time, who created Mendelian genes everywhere. These rather poor, today totally discredited, scientific contributions were used by political racists and eugenicists of the time for passing discriminatory immigration quotas. The increase in understanding of Mendelian genetics in experimental animals helped promote a healthier, agnostic attitude.

It was only in the late 1960s that the hereditarian tendencies surfaced again, this time in human psychological genetics, when A. Jensen tried to give a genetic label to the difference in IQ observed between U. S. blacks and whites. This famous attempt was based on the persuasion that the heritability of IQ is high (80 percent) and that therefore there is "likely" to be a genetic component to every observed difference—even when it refers to individuals raised in environmental conditions which can hardly be considered equal. The fallacy of this reasoning is clear to most experimental geneticists. But a rigorous answer could only come from observations of adopted children. Even if they often have flaws, they represent the only relatively safe means we have, today, of distinguishing cultural and genetic transmission and "heritability." Unfortunately, adoptions—especially transracial ones—are rare. We had to wait several years until the painstaking research of Scarr and Weinberg (1976) showed that black children adopted in good white homes reach an IQ practically indistinguishable from that of white children reared under comparable conditions, and much higher than that of black children raised in the poorer black environments. Very similar results were obtained in England by Tizard (1974).

Very recently, studies on French children of humble origin adopted in families of high socioeconomic standing have shown practically no differences in IQ and school achievement with children born and raised in the higher classes (Schiff et al. 1982). This seems to contribute no evidence in favor of the idea that the somewhat large differences in IQ and achievement observed in children from different socioeconomic classes have a genetic basis. The question was raised principally by Sir Cyril Burt, the distinguished British psychologist of strong hereditarian tendencies, whose scientific honesty was challenged not long ago.

To complete the fall of this hereditarian stronghold, the estimate of the genetic heritability of IQ has recently been revised by the groups engaged in path analysis of the inheritance of IQ (Rice et al. 1981; Rao et al. 1982) and has gone from 70 to 80 percent to around 30 percent. Genes are not stripped of all their importance in the determination of IQ, but are now being given a role that, at least to this writer, seems nearer to that suggested by common sense. A complete discussion should distinguish narrow and wide heritability, but this is beyond our present interest (see Bodmer and Cavalli-Sforza 1976).

New hereditarian fuel has come from some sociobiology (Wilson 1975; Lumdsden and Wilson 1981). The attempt has been made by these last two authors to narrow the meaning of the word "culture" by postulating a rather tight leash on genes. Now, it is clear that we can have cultural activity only if our brain is fully functioning, and the development of our brain depends on genes. But it is also clear that there are stringent requirements for being accepted in human society and that there are important constraints on the level of cognitive variation tolerated.

The constraints put by society on our minimum intellectual development must be very demanding. A proof is that in all human races only a small percentage of people cannot learn the skills of reading, writing, and arithmetic, although there has been no previous evolutionary experience of these activities that might have imposed a selection for being good specifically at them. It is the freedom of development of skills in a great number of different directions that makes cultural adaptations so powerful. In some behavioral traits the range of potential variation is limited, and genes have a powerful effect in influencing it. Such traits are usually recognized as "genetically determined." Confining the study of culture to them impoverishes the discipline and deprives the word "culture" of its most interesting aspects.

In a recent book with my colleague M. Feldman (1981; see also Cavalli-Sforza et al. 1982), we have taken a totally different approach, even if we still move from biological knowledge. We find cultural objects (varying from tools and technological products, which are all physical objects, to more abstract cultural creations like languages, sciences, law codes, and so forth) have the major quality of organisms: direction to specific aims, complexity, functionality, and interdependence of parts. They are transmitted, however, with rules entirely different from those of living organisms, being transmitted, in fact, *by* living organisms. Natural selection is here replaced by "cultural" selection, that is, *we* replace the "environment" as the agent that affects choices.

Changes are not as random as mutations, but often (not always successfully) directed towards an aim; still, they share with biological mutations some aspects like abruptness; moreover, a cultural equivalent of mutation—invention—is almost random in time and space, even if not in direction. Drift and migration are the evolutionary factors with equivalent effects in cultural and biological evolution. In table 5, I summarize these considerations, which have been repeatedly made independently (Gerard et al. 1956; Cavalli-Sforza 1971; Mundinger 1980).

We thought a theoretical analysis of cultural evolution, following some of the leads of the mathematical theory of biological evolution, and paying attention to differences, could be of use. A major difference

TABLE 5   Similarities and Dissimilarities of Biological and Cultural Evolution

	Biological Evolution	*Cultural Evolution*
Maintenance of type	Transmission from parent to child by mendelian rules	Many mechanisms of cultural transmission, including parents-children, social leader-followers, social-group pressure
Factors of transmissible change at individual level	Mutation	Invention, copy error
Factors of transmissible change at population level	Natural selection (choice of types fitted to environment)	Cultural selection (choice of types more acceptable to ego)
	Migration	Migration (cultural)
	Random genetic drift: (sampling accidents of magnitude determined inversely by population size, N)	Cultural drift: very powerful, especially in transmission with one transmitter and many recipients

L. L. CAVALLI-SFORZA

is in mechanisms of transmission, which are much more varied in culture than in biology.

I am giving in figure 9 three major types of cultural transmission that can be characterized by the number of transmitters per recipient. In the middle there is a mechanism involving one transmitter and one recipient, as typically found in the parent-child relationship. It is also the only one similar to that found in genetic transmission. At the two extremes are two mechanisms, with many transmitters per recipient, and with one transmitter for many recipients. The first is typical of social group pressure, and the second of teacher-students or social leader-followers relationships.

There is frequently, of course, cultural transmission by more than one mechanism. The combination of mechanisms generates other basic stereotypes of cultural transmission: biparental transmission, grand-parental transmission (and control by the elders), hierarchical trans-mission, and so forth (fig. 10).

Each transmission mechanism has its evolutionary properties. We have considered especially two effects which can be predicted theoretically: the variation between individuals of the same population (*within-populations*) and that *between-populations* (which do not communicate, or at least not freely). The latter predicts the spatial variation and also the variation in time; the former measures the cultural homogeneity of a population for the trait being studied. Table 6 shows the theoretical predictions.

It is of importance that the many-to-one and one-to-one transmissions are practically the only ones present in hunter-gatherers, who tolerate little evolution, leaving slight chance for the adoption and spread of inventions. It is practically only with the transition to agriculture that villages enlarge; towns originate; social structures, leaders, and hier-

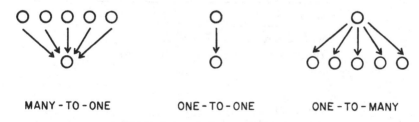

MANY - TO - ONE   ONE - TO - ONE   ONE - TO - MANY

Figure 9. Three major types of cultural transmission characterized by the number of transmitters per recipient.

 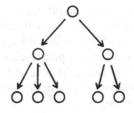

GRANDPARENTAL   HIERARCHICAL   BIPARENTAL
TRANSMISSION   TRANSMISSION   TRANSMISSION

Figure 10. Basic mechanisms of cultural transmission.

TABLE 6   Modes and Rates of Cultural Transmission

	Ratio of number of transmitters to number of receivers		
	Many to one	*One to one (Few to few)*	
Examples	Social class or caste influences	Vertical (Parent to child)	Horizontal (Child to child)
Rate of cultural change	Lowest	Intermediate	
Population Heterogeneity	Little acceptance of variants; between and within popu- lation, heteroge- neities are low	Persistence of variation (stable polymorphic equilibrium not uncommon); between and within population, heterogeneities are high	

	One to many		
	Social Hierarchies (Business)	Teacher/student, so- cial leaders, mass media	
	Highest		
	(Rapid flux) Between population, heterogeneity is high; within population, heterogeneity is low		

L. L. CAVALLI-SFORZA

archies develop. The one-to-many type of transmission now has a real chance to take on an important role and permit rapid extension of cultural changes. From the introduction of agriculture onward, cultural evolution will be faster and faster, taking the new pattern of vastly different cultures, each tendentially homogeneous, i.e., large between-variation and small within-variation which the one-to-many transmission makes more probable.

With the introduction of agriculture, a new factor of evolution has come in. Innovation is no longer acceptable, it is necessary. The possibility of feeding more people has shaken the balance of reproduction and death that caused no, or very moderate, increases of people among hunter-gatherers. There are now good reasons for having more children. Moreover, early agricultural techniques like slash and burn seriously and rapidly impoverished the soil. Thus, population growth and exhaustion of soil are powerful motivations for pushing in two directions: the search for new, virgin soil and the development of new technology to increase yields and permit continuous use of the soil by irrigation, fertilization, and so on. Both solutions are followed, the search for new soil—especially at the periphery—generating an expansion of people (a "demic" diffusion) from the original centers of agricultural development. Active technological progress is more evident, especially in the areas that had been under development for a longer time, nearer to the center of origin.

The feedback of technological development has accompanied us to the present day, and the velocity of development has been constantly increasing, despite fluctuations. A bacterial culture, made in a test tube of broth, soon starts multiplying and continues to do so at constant rate until it comes very close to exhaustion of nutrients. Then, fairly rapidly, the growth rate subsides and saturation is reached. From then on, numbers remain constant for an appreciable length of time until death begins. In the test tube Earth, humans have kept multiplying faster and faster as they approached saturation. We are only a few decades away from the inevitable end of growth and, in most parts of the world, the rate is far from declining and has recently increased. The mechanisms of birthrate regulation have obviously serious lags that do not permit fast major adjustments. It is of some interest to consider some of the things we know from ethnographic observations.

Spread of Farming from Centers of Origin. The statement I made earlier on the spread of farmers from the centers of origin of agriculture demands support. As I have been involved in work on this concept, it seems appropriate to give some information on it.

The existence of centers of origin of agriculture is fairly well documented, at least in some parts of the world—particularly the Near East from where domesticated wheat and barley, as well as cattle, sheep, goats, and pigs, spread to Europe and probably also to Asia and Africa. (The possibility that cereals may have been grown before in Egypt is now receiving some attention, however. See Wendorf and Schild 1980.) How did the diffusion from this center take place? There are two possibilities, and they are not mutually incompatible: spread by people (farmers) or by the diffusion of the idea of farming. We can call the two hypotheses demic and cultural. Archeologists have not shown interest in the problem, although they sometimes seem to adopt without discussion one or the other hypothesis. Archeological data, in fact, do not help in any direct way toward a distinction.

I conducted a quantitative analysis of the timing of the diffusion of agriculture to Europe with archeologist A. Ammerman, which showed that the spread was slow and regular. It is well known that it took some 4,000 years to cover the approximately 4,000 kilometers that are in between the nuclear area in the Near East and the most remote part reached by the spread, the north of England—computed as the crow flies (Ammerman and Cavalli-Sforza 1971, 1973, 1983).

The mathematical theory of the wave of advance by R. A. Fisher can be applied to these data. It predicts that the rate of advance here observed, about one kilometer per year, must be equal to about twice the square root of the product of the migration rate, m, and the growth rate a. The growth rate in the model is taken to be the initial growth rate of a logistic growth curve. We have estimates of migration rates of modern European farming communities, which tell us about possible orders of magnitude of m, from which we can infer that the growth rate a necessary for justifying the observed diffusion rate of 1 km/year is 0.7 percent per year or greater.

This value is not surprising at all. In fact, a should be estimated as the initial growth rate of a logistic curve, that is, the rate at the time the population is growing without constraints, almost exponentially. This period will not last long and is, in practice, confined to the region occupied by pioneers: the frontier of advance. Only a few hundred years after the first colonization of a new area, the population will be so close to saturation that the local growth rate must decrease and rapidly become close to zero. Whether this slowing down of the growth took place by increased mortality resulting from crowding, delay of marriage, or regulation of births is hard to say. But there are several historical examples showing that a rural population relocated in a new area grows exponentially for two or three centuries at a very fast rate, far closer to the maximum of the species, which is much higher than the minimum of 0.7 percent required above. For instance, the Bounty

L. L. CAVALLI-SFORZA

mutineers who occupied Pitcairn Island (a previously uninhabited Pacific Island), the settlers of Tristan da Cunha (another uninhabited island in the Atlantic Ocean), the Dutch settlers at Capetown, all developed at growth rates not far from 3 percent, doubling every twenty to thirty years. The Amish and Hutterites in the United States also multiplied, and to some extent still do, at rates considered the highest ever seen.

Whenever conditions of rural life are favorable, growth is initially very fast, but it must soon come to a stop or at least to a considerable slowdown. Some growth can continue after saturation, as the accumulation of new technological inventions permitting better soil exploitation increases the carrying capacity of the land. Also, whenever possible, part of the excess population emigrates to less crowded places. In the last 500 years Europe has kept growing slowly in numbers, sending at the same time many people to occupy remote continents.

When farmers from the Near East came to Europe, it was almost unoccupied territory. Hunter-gatherers were there at low densities, and they often favored a different type of soil, which made encounters with farmers even more rare. For certain, some of the European hunters did become acculturated, as the genetic picture will show. But in any case, local hunter-gatherers must have been no nuisance or at least no major one in most instances.

For example, it is clear that the occupation of Central Europe by the early Neolithic farmers (the "Bandkeramik" people) must have been quite peaceful. Reconstruction of their villages shows that they most often built isolated houses with no trace of fortifications. There is no reason why the first pioneers who moved in (and their children who kept pushing the frontier forward covering the ground from Czechoslovakia to Holland in some 1,000 years) could not have multiplied at the rate of the much later examples which I have cited. All of those historically known cases represented rural populations colonizing virgin areas, whose knowledge of hygiene was not superior and agricultural technology not much superior to those of early Neolithic farmers.

The evidence that the diffusion of agriculture was actually demic and not only cultural comes from other sources. What I have said so far only indicates that on the basis of ethnographic and archeological data it *could* have been demic. Genetic and skeletal data can, however, help us distinguish more clearly between the demic and cultural hypothesis. In particular, the genetic data, which are more robust toward confounding environmental effects than the skeletal data, should show a very special pattern, if there really was a diffusion of people from a center in the Near East and if a gradual admixture of them with local hunter-gatherers took place. This assumes some of them slowly acculturated and mixed with the advancing farmers.

Even though this process took place between 9,000 and 5,000 years ago, it turns out that it has in fact left its imprint on the gene-frequency patterns, confirming that the spread of agriculture was at least in part determined by the spread of the farmers themselves. This could hardly be seen in individual genes, but the pattern becomes visible if one adds in a suitable way the information from all genes. The analysis was made possible by making geographic maps of principal components of gene frequencies (Menozzi, Piazza, and Cavalli-Sforza 1978). The method showed built-in capacity to isolate independent migrations; the postulated one resulting from Near East farmers turned out to be the most important. Two other components, in order of importance, showed gene flow from Central Asia and from an area situated above the Black Sea. The former is in agreement with much historical information on mass migrations of people from Central Asia. The latter area could well be the region from which, according to many linguists, there was a spread of Indo-European-speaking populations, probably connected with the battle axe culture (see Goodenough 1970). It was quite rewarding that the first three principal components of modern gene frequencies indicated the existence of these three centers of diffusion, all of them clearly connected with previous, independent knowledge. We have also (Rendine, Piazza, and Cavalli- Sforza, in preparation) tested with computer simulations the power of the method of principal components in applications like the present one with satisfactory results.

Did the same phenomenon take place around other centers of origin of agriculture? It probably happened around centers of Southeast Asia and East Asia, in which rice, millet, and other crops originated, but the archeological and genetic information is much less adequate, and we can only make some educated guesses. Perhaps the best other case of a "demic diffusion" outside Europe is that of the Bantu expansion, starting from somewhere in Cameroon or Nigeria some 3,000 years ago (Hiernaux 1975).

Customs of Hunter-Gatherers Relevant to Reproduction. The transition to agriculture was a major break in most customs. It clearly did not take place overnight; it may have developed over hundreds or thousands of years, but it seems reasonable to assume that agriculture began spreading fast only when it had reached substantial development. In fact, it was a complex, mixed agricultural economy, which arrived through Turkey to the European mainland and spread to all the continents by way of land and sea.

In this economy, clearly, a high value was put on help for farm labor; children were probably considered very valuable, representing potential hands on the farm as well as supporters in old age. Children are very much loved in hunting-gathering societies, in particular the one in which I have worked—African Pygmies—but there they do not have such economic value. It is at first perhaps surprising to learn that hunter-gatherers have small families; there are on average a total of about five children born to a woman with completed fertility. In general, agricultural populations show higher fertilities. Under completely "natural" fertilities the number of children would be more than doubled. Our own observations will soon be published. For demography of Bushmen, see Howell (1979).

It seems very likely that the transition to agriculture involved a substantial increase in fertility, which made growth possible and spreading necessary. It is less clear if there was a contribution to growth from a decrease in death rates, but it might have been minor, considering that there is very little change in distributions of age at death from Upper Paleolithic to Neolithic (and until relatively modern times; see Hassan 1981).

Perhaps the most interesting consideration, at least for a "geneticist of culture" like myself (if I may use this potentially ambiguous term), is how this low number of children of hunter-gatherers is obtained. There are three important mechanisms of birth limitation in addition to infanticide. One is protracted lactation, which is probably not, however, a complete explanation of the four-year birth interval typical of hunting-gathering societies. Another mechanism is one that I would like to call "cultural menopause": a woman ceases to have children when her first daughter becomes pregnant (it would be a "shame" if she continued). This custom, although poorly studied, is widespread and not limited to hunter-gatherers. I suspect it is extremely old, but requires some understanding of the relation between sex and procreation, whose antiquity is unknown. (Is this knowledge the reason why Adam and Eve were chased from the garden of Eden?)   This custom may have two important consequences. If it is truly very ancient, it may explain the origin of physiological menopause, by lack of use of the female reproductory system, after forty to forty-five years of age, because of culturally induced termination of reproduction. It is useful to remember that humans are the only mammals in which females undergo menopause. The other consequence is that it effectively makes delay of marriage useless as a birth limitation procedure. In fact, if women start having children early, say at fifteen years of age, they will go into cultural menopause around thirty-six years of age on the average, for at that time it is expected that a daughter will start having children herself. This leaves her some twenty-one years time for reproduction, i.e., she will have about five children on average at four years per child. If women marry later they have time for more children, but there is not much freedom for taking advantage of the longer time before a daughter marries and has a child, if at forty-five years on average physiological menopause sets in. Under these conditions a woman can hardly have more than six children, which she can have by marrying at about twenty and continuing to have progeny, on average, until she is forty-five and either her first daughter has children or she becomes menopausal. These computations are rough and do not take account of stochastic variation. They are only made to explain by example the fact that, with cultural menopause, delaying marriage may increase rather than decrease fertility, up to a point determined by the age at physiological menopause.

The third mechanism, the postpartum taboo (Schoenmackers et al. 1981), is still widespread around the world: 16 percent of cultures of Murdock's ethnographic atlas have a taboo of twenty-five months or longer and 34 percent of thirteen months or longer. Not all hunter-gatherers have this taboo—the !Kung Bushmen do not, but they report infanticide of the last born if spacing fails. Pygmies do not practice infanticide, but a three-year taboo is common, even if not everybody observes it. According to Schoenmackers et al., the postpartum taboo was universal in sub-Saharan Africa, but has been eroded in some cultures: e.g., in cultures under Islamic influence (the Koran requests two years' lactation but only forty days of postpartum abstinence and, more exactly, as long as the bleeding lasts); in an East African area where coitus interruptus during lactation has apparently become an accepted practice; and in the lake region, where the postpartum taboo probably also existed and disappeared only recently, perhaps under the impact of Christianization, which viewed the postpartum taboo as potentially immoral because of the stimulus it offered to extramarital relations and polygamy.

From a survey by Caldwell and Caldwell (1981) on Yoruba women (Nigeria) the reason given for the postpartum sex taboo by the great majority of women interviewed is the health of the child—and for a minority, that of the mother. Only a few percent gave fertility limitations as the reason. The health of the child is also given by Central African Pygmies as reason for postpartum abstinence (B. Hewlett, personal communication). According to Gray (1981), there is some truth in the greater protection for the child of a longer interval between births.

Whatever its motivation, this practice has certainly important consequences in reducing birthrates. Even if this was an unwanted, or at least unplanned, side effect, it did most probably contribute to limiting the growth rates of hunter- gatherers. Today the postpartum taboo has been shortened or eliminated in many African populations, under various stimuli. This change is probably one of the main reasons why one observes considerable variation in fertility between different African people, causing growth rates to vary from zero to 3 percent per year.

Demographic Transitions. Customs regarding reproduction are of the greatest importance in determining the demographic behavior of a population. Hunter-gatherers even today have customs that would maintain a minimal population growth, if any. This must have been true for most of the history of man. It is also true that in the period starting some 100,000 years ago (approximately) there is likely to have been a slow but steady increase in numbers that made it possible for *Homo sapiens sapiens* to take over the earth. Wherever *Homo sapiens sapiens* had its strongest concentration at the beginning, it must have expanded after that time to America and Australia, perhaps some 30-50,000 years ago, as well as to Europe, where it supplanted *Homo sapiens neanderthalensis*, which had been there for more than 100,000 years. While there is little question, as long as present archeological evidence remains valid, that modern man arrived in America, Australia, and Europe most probably from Asia in the last period, the relation between African and Asian *Homo sapiens sapiens* is more obscure. However, some (weak) genetic data point to an origin of *Homo sapiens sapiens* in Asia, so that Africa might also have received *Homo sapiens sapiens* from Asia.

This expansion would require a very small growth rate, but even if there were only 10,000 people in the world about 100,000 years ago, and this had become 10 million at the time agriculture began, the growth rate over this long period may be misleading, since there are likely to have been important fluctuations in time and space.

Prior to the spread of agriculture, the population density in Europe was very low, similar perhaps to that of modern hunter-gatherers (one inhabitant every two square miles for African Pygmies). It is interesting that densities of hunter- gatherers are believed to be distinctly inferior (perhaps by a factor of three or four) to the "carrying capacity of the land" or the maximum density that would be possible with the particular economy of food collection (Lee and DeVore 1968). This is a very difficult estimate, but it is reasonable that a margin of safety is allowed to buffer possible fluctuations.

The population structure of African Pygmies shows an aggregation into clusters, or "camps," of twenty-five to thirty individuals on average. These are easy to count but unstable, their average duration being measured in months. The "hunting band" is probably a somewhat larger unit, and several camps often get together for hunting purposes. Net hunters, typically seven to twenty families (each owning a net), must get together for a satisfactory hunt, and hence a camp is an absolute minimum size of a cluster. For the convenience of net hunting, larger aggregates than the average camp would be preferable, but there are forces at work which act against larger groups. Some of them are seasonal and cause cycles. Not unimportant are the consequences of conflicts that tend to polarize the group into two quarreling factions: reconstitution of the peace is often obtained by fission.

Camps are mostly exogamous, and so are hunting bands. The social aggregate allowing marriage must therefore be larger. It has no specific boundary—tribal, linguistic, or geographic—but we have estimated its maximum size by studying distances traveled by individual Pygmies for any particular purpose (including that of finding a wife). An unacculturated Pygmy has visited places an average of fifty kilometers away from his birthplace or residence (B. Hewlett et al. 1982). This may take two days of travel, naturally on foot, the only possible transportation in the forest. The number of other Pygmies a man can meet in his travels is found to be on the order of 1,000 people. This represents probably a maximum size for the social network represented by all his relatives, friends, and acquaintances. It is among these people that a Pygmy also finds a spouse.

With groups of such size there is no need for a social hierarchy. In fact, Pygmies are an egalitarian society, which seems to have reached a nice balance between a reasonable degree of individual freedom, social cohesion, and mutual respect.

In spite of much exposure to agricultural practices, limited, however, to peripheral areas of the forest, many Pygmies have resisted the transition, and there are perhaps more than 100,000 Pygmies in the African forests still engaged primarily in hunting and gathering, even though they may have developed some partial agricultural activity. This is in spite of a prolonged exposure to agriculture, which may have been for a thousand or more years. A fairly complete transition is found only in areas in which the natural habitat of the Pygmy, the forest, has been destroyed beyond the point of no return. With it, game and the traditional way of life, which is an adaptation to the forest, have disappeared. Acculturation is then necessary for survival.

The attachment to the ancestral way of life is thus very strong, and the transition to agriculture through imitation of neighbors not a frequent phenomenon. Several explanations suggest themselves: change is always difficult; the imprinting of the forest environment may be especially difficult to cancel; the life of the hunter-gatherer is probably quite enjoyable (the garden of Eden). There are now some data (Bahuchet 1979) showing that, also for Pygmies, the situation is not dissimilar to that outlined by Lee for Bushmen: food collection is not a very time-consuming activity, and it leaves reasonable chances of enjoying life, probably more than for farming. There has been much discussion on whether the transition to agriculture was, or was not, determined by economic pressures (see Cohen 1977). Observations on Pygmies suggest that the transition happens only in areas in which traditional life has become truly impossible because of destruction of the natural habitat, and there is intense exposure to the agricultural way of life, as exemplified by neighbors.

After the transition to agriculture in Europe, densities may have been initially on the order of three to fifteen inhabitants per square mile (Ammerman and Cavalli-Sforza 1979) and rose almost regularly by almost two orders of magnitude. The agricultural modes of food procurement have now been extended to almost, but not all, the earth, and have caused a population increase of almost 1,000-fold in 10,000 years. People now cluster in aggregates of much larger size. Even though some farmers—for example in northern Italy and North America— still live in isolated houses, remote from their neighbors, the trend is toward greater and greater clustering into greater and greater cities, some of which contain as many people as there were in the whole world 10,000 years ago. Urbanization is perhaps one of the most dramatic changes that has taken place as a consequence of agricultural development.

Have we adapted to this new mode of life, and in what way? Has it been a biological (genetic) or cultural adaptation? The number of psychoanalysts in a block in Manhattan seems to indicate that our adaptation is, to say the least, incomplete. But man as he emerged at the end of Paleolithic and entered the Neolithic revolution was already capable of adapting to a great variety of niches. Perhaps he does not need much further genetic change for this adaptation. Moreover, even if we often don't seem to recognize it, cities are indeed created so that they can acceptably accommodate a great number of people. That their planning and development have been at least in part successful is shown by the fact that they have always kept increasing in size. Even if they are far from optimal in many respects, they are clearly acceptable to enough people so that they keep increasing. Cities are, in fact, cultural organisms whose growth is under cultural selection. They have a good chance of growing in size until there are no better alternatives. Cities will continue to be built at the minimum cost, which makes them still preferable to the other living conditions that exist and can compete with them. It is, thus, not so much we—or not entirely we—who must adapt to city life. The city keeps adapting to us as well.

City life, so profoundly different from that for which we were originally selected, is of course not the only challenge to our capacity for biological and cultural adaptation. There are a great number of other challenges, and we have recently come again to develop a feeling of nearing doomsday, as we did 1,000 years ago. It is unlikely that this is because the millennium is again to end. Probably what contributes to it more than anything else is the knowledge of what a nuclear war—or even an accident with a nuclear bomb—could do to humanity. Also, everything happens at an increasing rate: pollution, overpopulation, and exhaustion of resources. All of these things inevitably contribute to giving us a sense of impending disaster.

It seems to me, however, that there are elements of exaggeration in some of our current preoccupations about the future of our species. Not only pollution and overpopulation increase at a faster rate, but so do cultural and technological advances, and there are more solutions offered today to the old and the new problems than there could have been until some time ago. The rates of worsening and of improvement are both high because they come from the same source: the process of cultural evolution. Perhaps also because I have recently come from Europe and remember well its long history of appearances and disappearances of civilizations, of tragedies faithfully recorded in three or more thousand years, which may well obscure even possible future nuclear catastrophes, I have a little more faith in the capacity of the human species to come out of the worst disasters. The Black Death in the fourteenth century killed one-third of the European population—at least in some areas—a deed not easy to match even for many hawkish politicians of today. But even if the human species is very likely to survive, we may be heading for sorry times. If the unthinkable happens, recovery may take decades and leave deep scars.

Just to end on a more hopeful note, I would like to mention the books by Princeton physicist Gerard O'Neill (1981) on the future of space colonization. These books are convincing testimonials that we are very near to forming space colonies. The major interest, I find, is perhaps not—or not only—the possibility they offer of helping old Mother Earth with the population problem or with the energy problem. It is rather in the chance they give of forming new communities that can try social innovations. One can hope to come a little closer to the goal of generating social environments in which the development of one's personality can achieve greater harmony between one's inclinations and the needs of the rest of society.

References

Ammerman, A. J., and L. L. Cavalli-Sforza 1971. "Measuring the Rate of Spread of Early Farming in Europe." *Man* 6:674-88.

———— 1973. "A Population Model for the Diffusion of Early Farming in Europe." Pages 343-57 in C. Renfrew, ed., *The Explanation of Culture Change*. London: G. Duckworth & Co.

———— 1979. "The Wave of Advance Model for the Spread of Agriculture in Europe." Pages 275-93 in K. Cooke and C. Renfrew, eds., *Transformations: Mathematical Approaches to Culture Change*. New York: Academic Press.

———— 1983. *The Neolithic Transition in Europe: Demic Diffusion and Its Implication for Human Genetics*. Princeton, N. J.: Princeton University Press. In press.

Bahuchet, Serge, ed. 1979. *Pygmees de Centrafrique*. Paris: Selaf.

Bodmer, W. F., and L. L. Cavalli-Sforza 1976. *Genetics, Evolution and Man*. San Francisco: W. H. Freeman & Co.

Caldwell, P., and J. C. Caldwell 1981. "The Function of Child-spacing in Traditional Societies and the Direction of Change." Pages 73-92 in H. J. Page and R. Lesthaegh, eds., *Child-Spacing in Tropical Africa*. New York: Academic Press.

Cavalli-Sforza, L. L. 1971. "Similarities and Dissimilarities of Sociocultural and Biological Evolution." Pages 535-41 in F. R. Hodson, D. G. Kendall, and P. Tautu, eds., *Mathematics in the Archeological and Historical Sciences*. Edinburgh: Edinburgh University Press.

Cavalli-Sforza, L. L., M. W. Feldman, K. H. Chen, and S. M. Dornbusch 1982. "Theory and Observation in Cultural Transmission." *Science*, in press.

Cavalli-Sforza, L. L., and M. W. Feldman. 1981. *Cultural Transmission and Evolution: A Quantitative Approach*. Princeton, N. J.: Princeton University Press.

Cohen, M. N. 1977. *The Food Crisis in Prehistory*. New Haven, Conn.: Yale University Press.

Fisher, R. A. 1937. "The Wave of Advance of Advantageous Genes." *Annals of Eugenics VII* (pt. IV):355-69.

Gerard, R. W., C. Kluckhohn, and A. Rapoport 1956. "Biological and Cultural Evolution: Some Analogies and Explorations." *Behavioral Sciences* 1:6-34.

Goodenough, W. H. 1970. "The Evolution of Pastoralism and Indo-European Origins." Pages 253-65 in G. Cardona, H. M. Hoenigswald and L. Senn, eds., *Indo-Europe and Indo-Europeans*. Philadelphia: University of Pennsylvania Press.

Gray, R. H. 1981. "Birth Intervals, Postpartum Sexual Abstinence and Child Health." Pages 93-109 in H. J. Page and R. Lesthaeghe, eds., *Child-Spacing in Tropical Africa*. New York: Academic Press.

Hassan, F. 1981. *Demographic Archeology*. New York: Academic Press.

Hewlett, B., I. Van de Koppel, and L. L. Cavalli-Sforza 1982. "Exploration Ranges of Aka Pygmies of the Central Africa Republic." *Man*, in press.

Hiernaux, J. 1975. *The People of Africa*. New York: Charles Scribner & Sons.

Howell, N. 1979. *The Demography of the Dobe !Kung*. New York: Academic Press.

Lee, R. B. 1968. "What Hunters Do for a Living, or, How to Make Out on Scarce Resources." Pages 30-48 in R. B. Lee and I. DeVore, eds., *Man the Hunter*. Chicago: Aldine Publication Co.

Lumsden, C. J., and E. O Wilson 1981. *Genes, Mind and Culture: The Coevolutionary Process*. Cambridge, Mass.: Harvard University Press.

Menozzi, P., A. Piazza, and L. L. Cavalli-Sforza 1978. "Synthetic Maps of Human Gene Frequencies in Europeans." *Science* 201:786-92.

Mundinger, P. C. 1980. "Animal Cultures and a General Theory of Cultural Evolution." *Ethology and Sociobiology* I:183-223.

O'Neill, G. K. 1977. *The High Frontier: Human Colonies in Space*. New York: Morrow Publishing Co.

———— 1981. *Two Thousand Eighty-One: A Hopeful View of the Human Future*. New York: Simon & Schuster.

Rao, D. C., N. E. Morton, J. M. Lalouel, and R. Lew 1982. "The Pathological Analysis in Generalized Assortative Mating. II. American IQ." General Research of Cambridge. Vol. 39, pp 187–98.

Rice, J., C. R. Cloninger, and R. Reich 1980. "Analysis of Behavioral Traits in the Presence of Cultural Transmission and Assortative Mating: Applications to IQ and SES." *Behavior Genetics* 10:73-92.

Scarr, S., and R. A. Weinberg 1976. "I.Q. Test Performances of Black Children Adopted by White Families." *American Psychologist* 31: 726-39.

Schiff, Michel, Michel Duyme, Annick Dumaret, and Stanislaw Tomkiewicz 1982. "How Much *Could* We Boost Scholastic Achievement and IQ Scores: A Direct Answer from a French Adoption Agency." *Cognition 12*. In Press.

Schoenmaeckers, R., I. H. Shah, R. Lesthaeghe, and O. Tambashe 1981. "The Child-Spacing Tradition and the Postpartum Taboo in Tropical Africa: Anthropological Evidence." Pages 25-71 in H. J. Page and R. Lesthaeghe, eds., *Child-Spacing in Tropical Africa*. New York: Academic Press.

Tizard, B. 1974. "I. Q. and Race." *Nature* 247: 316.

Wendorf, F., and R. Schild 1980. *Loaves and Fishes: The Prehistory of Wadi Kubbaniya*. New Delhi: Pauls Press.

Wilson, E. O. 1975. *Sociobiology: The New Synthesis*. Cambridge, Mass.: Harvard University Press.

Commentary by John F. Eisenberg
Department of Zoological Research, Smithsonian Institution

In response to Cavalli-Sforza's essay I thought it might be useful to talk a bit about natural populations. By natural populations, I mean populations that are not regulated by human intervention. For this audience I will discuss nonhuman primates in nature. I am trying to bridge between biological and cultural, and wish to emphasize that the constraints on nonhuman primate populations are exerted through social interactions. Whether the behavior patterns are "hard-wired" or have a learning component is immaterial. The outcome certainly allows natural populations to regulate themselves within limits.

If you have a wild primate population, and if we assume that there is an optimal part of the year for birthing so that young tend to be pulsed in time, then our hypothetical population from year to year would exhibit fluctuations in density through time as shown in figure 11. The tendency to hover around a fixed value defines the carrying capacity for a given habitat.

Should you alter the carrying capacity—let's say by destroying some food resource base within the habitat, such as poisoning fig trees—then you will cause the population to reach a new equilibrium with a new carrying capacity. And should provisioning be provided, then at

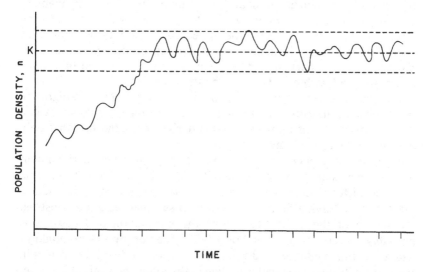

Figure 11. Changes in population density, n, in a hypothetical population having an annual birth pulse and increasing in density until it reaches its mean equilibrium density, K. The carrying capacity of the environment is defined by K rather than by the maximal n_2 or minimal n_1 values fluctuating about the mean K.

the rate at which it is provided the population will rise to a new carrying capacity at a higher level.

Births and deaths track the resources that are there. The only other way the population can change is due to immigration or emigration. If these tend to be equal, then the mortality and natality tend to cancel each other out from year to year. Mortality derives from such factors as disease, predation, and aging. Most mortality is socially influenced by limiting access to food through behaviors displayed by dominants toward subordinates.

Mortality may be expressed on an average basis at different rates when the sexes are compared. Mortality schedules reflect the mating system. That is, if the species is polygynous (multiple mating partners for each male), usually there is a strong asymmetry in the emigration tendencies. Namely, in polygynous societies it is generally the rule that the majority of males leave the natal troop, whereas the females are more prone to stay in the troop of their birth. On the other hand, in monogamous primate societies, the tendency to disperse is not asymmetric with the sexes, but tends to be equal. The emigrating sex is usually at a higher mortality risk before establishing itself in a new troop. If we calculate probability of dying and express this over time from infant to senile adult stage for a polygynous species, the probability of dying is high when an infant and falls rapidly. In males the probability of dying peaks again at puberty. The female curve does not show the peak at puberty (see fig. 12). The increased probability of dying at or shortly after puberty in males is a product of, on the one hand, their risk of injury or starvation during emigration and, on the other, the risk associated with efforts to enter into a new troop.

One will also note that accompanying the asymmetry in emigration rates reflected in differing mortality schedules for the two sexes is a built-in avoidance of inbreeding. This derives from the fact that only one sex tends to emigrate.

There is one aspect that comes out of all of this in pondering about nonhuman primate populations, and that is the regulation of the interval between births. There were three possible mechanisms discussed in the essay by Cavalli-Sforza. Suffice it to say that lactation anestrus, the failure of the cycle because of the stimulation of lactation, is probably the commonest method of delaying conception in monkeys and apes. For medium-sized primates, if a female loses her young in the year of its birth, she will conceive in the following year. Otherwise, if the young survives, she will skip and go through estrus every other year. Of course, in anthropoids the interval is even longer.

I guess what strikes me in all of this is how cultural methods of

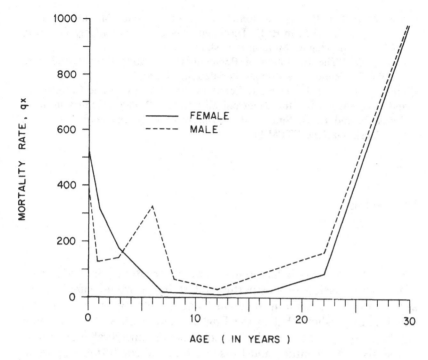

Figure 12. Age-specific mortality rate curves for male and female toque monkeys (after Dittus 1975).

regulating behavior, shall we say, mimic the more hard-wired solutions that we find in infra-human members of the order primates.

No vertebrate species other than man has invented agriculture. I have left out invertebrates since one could say that some ant species have developed a form of agriculture. If we consider the class Mammalia, of which we are members, then our unique passage from hunter-gatherer—the "Garden of Eden," as it were—to an agricultural society was a unique transition without any strict parallels.

The abuses of the land that followed this transition, of course, are a matter of historical record, in part, and it is that legacy that we live with. I personally do not have, as Cavalli-Sforza does, very much hope for interplanetary colonization as a way out of the mess that we find ourselves in. But I have no alternative solution any more hopeful or less.

References

Dittus, W. P. J. 1975. "Population Dynamics of the Toque Monkey, *Macaca sinica*." Pages 125-52 in R. H. Tuttle, ed., *Socioecology and Psychology of Primates*. The Hague: Mouton Publishing.

—— 1979. "The Evolution of Behaviours Regulating Density and Age-specific Sex Ratios in a Primate Population." *Behaviour* 69:265-302.

Eisenberg, J. F. 1979. "Habitat, Economy, and Society: Some Correlations and Hypotheses for the Neotropical Primates." Pages 215-63 in I. S. Bernstein and E. O. Smith, eds., *Primate Ecology and Human Origins*. New York: Garland STPM Press.

Discussion

DR. LASLETT: I would like to make two observations, one on Cavalli-Sforza's discussion of postpartum infertility. I would simply draw attention to a fact cited in the essay that I wrote for this conference, which goes to show that in the English population as far back as it can be traced—it can now be traced back, in considerable accuracy, to the sixteenth century and by some other uncomfortable approximations back to the thirteenth century—all of the indications are that fertility control was by the methods of nonmarriage, late marriage, and, what is interesting if one considers Cavalli-Sforza's observations, maximal breast feeding of children within marriage. This gave to English population history an extraordinarily constant fertility over many generations.

My other observation is simply this: I associate the European marriage and familial pattern with successful industrialization. Furthermore, all the lines of intensity, so to speak, of the European marriage pattern run just as Cavalli-Sforza's lines of transmission from southeast to northwest, from the northeast-southwest diagonal of Europe. On the one side, including the Balkan peninsula, most of the Iberian peninsula, most of central Europe, certainly all of eastern Europe, way into the Middle East, you have a non-European marriage pattern. You have a form of familial practice not associated with industrialization. In other words, that area which finally became most innovative and most capable of adaptation in the sense of industrialization doesn't, if I can be as speculative as you are, turn out to be that which received agriculture first, but that which received agriculture last, and which had, as I also hinted in what I have said, certain very interesting demographic similarities, not with early agriculture, but with the hunter-gatherers themselves.

DR. NEEL: I think one of the most eloquent proponents of space colonization in recent years has been Freeman Dyson in his book, *The Greening of the Universe*. And I guess I too would part company with you here, Luca. We have made such a mess in this one world—let's wait a little while before we take off to mess up some other planets! Regarding the control of fertility by lactation, there are some interesting physiological aspects I don't believe we understand, because certainly in some of the cultures among whom we have worked the women are lactating continuously from about age fifteen to perhaps age forty.

I tell the story—a little bit out of school—about the expedition in which my wife accompanied me into an area where there had been no white women before. We were surrounded in a village by a circle of completely nude Indian women. And since she was dressed in slacks and a shirt and had on a hat, they were not sure about her gender. And so within minutes, one of them dashed off and brought back a male child and pointed to the penis and then pointed to Priscilla. She shook her head, and that helped classify her. But then the other thing they did was to show her to a lactating woman and demonstrate how they could extract milk from their breasts. They asked her to join the competition, but she declined. They definitely had her catalogued within five minutes of our getting into the village.

DR. WIERCINSKI: From several dozens of blood-system genes it is possible to consider only two and obtain the coincidence with immigration routes regarding agriculture, as you have shown us. Did you study at the same time more different genes in that entire spatial distribution for these times? I am a little bit afraid of these systems in general. As you know, I base my research on the traditional traits.

DR. CAVALLI-SFORZA: I will address myself first to Dr. Wiercinski's intervention, as it bears more directly on some of the work we've done. The reason for using genetic markers was mostly that we were familiar with them, of course. But the other reason is that they are less sensitive to short-term environmental adaptations. If you eat more, on an average you have a bigger skeleton; you eat less, you have a smaller skeleton. And everything else follows. Not only the size, height, and weight, but other traits are affected.

That doesn't mean, of course, that there are not traits in the skeleton that can be inherited genetically. We always, therefore, are in doubt as to whether we are addressing short-term changes, which might take place in a very short time, or long-term changes, such as those in selection usually are.

Dr. Laslett mentioned the fact that breast-feeding based on recent findings must have been an old custom in England, at least prolonged breast-feeding. I suspect that this may be true in a wider context than England. I can offer at least an anecdote.

My grandfather grew up in northern Italy and was breast fed until he was over two years old. He was running after his mother with a little chair, so that he could climb to the breast. And his mother didn't know what to do to get him off the breast. So she decided to paint her breast black and that scared the child enough that he didn't want to breast feed any more. When I told this anecdote, which I believe is true, to a child psychologist, she was horrified. But I must say that my grandfather grew into a very normal person.

But prolonged lactation can hardly explain all of the actual delay in births that takes place, for instance, in hunter-gatherers. From data in African populations, after at most one and a half years following birth, menstruation and ovulation will resume. I presume this is one of the reasons why there are additional mechanisms of birth control among hunter-gatherers—among which the sex taboo is probably important and still widespread.

5. Biocultural Interaction in Human Adaptation

DONALD J. ORTNER

Curator of Anthropology, Smithsonian Institution

Editor's Summary. In his essay Donald Ortner stresses that culture must be viewed as an extraordinarily successful adaptive mechanism. While similar in some aspects to biological adaptation, the archeological and historical perspective indicates that cultural adaptation is far more responsive and flexible than biological adaptation, and can be transmitted from one individual to far more people than is possible for biological changes. Since cultural innovation is crucial to the future of human society, social conditions should be encouraged in which individual creativity can flourish.

It seems almost an obvious truth that human biology and culture interact in complex and important ways to produce what we call human beings. However, it is a curious fact that, despite millennia of speculation on the subject, we have made very little progress toward an understanding of the condition of being human. Today the myths of our past that provided a satisfactory cosmology for our ancestors are no longer acceptable, but they have not been replaced by acceptable new myths or their contemporary analogs. This has created a condition in which there is probably more controversy and, for many people, more confusion than at any time in human history. We are flooded with new information and new ideas at a time when our traditional perspectives about ourselves are being severely challenged.

Mankind's role in nature, or at least his perception of it, has always been of considerable consequence. (Passmore, 1974, has written an excellent book in which this subject is addressed.) The content of ancient philosophy and religion places great emphasis on clarifying this issue. In the Judeo-Christian tradition, for example, mankind was given a very special role in creation, which involved both stewardship and authority. This perception of mankind's role in the world implies a unique capacity that other creatures did not have. In the hunter-gatherer tradition there appears to have been a somewhat closer identification between mankind and the rest of nature. Among contemporary hunter-gatherers it is common to find a sensitivity to nature which we in civilized society are only rediscovering.

Mankind's role in nature traditionally has most frequently been related to his mythology and often to his perception of his relationship to deity. Much of Greek mythology focuses on man's quest for an understanding of his place in the physical and spiritual cosmos. The Greeks' vulnerability to events over which they had little or no control, and which were seemingly capricious, was undoubtedly a factor in their perception of their relationship with the gods. A recurrent theme in Greek mythology is that mankind was often a plaything for the amusement of the gods—a pawn in a cosmic chess match. This theme reveals that, despite the remarkable achievements of ancient Greek society, the Greeks remained subject to forces they did not understand.

The frustration over mankind's inability to control his destiny is revealed in more recent philosophical traditions in which perceptions of mankind range from a general doubt about human existence itself to the concept rather cynically expressed in John Gray's epitaph. "Life is a jest, and all things show it; I thought so once, but now I know it." Today an understanding of those qualities which make mankind unique in nature remains ephemeral, as if the essence of humanity were a soap bubble which bursts just as we begin to penetrate it.

Part of the problem is that the most difficult thing to understand in our cosmos is ourselves. It is a frustrating paradox that the qualities which make us human (such as individuality, creativity, and emotion) make understanding ourselves difficult. Indeed, the existentialist argues that the human experience transcends, to some extent at least, the capacity for a rational understanding of human existence. Even an alien intelligence, not subject to our inherent biases, would likely be frustrated in an attempt to achieve such an understanding, because of mankind's capacity for change, the complexity of human culture, and our proclivity for using deception as a mechanism for protecting our privacy. Despite the difficulty of understanding ourselves, the fact that an enormous amount of expense and energy has been and is being expended in trying to achieve such an understanding suggests that comprehending mankind is very important to us and to our future.

What are the qualities that make us human and how do they interact? These are questions which deserve our attention and are fundamental

DONALD J. ORTNER

for our entire symposium. First of all, we are animals, although I have heard at least one scholar at a scientific meeting vehemently deny this obvious fact. In the scientific lexicon, organisms are categorized as a plant or an animal. This distinction is not always clear-cut in some of the simpler organisms, such as euglena, which have both animal characteristics (e.g., motility) and plant characteristics (e.g., the ability to synthesize chlorophyll). Furthermore, we are eukaryotic animals, which means that we are an extraordinarily complex aggregate of cells, most of which contain the same genetic information but which—for poorly understood reasons—function in various but collectively helpful ways.

There has been, in Western society at least, a tendency to maximize the social and biological distance between mankind and the rest of nature. This tendency is partially responsible for the excesses which have brought us to the environmental crises of today. While a greater appreciation for our kinship with nature is important, it is essential to recognize that we are more than an aggregate of variously functioning cells which collectively go about the business of making more cells and tissues as well as new organisms. In other words, there is more to us than our biological nature. This additional quality, which consists of a complex behavior pattern we call culture, both arises from our nature and transcends our nature. At the same time, it adds enormously to our basic animalness. It is what makes being human different.

This is not to deny to other animals behavior which, in many respects, is similar to human culture. Such behavior is particularly evident in the higher vertebrates. The well-known learning capacities of other primates, particularly the chimpanzee, is an example of behavior which is properly considered incipient culture. However, it is equally clear that the human capacity for culture exceeds by several orders of magnitude the capacity of any other animal. Indeed, culture has become so important to mankind that it represents the most crucial factor in man's adaptive adjustment to the various environments in which he lives. Because of culture, mankind has the capacity to live below the sea, anywhere on land, and even in the airless environment of space. Also because of culture, mankind— today—has the capacity of doing unimaginable damage to itself and other members of the animal and plant kingdoms. It is the complex interaction between our biology and our culture (bioculture) that makes the human experience unique.

Dynamics of Adaptation. To understand the importance of culture it is perhaps useful to compare briefly the mechanisms of biological and cultural adaptation. Biological adaptation results from the interaction between the tangible biological (phenotypic) expression of an organism's genetic potential (its genome) and the environment. The latter includes other plants and animals and the resources necessary for life. New genetic potentials are created by changes (additions, subtractions, or modifications) of existing genetic potentials. These changes are called mutations. In the evolutionary model accepted by most scientists, the diversity of life which exists today is primarily the result of a dynamic interaction between groups of organisms and their environment in which adjustments to changing environmental conditions are possible because of mutations and the new genetic potentials they create.

This type of adaptation is by far the most common adaptation in both the plant and animal kingdoms. It, however, has rather obvious limitations, the most significant of which is that adaptation to a new environmental condition can only occur if the required genetic information exists in the collective genetic potential (gene pool) of a discrete group of organisms (a species) living in that environment. If the genetic potential for adapting to new environmental conditions critical to survival does not exist, the species will become extinct. While the fossil record is replete with examples of extinction, it is clear that most new challenges to a given species by the environment do not exceed the genetic potentials of the species. In general, changes in the environment occur slowly and well within the time framework required for most species to acquire the necessary genetic potentials.

If, however, major environmental change occurs very rapidly as, for example, the result of some environmental catastrophe, genetic potentials for the new conditions may not exist and the organisms will not have time for mutations and the process by which mutations are incorporated in the gene pool (natural selection) to provide the necessary genetic potential. One hypothesis regarding the major extinction of the large reptiles (e.g., the dinosaurs), which occurred at the end of the Cretaceous geological epoch (about 60 million years ago), invokes a scenario in which a large asteroid crashed on earth, throwing such a large amount of fine particulate debris into the atmosphere that the energy from the sun to the earth's surface was greatly diminished (Alvarez et al. 1980). In this hypothesis the reduction in temperature led to rapid changes in the environment for which the large reptiles were not adapted. Furthermore, there was inadequate time for the process of mutation and natural selection to provide the necessary genetic changes for adaptation and many species became extinct. Today, of course, the human species faces a major and unprecedented environmental challenge should nuclear war become a reality.

DONALD J. ORTNER

It is obvious, empirically, that genetically patterned adaptation has, by and large, been a very successful method of adapting. There is, however, another mechanism for responding to the environment and changes which may arise in it. An organism can create an artificial buffer between itself and the environment which attenuates or otherwise modifies the effect of the environment on the organism. In human society the buffer is culture.

What is culture? How does it compare conceptually with biology? How do culture and biology interact? These are questions which invoke some of the most heated debates in the scholarly world today. They represent a modern version of the ancient Greek's quest for an answer to the many imponderable realities of our existence. Books on the substance and nature of culture fill whole sections of most libraries and this topic will not be treated specifically in this essay. However, some insight into the nature of culture will be provided as I attempt to clarify the last two questions.

Analogies between the dynamics of biology and the dynamics of culture have been made by numerous scholars (e.g., Simpson 1962; Cavalli-Sforza and Feldman 1973). More recently, a subdiscipline has developed within anthropology in which biological evolutionary models have been applied to cultural dynamics (Harris 1979). Simpson's (1962) observation that culture is an adaptive mechanism having the effect of one or more biologically adaptive traits is, in many respects, both helpful and accurate. The stuff of culture—cultural traits, cultural complexes, and all the other categories of cultural activity—can be compared to phenotypic genetic traits. The cultural analog of the genotype is less obvious. In a sense, however, language can be thought of as the vehicle by which culture is created and transmitted. Innovations in culture are analogous to mutations in biology; they provide societies with the new ideas that allow them to adapt to changing conditions. Like mutations, the value and acceptance of innovations depends on both intrinsic and extrinsic factors.

The analogy is incomplete when we consider the dynamics of cultural and biological change. Mutations are random events whose frequency is a function of both cellular and extracellular factors, but which, in nature, arise in total independence of any conscious effort by the organism. Cultural innovations do indeed often have random and unplanned components, but they also are greatly influenced by the conscious efforts of people who respond to culturally defined needs by inventing ideas and things. Cavalli-Sforza and Feldman (1973) highlight an additional distinction of cultural evolution. The biological heritage of an individual is determined exclusively by his parents; an individual's cultural heritage is influenced by many people. Thus, a useful or necessary cultural innovation can be spread very quickly through a society.

Despite the cognitive aspects in cultural innovation, it is crucial to our understanding of cultural dynamics to realize that cultural traits need not be developed through some rational thought process in order to be beneficial. Suppose, for example, a myth exists that the gods have decreed that land should be uncultivated one year in seven. The fact that the myth did not result from a resolution emanating from a scientific conference on land management is not significant. As long as the myth results in a behavior pattern which confers an increased adaptive advantage on the society, the cultural innovation will be valuable in the society's interaction with its environment. Ross (1978) has provided a helpful statement on the functional significance of cultural traditions in his analysis of food taboos in Amazonian culture. Much remains to be learned, however, about the dynamics and mechanisms of this aspect of cultural evolution.

The analogy between biological and cultural evolution is perhaps most helpful in understanding the adaptive value of culture. Biological (phenotypic) traits provide the organism with the biological equipment needed to survive in a given environment. The success (measured in terms of survival and expansion) of a species depends on having the minimal and, ideally, the optimal biological traits needed to interact successfully with the environment and reproduce offspring in sufficient numbers to maintain the critical number of individuals for species survival. Modern biology is, in one sense, the story of those organisms which are successful.

The major problem with biological adaptation is that traits necessary for species survival must be or become part of the gene pool of the species. If conditions change, as they inevitably do, the existing traits may not be adequate for the new conditions. Depending on the genetic variability in the gene pool of a group of organisms, or on the introduction of new genes by mutations, or on gene flow from other groups of the same species which may occur during the transition to the new environmental conditions, a given group may or may not survive. One of the problems is that changes in the gene pool are largely limited to changes which occur in subsequent generations.

Cultural traits, like genetic traits, provide equipment for interacting with the environment, but changes in culture do not depend on the production of a subsequent generation. Furthermore, specialization and differentiation of culture have allowed mankind to create his own mini-environments, and thus permit him to range over the entire surface of earth. Mankind has done this largely without specific changes in his genetic equipment. Such adaptability is entirely unprecedented

DONALD J. ORTNER

in the biological history of life on earth. Thus, culture confers on mankind enormous flexibility to the point where this very success creates cultural potentials which threaten to destroy us or at least bring about profound changes in our way of life.

One need not be an alarmist futurologist to be concerned about the potential for disaster posed by nuclear warfare, continued population expansion, denudation of tropical forests, thoughtless exploitation of natural resources (including energy and water), and the loss of cultural diversity which, for me at least, is as threatening as the loss of genetic diversity that Neel (1970) projects as a product of modern geographical mobility.

The Relationship between Biology and Culture. Perhaps the most heated debate in the scholarly world today centers on the type and degree of interaction between culture and biology. Human biologists have long held that a feedback mechanism exists between biology and culture (Spuhler 1959; Washburn 1959; Dobzhansky 1962). The nature of the feedback mechanism has never been sharply defined. However, it is thought that biologically-based potentials for culture developed through genetic mutation early in mankind's evolutionary history. These biological adaptations included erect posture and great flexibility in manipulating objects with the hands. The full utilization of these biological traits, however, depended on increased cognitive abilities. Early hominids with greater intellectual abilities had a selective advantage over those who had less. Generally, the theory is that physical and mental capacities conducive to culture led to incipient culture, which gave adaptive advantages to the incipient culture-bearers. This adaptive advantage led to increased fertility of individuals with the appropriate physical and mental capacities. This led, in subsequent generations, to a greater frequency of culture-bearers.

The role of language in this process remains speculative. However, the crucial role of increasingly sophisticated communication in cultural evolution makes the development and utilization of language in the early stages of cultural adaptation very likely. The feedback loop between biological and cultural potentials continued until the mental potential to create culture equaled or exceeded the demand for additional biological ability to create culture. Once this point was reached, the buffering capacity of culture was so great and the existing potential for further creation of culture so unlimited that further adaptation was almost exclusively accomplished through cultural adaptation.

In this model, the development of culture transcends biology and becomes a buffer between mankind's biology and his environment.

Culture, rather than genetic phenotype, becomes the factor which confronts the environment. Although the capacities for culture have a genetic substrate, culture develops a dynamic of its own and develops apart from the generalized genetic substrate from which it arose—the biological potentials for culture are of a generalized nature.

Hamilton (1964) hypothesized a mechanism for the genetic basis of behavior which was readily accepted for many nonhuman animals. He proposed the concept of "inclusive fitness," in which "fitness" is an organism's ability to produce adult offspring, and "inclusive fitness" is an organism's own fitness plus its ability to favorably influence the survival of its relatives. Hamilton used the behavioral trait of altruism as a model and suggested that an organism may have greater genetic impact on subsequent generations by expending care and materials on its offspring already born rather than by reserving them for its own survival and further fecundity.

More recently, other scholars have extended Hamilton's concept to human behavior and thus to human culture (e.g., Barash 1977; Wilson 1978; Lumsden and Wilson 1981). Hamilton's hypothesis has been elaborated and utilized to formulate an integrated hypothesis relating behavior and biology for the entire animal kingdom, including man. This hypothesis has emerged as the newly proposed discipline of sociobiology.

It is important to emphasize that both the the earlier opinions regarding the relationship between human biology and culture and the hypotheses subsumed under the rubric "sociobiology" involve a genetic substrate underlying human behavior. The major difference between the two perspectives lies in the specificity of the relationship. The traditional concept involves the development of generalized potentials which gave mankind maximum flexibility to develop whatever culture traits are adequate to satisfy group needs. Sociobiology (as represented by Lumsden and Wilson 1981) is suggesting much greater specificity, in which specific genes control specific behavior. Indeed Lumsden and Wilson go so far as to suggest that a gene or gene complex may predilect for a certain culture. Thus a genome may exist that is more appropriate for Chinese culture than for European culture. Such a genome, if it exists, would be extraordinarily complex.

There do seem to be genetically based traits which could be related to cultural practices. An interesting example of such a trait is the ability of an adult to metabolize milk, and the relationship of this ability to the cultural tradition of using milk as a significant part of the human diet. Milk provides a valuable source of protein, calcium, and calories; however, the ability to digest milk depends on the presence of the enzyme lactase in the intestine. This enzyme is absent in a very high frequency of adults in many major human groups. Tests on

DONALD J. ORTNER

samples of Chinese suggest enzyme deficiency frequencies ranging from 83 percent to 97 percent and in American blacks frequencies from 30 percent to 100 percent, with most samples exceeding 70 percent (McCracken 1971). In areas of Europe where milk is an important part of the diet, the frequency of lactase deficiency is low (typically under 20 percent).

There is some debate about the cause of the deficiency (Harrison 1975), but the evidence favors a genetic basis. McCracken (1971) suggests an association between the cultural subsistence base and the use of milk in the diet. He suggests that a strong cultural dependence on milk in the diet of a society would create selective pressure favoring adults having the enzyme lactase. Conversely, modern attempts to introduce milk into the diet in societies which have high lactase deficiency are often unsuccessful since digestive disturbances are a common result. In the case of lactase deficiency, a presumed genetic trait has a direct effect on the cultural content of a society— in effect eliminating a potentially significant source of high quality food.

As culture assumes a more significant role in mankind's evolutionary history, the relationship between genetic potentials and culture is looser (Lumsden and Wilson's "leash principle"). Nevertheless, the specificity of the relationship between biology and culture in the sociobiology model is much greater than in the more widely held model. When applied in the context of human behavior, sociobiology challenges one of the most strongly held affirmations in the social sciences; that is, we are culturally neutral when we are born and have the potential to learn virtually any culture—provided we are intensively exposed to it early enough. Because of the genetic determinism inherent in some aspects of sociobiology, many scholars view it as neoracism and react strongly against it.

There are really two problems in any discussion of human bioculture: one involves matters of fact, the other involves matters of professional ethics and social philosophy. (Since our biology preceded our culture, I prefer a term, bioculture, that symbolizes this phylogenic relationship.) It is conceivable that one could be factually correct, but highly imprudent ethically, in promoting a hypothesis. On the other hand, it is important for us, as scientists and scholars, to evaluate the matters of fact regarding sociobiology objectively. There may indeed be ethical social issues implicit in sociobiology but, as Midgley (1978) cogently argues, assigning a social or ethical value to a scientific hypothesis is a complex matter, and the same hypothesis can be viewed as good or bad depending on one's perspectives and assumptions.

Diversity of Culture. One of the most critical issues to be dealt with in a model of the relationship between genes and culture is the diversity of culture. It is one thing to be able to define culture, it is quite another to define *a* culture. Culture can be defined as the traits, complexes, and patterns of relationships between and among people that characterize a society. Coming to an understanding of *a* culture, particularly a complex one, is almost by definition impossible. Even the most comprehensive ethnography of the simplest human community can be no more than a feeble approximation of what really constitutes the community.

Thus, for example, when Lumsden and Wilson say we are genetically preadapted to a specific culture, what is it that we are preadapted to? Is an American preadapted to American culture? This question immediately begs another: what is American culture? The facility with which many foreign-born people adjust to the freedom of American society suggests that at least some aspects of American culture are compatible with almost universal human cultural potentials.

Mead (1956) has documented the ease with which the Manus people in the Admiralty Islands transformed their society, within a single generation, from a Stone Age culture to a society based on modern concepts of social organization and technology. This experience is but one of many which highlight the enormous flexibility in mankind's relationship with culture that exists in all human groups.

The diversity and flexibility of culture suggest that a concept of genetic preadaptation for a culture must first confront the fact that culture is extraordinarly complex, with an enormous number of more or less equally adaptive options available and great potential for inventing and adopting even more. Any single individual in a society is able to, indeed must, choose among the various options available to him.

It is of more than academic interest to note that our understanding of biocultural dynamics depends on the degree to which one or the other of two models about the relationship between biology and culture is correct. If the sociobiology model is correct, we would have less flexibility in changing the directions of the human society, since significant amounts of collective behavior are the result of the expression of genetic potentials. For major change to occur, we would need a change in our genetic potential. If, for example, the aggressive relationships which traditionally have existed between competing human societies are based on genetically patterned behavior, our potential for developing less destructive methods of resolving conflict may be limited. On the other hand, if aggressive social behavior is culturally patterned behavior, with only a generalized genetic substrate, we can bring about whatever changes are deemed necessary by changing our culture. We need not wait for genetic mutation and

DONALD J. ORTNER

natural selection to work their course. Culture, itself, is in effect the phenotype, and we can change it through cognitive processes.

Probably most people raised in the Western tradition would prefer to have the greater flexibility inherent in the more traditional model of biocultural relationships, in which culture transcends biology and the relationship between biology and culture is more generalized rather than more specific. Wishing it, however, does not make it so. The conflict between the two models will have to be resolved in the crucible of scientific debate. Being an optimist by nature, I prefer to believe that our genetic heritage does not seriously limit our present or future cultural options or, at least—if there are genetic limitations to change— that we will be able to override them by utilizing cultural mechanisms. Clearly, if there is a specific relationship between cultural behavior and the gene pool of human society, then our options for adapting culture to developing realities is limited and the future prospects for humanity could be rather grim.

Returning to our example of aggressive behavior, we do have the option of allowing our future to be determined in the crucible of hostile competition between world societies. The risk in terms of individual human suffering and disruption is great, particularly if the competition degenerates to the physical level. This, of course, would not be something new in human history. Much of human history is a litany of strong societies conquering and dominating weaker societies and, in turn, being overthrown and dominated by other stronger societies. In these conflicts, little thought appears to have been given to the problems they create for the lives of countless people killed or adversely affected by the conflict and the impact social disintegration has on cultural innovation and the development of human society. I was appalled recently to hear on a Washington, D.C., radio talk show a well-known political figure express the opinion that we needed larger population sizes in Western countries in order to have the manpower to fight wars. In the context of that statement, I cannot think of a better argument in favor of population control.

Strategies in Biocultural Interaction. The problem really is whether a policy of biological and cultural laissez faire is appropriate to the realities which confront international human society today. I think not. I further think a comparison of past conditions with conditions today in areas such as resolution of conflict and world population growth will demonstrate that such a policy could have disastrous consequences.

Let us consider, for example, the matter of world population growth. Throughout most of human history, high infant mortality and short life span were realities which placed a premium on maximum fecundity in human groups. It is thus not surprising to find that the behavioral baggage we carry from our past, perhaps from our genes and certainly from our culture, favors reproduction rates that are completely out of phase with contemporary realities.

The problem is that human society has invented, through cultural innovation, ways to control many infectious diseases and thus minimize infant mortality and increase longevity. This result, coupled with continued high fecundity throughout the world, has led to a dramatic increase in human population during the last 100 years to the point where sheer numbers of people pose a serious threat to the stability of civilization.

While it is difficult to overstate the hazards of this contemporary version of the Damoclean sword, it is important to consider the benefits which have accrued to human society because of the cultural advances which have been the major contributing factor to the population explosion. Increased longevity has undoubtedly enhanced the potential for human society to benefit from the wisdom of human experience embodied in senior citizens. One might even hope that, as the life span increases throughout the world, greater wisdom and reason will prevail in international relationships. To the social benefits of increased longevity, I would add that improved health has probably allowed a broader spectrum of people to survive, resulting in greater biological and cultural diversity and potential for adaptation.

However, the benefits associated with the population explosion are overshadowed by known and unknown problems of enormous magnitude. Witness the starvation of tens of thousands of people every year, particularly in Third World countries. If even the most favorable scenario projected by the recent Global 2000 Report (Barney 1981) is true, millions more will be condemned to malnutrition and starvation in the twenty-first century. The geopolitical problems created by the demand for food and water resources would almost undoubtedly adversely affect even those countries with adequate resources. Witness the irresistible economic forces that have resulted in millions of illegal immigrants coming to the United States from Mexico.

Assuming that we have both the wit and the will to provide adequate food and water for all the people born in the twenty-first century, we are faced with additional problems. While the adverse effects of human crowding are poorly understood and indeed have been questioned (compare Freedman 1980; Baron 1980), it seems likely that unknown but serious problems await human societies when we have too many people in a limited amount of space. From the standpoint of improved technology and agriculture, it is conceivable to feed several times the number of people that now exist on earth. However, our emerging

DONALD J. ORTNER

knowledge about the effects of crowding on both man and animals suggests at least the potential of social psychological problems on a massive scale, whose solution might involve Draconian measures of social control which could make Orwell's 1984 seem like Paradise.

In the context of this potentially grim scenario it is most appropriate to ask questions about our biocultural heritage. We need to know about the cultural baggage we are carrying with us today and the extent to which our biocultural heritage provides adequate mechanisms for adapting to modern social conditions. Human history is filled with examples of cultural changes which have adversely affected human society. Most often these effects were not anticipated. A few examples will make the point.

The earliest roots of what we recognize as Western civilization go back to early experiments in urban living during the Neolithic period in the Near East. The major impact of urban culture, however, developed during the Chalcolithic (fourth millennium B.C.) and became the dominant social force in the Near East during the Early Bronze Age (third millennium B.C.). City life brought about a profound change in the relationship between man and the environment. Hunter-gatherers and nomadic-pastoralists tend to move about a fairly large geographical area in fairly small groups (about 50-150 people). City life, however, is based on agriculture and inevitably means a much more sedentary society. Urbanism, even with its increased population density, was certainly an attractive cultural innovation. Associated with urbanism, however, were increased economic vulnerability and severe health problems. There is fairly convincing evidence of a major increase in infant mortality involving an order of magnitude of about four times in urban culture (Angel 1971, 1976; Ortner 1976). Increased population density also brought increased contagion from infectious diseases. In all probability the problems of the early cities were not significantly different from medieval cities which, as McNeill (1976) emphasizes, were population sinks. By this he means that death rates were so high in cities that only high rural fecundity and continuous, major migration to cities could maintain the city population at levels known to exist.

Despite the attraction of city life, the archeological record provides ample evidence of the vulnerability of city life. At the end of the Early Bronze Age in the Near East there was widespread abandonment of cities which had been built with such prodigious effort (Kenyon 1979:117). The reasons for this are still poorly understood and speculative, but help to place our contemporary problems in historical perspective. Some of the major man-influenced reasons may include:

(1) Denudation of contiguous forests which supplied the energy for urban technology (e.g., pottery and metallurgy);

(2) Warfare, which cut the tenuous agricultural and trade umbilicus vital to survival of cities;

(3) Disease (e.g., cholera);

(4) Hydrologic changes resulting from human intervention (e.g., burning, deforestation);

(5) Misuse of land, resulting in diminished fertility of the soil;

(6) Population pressure;

(7) Overgrazing of domestic animals;

(8) Social decadence of some city cultures.

The abandonment of the ancient cities provides mute testimony regarding the vulnerability of urban society. It would be foolish to assume that modern society is substantially less vulnerable to the problems which confronted ancient societies and which were resolved with varying success. Those of us who live in cities subject to disruptive snowfall know how quickly the shelves of food stores are emptied of supplies. The problem is that today we are not talking about a single city or region but massive disruption throughout the world should some of the problems not be resolved.

Another example of unanticipated problems arising from cultural innovation is apparent in the improvements associated with early agriculture, which brought about greater potential for food production. A major innovation was irrigation, which goes back to at least the fourth millennium B.C. and which provided greater stability in crop production. However, with certain forms of irrigation one finds an increase in water-borne diseases, such as schistosomiasis, in which the vector is a snail living in polluted water.

Health problems arising from the development of urban society were not limited to ancient human society. The Industrial Revolution brought about an expansion in the size of cities unprecedented in human history. With this expansion came factories belching smoke, massive expansion of construction, narrow dark streets, malnutrition and disease. Metabolic diseases such as rickets and osteomalacia (both involving a deficiency of Vitamin D resulting from poor nutrition often associated with reduced exposure to sunlight) became common medical problems for children and lactating women (Wells 1964). Furthermore, the crowding and poor sanitation associated with crowded cities provided ideal conditions for the spread of infectious diseases such as plague.

Conclusions. I would argue that an international condition of biological and cultural laissez faire, such as characterized most of human history, could have disastrous consequences, because cultural innovation has created realities unprecedented in their capacity to affect world human society. The question remains: can we influence the development of our world culture in ways that effectively relate to current and future conditions? We have seen that the development of culture can have unforeseen and adverse effects on individuals and societies. The capacity to respond creatively to these challenges will be crucial to future biocultural adaptation. In this context, an important decision confronting human society is the emphasis future society will give to individual versus social requirements. A greater emphasis on the conceptual dichotomy between individualism and collectivism is, I think, a more important endeavor than the contemporary popular preoccupation with the ideological confrontation between capitalism and communism.

What we know about contemporary hunting societies and, indeed, many preliterate agriculturists—as well as what we can infer about our early hunter-gatherer ancestors—suggests that many of the qualities associated with individualism were important to the success of such societies (Lee 1959). However, with the emergence of settled communities and urban societies an inevitable tension must have developed between the needs of the individual and the requirements of social living in which at least some conformity and attenuation of individual expression was essential to the survival of urban society. From that day to this, one of the great issues confronting civilized society has been to achieve an optimal balance between the demand for individual expression, initiative, and creativity, and the wishes of rulers to insure reasonable levels of predictability in the behavior of its members.

The fulcrum has been fixed at different points by different societies. In some complex civilized societies, rulers have been very repressive; in other societies, rulers have encouraged and permitted considerable individual expression. For me, one of the bothersome human paradoxes is that some of the greatest expressions of human creativity and innovation have arisen in what can only be called very autocratic societies. A partial explanation of this paradox is undoubtedly that individualism was encouraged in at least some segments of such societies and the qualities of creativity were rewarded in exceptional people, even when they were not of the appropriate class, as long as their creativity was kept within boundaries defined by the social group with which they were identified.

The point, of course, is that innovation and creativity are essential for cultural adaptation and survival of a society. I suggest that, as the problems of human society become more complex—which they are certainly doing—the need for individuals and groups with these qualities becomes even greater and thus even more important to the future of civilization. I do not accept, on the basis of current evidence, the assumption that our genetic heritage has significantly limited our cultural potential. It seems to me increasingly important to create a world society in which individual and collective genius is given even greater encouragement to achieve creative solutions to the problems confronting mankind. To do this, we must create a world society in which individual freedom is seen as an essential factor in human survival, a world society in which individual and cultural diversity is seen as the crucible from which innovations essential to future biocultural adaptation can emerge.

As Midgley (1983) argues elsewhere in this symposium, there should be reasonable limits on the expression of individualism. Perhaps the problem today is that there is too much individualism in some societies and not enough in others. Furthermore, irresponsible and uninformed individualism could easily be a monstrous curse. Passmore's (1983) eloquent plea for an educational program which provides a broadly based education for citizenship rather than for technology seems, to me, an essential element of future society. Somehow we must create for all people, everywhere, the educational system that will respect cultural diversity, but equip most world citizens to make rational individual and collective choices between the alternatives offered them by the products of individual and collective genius. Obviously, we must also create the sociopolitical environment in which such individuals can make these choices in freedom.

NOTE BY DR. ORTNER:
Without usurping the insight and clarification of those who commented on my essay, I should like to clarify some concepts that are crucial to understanding what I am attempting to say. A distinction needs to be made between social planning that is dependent on elitist decision-making and social planning that involves a broadly based process of decision-making, in which an educated population reaches a consensus about appropriate adaptive policies, which are then implemented by a responsive leadership. The former immediately invokes questions raised in other essays and commentaries about who the "we" are who decide how we should adapt. The latter, in effect, says that the "we" are all of us and that we must all be involved and respond to identifiable issues in a creative way to best ensure adaption for all. This may seem utopian and unrealistic, given contemporary problems in reaching national and international agreements and consensus. For it to work,

DONALD J. ORTNER

narrowly defined, short-term self-interest will need to be replaced or, at least, supplemented by a more broadly defined, long-term self-interest.

This certainly implies the development of greater economic linkages between geographical and social components of world society. It also implies greater levels of tolerance and an international perspective which is enhanced through greater literacy and education throughout world society, as well as the creation of the economic potential necessary to achieve and utilize the potential created by education. It is obvious that for someone who is starving, the most immediate need is not a better education but more food, but the quest for that food may take more adaptive directions, if the individual is more aware of the various options for getting it through constructive means. Recent experience suggests that improvement of economic, social, and educational conditions may result in the welcome bonus of a reduction in population growth.

A further point of clarification is needed regarding the capacity of human societies to make major changes in their culture. Reference has been made in some of the essays and during the commentary and discussion to the fact that culture is singularly resistant to change imposed by government. Here, I think, it may be helpful to make a distinction between flexibility and governability. Empirically, some societies have shown remarkable ability to make even major adjustments in response to social and technological innovations. The experience of the Manus, which I cited in my essay, is only one of many such examples of this flexibility. On the other hand Masters, in his commentary on my essay, is correct in emphasizing just how resistant most people are to decisions and policies imposed by government. Among the Manus the decision for change was the result of a broadly based consensus. What we need in large complex societies may be the development of more effective mechanisms for creating such a consensus.

A final conceptual distinction is that when we talk about a future for the human race we tend to think, rather narrowly, of survival of the society and the cultural traditions to which we belong. Thus we may think of a future for the society representing the American tradition. The biologist, of course, tends to think more broadly in terms of the future of the entire species. This distinction is important in the context of projections about the result of various calamities such as nuclear war. The future of civilized society is certainly threatened by such an event. The threat, however, to the human species is very much less. Thus what we tend to talk about in projecting some of the potentially disastrous situations is the destruction of cultural elements which we have come to value, but not necessarily the destruction of the human species.

References

Alvarez, L. W., W. Alvarez, F. Asaro, and H. V. Michel 1980. "Estraterrestrial Cause for the Cretaceous-Tertiary Extinction." *Science* 208:1095-108.

Angel, J. L. 1976. "Early Bronze Age Karatas People and Their Cemeteries." *American Journal of Archaeology* 80:395-91.

Barash, D.P. 1977. *Sociobiology and Behavior*. New York: Elsevier North-Holland, Inc.

Barney, G. O. 1981. *The Global 2000 Report to the President*. Vols. 1-3. Washington, D. C.: U. S. Government Printing Office.

Baron, R. M. 1980. "The Case for Differences in the Responses of Humans and Other Animals to High Density." Pages 247-73 in M. N. Cohen, R. S. Milpass, and H. G. Klein, eds., *Biosocial Mechanisms of Population Regulation*. New Haven, Conn., and London: Yale University Press.

Cavalli-Sforza, L., and M. W. Feldman 1973. "Models for Cultural Inheritance 1. Group Mean and Within Group Variation." *Theoretical Population Biology* 4:42-55.

Dobzhansky, T. 1962. *Mankind Evolving*. New Haven, Conn.: Yale University Press.

Freedman, J. L. 1980. "Human Reactions to Population Density." Pages 189-207 in M. N. Cohen, R. S. Milpass, and H. G. Klein, eds., *Biosocial Mechanisms of Population Regulation*. New Haven, Conn., and London: Yale University Press.

Hamilton, W. D. 1964. "The Genetical Evolution of Social Behavior, I and II." *Journal of Theoretical Biology* 7:1-52.

Harris, M. 1971. *Cultural Materialism: The Struggle for a Science of Culture*. New York: Random House.

Harrison, G. G. 1975. "Primary Adult Lactase Deficiency: A Problem in Anthropological Genetics." *American Anthropologist* 77:812-35.

Kenyon, K. M. 1979. *Archaeology in the Holy Land*. 4th ed. New York: W. Norton and Company.

Lee, D. 1959. *Freedom and Culture*. Englewood Cliffs, N. J.: Prentice-Hall.

Lumsden, C. J., and E. O. Wilson 1981. *Genes, Mind, and Culture*. Cambridge, Mass.: Harvard University Press.

McCracken, R. D. 1971. "Lactase Deficiency: An Example of Dietary Evolution." *Current Anthropology* 12:479-517.

McNeill, W. H. 1976. *Plagues and Peoples*. Garden City, N. Y.: Anchor Books.

Mead, M. 1956. *New Lives for Old*. New York: William Morrow and Company.

Midgley, M. 1978. *Beast and Man*. Ithaca, N. Y.: Cornell University Press.

_____1983. "Toward a New Understanding of Human Nature: The Limits of Individualism." Pages 517-33 in D. J. Ortner, ed., *How Humans Adapt: A Biocultural Odyssey*. Washington, D.C.: Smithsonian Institution Press.

DONALD J. ORTNER

Neel J. V. 1970. "Lessons from a 'Primitive' People." *Science* 170:815-22.

Ortner, D. J. 1979. "Disease and Mortality in the Early Bronze Age People of Bab edh-Dhra, Jordan." *American Journal of Physical Anthropology* 51:589-97.

Passmore, J. 1974. *Man's Responsibility for Nature.* New York: Charles Scribner's Sons.

_____1983. "Education and Adaptation for the Future." Pages 457-76 in D. J. Ortner, ed., *How Humans Adapt: A Biocultural Odyssey.* Washington, D.C.: Smithsonian Institution Press.

Ross, E. B. 1978. "Food Taboos, Diet, and Hunting Strategy: The Adaptation to Animals in Amazon Cultural Ecology." *Current Anthropology* 19:1-36.

Simpson, G. G. 1962. "Comments on Cultural Evolution." Pages 104-8 in H. Hoagland and R. W. Burhoe, eds., *Evolution and Man's Progress.* New York: Columbia University Press.

Spuhler, J. N. 1959. "Somatic Paths to Culture." Pages 1-13 in J. N. Spuhler, ed., *The Evolution of Man's Capacity for Culture.* Detroit: Wayne State University Press.

Washburn, S. L. 1959. "Speculations on the Interrelations of the History of Tools and Biological Evolution." Pages 21-31 in J. N. Spuhler, ed., *The Evolution of Man's Capacity for Culture.* Detroit: Wayne State University Press.

Wells, C. 1964. *Bones, Bodies and Disease.* London: Thames and Hudson.

Wilson, E. O. 1978. *On Human Nature.* Boston: Harvard University Press.

Commentary by Mark N. Cohen
Department of Anthropology, State University of New York

My first comment has to do with a statement that Don Ortner made toward the end of his essay in which he implies that optimism and pessimism toward the future should be linked to whether human behavioral traits turn out to be genetically governed (as the sociobiologists argue) or to be more purely cultural (and therefore more malleable).

One of the interesting things about our own culture is that it has a funny blind spot. We assume that if a trait is genetic we cannot "fix" it, whereas if it is cultural we can. In fact, there is much more evidence that we know how to begin to fix genetic disorders than that we know how to fix cultural ones. Grant that cultural systems change more rapidly than genetic ones, and appear to be more flexible in their meanderings. It is not at all clear, however, that cultural systems are more tractable than genetic ones for purposes of *planned* change. If we were to discover that "aggression" was a product of genes—and therefore correctible like any metabolic disorder—I suspect we would be closer to a solution to the problem, not farther from one.

I would like to discuss, at greater length, a point that Don made about the relationship between human numbers and crowding phenomena as they relate to potential future psychological stress. I choose to discuss this point because crowding phenomena provide an interesting example of the interaction of biology and culture, an example in which the relationship between the two can be mapped fairly precisely. Further, the issue of future crowding allows me to discuss another more general theme which underlies many of the presentations in this symposium: the question of whether and to what extent projected future stresses are "unprecedented."

Clearly, our current and future problems are unprecedented in several respects. Our technology is more powerful and dangerous and our world is changing more rapidly than ever before. But I maintain that there are good historic and prehistoric precedents for some of the problems we face, including the stress of crowding. These precedents are important for two reasons. One, they can give us some kind of hope and confidence. Two, they can show us the form of possible solutions. We can gain some of Dr. Dubos's optimism by looking at the resiliency that we have shown in the past. More important, I think, we can see some very specific solutions that have operated in history to solve problems like the ones we now face.

One thing that is *not* unprecedented is the need for the human population to compromise some existing values and standards in the face of growing populations. Part of our frustration today comes from a sense that, until very recently, things had been "getting better" and that only now is "progress" threatened. Yet there is evidence to suggest that we are already living with some old economic and social compromises resulting from past population growth. We have in fact been compromising for a long time, and we will probably be able to live with the next round of compromise, too.

Human populations have been outgrowing their existing strategies for a long time. I don't believe in the model of rash population growth and crash that Dr. Hassan attributes to me in his essay (chap. 7), but I do think that human populations constantly push at the fringes of their environments or at their "carrying capacity." I think this has resulted in economic change and in the compromise of existing standards. Like all animals, we can and do make lots of babies. Most species eliminate the excess of their offspring by discriminating against them socially or destroying them or forcing them out to the boundaries of the niche (where they are almost certain to die). The result of this

DONALD J. ORTNER

combination of social discrimination, forced emigration, and death is that most animal populations "stabilize," but only while constantly probing their niche margins. Occasionally, these marginal animals discover a whole new environment and the species expands. Most often, the marginal animals simply fail to thrive because they lack necessary behavioral and dietary flexibility.

Among human beings, two other factors operate. First, we have the ability to regulate population intentionally in anticipation of threatening the margins of the niche, but we do so imperfectly. Second, we have some ability—and some perception of our ability—to alter the carrying capacity of our environments and expand the margins of the niche. I think that what we have seen through human history is the expansion of technology and culture necessary to offset negative feedback at niche margins.

The tendency for human populations to push against our margins and limits has pushed us toward cultural means of extending our limits. Often we pay a price for that extension. The origins of agriculture were such an episode. There is now fairly good paleopathological evidence suggesting that the beginning of agriculture was a real low point in terms of the quality of human life. Apparently farming fed more people than did hunting and gathering, but it didn't feed them as well. A period of adjustment was required before cultural solutions to new problems were found and levels of well-being were restored.

Let me now return to the question of crowding and what it has to do with future social design. You are all probably familiar with the famous Calhoun experiments done some years ago. These experiments demonstrated that if one provided rats in a cage with an infinite food supply and let the population grow free of disease, the population eventually stabilized as a result of behavioral interactions among the rats, which produced glandular changes, reduced fertility, reduced disease resistence, promoted antisocial behaviors of various kinds, and distorted sexual as well as parental behaviors. Some of the effects, of course, were the result of the artificial cage, which prohibited emigration as a natural solution and heightened crowding. But there are hints that the same mechanisms operate in a range of free living organisms. In such populations, aggressive behavior, social spacing, and the inhibition of growth and fertility within the social group are combined with the emigration of excess individuals so that, again, populations stabilize even though the limits of the food supplies may not have been reached.

There are indications, reported by John Christian some years ago, suggesting that most of the glandular mechanisms involved in this phenomenon are shared by human beings as well as other animals.

Ever since this was pointed out, however, two major questions have remained. First, why don't human populations "asymptote"—that is, level off—at some high density even if the food supply can still be expanded? Second, as various social psychologists have pointed out, why isn't there any clear relationship between high population density and social pathology in human cities?

The answer to both questions has to do with defining exactly what is stressful about high density. Even among animals it is not the number of organisms per se that counts. It is the dynamics of the interaction among them. We know, for example, that dominant animals don't feel the stress as much as subordinate ones do.

The study of human subjects suggests that crowding is stressful in particular ways. First, individuals lose control of their own actions. That is, they can no longer choose a course of action that carries them unobstructed to a desired end. The sense of loss of control generates stress. The more "others" there are, particularly if one is subordinate and must get out of the way, the more stressful the situation becomes. Second, there is a condition called "privacy," which refers to one's ability to control and limit the access of others to oneself. The more other organisms there are and the more that each must interact with others against its will, the worse the problem becomes.

The third stressful element is information-load. The number of other organisms present is only indirectly important. What *is* important is the number of contingencies one has to play with, and the number of variables one has to juggle, in choosing a course of action. The more "others" there are—other things being equal—the more contingencies each will have to consider. In addition, of course, crowding is stressful if resources are scarce or if access to them is congested.

What I have described so far is the biological heritage that human beings share with other animals: a system of glandular responses to behavioral cues and a set of evaluative categories in which stressful cues are measured.

The interesting point is that human beings have developed a series of organizational techniques to offset sources of stress. It is at this point that we can view the cultural overlay on the biological pattern. If we trace the evolution of human social forms as the scale of society grows—following the prehistorian's simple threefold division of societies from hunter-gatherers through small-scale farmers to civilization, or what Morton Fried speaks of as egalitarian, rank, and state levels of organization—what we see is a gradual elaboration of mechanisms to cope with larger and larger aggregates.

Hunter-gatherers, or egalitarian groups, are often described as lacking various kinds of formal political apparatus. What I suggest is that they lack the formal apparatus that the rest of us use to get around

DONALD J. ORTNER

each other, to reduce information-load, to avoid problems of privacy, and to avoid problems of lack of control. For example, we all know that hunter-gatherers build very flimsy shelters, and it is usually assumed that this results from a lack of technological skills and from a lack of incentive related to the fact that they will soon move on. What I suggest is that it is also because they don't *need* shelter to define privacy and limit information-load in the way the rest of us do. Similarly, they don't *need* the formal mechanisms of subdivision, scheduling, and control that the rest of us use to prevent "others" from overwhelming us.

At the other extreme, one of the things that states do, as political systems, is limit the ability of others to impinge on one's behavior. Thus they attempt to restore the kind of control that each individual used to have as a hunter-gatherer. People who live in states also employ a number of classificatory devices, such as stereotyping ethnic groups, to simplify their information-processing problems. One deals only with the group as an aggregate and therefore need not concern oneself with all of the individual ways in which some group members might impinge.

Very generally, what I think has happened throughout human history as populations have enlarged—in addition to the technological extension of niche boundaries and resource supplies—has been the gradual elaboration of mechanisms to offset potential crowding stresses. These include increasing compartmentalization and scheduling of human activities around one another; the differentiation of groups into categories which further separate spheres of activity; and the elaboration of physical barriers to communication, such as housing and clothing, which reduce information-load and increase privacy.

Stress, then, is likely to occur, or at least to peak, when populations outgrow the existing mechanisms or when these mechanisms fail for other reasons. Indeed, I think that we may be hearing such a message in the political restiveness of the "New Right." The historical perspective is important, I think, because it helps to predict the kinds of social design features that an even more crowded world will require. There are thus some specific guidelines to follow from our past experience. There is precedent also for the fact that a successful solution will require the selective sacrifice of some valued social forms.

Commentary by Roger Masters
Department of Government, Dartmouth College

Ortner's essay is a welcome advance beyond the hackneyed "nature-nurture" dichotomy which persists in popularized discussions of biocultural evolution. As he points out, an evolutionary approach to human social behavior can focus *either* on "the analogy between biological and cultural evolution" *or* on the "feedback mechanism" relating biology to culture. In the former, which is generally less controversial, the "adaptiveness" of cultural practices is analyzed on the assumption that individually learned or socially transmitted behaviors, as well as bodily structures, can be subject to natural selection. Most debate has focused, therefore, on mechanisms of biocultural interaction.

Ortner is surely correct in suggesting that "people raised in the Western tradition" seem to prefer the "traditional model of biocultural relationships, in which culture transcends biology"; this apparently provides us with more "flexibility" than the presumed determinism of genetic predispositions or preadaptations to specific cultural practices. Oddly enough, however, this widely shared presumption may be incorrect.

Modern political systems have extreme difficulty in governing, in good part because of the resistance to cultural change. Were we to discover genetic factors controlling social behavior, the possibilities of conscious human manipulation (for good or ill) would probably be *increased*. Now that the technologies described in *Brave New World* have become a practical reality, isn't it time to reconceptualize biocultural evolution without imagining that its political implications depend on the extent to which biology explains human behavior (Masters 1982)?

Ortner speaks of sociobiology primarily with reference to models of genetic preadaptation or influence on culture (e.g., Lumsden and Wilson 1981). But the concepts underlying Hamilton's (1964) genetic models of natural selection can also be used to analyze cultural mechanisms that are *not* genetically transmitted from generation to generation (Campbell 1965, 1972; Masters 1981a, 1981c, 1982). In so doing, our entire approach to cultural evolution can be improved and made more consistent with biological theory.

For example, Ortner speaks of the analogy between biological and cultural evolution as if selection operates only at the level of the entire group:

> . . .*cultural traits need not be developed through some rational process in order to be beneficial. . . . As long as a myth results in a behavior pattern which confers an increased adaptive advantage on the society* the cultural innovation will be valuable in the *society's* interaction with its environment [italics added].

Without denying the importance of group selection in specialized

DONALD J. ORTNER

circumstances—especially for humans (Alexander and Borgia 1978; D. S. Wilson 1979)—Ortner's formulation reminds one of Wynne-Edwards's generally abandoned views. Today, following Hamilton (1964), Williams (1966), Wilson (1975), and others, inclusive-fitness models generally begin from the assumption that kin-selection or nepotism will be a powerful force limiting the scope of cooperation among reproductive rivals.

As has been pointed out elsewhere, inclusive fitness is essentially a cost-benefit theory, formally isomorphic with game theory (Maynard-Smith 1978; Axelrod and Hamilton 1981), economic theory (Hirshleifer 1978; Chase 1980), rational choice or decision theory (Margolis 1982), and—in traditional political philosophy—social contract theories from the pre-Socratics to Hobbes, Locke, and Rousseau (Masters 1982).

While such cost-benefit models of social aggregates are a useful way of studying natural selection and adaptation, other biological approaches focus on holistic or hierarchical systems, much as sociological analysis begins from the group as a whole whereas economic models start from individual decision-making units (Barry 1970). But the importance of linking inclusive-fitness theory with similar "rational actor" models of human social behavior should not be underestimated as a potential bridge between the natural and social sciences (Axelrod and Hamilton 1981; Masters 1981b).

In my own work, for example, it has become apparent that the "four types of behavior" distinguished in Hamilton's original paper introducing the concept of "inclusive fitness" (1964) are fundamental in human cultural behavior as well as in the adaptation of other species. The Prisoner's Dilemma (see fig. 14), widely used to illustrate the limits on cooperation between competing actors, illustrates the biological problem of extending cooperation to those who could further increase their gains by appearing to conform while actually defecting or cheating. From this perspective, it becomes obvious that the law, governmental systems, and religious taboos have played an incalculable evolutionary role by generating self-sacrificial cooperation in large-scale, frequently anonymous human societies. And while biologists have usually spoken of "altruism" when dealing with this phenomenon, such a motivational term is unnecessarily confusing and should be replaced by more objective behavioral concepts like *sociality* (Alexander 1978) or *social virtue* (Masters 1981a, 1983).

The utility of this kind of cost-benefit approach to human evolution has hitherto been primarily directed to theoretical matters. Very recently, however, we have begun to see empirical findings that demonstrate the importance of natural selection in explaining cultural myths, taboos, or practices. Perhaps the most important of these is

Durham's study of the New Yam Festivals in West Africa, showing how a powerful taboo against eating newly ripened yams is adapted—in specific ecological conditions—to the needs of carriers of the sickle-cell gene (Durham 1981). Since hungry adults have been known to starve to death for fear of violating this taboo, it provides an archetype of self-denying or socially virtuous behavior of a sort rare in other vertebrate species, but essential for the survival of human cultures (Campbell 1972).

Ortner concludes by underlining the evolutionary dangers of a policy of biological and cultural "laissez faire." Rational choice models of collective goods, like Hardin's "Tragedy of the Commons" (1968), indicate why this is so. Virtuous restraint in the short run, in order to benefit the group over the long run, is unlikely unless it is probable that the defectors will suffer and the virtuous will not be mere "suckers." In the absence of mechanisms that can balance the nepotistic urges selected over more than 3.5 million years of hominid evolution, every known civilization seems to have overexploited its resource base and collapsed.

Can we hope to avoid this fate without benefit of the lessons which a deeper understanding of biocultural evolution promises to teach us? And if *Brave New World* is the only answer, dare we take the risk of using genetic engineering to make our descendants into a cooperative, but genetically uniform, species? Clearly we have a moral obligation to know more about the evolutionary processes that formed human nature.

Contemporary approaches to natural selection suggest a general theory of human behavior that reinforces the concern for the future. Ortner points out that as soon as we look at any evolutionary model of human life, the future is somehow more challenged than we would like to admit; and I think that is going to continue to be a basic issue in the remainder of our deliberations. My reason for saying that is apparent in figure 13, which shows how evolutionary theory, particularly the inclusive-fitness models, reminds us of why the problem of the future is so intractable.

What you see is a simple fourfold table. A and B are any two organisms, and if A's behavior produces a gain for A in terms of inclusive fitness and B's behavior produces a gain for B, we can speak of that as mutual benefit or reciprocity. If A gains and B loses, then the outcome from A's point of view is what social biologists call nepotism or kin selection. If they both lose, we can speak of mutual harm. If A loses but B gains, I suggest we should call that, with Dick Alexander, sociality or virtue (Masters, 1983).

You find a matrix like this in Hamilton's classic essay of 1964 and

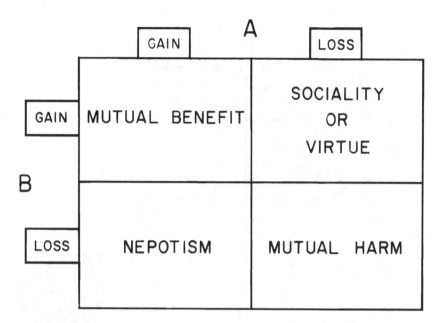

Figure 13. Basic types of social behavior (adapted from Hamilton 1964 as reprinted in Caplan 1978:207).

in other works by sociobiologists, but it doesn't mean anything like a theory of the "selfish gene." Humans have a peculiarity in that we exploit at various times all of those categories of behavior. Other species sometimes do so also, but not quite as markedly as humans; that is, our specificity is the variety of strategic responses of which humans are capable.

This theory is very nice because it permits us to link evolutionary biology with economic theory, game theory, rational choice theory, and the like. Let me give you a quick indication of one way that goes.

Figure 14 illustrates the very famous example known as the Prisoner's Dilemma, which I mentioned earlier. Although I won't go through it in detail, here is just one little thing. If you look at the payoffs from A's point of view, A will always prefer to talk, that is, to "rat" on his fellow prisoner, even though staying silent is doing the honorable thing. And in the process B will do the same, and so they'll both talk, ending in the worst of all possible situations. The Prisoner's Dilemma has often been used as an indication of the Hobbesian character of human nature if we calculate cost and benefit. Competition thus seems more "natural" than cooperation. But there is an interesting thing that

Non-Kin (a)

A's Strategy

B's Strategy	Silent	Talk
Silent	A = +9 B = +9 (Mutual Benefit)	A = +10 B = -10 (Nepotism)
Talk	A = -10 B = +10 (Sociality or Virtue)	A = -9 B = -9 (Mutual Harm)

Kin (b)

A's Strategy

B's Strategy	Silent	Talk
Silent	A = 9 + 4.5 = +13.5 B = 9 + 4.5 = +13.5 (Mutual Benefit)	A = 10 - 5 = +5 B = -10 + 5 = -5 (Nepotism)
Talk	A = -10 + 5 = -5 B = +10 - 5 = +5 (Sociality or Virtue)	A = -9 - 4.5 = -13.5 B = -9 - 4.5 = -13.5 (Mutual Harm)

Figure 14. The Prisoner's Dilemma—with non-kin (a) and kin (b). Assumptions include a two-person game in which pay-offs are transitive and commensurable; communication or behavioral coercion between A and B is impossible. Nepotism (private self-interest) and sociality (virtue) are defined in terms of A's pay-offs. Kin is defined as either full siblings or parent-offspring (technically a coefficient of relatedness or $r = 0.5$); hence, in each cell, A adds 0.5 of B's pay-off in figure 14a to his own pay-off, and so forth (see Masters 1983).

DONALD J. ORTNER

happens with the Prisoner's Dilemma that is not discussed very much. If A and B are brothers or father and son, as in figure 14b, according to inclusive fitness theory, their coefficient of relatedness (R) equals .5: hence A has to include as part of his payoff half of B's payoff and vice versa. If you include these payoffs in the matrix, it turns out that being silent or cooperating is the outcome for close kin. So that if you take rational-choice theory, you immediately see why hominids would have a tendency to cooperate in kin groups. In fact, for more than 3 1/2 million years hominids consisted of extended kin-group bands; that is, bands that have an R which is considerably greater than zero, as Neel's data show, and in which cooperation has been very highly developed, especially compared to the other primates. Food sharing is a good example of this cooperation in humans as the Lancasters (1983) show.

Still, there is a difficulty with using this kind of model to explain all of human history, and it is very simply indicated by our presence here. We no longer live in societies of extended kin which number up to about a hundred. From a biological point of view, one of the really interesting questions is where a society of 200 million non-kin comes from, and why on earth should we sacrifice for it? Why should we pay taxes?

Now, rational choice theory has already considered this (for example, Olson's *Logic of Collective Action* and Down's *Economic Theory of Democracy,* to give you a couple of examples). This is a real problem, which can be illustrated with a model that many of you will know as Garrett Hardin's "Tragedy of the Commons." Figure 15 formalizes the model in a way parallel to game theory and has the same categories that came out of Hamilton's paper.

I'm sure you all know the "Tragedy of the Commons" model, represented in figure 15 as an n-person game. You have a bunch of herdsmen. They each have the option of putting another cow on the commons, and if they do, in the short run each gains. (The matrices are based upon the notion that each cow gives a marginal utility of + 1.)

Under those circumstances in the short run either A or B would be foolish not to add cows to the common resources. The difficulty is that in a specific set of situations, there are common resources that at some margin will be overgrazed. In the commons on which the cows feed, you suddenly get a situation like that illustrated in figure 15b. Assume that there is an Xth cow (but nobody knows what the X is) such that there is enormous negative payoff for all.

Without going through all of the details, this kind of matrix presents a Hobbesian argument. William Ophuls has shown very well how, as

Short-Term Calculus (a) (Add 1 cow)

		Strategy A	
		Don't Add Cow	Add Cow
Strategy B	Don't Add Cow	A = -1 B = -1 (Mutual Harm)	A = +1 B = -1 (Nepotism)
	Add Cow	A = -1 B = +1 (Sociality)	A = +1 B = +1 (Mutual Benefit)

Long-Term Calculus (b) (Add Xth Cow)

		Strategy A	
		Don't Add Cow	Add Cow
Strategy B	Don't Add Cow	A = X - 1 B = X - 1 (Mutual Benefit)	A = X + 1 B = X - 1 (Nepotism)
	Add Cow	A = X - 1 B = X + 1 (Sociality)	A = X - 1000 B = X - 1000 (Mutual Harm)

Figure 15. The Tragedy of the Commons: a short-term calculus (add one cow) and a long-term calculus (add xth cow). Assumptions include an n-person game in which all actors know that only two strategies exist—and that others are free to choose either. In long-term calculus, commons can support an xth cow if added by *either* A strategists *or* B strategists, but not by both. All cows have marginal utility equal to +1; the value of x is unknown. Nepotism (private self-interest) and sociality (virtue) are defined in terms of A's payoffs (see Masters 1983).

DONALD J. ORTNER

a deduction from the "Tragedy of the Commons," the way you avoid overharvesting the resources is to delegate authority to one group— the government or Hobbes's sovereign—which says you may not add more than X minus one cows to your herd (Masters 1983).

I believe that this logic explains the origin of the centralized state. But in our circumstances, it is extremely plausible that one solution people will want to adopt in order to avoid the disaster of a "Tragedy of the Commons" is "Brave New World."

That the "Tragedy of the Commons" remains a concrete outcome that we have to consider is very obvious in the daily newspaper. To me it's symbolized by that sinkhole in Winter Park, Florida. In our civilization, there is a common resource basis—water under the ground. Everybody says, well, I might as well go out and pump a little water; it's just like adding another cow in Hardin's model. If I don't do it, my neighbor will do it anyway. The only trouble is that each individual keeps on pumping water, and all of a sudden, as illustrated in figure 15b, the house disappears.

I would remind you that every known civilization has gone this way, whether you use Marvin Harris's approach or any other approach that I know of. If you look at the desert of Saudi Arabia, the Tigris and Euphrates, Rome—whatever civilization we look at, every prior one has ended up in that situation described as mutual harm in the game matrix.

And in order to indicate the seriousness of this problem, consider the risk to the environment of a global population of six billion human beings, and the possibility of an awesome choice between "Brave New World" and a nuclear war. I think we have got to begin to seriously figure out how we stay out of that outcome known as mutual harm.

References

Alexander, R. D. 1978. "Natural Selection and Societal Laws." Chap. 7 in T. Englehardt and D. Callahan, eds., *Morals, Science and Society*. Vol. 3. Hastings-on-Hudson, N. Y.: Hastings Center.

Alexander, R. D., and G. Borgia 1978. "Group Selection, Altruism, and the Levels of the Organization of Life." *Annual Review of Ecology and Systematics* 9:449-74.

Axelrod, R., and W. D. Hamilton 1981. "The Evolution of Cooperation." *Science* 211:1390-96.

Barry, B. 1970. *Sociologist, Economists, and Democracy*. London: Macmillan.

Campbell, D. T. 1965. "Variation and Selective Retention in Sociocultural Evolution." Pages 19-48 in H. R. Barringer, G. I. Blanksten, and R. W.

Mack, eds., *Social Change in Developing Areas*. Cambridge, Mass.: Schenkman Publishing Co.

Campbell, D. T. 1972. "On the Genetics of Altruism and the Counter-hedonic Components in Human Culture." *Journal of Social Issues* 28:21-37.

Chase, I. 1980. "Cooperative and Noncooperative Behavior in Animals." *American Naturalist* 115:827-57.

Durham, W. H. 1981. "Coevolution and Law: The New Yam Festivals of West Africa." Paper presented to Hutchins Center—Goethe Institute Conference on Law and Behavioral Research, Monterey Dunes, California (Sept. 25-27, 1981).

Hamilton, W. 1964. "The Genetical Evolution of Social Behavior." *Journal of Theoretical Biology* 7:1-16. Reprinted in Arthur Caplan, ed., *The Sociology Debate*, pp. 191-209. New York: Harpers.

Hardin, G. 1968. "The Tragedy of the Commons." *Science* 162:1243-38. Reprinted as chap. 6 in H. E. Daly, ed., *Toward a Steady-State Economy*. San Francisco: W. H. Freeman.

Hirshleifer, J. 1978. "Natural Economy vs. Political Economy." *Journal of Social and Biological Structures* 1:319-37.

Lancaster, J. B., and C. S. Lancaster 1983. "Parental Investment: The Hominid Adaptation." Pages 33-56 in D. J. Ortner, ed., *How Humans Adapt: A Biocultural Odyssey*. Washington, D.C.: Smithsonian Institution Press.

Lumsden, C. J., and E. O. Wilson 1981. *Genes, Mind, and Culture*. Cambridge, Mass.: Harvard University Press.

Margolis, H. 1982. *Selfishness, Altruism, and Rationality*. Cambridge: Cambridge University Press.

Masters, R. D. 1981a. "Social Biology and the Welfare State." Paper presented to Conference on the Future of the Welfare State, St. Pierre des Canons, France (Oct. 5–9, 1981); to appear in *Comparative Social Research Annual* 1983 (Greenwich, Conn: JAI Press). In Press.

—— 1981b. "The Value—and Limitations—of Sociobiology." In Elliott White, ed., *Sociobiology and Human Politics*. Lexington, Mass.: Lexington Books.

—— 1981c. "Evolutionary Biology, Political Theory, and the Origin of the State." Paper presented to Hutchins Center–Goethe Institute Conference on Law and Behavioral Research, Monterey Dunes, Calif. (Sept. 25–27, 1981); to appear in *Journal of Social and Biological Structures*.

—— 1982. "Is Sociobiology Reactionary? The Political Implications of Inclusive Fitness Theory." *Quarterly Review of Biology* 57:275-92.

—— 1983. "The Biological Nature of the State." *World Politics* (Jan.). In Press.

Maynard-Smith, J. 1978. "The Evolution of Behavior." *Scientific American* 239:76-92.

Williams, G. C. 1966. *Adaptation and Natural Selection*. Princeton, N. J.: Princeton University Press.

Wilson, D. S. 1979. *The Natural Selection of Populations and Communities.* Menlo Park, Calif.: Benjamin Cummings.
Wilson, E. O. 1975. *Sociobiology.* Cambridge, Mass.: Harvard University Press.

Commentary by James F. Danielli
Danielli Associates, Worcester, Massachusetts

I would say that generally speaking human beings operate in terms of programs. For example, if I walk across a room like this, I am not making individual decisions about what each of my muscles should do; I am using a program for walking across the room. And if you look at the way in which you behave, I'm sure you will find for practically everything you do—except innovation (and even there there may be programs for innovation)—what you do is use a program. Therefore, if we want to develop models of how human beings operate, I think we might very well consider analyzing their programs.

We evidently have a substantial number of programs (P_G) which are essentially genetic in nature and which are the substance of what sociobiologists have dealt with. Second—I'm simplifying this a bit— we also have a number of programs which are social in nature (P_S) and which are derived from the interactions between human beings, and incidentally, between human beings and libraries and things like that. However, we can say of the social program that it will consist of a genetic element, plus a socially derived element, so that we end up with a set of programs, some of which are purely genetic, some might be described as purely social, but many contain components of both.

What takes the social program away from being dominated by genetics and enables us to operate with some degree of freedom, I think, is the internal reward system which, as I see it, consists of the following scenario. We have learned recently from neurobiology that opiates and various other substances are released in the brain. I am postulating that when we carry out certain of these programs, these opiates and so forth are released in the brain and give us an internal reward, which enables us to feel happy about doing things that may otherwise look unrewarding, such as altruistic actions.

Now, if that is so, you can see that as one becomes incorporated from childhood into the social program, if the internal reward system can be conditioned so that when we repeat in adult life the things which we have learned are approved of in childhood, we get this internal reward and become a satisfied human being. Thus the social program attains a very large measure of its independence from the genetic program as a result of the development of the internal reward system.

The next comment I would like to make is that the greatest adaptation human beings are likely to make in their lifetime is the adaptation to the culture into which they are born. For example, the culture which determines whether one grows up, shall we say, as a typical French citizen or a typical Chinese citizen. That is the biggest adaptation any human being can make, and it is probably irreversible.

The reason I mention this is that I feel that if we are going to talk about adaptations and how they are made, maybe one of the best sources for information about adaptation lies in the study of the adaptations children make to their environment as they are growing up. Adaptations, it seems to me, fall, if I can put it in a very crude way, into two classes: those which we make to a culture as we grow up and those which we make to vicissitudes for which we have not been prepared by the system in which we grew up.

To illustrate those adaptations we make to vicissitudes, I mention three things: first of all, conversion to a religion or ideology, which gives us a way of coping with vicissitudes, even if it is ridiculous and counterproductive. Another way of coping with vicissitudes is the use of drugs, which we all know is ridiculous and counterproductive, but we also know that many of us can't do without it. The third way of coping which I might mention is the improvement in management capacity, upon which, after all, the survival of the civilization depends. I would say at the moment, for example, the vicissitude of having an immense body of data to deal with, which has confronted our society in recent years, has been solved by the invention of electronic computers, which enable us to develop management systems for handling that data. And if it were not for that invention, that way of coping with the vicissitude—that adaptation, if you like—the future of our society would look totally different from the way it looks right at this moment.

What I would say in conclusion is that if we wish to change adaptability, we're not going to do very well by trying P_G itself, because that would mean genetic engineering, which is not on the agenda for the time being. But although, as one of the speakers said earlier, it is difficult to change the social program, that is where we have to look; that is, we need changes in social programs if we are to adapt to the changes taking place in our society.

DONALD J. ORTNER

We can write social programs which only elicit those parts of the genetic programs favorable to our endeavor. In other words, because there are certainly elements in the genetic program which are very unfavorable, we should choose carefully *which* elements of the genetic program are in fact elicited by the social system we are trying to develop.

6. Explaining the Course of Human Events

BETTY J. MEGGERS

Research Associate, Department of Anthropology, Smithsonian Institution

Editor's Summary. Betty Meggers focuses on the fundamental problem of the reference point in our concept of human society. In brief: Who are the "we" who should decide what the components of human society should do and, indeed, is it possible to control the course of human events? With respect to the latter question Meggers states that any perception that we can exercise such control is a delusion. But even if we could, there remains the problem of which component of society should do the deciding and how should that decision be made? Meggers emphasizes that humans should continue to innovate and try new ideas with the assurance that the best ideas will win out in the future as they have in the past.

More than one hundred years have passed since Darwin knocked the pedestal of special creation from beneath our feet and dumped us among the other vertebrates, yet acceptance that our behavior is explainable in terms of evolutionary theory still meets great resistance among both scientists and laymen. It is instructive to consider this antipathy in the context of present knowledge of the origins of the universe and of life.

Astronomers now believe that an explosion between 10 and 20 thousand million years ago created all the matter and energy incorporated into the countless celestial bodies hurtling through space. Some 4,600 million years ago, our planet coalesced at just the right distance from a star of the right intensity to provide conditions compatible with the emergence of life. After another 1,000 million years, the surface of the earth achieved a state appropriate for the survival of self-replicating molecules, but additional eons elapsed before these combined into single-celled organisms and initiated an evolutionary dialogue that transformed the biosphere and produced millions of kinds of creatures, whose diversity in appearance masks their chemical uniformity. The biota familiar to us constitutes the latest chapter in a long, complicated, and rambling epic. Although many details remain obscure, it is clear that our existence is no more inevitable and our persistence no more probable than those of any other species that has ever lived.

Compare this panorama with the message expressed or implied in daily news reports, political speeches, advertisements, books—in fact, almost every type of popular and scholarly medium—that our species has not only overcome the constraints of natural selection, but has achieved control over the environment of the biosphere. Overexploitation of renewable resources and depletion of nonrenewable ones, pollution of the atmosphere and the ocean, explosive human population growth, extinction of other species whose habitats we destroy—these and other accelerating processes are seen as readily remediable, when and if we choose to take action. Does not the very fact that we alone have exposed structures and events too small, distant, and ancient to be observed directly—and from these have deduced the laws regulating galaxies and atoms—prove that we have sundered the ecological bonds that limit the freedom of all other kinds of organisms?

Presented with these two interpretations, some of us find the evolutionary view more acceptable. We find it inconceivable that our species has the capacity consciously to rechannel—much less bring to a halt—processes that have operated on a cosmic scale for thousands of millions of years. Furthermore, beneath the cultural veneer, human behavior is so similar to that of other animals that it seems best explained by general evolutionary principles. This paper will call attention to some of the unobtrusive constraints on our freedom of choice and examine their implications for the future course of human events.

First, a brief digression is made necessary by the anthropocentric outlook that not only dominates popular thinking, but also prevails among social scientists. The "gut reaction" of most anthropologists is to reject the possibility that aspects of the environment are beyond our control. This attitude pervades even textbooks on "ecological anthropology." One authority tells us that "we may . . . construct models of the social process that contain many features reminiscent of natural ecosystems, but we can, if we choose, remain agnostic on the question of whether this parallelism makes it necessary to view social systems *as* ecosystems" and that "the rational or purposive manipulation of the social and natural environments constitutes the human approach to Nature" (Bennett 1976:19-20, 3). Another specifies that "we shall use *ecological system* to avoid the biological bias associated with the name ecosystem" (Hardesty 1977:14, emphasis in the original) and asserts that "cultural evolution undoubtedly involves some kind of 'selection' process, indeed probably several kinds, but it is unlikely that something analogous to natural selection is common" (Hardesty 1977:39). That statements such as these are attributable in part to misunderstanding of biological principles becomes clear when circumstances anthropologists consider distinctive to culture are compared with statements by biologists (table 7).

The impression of human uniqueness is reinforced by the complexity and variety of human behavior. As animals, we have ranges of tolerance for temperature and pressure, abilities to subsist on a vast variety of foods, and many other characteristics determined by our genetic heritage. We also have a social dimension, which is not exclusive to our species or even to the primate order. Finally, we have culture. Whether or not we are unique in this respect depends on how culture is defined (e.g., Bonner 1980), but the degree of elaboration we have achieved is certainly without precedent or parallel. We exult in our ability to probe the depths of time and to expose the structure of the atoms, pausing only occasionally to wonder whether our perceptions of "reality" are "true." Whereas an erroneous interpretation of the history of the universe is harmless, except perhaps to the ego of its proponent, an erroneous estimate of our capacity to control our environment can be disastrous. It thus behooves us to look carefully at our situation. What are the relative inputs of our biological, social, and cultural heritages? Do biological limitations on cultural behavior equal or exceed the effects of culture on biological processes? Under what circumstances and to what extent are we really able to ordain or even channel the future evolution of culture?

Another factor inhibiting scientific understanding is the bias inculcated in us by the very phenomena we are attempting to study. We believe that possession of "cognitive awareness" sets us apart from other organisms and that we alone can recognize and solve "problems." Thus, when hunting and gathering no longer supplied enough food, our ancestors solved the problem by domesticating plants and animals; when food production fell below the requirements of growing populations, they bred higher-yielding grains or improved the conditions for their growth; when pottery was needed, it was invented.

The anthropocentric view that "a beneficial response to an environmental problem cannot be made unless the organism is aware that a

TABLE 7 Similarities between Cultural and Biological Processes (Seen as Differences by Anthropologists)

Anthropologist	*Biologist*
"Cultural variations . . . are not capable of precise replication and transmission from parents to offspring as are genetic variations; rather they are susceptible to infinite blending and reinterpretation. It is this characteristic of cultural variation that makes a process analogous to natural selection questionable as responsible for evolutionary 'sorting' " (Hardesty 1977:38).	". . . sexual reproduction, which probably occurred early in evolution, compels reassortment of genetic programs in interbreeding populations. As a result, every genetic program (that is, every individual) is different from the others. This permanent reshuffling of genetic elements provides tremendous potentialities of adaptation" (Jacob 1977:1166).
"While human sociocultural history, like biological history, involves processes which are general and predictable, given specific conditions, its actual course involves immensely complex interplay of processes, sociocultural and ecological, which is, in its full concreteness, unpredictable and nonrepeatable" (Fallers 1974:140).	"The more I study evolution the more I am impressed by the uniqueness, by the unpredictability, and by the unrepeatability of evolutionary events" (Mayr 1976:317).
"Any theory of human or cultural ecology that is based on the proposition that Man's relations with Nature can be understood on the basis of methods and concepts derived from biological ecology, tends to neglect the variability and openness of the human behavioral process. . . ." (Bennett 1976:245).	"The most exciting aspect of biology is that, by contrast with physics and chemistry, it is not possible to reduce all phenomena to a few general laws. Nothing is as typically biological as the never-ending variety of solutions found by organisms to cope with similar challenges of the environment" (Mayr 1976:424).

BETTY J. MEGGERS

problem exists" (Hardesty 1977:28) contrasts with the biological principle that "anything adding to the probability of survival and reproductive success will automatically be selected for" (Mayr 1976:38). Can we safely ignore the possibility that patterns of cultural behavior originated and persisted through natural selection rather than conscious choice? Or that goal-directed choices constitute a means of increasing the random variation on which selection can operate (Dunnell 1981:210)? Can we be sure that cognitive awareness is not an illusion fostered by natural selection as a mechanism of adaptation?

When we look at ourselves from the perspective of natural selection, we can see two principal ways to examine the question of how humans adapt. One is the interface between biology and culture: to what extent does our behavior have a genetic basis and what are the biological effects of cultural practices? The other is the process of evolution: to what extent are the causes of biological and cultural change similar and what are the consequences of differences in methods of transmission of innovations?

The biocultural interplay is manifested in the effects of diet, medical knowledge, values, social relations, and other cultural variables on stature, life expectancy, frequency of specific pathologies, fertility, and other biological differences among individuals and populations. More subtle correlations also exist between physiological attributes and cultural practices. The genetic consequences of "self-domestication" are becoming sufficiently obtrusive to arouse concern about their long-term implications (see Neel, this volume). The complementary aspect of the biocultural interface—the extent to which cultural behavior is channeled biologically—is being explored by sociobiologists. Clearly, ours is a "simian world," as it was labeled half a century ago (Day 1936), and our behavior and institutions would be different were we felines or bovines rather than primates. Whether specific cultural expressions reflect biological differences between human populations rather than the operation of selective processes similar to those responsible for biological configurations remains to be established.

I wish to apply the perspective of evolutionary theory to the question of how humans adapt. If we look back over the path our ancestors followed, we can see that genetically programmed behavior was gradually supplemented and then incrementally supplanted by learned behavior as the means of articulating individuals with their environments. Culture, the culmination of the trend and the dominant mode of adaptation by *Homo sapiens*, is a specialized type of learned behavior. Individuals who could improve what they were taught and transmit a larger and more reliable store of information to their contemporaries and descendants gave the latter a better chance of survival. The more that cultural practices were elaborated, the better those individuals and groups that possessed them were able to ameliorate previously devastating crises. Again, the most successful survived and multiplied.

Although the primary focus of natural selection has moved gradually from biological strength and agility to cultural tools and knowledge, the processes initiated when life began have remained (apparently) unchanged. The same principles underlie the biological progression from single-celled organisms to higher mammals and the cultural progression from family bands to "superpowers" (Bonner 1980).

Viewing cultural behavior through glasses tinted (or tainted) by exposure to the theories and methods of natural science suggests that much of the confusion, discord, uncertainty, and general unease prevailing among social scientists reflects failure to achieve two goals fundamental to scientific inquiry: (1) to develop a general theoretical framework useful for recognizing significant kinds of cultural data and for generating hypotheses to explain them, and (2) to free ourselves sufficiently from the biases instilled by culture to examine human behavior objectively. These two threads are interwoven: until we achieve objectivity, we cannot increase our scientific understanding, yet the acquisition of understanding requires greater objectivity than we now possess.

Our pursuit of understanding is further handicapped by the fact that our tools not only are part of our subject matter, but have developed in the context of a particular variety of culture. The strength of this obstacle is evident from the minimal progress toward its removal. The first step is a clear recognition of its existence; the second is assembling the clues we have that are relevant to explaining human behavior and examining how they might be integrated and augmented.

In the pages that follow, I shall present some of the aspects of evolutionary theory that seem to me to account for cultural phenomena that are otherwise unintelligible. The following propositions will be assumed to be valid:

(1) evolution is a universal, continuous process that operates today essentially as it did when life began;

(2) culture is a form of behavioral adaptation that gives *Homo sapiens* a unique capacity to respond to environmental pressures rapidly and variably and to affect the global environment drastically and irreversibly;

(3) "the diversity and harmonious adaptation of the organic world [is] the result of a steady production of variation and of the selective

BETTY J. MEGGERS

effects of the environment'' (Mayr 1963:1), and the diversity of the cultural world is attributable to the same kind of interplay.

After describing similarities in the sources of diversity and mechanisms for differential preservation of biological and cultural innovations, I shall review some of the consequences of their different methods of transmittal. Finally, I shall examine some of the implications of this point of view.

Sources of Diversity. Biological and cultural changes are strikingly similar in their genesis and implementation. Mutations, which create new genes, are comparable to inventions and discoveries. Genetic recombination, which alters the sequences of genes in chromosomes, produces new phenotypic effects. Similar results follow new combinations of cultural traits, as when domestic animals were hitched to chariots and plows. Gene flow spreads variations from one population to another, providing the opportunity for novel combinations. Its cultural counterpart, diffusion, disperses ideas and objects among human populations.

Sampling bias, also known as the "founder effect," causes unequal representation of genes among two or more formerly interbreeding populations, leading to their diversification. Similar divergences are observable in the cultural behavior and language of human groups whose communication has been reduced or terminated. Finally, genetic drift, which gradually changes the representation of alleles within a population, is homologous with cultural drift, which produces unobtrusive alterations in cultural behavior.

Biological and cultural variability share other characteristics. Most biological innovations either convey no immediate advantage or are deleterious to their possessors (Mayr 1976:522; Blute 1979:56). The ethnographic literature attests to the repression and ostracism of deviant individuals whose behavior is believed to threaten the security of the community. Although greater internal diversity is tolerated by complex societies, we still penalize nonconformists (e.g., drug users and homosexuals), who appear to challenge the validity of dominant values and institutions.

The continuous and random production of both biological and cultural innovations provides the potential for rapid adjustment when formerly successful behavior becomes obsolete. The speed with which certain kinds of insects have developed tolerance for pesticides is a dramatic example of the importance to a species of maintaining biological heterogeneity. Minority cultural practices have played similar roles during periods of crisis. The dominant world religions are traceable to insignificant local cults, whose values were preadapted to the maintenance of order under changed sociopolitical and economic conditions.

Occasionally, an unusual innovation initiates a new line of evolution. Among animals, insects and vertebrates appear to have arisen from separate ancestral species that developed peculiar specializations (Mayr 1976:522). Among cultures, unpredictable and pervasive consequences followed the invention of the steam engine. Today, our lives are being drastically altered by the explosive ramification of microelectronics.

Differential Perpetuation of Innovations. The mechanisms for differential perpetuation of biological and cultural innovations are also similar, as would be expected if the evolutionary process is a universal one. Many social scientists object to applying the term "natural selection" to cultural phenomena and have proposed substituting "cultural selection" (e.g. Durham 1976:91). Because the concept of natural selection predates the discovery of the genetic means of transmittal of biological variations and because the same kinds of processes can be observed among biological and cultural phenomena, I align myself with those who prefer a nongenetic definition of natural selection (e.g. Richardson 1977:14).

Among the expressions of selective processes shared by biological and cultural phenomena are adaptive radiation (Kottak 1977; Linares 1977), niche specialization (Despres 1969), mutualism (Peterson 1978), competitive exclusion (Margolis 1977), phenotypic convergence (Rhodes and Thompson 1975; Adams 1966; Meggers 1972), and equilibrium area-diversity ratios (Terrell 1977). Space does not permit exemplifying all of these—and some, particularly competitive exclusion, are obvious. Most individuals experience it in finding a mate, gaining admission to college, getting a job, obtaining funds for a research project or a loan to buy a house, and in innumerable other situations. Smaller and simpler societies lose out to larger and more advanced ones, a process underway in many parts of the world today. Biological and cultural examples of convergences and area-diversity ratios illustrate their similarity.

Convergence—the emergence of similar forms from dissimilar antecedents—is one of the most fascinating expressions of the evolutionary process. It may manifest itself biologically in the independent development of structures that look different but perform the same function, such as the wings of bats and birds and the bifurcated tails of whales and fishes, or it may lead to astonishing morphological resemblances, such as the membranous wings of bats and pterodactyls and the "driptip" leaves and tapered buttresses of rain forest trees.

Unrelated species that occupy equivalent niches on different continents may resemble one another more closely than either does its biological relatives. Among examples from the mammalian fauna of tropical Africa and America are pangolins and armadillos, pigmy hippopotamuses and capybaras, royal antelopes and agoutis (Bourlière 1973: fig. 1). On a larger scale, communities of plants may look so much alike that only a specialist can tell whether a given photograph depicts, for example, a desert landscape in the province of Catamarca, Argentina, or a "forest" of saguaro cacti in southern Arizona.

Cultural convergences are equally striking. Prehistoric traits and complexes from widely separated regions with similar environments are often remarkably alike. At the time of European contact, the forests of eastern North America and Amazonian South America were inhabited by slash-and-burn agriculturalists who lived in pole-and-thatch communal dwellings, often surrounded by defensive palisades; contemporary groups in the southwestern United States and northwestern Argentina practiced dry farming or irrigation, lived in multiroom "pueblos," and decorated their pottery with identical geometric motifs traced in black on a white background (Meggers 1972). The traditional ways of life in highland Switzerland and the Himalayas share details of material culture, subsistence techniques, land tenure, and sociopolitical organization, including such specific features as late marriage, high proportion of celibacy, and low birthrate (Rhodes and Thompson 1975).

The evolution of urban society in pre-Columbian Mexico followed that in Mesopotamia by several millennia, but the kinds of sociopolitical and religious institutions, and their sequences of occurrence in the two regions, are very similar. Adams, who made a detailed comparison, concluded that "we have dealt . . . with independently recurring examples of a single, fundamental, cause-and-effect sequence" that "involved not the reenactment of a predetermined pattern but a continuing interplay of complex, locally distinctive forces whose specific forms and effects cannot be fully abstracted from their immediate geographical and historical contexts" (1966:173). A biologist could not have phrased it better.

Cultural convergences are often interpreted by anthropologists as proof of human inventiveness rather than consequences of natural selection. Other kinds of patterns shared by biological and cultural configurations are more difficult to dismiss in this fashion, however. Consider, for example, the correspondence between the area-diversity ratios exhibited by bird fauna and indigenous languages on the Solomon Islands (figs. 16 and 17; table 8) in western Melanesia (Terrell 1977). Biogeographers have found that the number of species of land and freshwater birds encountered on each of the main islands of the

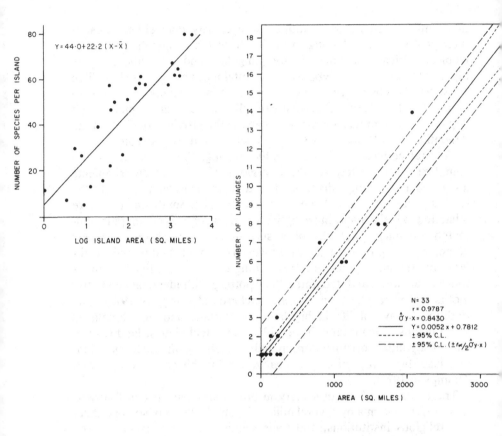

Figure 16 (left). Relationship between island area and the number of land and freshwater species of birds on the Solomon Islands of Melanesia (after Terrell 1977, fig. 3a).

Figure 17 (right). Relationship between island area and the number of indigenous languages spoken on the Solomon Islands (after Terrell 1977, fig. 5).

archipelago is so closely correlated with the land area that the equilibrium number of species on an island can be predicted if its size is known. Environmental factors such as increasing diversity of niches with increasing island size underlie this "rule," but the rigidity of the ratio is remarkable.

More remarkable still is the discovery that the number of languages

172   BETTY J. MEGGERS

TABLE 8 Area-Language Relationships for Solomon Islands

Island	Area (Sq. miles)	Expected Languages	Estimated Languages	Deviation
Rendova	147	1.55	2	0.45
Vanugu	210	1.88	3	1.12
San Cristobal	1193	7.00	6	1.00
Choiseul	1145	6.75	6	0.75
Isabel	1581	9.02	8	1.02
Malaita	1663	9.45	8	1.45
New Georgia	789	4.89	7	2.11
Guadalcanal	2039	11.40	14	2.60
Bougainville	3317	18.06	18	0.06

Source: After Terrell 1977: Table 1.

spoken on a given island is also closely predictable from its area. Languages have no apparent ecological dimension and one might expect monolinguality to be advantageous. There is no obvious explanation for this correspondence other than the existence of similar fundamental selective forces acting both on the biological and cultural systems. If this be granted, we must then ask how much of our behavior is governed by such unperceived imperatives? How much control do we really have over the basic processes that enmesh us?

Transmission of Innovations. The differences in methods of transmittal of biological and cultural phenomena have significant evolutionary consequences. Biological innovations are perpetuated principally by reproduction, which is a one-way street. Offspring inherit from parents, but cannot reciprocate. Furthermore, each individual's genetic makeup is fixed at conception. Although novelties with adaptive value can spread rapidly in fast-breeding populations, they are confined within species. Cultural innovations, being transmitted by learning, circumvent the barriers of kinship, age, sex, generation, race, language, and even physical proximity. Learning also eliminates the need to repeat developmental sequences and inventions. The number and variety of mechanisms for facilitating the diffusion of knowledge testify to the benefits of this short-cut (Meggers 1980) and the speed with which innovations can spread is impressive. The appearance of stone projectile points at the southern tip of the Western Hemisphere only a few centuries after their adoption by North American paleo-Indians is a particularly striking example. New technologies can be employed by groups who obtain the finished products by trade, rather than learn the methods of their manufacture. Individuals and communities can move directly from the Stone Age to the Nuclear Age in a few weeks or even hours (fig. 18).

Another important by-product of different methods of biological and cultural transmission is the potential for accumulating variations. Although the amount of information stored in each biological organism is phenomenal, changes involve substitutions rather than additions (except when the number of chromosomes increases). Cultural variation is not similarly restricted. The knowledge acquired by each generation can be added to that previously accumulated by the same group or by another group. Innovations originating at different times and in different places can be combined, as when steam engines were put on wheels. Abandoned techniques can be resurrected, as occurred in Israel where

Figure 18. "So by a vote of 8 to 2 we have decided to skip the Industrial Revolution completely, and go right into the electronic age." © 1981 by Sidney Harris—*American Scientist* magazine.

BETTY J. MEGGERS

an ancient method of trapping the moisture produced during nightly condensation has been used to reclaim the desert.

The advantages of cultural over biological transmission of innovations are offset by potentially serious disadvantages. On the biological level, unimpeded gene flow would be deleterious because it would dissolve adaptive combinations as rapidly as nonadaptive ones. The genetic barriers between species are a compromise that permits both intraspecific variation and interspecific diversification (Mayr 1976:19, 519). Successful combinations can be perpetuated and inferior ones eliminated by natural selection. On the cultural level, the tremendous capacity for dispersal of ideas and inventions that conveys adaptive superiority is also incompatible with the development and maintenance of viable social configurations.

There is abundant evidence that barriers comparable to those impeding the flow of genes between species prevent the wholesale diffusion of cultural innovations. Among these are the widespread attitude that members of other groups are inferior and unworthy of imitation; the view that traits of culture are as inalienable as biological features and hence impossible to acquire; the belief that outsiders are hostile and intercourse with them is to be avoided. Mechanisms impeding cultural flow are coupled with methods of affirming in-group identity, such as patterns of body painting, costumes, hair styles, flags, dietary restrictions, ceremonies, myths, slang, and innumerable other kinds of "emblems." The importance of these isolating mechanisms is vividly attested by the demoralization and disintegration that results when they prove inadequate.

A recent change in attitude among the population of the United States is interesting in this connection. Until a few decades ago, the goal of minorities was to assimilate, to leave behind behavior, values, and material evidence of their antecedents. Now, the emphasis is on preserving ethnic and racial identity. Has cultural homogenization gone too far? Instantaneous communication by radio and television, rapid movement of people and goods, and nationwide standardization of food, clothing, transportation, and entertainment give a veneer of sameness on an unprecedented scale. We are learning the danger of monoculture among domesticated plants; is it as hazardous culturally as it is biologically? Are deep-seated mechanisms of natural selection working to preserve cultural heterogeneity?

Culture not only affords the opportunity to adopt techniques and ideas invented elsewhere, it also permits moving large amounts of goods from one place to another. This capacity too can be a mixed benefit. As long as survival depended on long-term success in exploiting the immediate environment, a group that outgrew its food supply or experienced shortages from natural causes was forced immediately to correct the imbalance by temporarily increasing its death rate, altering its subsistence pattern, or emigrating. Over the millennia, procedures that minimized subsistence stress and environmental degradation were perpetuated and institutionalized.

In Amazonia, villages were small, transient, and dispersed; food was obtained by hunting, fishing, collecting wild plants, and raising crops in temporary clearings in the forest. Various customs, taboos, and attitudes kept population size compatible with carrying capacity. Humans were integrated into the ecosystem, affecting it little more than other components of the biota. At the opposite extreme, the natural landscape of northern China was slowly remodeled into a stable artificial ecosystem that can support a dense human population indefinitely. In other times and places, apparently successful cultural configurations evolved and then disintegrated.

Did climatic change cause subsistence failure? Did the population overshoot its food resources? Or did cultural factors, such as warfare, insurrection, unequal distribution of wealth, and ideological transformation upset the balance? The difficulty in assessing the evidence is attested by the continuing debate over the reason for the collapse of Mayan civilization a few centuries before European contact.

Cultural remedial measures, such as movement of food, raw materials, and finished goods from the region of origin to increasingly distant consumers, make possible larger and more stable adaptive configurations. They have also permitted our species to expand temporarily or permanently into habitats where self-sufficiency is precluded. Now, they are able to buffer the immediate effects of local environmental change on an unprecedented scale, not only saving populations that overrun their food supplies from extinction, but allowing them to continue to grow. The size and composition of our planet are fixed, however, as are the parameters within which life as we know it can exist. We are changing these parameters so rapidly that many other kinds of organisms are unable to adjust. Whether our species will survive is a question we cannot yet answer.

The Cultural Context of Perception. One of the reasons we cannot evaluate the consequences of our actions is that our perceptions are channeled by both culture and language. Although physicists and astronomers have been obliged to recognize that traditional concepts of time and space bear no necessary relationship to the behavior of the cosmos, the assumption is generally made that understanding natural and cultural processes is within our capacity and depends primarily on developing appropriate instruments and identifying key relationships. The extent to which our imaginations are constrained by the grammar of our languages is seldom considered.

The fallacy of believing that the laws of logic are the same for all observers regardless of what language they speak was pointed out by Whorf more than forty years ago. He noted that the world view prevailing today arose in the context of Indo-European languages, which treat time as an entity that can be measured and counted and—consequently—spent, wasted, and saved. Other languages deal with time as a process, a continuum, or a cyclic phase. One cannot talk about ten days in the same fashion as ten books or ten houses. Summer is not a unit, but a period with certain climatic characteristics. In Whorf's opinion:

Whether such a civilization as ours would be possible with widely different linguistic handling of time is a large question. . . . We are of course stimulated to use calendars, clocks, and watches, and to try to measure time ever more precisely; this aids science, and science in turn, following these well-worn cultural grooves, gives back to culture an ever-growing store of applications, habits, and values, with which culture again directs science (1941b:89).

When we have difficulty in imposing our attitude toward time on peoples whose languages and cultures treat it differently, we accuse them of indolence, stupidity, and obstinacy. Because languages with very different grammars can be spoken by the same individual, we incorrectly assume that thought processes and behaviors are similarly translatable.

Another feature of Indo-European languages that colors our perceptions is the construction of sentences by subject and predicate. Someone or something performs an action. When we analyze events from this perspective, we are likely to simplify them. For example, the fact the crocodillians eat fish implies that their removal will make more fish available for human consumption. Eliminating them from one of the tributaries of the Amazon, however, had the opposite effect. There was a pronounced decrease in the abundance of fish. An ecologist studied the food web and found that the crocodillians contributed more in nutrients to the local aquatic regime than they extracted in food. Their contribution, in fact, was essential to the maintenance of the food chain (Fittkau 1970). No one knows whether the structures of indigenous Amazonian languages, which are very different from Indo-European, facilitated adaptation of their speakers by preventing similarly mistaken perceptions, but such an investigation would be worth making.

The great variety of ways of perceiving built into the hundreds of languages that have been recorded is a resource we have seldom recognized, perhaps because we have been taught that all languages can be translated into all others. While this is true in gross terms, subtle distinctions are lost even among languages that are closely related. The Portuguese facility for transforming nouns into verbs permits nuances that cannot be preserved in English translation. Far more profound perceptual differences are concealed by translating languages with distinct grammatical structures, ways of classifying events, and forms of logic (figs. 19 and 20). Whorf suggests that

examination of other languages and the possibility of new types of logic that has been advanced by modern logicians themselves, suggest that this matter may be significant for modern science. New types of logic may help us eventually to understand how it is that electrons, the velocity of light, and other components of the subject matter of physics appear to behave illogically, or that phenomena which flout the sturdy common sense of yesteryear can nevertheless be true (1941a:20).

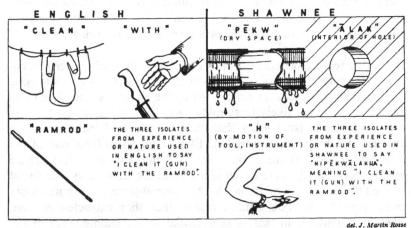

del. J. Martin Rosse

Figure 19. The different isolates of meaning (thoughts) employed in English and Shawnee (a North American Indian language) to report the same procedure—that of cleaning a gun by running a ramrod through it—reflect distinct kinds of logic. The pronouns "I" and "it" in English have the same meaning as "ni" and "a" in Shawnee and are not shown (after Whorf 1940, fig. 1).

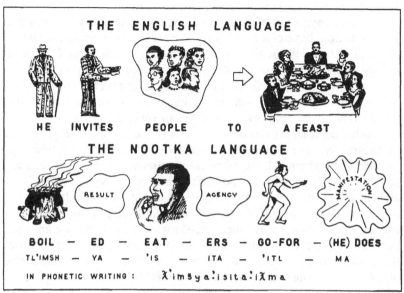

Figure 20. Distinct concepts are employed in English and Nootka (a North American Indian language) to describe the same event. The English sentence is divisible into subject and predicate; the Nootka equivalent is one word, consisting of the root "tl'imsh" with five suffixes. Although difficult for a speaker of English to comprehend, the Nootka formulation is complete and logical (after Whorf 1941a, fig. 2).

He emphasizes the importance of preserving linguistic diversity to escape the ethnocentric trap, saying,

I believe that those who envision a future world speaking only one tongue, whether English, German, Russian, or any other, hold a misguided ideal and would do the evolution of the human mind the greatest disservice. Western culture has made, through language, a provisional analysis of reality and, without correctives, holds resolutely to that analysis as final. The only correctives lie in all those other tongues which by aeons of independent evolution have arrived at different, but equally logical, provisional analyses (1941a:23).

Some Implications. Acknowledging ourselves to be products of evolution obliges us to confront several facts we would prefer to ignore. One is that natural selection is opportunistic and amoral. Pity and charity are alien concepts; "ought" and "should" are irrelevant verbs. Many more species have become extinct than are now alive. Whole landscapes, composed of flora and fauna as unlike today's as those in science fiction, have come and gone. We are unimportant in the history of our planet, which got along very well without us for several thousand million years. If we lose the flexibility to adapt, we too will become extinct. Other species will take our place, fill our niche, and carry on the evolutionary process—unless we, in passing, so alter the conditions for life that no existing organic forms can survive.

Another critical factor is our ignorance of the fundamental processes of nature. We are beginning to see that biological interactions are channeled by chemical and physical reactions, and these in turn are governed by tremendously powerful physical forces. Physics is an ancient discipline, yet physicists are still discovering novelties among the atoms and the stars that defy explanation. Biology is much younger and has come a long way since pre-Darwinian days, but we are only beginning to appreciate the marvelous intricacy of the genetic code, the astounding diversity of living things, and the amazing complexity of ecosystems. The social sciences have barely entered the complicated labyrinth of cultural structures and interactions, impeded even more than the physicists and biologists by the biases inculcated by our particular cultural and linguistic heritages.

Given the level of uncertainty within each of these scientific fields, it is not surprising that the interactions between physical, biological, and cultural phenomena are even more difficult to define. Ecologists, agronomists, climatologists, and others who attempt to synthesize knowledge about climates, soils, crops, and fuels, and to predict the impact of technological manipulations and inputs can only issue warnings because, although all organisms affect their surroundings, none before us appears to have altered conditions irreversibly on a planetary scale at so rapid a rate.

A third implication is that our ability to control the course of human events is a delusion. Confidence is adaptive, and many cultural practices have as one of their functions the maintenance of confidence. Myths tell us we were created to rule the earth; rituals ensure our success in hunting, the bounty of the harvest, and the favor of the gods. Confidence in the effectiveness of a treatment can cure an illness; confidence in the justice of a cause can win a war. But confidence, like other attributes, can become maladaptive. This seems likely to be true of our confidence that burgeoning human population, substantial modification of the composition of the atmosphere, significant depletion of the natural biota, incremental exploitation of nonrenewable resources, massive injection of toxic materials into the soil, the sky, and the sea, and large-scale preemption of food-producing land for other uses are temporary "problems," which "we" will somehow solve when it pleases us to do so.

But who are "we"? The species, *Homo sapiens*? The small success of members of the United Nations in obtaining agreement even on matters of minor significance eliminates this definition. Are "we" the Chinese? The Brazilians? The Saudis? The Navajos? The !Kung Bushmen? If I judge correctly, "we" are none of these; rather we are the intellectual products of Western civilization, represented by the participants in this symposium. In the context of the world today, much less of natural selection, the likelihood that "we" will decide the future of the planet is vanishingly small. Even if we had the power, what destiny might we choose? What form of government? What values? And, were we able to agree on these, how would we implement our decisions? To do so, we would need to know far more than we do about how and why culture changes. What is happening throughout the world reflects the operation of evolutionary principles still largely beneath our awareness. This applies to changes in culture, as well as in other organic and inorganic components of the biosphere. The blinders of anthropocentrism may be an essential aspect of our relationship to our cultural cocoon. If so, we may never be capable of understanding how humans adapt.

Does this mean we should sit back and let events take their course? Are those who are vigorously espousing causes and laboring to make converts to their way of thinking wasting their efforts? The thesis I have presented implies they are not. Although cultural behavior appears to have become our principal means of articulating with our surroundings, this behavior is inseparable from our biology. Both our biological capacity for teaching and learning and our cultural capacity for improving, accumulating, and augmenting knowledge have been favored by natural selection. It follows that those most successful in transmitting their attitudes, ideas, and information will continue to chart the course of our biocultural odyssey in the future, as they did in the past.

References

Adams, R. McC. 1966. *The Evolution of Urban Society: Early Mesopotamia and Prehispanic Mexico.* Chicago: Aldine.
Bennett, J. W. 1976. *The Ecological Transition: Cultural Anthropology and Human Adaptation.* New York: Pargamon Press.

Blute, M. 1979. "Sociocultural Evolutionism: An Untried Theory." *Behavioral Science* 24:46-59.

Bonner, John T. 1980. *The Evolution of Culture in Animals.* Princeton, N.J.: Princeton University Press.

Bourlière, François. 1973. "The Comparative Ecology of Rain Forest Mammals in Africa and Tropical America: Some Introductory Remarks." Pages 279-92 in B. J. Meggers, E. S.Ayensu, and W. D. Duckworth, eds., *Tropical Forest Ecosystems in Africa and South America: A Comparative Review.* Washington, D.C.: Smithsonian Institution.

Day, Clarence. 1936. *This Simian World.* New York: Knopf.

Depres, Leo A. 1969. "Differential Adaptations and Micro-cultural Evolution in Guyana." *Southwestern Journal of Anthropology* 25:14-44.

Dunnell, R. C. 1981. "CA Comment." *Current Anthropology* 22:210.

Durham, W. 1976. "The Adaptive Significance of Cultural Behavior." *Human Ecology* 4:89-121.

Fallers, Lloyd A. 1974. *Social Anthropology of the Nation-State.* Chicago: Aldine.

Fittkau, E.-J. 1970. "Role of Caimans in the Nutrient Regime of Mouth-lakes of Amazon Affluents (An Hypothesis)." *Biotropica* 2:138-42.

Hardesty, D. L. 1977. *Ecological Anthropology.* New York: Wiley.

Kottak, C. P. 1977. "The Process of State Formation in Madagascar." *American Ethnologist* 9:136-55.

Linares, O. 1977. "Adaptive Strategies in Western Panama." *World Archaeology* 8:304-19.

Margolis, M. 1977. "Historical Perspectives on Frontier Agriculture as an Adaptive Strategy." *American Ethnologist* 4:42-64.

Mayr, E. 1963. *Animal Species and Evolution.* Cambridge, Mass.: Harvard University Press.

———1976. *Evolution and the Diversity of Life: Selected Essays.* Cambridge, Mass.: Belknap Press of Harvard University Press.

Meggers, Betty J. 1972. *Prehistoric America.* Chicago: Aldine.

———1980. "The Evolutionary Significance of Diffusion." Paper presented at the plenary session of the American Anthropology Association, December 1980. Manuscript.

Peterson, J. T. 1978. "The Ecology of Social Boundaries; Agta Foragers of the Philippines." *Illinois Studies in Anthropology* 11. Urbana: University of Illinois Press.

Rhodes, Robert E., and S. I. Thompson 1975. "Adaptive Strategies in Alpine Environments; Beyond Ecological Particularism." *American Ethnologist* 2:535-51.

Richerson, P. J. 1979. "Ecology and Human Ecology: A Comparison of Theories in the Biological and Social Sciences." *American Ethnologist* 4:1-26.

Terrell, J. 1977. "Human Biogeography in the Solomon Islands." *Fieldiana, Anthropology* 68:1-47. Chicago: Field Museum of Natural History.

BETTY J. MEGGERS

Whorf, B. L. 1940. "Science and Linguistics." *The Technology Review* 42:April.
———1941a. "Languages and Logic." *The Technology Review* 43:April.
———1941b. "The Relations of Habitual Thought and Behavior to Language." Pages 75-93 in L. Spier, A. I. Hallowell, and S. S. Newman, eds., *Language, Culture, and Personality*. Menasha, Wis.: Sapir Memorial Publication Fund.
———1950. *Four Articles on Metalinguistics*. Washington, D. C.: Foreign Service Institute, Department of State.

Commentary by John C. Ewers
Department of Anthropology, Smithsonian Institution

Two centuries ago the French dramatist Pierre Beaumarchais wrote a very popular comedy, *The Marriage of Figaro*, in which one of his characters said, "We drink without thirst and make love any time. Only that distinguishes us from the animal." There are, of course, many other differences between man and the other animals. I am inclined to believe that the most striking difference may be man's insatiable curiosity, not just about that part of his environment that lies within the life zone he normally occupies, but about everything in the universe. Man doesn't bay at the moon; he sends members of his species to investigate it. Through microscopes he examines organisms so small he can't see them with the human eye. Through telescopes he becomes familiar with galaxies that are light years away, that cannot be seen with the human eye.

The American ethnologist Robert Lowie wrote seventy-four years ago: "Culture is a matter of exceedingly slow growth until a certain threshold is passed when it darts forward, gathering momentum at an unexpected rate." I think a very important, and perhaps a most important threshold, was man's invention of writing, which enabled him to use language in such a way as to transmit what he had learned far beyond the range of his own voice to his own contemporaries and to preserve a record of that knowledge for future generations. Certainly the point of view of every scholar is conditioned by his family background, childhood experiences, education, and research experiences. Betty Meggers and I have been associated in the Department of Anthropology at the Smithsonian for three decades, but we learned a great deal before we came here and some of the things we learned were not the same.

Our backgrounds and research involvements have been different. So has our education. At the University of Michigan Betty came under the influence of Leslie White, a leading exponent of the Evolutionary School of Anthropology. I had as my mentors at Yale an outstanding group of anthropological scholars from an earlier generation who reacted rather strongly and negatively to a theory of evolution which hypothesized evolutionary stages from savagery to civilization. That theory went back to the eighteenth-century English philosophers and was represented in this country through the works of the father of American anthropology, Louis Henry Morgan, and many of his nineteenth-century contemporaries.

My mentors were wary of applying biological theory to the development of human cultures. With Alfred R. Kroeber, they tended to think of culture as something superorganic and the processes of cultural change as different from those of natural selection in biological evolution. They came to be lumped together as members of an American Historical School, for they insisted that to know the development of any culture one must study the history of that particular group and its relations with other groups of human beings.

Clark Wissler, who steered me in the direction of Plains Indian research, was himself a leading figure in that field. He was also much concerned with problems of culture content and culture change. His comparative studies of numerous cultures led him to recognize a considerable number of traits or aspects of culture that could be found in every human culture, regardless of the relative simplicity or complexity of its economy or its technology.

George Peter Murdock, a colleague of Wissler's at the Institute of Human Relations at Yale (and another of my teachers), expanded and refined Wissler's universal culture pattern into an Outline of Cultural Materials which has served Murdock and his associates as a basis for comparative studies.

These studies tend to demonstrate that people elaborate the different aspects of their universal culture patterns differently and to a greater or lesser degree. Murdock's major interest, of course, was in the comparative study of social organizations, and he found that the most complex of all known human kinship systems was that of the Australian aborigines, a system so complex that it has proved the despair of most students. And yet on a scale of economic and technical advancement the Australian aborigines must rank very low.

Franz Boas, under whom I studied at Columbia, is remembered for many significant contributions to different fields of anthropology, one of which was the study of the languages of the native peoples of North America. Boas became greatly impressed by the complexity of the

Eskimo language. Not only did it have a highly specialized vocabulary, but its grammar was more complex than that of either Chinese or modern English.

In a lecture delivered at Columbia University as early as December 18, 1907, and later published, Boas declared that all aspects of culture have not developed from the simple to the complex; that in language it has often been from the more complex to the more simple.

Anthropologists of the evolutionary persuasion may find that statement rank heresy. Linguists do not. Boas challenged the evolutionists in their consideration of another very important aspect of culture, art. In his well- known book *Primitive Art*, Boas went after A.C. Haddon's work, *Evolution in Art*, showing that the examples Haddon chose to demonstrate evolution were in fact contemporaneous variants and did not illustrate a historical sequence. Ruth Bunzel, a student of Boas, later referred to Haddon's theory as one "based upon the fallacy inherent in all evolutionary arguments, namely the assumption that universal sequences of culture forms can be established."

My own years of study of the history of the graphic and plastic arts lead me to believe that the concept of evolution in art is a myth. Art has been universal, and it is very difficult to compare the artistic achievements of different cultures. If called upon to compare a painted bison from the cave of Altamira, a selected animal effigy pipe from Moundville, Ohio, carved by an American Indian at the very beginning of the Christian era, and a relatively abstract painting by Pablo Picasso of the early twentieth century, the art critic might very well be inclined to award Picasso third prize, especially if he was aware that Picasso, in his search for significant form, had been greatly influenced by the creations of West African sculptures. His style therefore was actually derivative; it was not original.

We know that evolutionary theory tends to minimize, if not to deny, the influence of the individual in cultural change and stability. Yet many years ago Alexander Goldenweiser, a student of Boas, observed that "a civilizational stream is not merely carried, but is unrelentingly fed by component individuals."

As long ago as 1917, another of my mentors, Edward Sapir, reminded us that "the determining influence of individuals is much more easily demonstrated in the higher than in the lower levels of culture."

Need I remind you of the behavior of millions of people in all parts of the world, still strongly conditioned by the teachings of Jesus in the Sermon on the Mount some 2,000 years ago, while the Ten Commandments, transcribed by Moses at an earlier time, continue to exert an influence upon the actions of large numbers of Jews and Christians.

The persistent influence of Charles Darwin upon the thinking of late-twentieth-century scientists and philosophers should be apparent to anyone who reads the proceedings of this symposium. And is not the persistent practice of scholars of citing by name those individuals whose ideas they have accepted, an acknowledgment of the importance of the individual contribution to human thought?

Betty Meggers has referred to recent efforts on the part of minorities to preserve their ethnic identity. My experiences lead me to doubt the long-range success of many of these efforts.

It was my privilege to have lived for three and a half years on the Blackfeet Reservation in Montana nearly four decades ago. At that time there were some thirty Indians on the reservation who were old enough to have been brought up back in buffalo days, who experienced life in the camps of nomadic hunters, including the buffalo hunt and intertribal warfare. These old people knew they were the last generation of Blackfoot Indians to have lived that life, and they were eager to tell me what they remembered of it so that it could be recorded.

In those days I used to wonder what Blackfoot life would be like when these fine people were gone, when they had "gone to the Sand Hills," and when younger generations of the tribe lost all living ties with their way of life in prereservation days. The last of those old Indians passed away during the 1950s.

Their descendants still celebrate their annual Indian days. They occasionally dress Indian. They dance their war and rabbit dances. But they no longer gather around the fire on cold winter nights to hear the old people tell their rich experiences in buffalo days, when members of their tribe were the mighty hunters and awesome raiders of the Northwestern Plains.

Now, I often wonder why I had not had the wit to ask this or that question of Weasel Tail or Julia Wades-in-the-Water. No one knows the answers to my questions about the old ways of life any more. That is what happens when a living culture dies. I know that some of my books have become required reading in the new community college on the Blackfeet Reservation. Just last month my wife and I received a letter from Earl Old Person, the present chief and chairman of the Tribal Council on the reservation. We knew Earl as a small boy. In fact, he was a Boy Scout when we lived on the reservation. In his letter he thanked us for what we had done to preserve the knowledge of the history of his tribe.

I sincerely hope that what I wrote was accurate, for on some aspects of tribal culture and history I am very much afraid it is literally the last word.

BETTY J. MEGGERS

Commentary by Roger Masters

Department of Government, Dartmouth College

We have had an interesting juxtaposition of two views of the way evolutionary theory might or might not relate to human cultural behavior. I am so thoroughly in agreement with what Betty Meggers said—and so thoroughly in disagreement with the conception of evolution implicit in Dr. Ewers's otherwise moving and humane remarks—I will simply leave it at that.

I want to make one comment that might advance our discussions. It has to do with the mechanisms of cultural change, making it possible for humans to have an information store upon which selection can operate, as Betty Meggers has argued.

What seems to be unique about humans is verbal communication. This is an old idea that goes back to the Greeks (Aristotle, Politics, I, 1253a). But it has implications, in the light of contemporary linguistics and evolutionary biology, that may help us to understand cultural adaptation.

If you consider DNA as a system of communication, it has some interesting properties. DNA molecules are very long strings of nucleotide bases which form triads, i.e., the so-called "genetic code." Without going into the complexity of the latest work on the "grammar" of the genetic code, this structure of DNA is itself remarkably parallel to the structure of human speech. Using the conventional symbols, consider the example of a string of nucleotide bases forming a DNA string: (1) . . . TAT CAT ATT. . . . The written symbols of our alphabet obviously form similar strings, composed of discrete subunits: (2) . . . THAT CAT AT. . . . In the first example, of course, each written symbol stands for one of the nucleotide bases, whereas in the second each symbol— i.e., each letter—represents a phonemic sound contrast. There is, however, a third mode of written communication—ideograms such as a line drawing of a cat.

The interesting thing about these three forms of laying traces is that the first two share something that is different from the third. While the first two communicative systems can be compared to speech, the third is akin to nonverbal cues. What you find in primates is a very rich nonverbal repertoire in both the vocal communications system and especially in the facial-bodily gestures, as well as some other channels of communication. Hence this comparison provides some insight into what is special about human speech as contrasted to earlier forms of animal communication.

The third kind of system, of which primate and human nonverbal communication is an example, generally tends to be analog, not digital; it tends to have a relatively limited and closed repertoire. It is true that when you look at writing systems of this sort, such as Egyptian, Chinese, or Mayan ideographs, there can be large numbers of characters. But ideographic writing is more limited than alphabetic systems, just as the number of things that you can communicate with nonverbal cues is less than the range of verbal communication. Why? The key seems to be that in our speech patterning, there are phonetic contrasts which don't bear meaning, but which then form morphenes which do carry meaning.

I have used writing, of course, because alphabetic writing mimics the structure of speech: alphabetic writing has letters which, like the phonetic contrast, are discrete and do not carry meaning, but form strings where subassemblies have meaning, and then longer assemblies have more meaning. This structure is, of course, similar to that of DNA.

Hockett, when he introduced this argument about human speech, spoke of dual patterning, but clearly that is incomplete. There is at least triadic patterning. For a fuller discussion of this argument, see my paper, "Genes, Language, and Evolution," *Semiotica*, vol. II (1970), pp. 295-20. First, there must be the patterning of the phonetic contrasts, the distinction between the "ah" and the "oh" sound. Second, these contrasts must be combined: if I want to say the "cot," that is different from the "cat." Hence, the meaning unit is an assemblage of phonetic contrasts that don't carry meaning, and the same thing happens with DNA. Third, the minimal meaning unit—codon in DNA morpheme or word in speech—must be combined into a longer string, be it a chromosome or a sentence.

The systems using "triadic patterning," unlike nonverbal cues or ideographs, generate tremendous possibilities for what is technically called "openness" in linguistics. This is particularly important for evolution because information stores based on "triadic patterning" can have meaningless stretches of information, which is something that we are reminded of in a symposium like this—I'm giving you one right now, I suppose. But those meaningless stretches can then be preadaptations of material upon which selection can work. Hence it is possible for humans to say things which are not completely determined by the context or completely adaptive at one time, but will subsequently become useful resources. In the same way, recent work in genetics shows that the duplication of stretches of DNA makes possible subsequent mutation and evolutionary change.

Such a mechanism could be crucial in making possible a form of cultural evolution in humans that is not found in other species. I think it is useful to reflect on this, especially because it indicates that, *as a species*, humans do indeed have something that makes us rather different in our capacities for adaptation to ecological settings.

That being said, I want to conclude with stress on one very nice thing in Betty Meggers's paper. At the end, Betty asks a key question: who is the "we," when we speak of the adaptive character of "us"? She very properly emphasizes a common error by pointing out the importance of defining the referent of that particular word, "we."

It is very easy to make a very big mistake in this regard. To start some sparks flying, I'll give one concrete example so that we can have a good argument about this. Consider the following sentence in Professor Cavalli-Sforza's paper: "Culture is one other method of adaptation that makes *us* more flexible, more versatile, and more ready to occupy successfully new niches, so much so that *we* are now close to having saturated *our* own planet." That is a proposition which is clearly true if "we" means the species *Homo sapiens*. It is clearly false if it refers to a particular society. While individuals learn, and the species has evolved a variety of cultural norms, *within* a society, culture normally works to *reduce* flexibility by increasing the predictability and conformity of behavior. Hence what is true with regard to the species—or even at the individual level—can be just plain false with reference to social groups.

Commentary by Bruce D. Smith

Department of Anthropology, Smithsonian Institution

In reading through the papers prepared for this symposium, I was somewhat surprised by the approach/avoidance attitude that many of the essayists displayed as they worked their way around what seems to me to be a rather fundamental question. I would like to raise this fundamental question directly and reiterate some of the various positions taken in regard to it by the distinguished essayists and suggest that it might be an interesting topic of discussion here and there through the rest of the symposium.

The question itself is fairly straightforward: Can the present potentially disastrous course of events be successfully altered by the intervention of humankind, or are we now relegated to a somewhat blustery and bumbling secondary role in the unfolding drama of our own demise?

At one end of the spectrum of positions taken on this question, Stephen Toulmin unequivocally and confidently champions mankind's ability to rise to the occasion. With spirit somewhat uplifted by this upbeat optimistic message, I read the essay by Dr. Meggers, whose work I have always admired, only to read that Toulmin's confidence may well be of the maladaptive variety.

Other essayists fall between Toulmin and Meggers on this question, touching it briefly and somewhat apprehensively. James Neel almost qualifies for inclusion in Toulmin's category of "prophet of ecological doom," clearly displaying early symptoms of what Toulmin calls "premature despondency." Professor Scrimshaw, in an illuminating conclusion to his paper, echoes Neel's identification of the necessity of the sociopolitical framework/global cooperation solution to the problem. Similarly, Hassan voices the solution of "a worldwide mangement of resources and technology, of labor management, in a spirit of cooperation."

Although not explicitly denying the possibility that such a framework of global cooperation could be established from the full spectrum of the world's governmental structures, currently afloat in swirling cross-currents of conflicting objectives, attitudes, and policies, Richard Barnet's essay (chap. 15), and that of McNicoll and Nag (chap. 11), did little to drive away my growing feeling of apprehension.

McNicoll and Nag, for example, point out our rather chilling inability, even as recently as the 1960s, to forecast with any degree of accuracy future world economic growth. So, in conclusion, I must confess to a certain lack of faith, a certain bunker mentality. I have not yet started to stockpile canned goods in my mountain retreat, but I do ask the question, "Where are the examples, the prototypes in today's world for this spirit of global unification to resolve a common problem?" Is it in the recent economic summit in Mexico or the treaty covering exploitation of undersea minerals? Or perhaps the international co-operation in setting quotas on the harvesting of whales and other sea mammals? Or the cooperation between Canada and the United States in solving the problem of acid rain? Or, finally, and closer to home, the cooperative public/private effort to preserve the Ogalala aquifer?

7. Earth Resources and Population: An Archeological Perspective

FEKRI A. HASSAN

Associate Professor of Anthropology, Washington State University

Editor's Summary. Fekri Hassan points out that the agricultural revolution created greater control over food resources at the cost of harder work and loss of individual freedom as compared to the earlier hunter-gatherer subsistence pattern. As the world becomes more crowded, migration—the relief valve for rapidly expanding populations in agricultural centers in the past—is no longer an option. He suggests that many of the world's problems can only be solved by a significant transfer of capital, technology, and industrialization to the Third World.

Just as a boat cut loose from its moorings drifts aimlessly hither and thither at the mercy of the wind and the waves so do we drift when we lose touch with the past. KARL JASPERS

The problem of overpopulation today can be clarified by an examination of the changing relations between people and resources through time. From the dawn of human existence, cultural responses were developed to ensure human survival. Under hunting and gathering, a mode of food procurement that lasted for 2 to 3 million years, the interaction with the environment was based on the efforts of mobile bands of three to ten families, who shared information and food. The growth of their numbers was limited by the natural yield from wild resources, which did not support more than 12 million persons for the whole world. Accordingly, behavioral population regulation mechanisms were adopted and a strategy favoring few offspring invested with care was selected.

A combination of ecological, biological, demographic, and cultural factors by the close of the last Ice Age—about 10,000 years ago—facilitated the emergence of agriculture. It was the great potential of agriculture for economic growth, coupled with the changes in the organizational structure of incipient (Neolithic) agricultural societies and the attendant economic and demographic changes, that led to a spiral of agricultural developments, trade, and population increase. The expanding economy placed a premium on human labor and nonsubsistence goods. At the same time, the emergence of managers, craft specialists, artisans, soldiers, priests, scribes, and so forth, led to a disproportionate increase in the ratio of consumers to foodproducers. In addition, the standard of living shifted from one related to subsistence to an emphasis on nonsubsistence goods for an increasing proportion of the population.

These Neolithic developments continue to the present: the expansion of non-food-producers includes in some cases 80 percent of the population in some contemporary European societies and, with the addition of fossil fuels for agricultural and industrial production, a very sharp increase in nonsubsistence goods. Industrialization in Europe over the last 160 years has led to a decrease in mortality, which was balanced by a drop in fertility. However, contact between nonindustrial societies and industrial technology has led to a reduction in mortality without a concomitant reduction in fertility.

This anomalous situation has led to an excessive rate of population increase that places the survival and well-being of a great many of the world population in jeopardy. The increase in population is worsened also by rising expectations. The developments in transport and international trade since medieval times have also predicated economic stability on the flow of resources from distant places. The disparity between national interests and international dependency, concomitant with differential access to wealth, is undoubtedly the key problem of our present situation. This problem is endemic in agricultural and agroindustrial economy and has now reached a global dimension because of telecommunication, advanced transport, and international trade. A remedy is urgently called for if discontent, revolts, acts of terrorism, strikes, and international strife are to be averted. Transfer of technology, tight international integration of resource management and allocation, and rechanneling of growth (from one oriented toward superfluous consumption and wasteful expenditures to one oriented toward economic and technological growth geared to enhancing health and intellectual/artistic opportunities) are some of the possible alternatives to misery, starvation, and potential extinction.

Overpopulation, Julian Huxley has remarked, is the world's greatest problem today. Many people view this as a strictly demographic problem that can be overcome if one method or another of dampening fertility is widely adopted. The problem of overpopulation, however, is inseparable from economy and culture, and the seriousness of the population dilemma can only be appreciated within a broad perspective that goes beyond proximate demographic causes. In this essay, I will attempt to delineate the relationship between economy and population from the beginning of the prehistoric past to the emergence of civilization, and to show that with the emergence of agriculture about 10,000 years ago the changes in the dynamic relations between people and resources and the attendant modification of the sociocultural system have fostered a demographic condition and an economic pattern that underlie some of the problems of the world today.

Resources and Population: The Cultural Connection. About 3.5 million years ago, the ancestral forms of human beings were to be found in various places in East Africa. These creatures were primates who happened to undergo an evolutionary trajectory that emphasized the acquisition and transmission of learned behavior at a scale much greater than that of other animals. Although other primates show advanced learning abilities based on longer recall and some form of symbolic information storage (Dethier and Steller 1961:105), the feeding strategy of the early hominids—in a terrestrial, savanna habitat associated with the emergence of the nuclear family (Bartholomew and Birdsell 1953; Devore 1971; Lovejoy 1981)—placed them on an evolutionary track that had potential for development in cognition. Localized female food-collecting and wide-ranging male food-gathering, scavenging, and hunting would have been advantageous in maximizing food gain from the scarce and dispersed sources of protein in the dry, savanna environment.

Female-male bonding, resulting in part from the loss of estrus and development of year-round sexual receptivity (Fox 1967:38), was selected for because of the greater survival potential of offspring who were fed, protected, and cared for by a parental couple. Groups of hominids consisting of several families with a food-sharing pattern also had a greater probability of survival and a greater chance of reproductive success compared with solitary families. This is because group activity ensured greater protection from predators and greater access to food by each individual at all times, with sharing of information about food locations as well as actual sharing of food produce. The ability to carry food to a homebase was inseparable from the ability to share. The freeing of hands and the ability to walk on two limbs (bipedalism) were strongly selected for (Lovejoy 1981).

Information sharing, a greater awareness of the sparse resources, an emphasis on parenting and sociality benefit from greater cognitive abilities. Evidently a dynamic feedback relationship between further developments in food-getting activities and familial/social interaction and cognition was established. Significant changes in the brain, consisting of an expansion of cerebral cortex, are dated at about 2 to 1.5 million years ago. This is documented by the cranium from Lake Turkana in Kenya, known as 1470 or *Homo habilis* (Lewin 1981:807). Toolmaking was one of the key developments in the early stages of hominid evolution. It was advantageous in warding off predators, increasing the probability in capturing game, and extracting roots and other food items. The oldest tools from Ethiopia seem to date back to 2.1 million years or perhaps to 2.5 million years (Lewin 1981:806,807).

The aspects of early hominid culture or protoculture that evolved in response to the interaction between our earliest ancestors and their habitat along the trajectory that has ultimately led to our own emergence as a species thus consist of: biparental family, structured social groups, some form of communication, learned behavior, and toolmaking. These are the dimensions of the human culture as outlined by Hallowell (1956)—dimensions that are crucial in our understanding of the essence of the human condition and the nature of culture (see also Reynolds 1968).

The relationship between people and resources from the early dawn of hominid prehistory was, in the light of the above remarks, subject to the intervention of social and cultural mechanisms. It is indeed these mechanisms that have undergone dramatic changes throughout the human past and are responsible for the remarkable developments in human culture, the expansion of the human population, and the indelible impact of people on the ecology of the earth. These social and cultural mechanisms involve energy input for food production, efficiency of energy conversion, organizational pattern of production, and distribution and consumption. I will focus on the changes in these elements in the following sections.

Stone Age Hunters and Gatherers. From the inception of human society to about 10,000 years ago, a period of 2 to 3 million years, wild game, plants, and other naturally available wild-food resources were the source of human nourishment and sustenance. This mode of subsistence entailed some strict and inescapable consequences for human culture and demography. It was inevitable, given this dependence on the yield of wild resources, that the human population could not climb beyond a small predetermined limit unless such dependence was broken. I have estimated that the limit for the whole world was about 8.6 million persons if a generalized hunting-gathering subsistence was pursued, and probably as many as 12 million persons with the addition of fish resources, sea mammals, and small animal game and fowl (Hassan 1981:204). By 10,000 years ago, archeological evidence indicates that the actual world population was not greater than that figure—the world population at the twilight of the Paleolithic period was comparable to the population of Greater New York City or Tokyo.

On a local scale, the size of hunting-gathering human communities (called bands) was as small as 15 to 50 persons on the average, with regional units of 500 to 1,000 (Wobst 1974; Hassan 1981). The small community, the band of hunter-gatherers, had to be tuned to the seasonal changes in food availability and the periodic fluctuations of climate affecting the location, quality, and quantity of food (Hassan 1979). The small numbers of people reflect the dependence on wild resources and the small area that can be routinely covered from a home base in the course of daily activity, an area often less than 10 kilometers in radius. The number of people, however, could not fall below twenty-five for the band and 500 for the regional unit without endangering the biological viability of the group because the probability of extinction from random (stochastic) fluctuations in mortality increases as the size of the group decreases.

Thus, under the mercy of fluctuating and often scarce resources, successful populations had to maintain their numbers above a minimum of twenty-five persons, but the numbers had also to be regulated below that of the maximum yield from wild resources. This has favored a population potential and growth pattern that consisted of the reproduction of a limited number of offspring that were invested with parental and social care to maximize their survivorship, a pattern referred to by ecologists as the "k-strategy." It is a pattern, unlike that of other organisms, that produces a large number of offspring, with little investment in rearing, and in which regulation is dependent on the availability of food. This strategy, which is referred to as the "r-strategy," would have been highly disadvantageous to the very elements of human existence: sociality, continuity of family, and the transmission of learned behavior from one generation to the next.

Although a pattern of rash overpopulation, followed by an inevitable crash, has been posited for prehistoric hunters and gatherers (Cohen 1977), the continuity of human culture and the evidence for population control in prehistory—including such extreme measures as infanticide

(Divale 1972)—suggest that the low rates of population growth estimated for the Pleistocene (generally less than 0.01 per year) do not represent an alternation of episodes of rapid growth (exceeding 0.5 percent) followed by depopulation. The increase in world population during the Old Stone Age was slow and intermittent (see fig. 21). Initially (about 2-3 million years ago) the world population was restricted to open grasslands or parklands of East and South Africa (the age of early hominids from Southeast and East Asia is uncertain, but is somewhere between 1.9 and 0.5 million years ago). By about one million years ago, the early hominids had dispersed to North Africa, the Near East, southern France, and East Asia. The newly occupied territories were grasslands, wooded steppes, and woodlands (Butzer 1971).

By half a million years ago, the northern latitudes were invaded. The Choukoutien site in northern China, which yielded the remains of a close ancestral form (*Homo erectus*), was inhabited when a boreal coniferous forest covered the region. By about 40,000 to 20,000 years

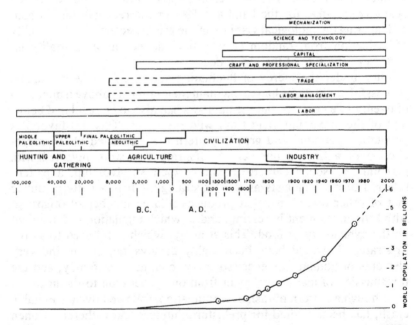

Figure 21. Stages in the development of human society, shown in association with the estimated increase in world population.

FEKRI A. HASSAN

ago, areas that were still colder than northern China were inhabited, including the cold steppes of Russia, Poland, and Germany. Tropical and desert regions were also penetrated during that time period, and the transcontinental migrations that brought people to Australia and the New World were under way.

The Prelude to Agriculture. The dispersal of human population was one way of diffusing the danger of local overpopulation. The dispersal rate, however, was slow and does not seem to reflect a response to rapid population increase which would otherwise have meant that the world would have been saturated much earlier. The peopling of the world throughout the Ice Age was mostly a function of a very slow increase in world population and an increasing adaptability through cultural developments and biological evolution. Changes in the human brain are evident throughout the Pleistocene Ice Age. The modern form of human brain dates back at least 40,000 years with a similar ancestral form associated with the Neanderthals (*H. sapiens neandertalensis*). Greater cognitive ability was reflected in greater sophistication in toolmaking. Greater versatility, variability, and specialized techniques are the hallmark of the later part of the Stone Age (the Upper Paleolithic).

Art and mathematical thinking are also recognizable at that time. The widespread use of fire during the Late Paleolithic was also one of the major innovations, because it allowed people to inhabit colder regions and to have access to many food resources. Cooking destroyed toxic substances in many foods and also helped to tenderize many others. Fire might also have been used in hunting and regenerating vegetation (Butzer 1971:482-3).

The final Paleolithic Age (20,000-10,000 years ago) is a period of special interest in the relations between resources and people. It was a period of rapid technological innovation, experimentation, and ferment. This is in part a result of the emergence of our own subspecies (*Homo sapiens sapiens*), with highly cognitive abilities, the presence of many people expanding the pool of innovators; and greater proximity of human groups, permitting a fairly rapid cross-cultural transmission of ideas. This period in human prehistory coincided with the climatic fluctuations associated with the end of the most recent major glaciation in the history of the earth. These fluctuations undermined the stability of the specific patterns of subsistence by making the location and yield of wild food resources less predictable. Yield could not be anticipated from one year to the next, and the knowledge of where certain foods were located became less certain.

Consequently, worldwide changes in subsistence emerged at that time in places that are very different in location and in the kinds of resources utilized. In northern Europe, coastal and riverine resources—once less emphasized—became the basis of subsistence. At that time, specialized hunters, who pursued major herds of game, also emerged and, in the Near East, the catch was expanded to include birds, fish, snails, tortoises, and weedy cereal grasses.

The terminal Pleistocene (about 20,000 to 10,000 years ago) was one of the most dynamic periods for cultural development since the emergence of early hominids. It signaled a new phase in the relations between people and resources. With the expansion of exploitable resources to include a great variety of species, people might have inadvertently begun to destroy their habitats by specialized predation and by the use of fire. For the most part, hunters and gatherers are concerned about the conservation of natural resources. They live constantly at the mercy of the delicate balance of the ecosystem. With the use of mythology, by endowing the wild game with spirits, and by taboos and totems, they developed numerous safeguards to prevent the overexploitation of resources. However, hunters—perhaps no more than ourselves—cannot foresee the long-term consequences of their actions, nor could they perceive and hence respond in time to long-term changes in climate.

During the terminal Pleistocene, the ability on the part of people to innovate and intensify their assault on the environment, along with the changing nature of the resources of the earth at that time, were responsible for further spiraling intensification of human exploitation of resources. Specialized hunting might have destroyed large populations of mammals. Use of fire also led to many accidental fires that swept grasslands and forests. In Egypt, for example, fire swept the Nile floodplain for hundreds of kilometers about 12,500 years ago, as shown by reddened, fire-baked silt in many places along the Nile.

The changes during the final Paleolithic period were only a prelude to the most dramatic event in human prehistory: the advent of agriculture. The changes, in sum, were:

(1) the diversification of and specialization in hunting, gathering, and food-processing technology (e.g., grinding stones to exploit cereal grains and nuts, bows and arrows, cooking techniques);

(2) the diversification of food extractive strategy—very generalized, specialized, riverine, maritime, lacustrine;

(3) the increasing impact of people on their habitat.

The Origin of Agriculture: The Remaking of the Earth. Among the

FEKRI A. HASSAN

many changes in the subsistence strategy of hunter-gatherers at the close of the last Ice Age, one was destined to change the world forever. It was the incorporation of wild cereal grasses in the diet. These grasses were previously less attractive to the Paleolithic peoples, presumably because they required extensive processing and were most likely less palatable than meat and legumes. In the Near East, where violent fluctuations in climate made the life of the hunters full of uncertainties and risk, the fields of wild cereals acquired a new value. The availability of grinding equipment (perhaps initially "invented" to prepare ornamental or ceremonial pigments for body paint), combined with the spread of grasses as a result of changes in vegetation under climatic influences (Wright 1977), as well as the probability of fire-made cereal grasses by 16,000 years ago, made this food source an integral part of the diet of the hunters and gatherers. The cereal grasses had a tremendous potential for generating a variety of responses on the part of their exploiters (Hassan 1977). They occurred in large fields, they matured in short time, required processing, and were storable.

To increase the efficiency of exploiting cereals, people gathered together during the harvesting season, and a considerable amount of time was spent in parching and grinding the grain. Storing was practiced perhaps first by heaping cereals above ground and later in pits and protective structures. These responses meant that larger groups of people (50 to 100 compared with the 15 to 50 of the Paleolithic) could reside in one place a good part of the year. Eynan, one of the earliest village sites in Palestine, contained about fifty houses. The permanency of settlement, marking a major break with the peripatetic pattern of the hunter-gatherers of the past, is indicated by a succession of three occupations at the village and intramural burials, including those of children and infants. One of the burials was that of a chieftain (Mellaart 1967), which is of great significance since one of the major developments in human culture later on was the emergence of a managerial segment of the population, of which the modern state is the ultimate product.

Among hunters, small in number and with little or no material possessions, the incentive for formal leadership was nonexistent. Status and informal leadership are recognized as a result of prowess in hunting, greater knowledge of the environment, greater intelligence, or individual skills. The emergence of large aggregates of people who resided together for a long part of the year or year-round, and the advantages to be accrued from collective harvesting, storage, and defense of the storable "wealth," encouraged the emergence of social integrating mechanisms, including that of chieftains.

The increasing dependence on cereal grasses entailed a hidden danger that could not have been realized at the start. Cereals are lacking in certain essential amino acids and must be supplemented with meat or legumes. Moreover, permanent occupation of one place by a large number of people soon leads to a depletion of available wild foods. To ward off a crisis, the game and legumes had to be "managed." They had to be brought under control—a step that marks the beginning of domestication both of animal game and plants and the beginning of the modern world order (Hassan 1977).

Preindustrial Agriculture—No Limits to Growth. Agriculture and stock-raising led to the eclipse of the world of hunters and gatherers, and initiated a new pattern of interaction between people and resources and among individuals. It was more than a change in the way people procured their food. It was the beginning of the human mastery of its habitat—the beginning of the dichotomy of man vs. nature. It was also the beginning of law, order, organization, and the reshaping of the human spirit to yield to the force of social institutions and collective will. Agriculture had an inherent potential for fabulous economic growth. It also entailed a simplification of the ecological network by focusing on a few domesticable resources, and thus reduced ecological stability.

The unforeseen consequences of agricultural practices such as salinization, loss of fertility, and soil erosion, added to the destabilization of the human habitat. Agriculture was not "paradise found." It was hard work: droughts, pests, uncertain yields, and an increasing subservience of the individual to the chieftain or headman, who mediated conflicts but also supervised the intracommunal sharing. The relatively voluntaristic sharing of the hunters and gatherers had to be replaced by a formal order. The freedom to come and go characteristic of the unbound hunters was intolerable in the new social order. Formal relations with neighbors were also favored as a means of regional integration of resources. Trade in foodstuffs and other commodities became a characteristic of this episode. Although trade in exotic shells, flint, and other substances was practiced during the later part of the Paleolithic, grain, fish, oil and wine were the predominant items of trade and exchange.

Sedentarization meant that in times of famines and scarcity, resources moved to where people were located: movement of goods replaced movement of people. Modes of transport such as the ass, the wheel-drawn carts, and boats became a necessity of the new order. The spatial boundaries of food-producing territories limited to small areas were broken, making it possible later on for major settlements to emerge in habitats that could hardly support a scant population. Places at the nodes of transportation on rivers, mountain roads, and coasts acquired a special prominence that is now reflected in the location of major cities and towns of the world. Two of the earliest cities in Egypt (Buto and Hierakonpolis) are at the peripheries of the Nile Valley and owe their prominence to the connection between Egypt and neighboring countries (Wilson 1955).

Changes in the mode of subsistence from food extraction to food production thus entailed cultural and demographic changes. Demographic changes consisted of the emergence of large population units, a reduction in spatial mobility of residential groups, and sedentarization. The most notable element of agriculture was not simply its greater productivity compared to that of hunting and gathering but its potential for economic growth. Agricultural yield can be increased to magnitudes that, at one time, could not have been contemplated. Even today, 10,000 years after the initial attempts at agriculture, we continue to increase agricultural productivity by developing new cultigens (such as miracle rice), mechanization, collective farming, and the use of fossil fuels.

The Emergence of Civilization. The elasticity of the agricultural mode of subsistence has fostered a dynamic relationship between resources and culture. The major cultural changes during the preindustrial stage consisted of

(1) agricultural innovations (e.g., ploughing, fallowing, fertilizing, water control, land reclamation, genetic selection, and multiple rotation of crops), thus, changes in cultigens, land, and water;

(2) changes in the size of the labor input; and

(3) changes in the distribution of food substances.

These three categories of change were selected for by the metastable condition of proto-agricultural systems. Agricultural innovations increased yield. An increase in size of the labor force not only ensured greater yield, but also greater efficiency. Intracommunal and intergroup patterns of distribution ensured interregional equalization and integration of resources. The simplification and the consequent destabilization of the ecological system accompanying agriculture were thus combated by cultural complexity. Management of intra- and interregional distribution of food and of labor created the need for specialized managers, who by virtue of their control of "wealth" acquired greater political eminence and with it the means to enforce their regulations and edicts. As managers they also guaranteed the emergence of specialized artisans to provide the specialized tools and implements of agriculture.

Writing was also developed to document transactions and mathematics to calculate equivalencies of goods, astronomy to provide an almanac, and priests to reduce the anxiety of an uncertain agricultural world and to express the new order of man-land relations and the emerging social enactment. The rise of managers and craft specialists and others dependent on the fruits of agricultural labor was an incentive for intensifying labor and agricultural innovations.

To suggest that availability of more food encouraged population increase that led to agricultural innovation is to miss the whole essence of the managerial revolution that is at the heart of the first major transformation in human economy. The demands for labor can readily explain the increase in early populations. That increase, however, was not on the average greater than 0.1 percent per year (Hassan 1979). It was, however, not a sustained increase. It consisted of episodes of fast increase followed by stabilization. In addition to population increase, the shuffling of populations (internal migration) was a rapid method for deploying labor whenever it was needed. The population of early farming communities was responding to the vagaries of agricultural production, to the new niches opened by new innovations, and to the differential success of adjacent communities, resulting from either random factors or differences in the efficacy of agricultural management or differences in the potential of the land for food production or commerce.

The Nonutilitarian Economy. The demands for labor for agricultural produce were also at that time associated with a remarkable development—the demand for nonsubsistence goods, not just for the raw materials and finished products that serve as implements of production but, in addition, the nonutilitarian items of luxury, prestige, and power. The changes in subsistence strategy from food extraction to food production were thus accompanied not only by changes in distribution but also in consumption. More people who were not farmers had to be fed and more food had to be produced to exchange for nonfood goods. It was this pattern that characterized the interrelation between people and resources until the Industrial Revolution.

The Roots of Destruction—Agriculture. The spiral of agricultural developments took its toll on the environment. Perhaps one of the major activities that led to environmental destruction was the denudation of landscape. In Egypt, for example, the demands for fuel and wood for construction and tools, coupled with overgrazing, soon turned the land surrounding the Nile Valley into a barren desert and eliminated the trees and grasses that could still survive under arid conditions. In Palestine, land clearing and demands for timber during the Early

Bronze Age led to considerable deforestation, which was followed by erosion. For example, the denudation of the fields that sustained one of the earliest agricultural settlements in Palestine, Jericho, undermined its subsistence base. The occupants dwindled in numbers and energy. The last of the Early Bronze Age buildings show definite signs of the end. They were hastily put together from old and broken bricks. Then the town was abandoned and left to neglect and further destruction (Kenyon 1971:133-134).

Man was working against himself. Erosion destroyed agricultural land. Thus, further developments had to be undertaken, which also had their unforeseen negative side effects. A spiral of destruction was initiated, a monster that feeds on progress and one that spews progress—a paradox of man-land relations that has not ceased to plague humanity and threatens to drive it to the very edge of survival, with no room for progress that could save the earth from the vicious circle of technological advancement and environmental destruction.

In the eighteenth century, Francois René, Vicompte de Chateaubriand, remarked, "Les forêts precedent les peuples, les déserts les suivent." (Cited by Hardin 1964:101.) This was the fate of Jericho and many other places since. The erosion of the landscape was only rivaled by that of cities and civilization from warfare. Systematic warfare was a development of the Neolithic, and it has become so prevalent that some have presumed that it is rooted in instinctive aggression. The causes of war are for the most part related to socioeconomic conditions, power, and politics (Holloway 1967).

War claimed cities like Ur, about 2004 B.C., and Troy VII between 1209 and 1183 B.C. (G. Clark 1969:155). The Empire of Ur in Mesopotamia included Sumer, Akkad, and northern Mesopotamia. It was a civilization that boasted extensive building projects, efficient administration, flourishing temples, and active commerce. About 2004 B.C., Ur was overrun by Elamites. The ruins of the city, and the surviving lamentation over its destruction, are testimonies to the devastation and savagery by which Ur was laid waste (Kramer 1940):

In all its streets where they [the people of Ur] were wont to promenade
　　dead bodies were lying about;
In its place where the festivities of the land took place
　　the people were *ruthlessly laid low*
The blood of the land like bronze and lead. . . .
Its dead bodies, like fat put to the flame, of themselves melted away. . .
The city they make into ruins.
Its lady cries: "Alas for my city,"
Cries: "Alas for my house."

The accumulation of wealth, the emergence of a power elite, the need to secure trade routes and markets, the need to recruit labor and to stabilize the subsistence base by formal integration of adjacent regions, favored the emergence of military power. The causes of war were institutional not instinctive. Although the hunters and gatherers may engage in feuds and violent confrontations, the emergence of systematic warfare, fortifications, and weapons of destruction follows the path of agriculture and the concentration of wealth. From the beginning of civilization, walled towns, maceheads, axes, and sling-stones joined the plowshares and underwent a parallel technological advancement that has ultimately led to nuclear warheads side by side with farm machinery.

Toward Industry: A Historical Archeology. During the last 10,000 years, humanity has survived on the produce of agriculture. It was not until about 200 years ago that agriculture took a dramatically new form. It was also during this last 200 years that mechanized technology ushered in a major transformation in the relation between people and the resources of the earth and in the realm of societal relations. This was the greatest transformation since agriculture first was practiced and is truly a revolution inasmuch as revolutions entail a radical transfor-mation of a preexisting order.

These cultural revolutions are not sudden upheavals or instantaneous developments; they are the result of cumulative change, which leads to a threshold as the system is "suddenly" and "catastrophically" transformed and propelled to a new state. From 10,000 B.P. (years before present) to the eighteenth century, the world was characterized by regional political units that were in continuous amoebic flux, expanding and shrinking, coalescing and fragmenting, in response to environmental calamities, political ambitions, trade, and economic gains. Although in some cases the political flux reflected attempts to stabilize the means of subsistence, it was often a response to a new standard of living and the demands for nonfood goods. This pattern has not changed over the last 200 years. If anything, it has now reached a far greater dimension. More people today desire more nonfood goods. Disparity within and among nations has widened, and the introduction of machinery has made more goods available and thus encouraged the potential consumers to produce other goods in exchange—and a new cycle of nonutilitarian production has been generated. It is against this background that the modern era in man-land relationship must be understood.

Overpopulation today is not a problem of an overabundance of people relative to available or potential food that can be produced by *existing* technology. In spite of mass starvation, famine and appalling malnutrition, we already possess the technology and means to feed and sustain the world population (Clark 1970:vii; Marstrand and Pavitt 1973:65). But to suggest that the world population problem could be solved by a new system of food redistribution would be foolish because it would lower the standard of living everywhere and retard enlightened economic growth. From the beginning of human society, the emphasis on the quality of human life is undeniable—and it is the quality of life that we ought to value.

Parental care and mutual social support invested in the quality of a few offspring to secure for them a long and a good life have been human practices since the dawn of prehistory. The advent of agriculture made it possible for some members of the human society to prosper and to enjoy the good life to a level previously unknown to the hunter of the open fields. Nothing perhaps will catch the spirit of this transformation better than these lines from the Gilgamesh epic, which was probably composed 4,000 years ago, soon after the emergence of agricultural city life. Enkidu, a hunter of wild animals, who was brought to Uruk, now experiences the comforts of civilization (McNeill and Sedler 1968:78ff):

The milk of wild animals
He was accustomed to suck
Bread they placed before him;
He felt embarrassed, looked
And stared.
Nothing does Enkidu know
Of eating bread
[And] to drink strong drink
He has not been taught.
The Courtesan opened her mouth,
Saying to Enkidu:
"Eat the bread, O Enkidu,
[It is] the staff of life:
Drink the strong drink, [it is] the fixed custom of the land."
. . . .
His soul felt free [and] happy
His heart rejoiced,
And his face shone
He rubbed. . . .
His hairy body;

He anointed himself with oil,
And he became a human being.

Europe—from Barbarism to Civilization. The most ancient civilizations emerged in the wake of the Neolithic agricultural life in Mesopotamia and Egypt about 3,500 and 3,150 years B.C., respectively. These were followed by the civilization of the Hittites in Anatolia, about 1800-1700 B.C., the Phoenician civilization in the first and second millennium B.C., and the Minoan civilization shortly after 2000 B.C. (Wenke 1980). The classical world, with which the Western world today identifies itself, came into being in Greece about 750 B.C. and in Italy around 295 B.C. Most of Europe at that time was occupied by farming communities.

Initially, agriculture spread to Middle Europe from the Middle Danube and to the Rhine from the Aegean and the South Balkans during the fifth millennium B.C. During a second phase of expansion (mostly during the third millennium B.C.) agriculture spread to the rest of the temperate zone as far north as the northern limit of the deciduous forest (Clark 1977:116-18). Beyond the limits of the Greek world, the Barbarian world of Europe started to acquire iron by the seventh century B.C., and the communities were grouped into "chiefdoms." A warrior class is distinguished archeologically by grave goods which indicate lavish consumption of luxury goods. During the fifth century B.C., fortified townships emerged in Gaul and South Germany.

In Italy, a play of Greek and probably Phoenician influence on an indigenous foundation led to the emergence of the Etruscans, who were skilled in working copper and iron. The Etruscans were organized in self-governing city-states. It was from the Etruscan foundation that Rome, a tiny republic, emerged to dominate the Mediterranean world from 295 B.C. to 30 B.C. The Romans also subjugated the Barbarian chiefdoms of Europe. By A.D. 117 the Roman Empire reached the Danube-Rhine line. The fragmentation of the Roman Empire by A.D. 300 was followed by the emergence of Barbarian kingdoms around A.D. 500, but it was not until the tenth century that more or less settled conditions prevailed in Central Europe (see figure 22).

In the East, Islam spread from Arabia in a sudden eruption that soon engulfed the basin of the Mediterranean by the seventh century. With the rise of the Moslem Empire, Europe suffered a serious setback, which might be related to the disruption of the Mediterranean trade (Pirenne 1937). Commerce was limited, interregional integration was disrupted, cities collapsed, ruralization prevailed, and illiteracy spread. The aspect of the early Medieval period was that of localization and ruralization, or deurbanization, except for cities where bishops resided.

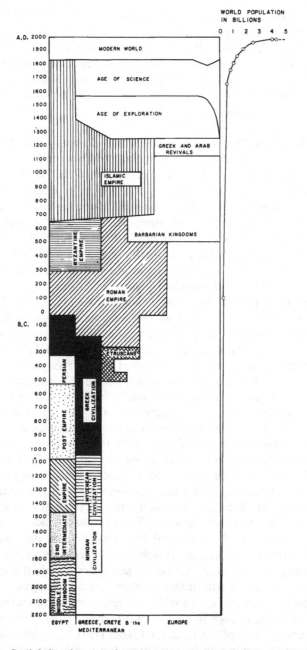

Figure 22. Social development in Egypt, the area around the Mediterranean, and Europe from 2200 B.C. to the present, shown in association with the estimated increase in world population.

The emphasis was on local economy, local markets, estate agriculture, feudalism, and aggrandizement on the part of the Church because of its land holdings. Ironically, the Moslems maintained and added to Greek science and philosophy during the time when Europe was in the "Dark Ages." The rise of Europe in the high Middle Ages would not have been possible without the revival of the Greek and Arab studies in the twelfth century.

The High Middle Ages—the Resurrection of Europe. From the seventh century a new cycle of development was under way in Europe, coming to a definite turning point by the recession of the Moslem Empire, which began in the latter part of the eleventh century. There are various contrasting views on the changes in the cultural landscape of Europe during the high Middle Ages (A.D. 1000-1300), which led to the prominence of the countries of northern Europe. Resumption of trade with the Mediterranean markets and beyond following the disintegration of the dominion of the Moslem Empire was proposed by Pirenne (1937), technological change is favored by the Marxist school (Engels 1942; Childe 1947), and population growth by North and Thomas (1973). The changes that became manifest in Europe by the eleventh century, however, were most likely a result of a combination of trade, technological advancements, and population increase, among other factors that were triggered by the reorganization of the cultural system following its disruption by the Moslem invasion.

North and Thomas (1973) build their case for the role of population increase without citing any figures on the rate of that increase. Assertions about the independence and exogenous force of population increase as the antithesis of the Malthusian principles provide an attractive, simple explanation of economic change. The trouble with the Malthusian thesis and its antithesis lies mainly in neglect of the reciprocal relationships between economic growth and population. They also suffer from the obsession with "first causes" and "prime movers," which leads to gross simplifications and a lack of regard for the causal network of events and the dynamics and mechanisms by which the process of change is achieved. Variations in the rates of population growth through prehistoric and historic times (table 9) show a distinct association with major technoeconomic changes.

Human populations are capable of rapid population increase even under conditions of high infant mortality, short life expectancy, and a child-spacing period of three years (Hassan 1981). The potential fertility is dampened behaviorally to equilibrate the population with available resources and socioeconomic conditions. When opportunities for labor become available as a means of optimizing returns or maximizing gain,

TABLE 9   World Population from Lower Palaeolithic to Present

Chronology	World Population in million persons	Population Growth Rate (percent per year)
Lower Pleistocene	0.8	0.00007
Middle Palaeolithic	1.2	0.0054
Upper Palaeolithic	6.0	0.0100
Final Palaeolithic	9.0	0.0033
Neolithic	50.0	0.085
B.C./A.D.	300.0	0.046
1300	400.0	0.022
1650	553.0	0.37
1750	800.0	0.44
1800	1000.0	0.52
1850	1300.0	0.54
1900	1700.0	0.79
1950	2500.0	1.74
1977	4300.0	2.01

Source: Data from Coale 1974, Hassan 1981, Thomlinson, 1965.

the population controls can be relaxed to accommodate the new situations. By contrast, when food becomes scarce and the opportunities for expanding the economy are lacking, growth is dampened. There are also other factors that contribute to population increase, such as inadvertent changes in fertility or mortality as a result of health-related matters (e.g., prolonging the child-spacing period to ensure good health for the mother and infant), cost of children (related to availability of weaning foods, costs of education), and benefits of children (opportunities for child labor). Changes in the average life span during prehistoric and historic times do not thus seem to have been a major determinant in the overall long-range pattern of population growth (table 9).

The cybernetic (circular) relationship between production and reproduction must be born in mind in the historical analysis of economic change. The increase in population and technological development is inherent in the agricultural mode of production. The changes are autonomous. Collapse of agricultural systems as a result of technological or managerial failure and decimation of populations by epidemics or war also lead to displacement of populations and thus create "uprooted" laborers (i.e., excess population). In addition, populations suffer from stochastic changes that may be of a smaller magnitude among the larger populations of the agricultural peoples than among hunter-gatherers, but they do lead to a large number of individuals.

The agricultural system thus cannot be characterized by a static demographic pattern, but by a fluctuating demand for labor, occasional overpopulation or depopulation, and population dispersal and internal migrations.

During the Neolithic revolution, the rise in the rate of population increase was a result of the relaxation of the measures of population control in response to the opening of work opportunities related to the change and increase in the optimum size of the economic population unit (more people lead to greater output per unit of capital), as well as a reduction in the cost of children because of the availability of baby food and the availability of opportunities for child labor.

The agricultural mode of life also fosters a growing demand for labor as a result of the fluctuations in yield, which may be due to weather, pests, unforeseen negative effects of agricultural practices (loss of soil fertility, soil erosion, salinization), epidemics (made more common under agriculture because of the increase in the number of hosts), or stochastic factors. The fluctuations can be combated by increasing labor input per capita, as well as by an increase in the number of laborers. This may be also associated with agricultural extensification (reclamation, dispersal, and so forth). The fluctuations also lead to agricultural intensification through technological innovations (plowing, fertilizing) as well as by interregional integration of resources (via trade or war). The technological innovations, trade, or war led to the creation of craft specialists, managers, soldiers, tradesmen, and other personnel not directly involved in food production. The increase in this sector expands the need for labor and thus promotes population increase, immigration, and warfare for slavery. The rise in the standard of living has also a similar effect on labor demands.

The medieval period was in essence a repetition of the old cycles of the preindustrial agricultural systems: rural development is associated with the rise of trade and craft production, accumulation of capital and investment, and the emergence of the elite and middlemen. The collapse of large-scale trade following the Moslem expansion in the seventh century increased the vulnerability of agricultural communities to the impact of severe climatic fluctuations and crop failures. A bad harvest meant a reduction in available work and unemployment. The collapse of the preexisting order and the demise of many once-flourishing cities created uprooted farmers who reclaimed new lands and created *villes neuves*. This "free" labor also found occupation in the trades and the crafts. By the eleventh century the face of Europe was changed as a result of land reclamation. The opportunities for agricultural work stirred population growth, which in turn added to the force of labor engaged in farming, trade, and the crafts.

FEKRI A. HASSAN

Industrialization: Capital, Technology, and Energy. Trade within continental Europe and on the North Sea, the Baltic Sea, and the Mediterranean (which was regained after the defeat of the Moslems and their retreat in the eleventh century) was associated with the rise of a new segment of society—the middle class, which expanded at the expense of the church and the nobility. The new regime fostered secular and practical knowledge, thus facilitating the emergence of new industrial technology. It also stimulated development of the commercial companies which in the fourteenth and fifteenth centuries grew very rapidly and accumulated huge capitals. It was capital and advanced technology that led to further advances in agriculture and "industry."

Commercial crops based on specialized farming (flax, tobacco, hops) replaced subsistence farming in many places in Europe by the seventeenth century. "Enclosures," large tracts of land under great landowners, emerged in England (as many as 5 million acres were enclosed from 1760 to 1800), and technical treatises on farming became widely available, beginning in 1663 with an English treatise on the cultivation of clover (Anderson 1977:65). By the 1780s, complex machinery and the development of the steam engine ushered in the stage of mechanized agriculture and industry. The steam engine was used commercially in 1785, but became widely used after 1820 (Cipolla 1970:52). Electricity was discovered in 1822 and generators were made available by 1870. Coal and electricity were new sources of energy that opened up a new world and led to a new round of interaction between the earth and us. The stage we now live in is characterized and differentiated from the previous preindustrial stage by the application of machinery and fossil energy toward food production and by the expansion of trade as a result of mechanical modes of transportation (see fig. 22). It is also characterized by a large world population, the majority of which is non-food-producing, and one that is dominated by the need for both agricultural production and the production of nonsubsistence goods (see fig. 23).

Population—An Economic Problem. The marvelous advances over the last 10,000 years have fostered a demographic balance that was based on the erratic demands of agricultural production for labor. Given a high rate of mortality and the advantages of children (cheap labor, insurance against old age, and social support), the agricultural mode of life also encouraged higher rates of fertility than previously possible. The malleability of agricultural production, and its response to economic growth, have fostered a dynamic relationship between population growth and agricultural intensification. The advent of advanced medical

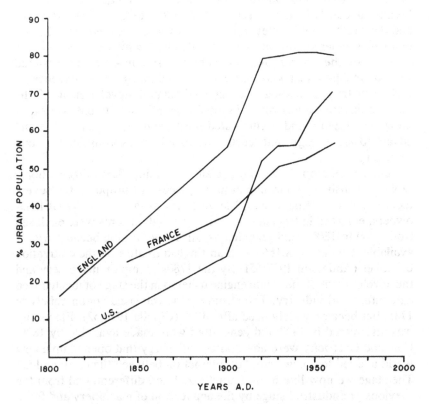

Figure 23. Changes in the percentage of urban population of England, France, and the United States from A.D. 1800 to the present.

hygiene, improvement in life conditions, and availability of preventive and other kinds of medicine led in Europe to a reduction in mortality, which was balanced by a decrease in fertility as the cost of children increased (as a result of the rise in the desired standard of living) and other urban and industrial changes took place. The transfer of medical technology, without industrialization, to other nations had led to a decrease in mortality, greater survivorship, without the drop in fertility associated with the socioeconomic changes in industrial Europe.

Resources—Conservation for Life. In addition to "overpopulation," the earth's resources are now threatened by the erosion of the hunter's conservation ethic and the emergence of the arrogant ethic of the Neolithic compounded by that of industry. This is well portrayed in

FEKRI A. HASSAN

the encounter between Gilgamesh, the strongman from Uruk, and Humbab, whom Enlil (the god of thunder who brought abundance and prosperity) appointed to preserve the cedar forest.

To preserve [the cedar forest]
[Enlil has appointed him] as a sevenfold [?] terror
To preserve the cedar [fore]st
Enlil has appointed him as a terror to mortals.
Humbaba—his roaring is [like that of] a flood-storm, his mouth is fire, his breath is death!
. . . .
Gilgamesh opened his mouth
[And] said to [Enkidu]
. . .[I will] put [my hand to it]
And [will cu]t down the cedar.
An everlasting [name] I will establish for myself.

Hunters and gatherers had a very negligible impact on the environment because of their small numbers, because of their spatial mobility, and because of their simple extractive technology. Under certain conditions, however, they might have contributed to local depletion of resources. Their greatest impact appears to have been during the end of the last glaciation, when fire and specialized hunting were commonly practiced.

The emergence of agriculture accelerated the impact of man's adverse effect on the resources of the earth as a result of localized overexploitation of wild game by large, sedentary populations, by the impact of agricultural practices destroying native vegetation and accelerating erosion, destruction of top soils, water-logging, and salinization.

The impact of the industrial man, however, is unmatched. Mechanized agriculture and industry, use of fossil energy—given the volume of world activities—threaten the quality of the air, water, and food resources. The greatest danger stems from changes in the amounts of carbon dioxide and other materials (e.g., dust) that are likely to affect global climate (Sagan et al. 1979). A slight change can cause serious problems in agricultural productivity and upset the world economy.

Concluding Remarks—The Past Is Only Prelude. The answers to world problems today are not simple, but the past illuminates our path. We are caught in a spiral of rising numbers of population, rising expectations, increasing demands for food, and increasing demands for energy. The world also is divided into separate political units with vast differences in economy, ideology, population, and military power.

In the prehistoric and historic past, wars were a means of securing goods, labor, and markets. With nuclear weaponry and the threat of ultimate destruction, the prospect of war is grim. Differential wealth and the worldwide awareness of the high living standards of some societies can only breed misery and discontent. Revolts and revolutions are already evident signs of pressures that are perhaps the greatest danger to the stability of the world today. Acts of violence can spread, and the despair that led in the past to the revolt of peasants can now lead to sabotage of resources of vital interest and even the deployment of small nuclear bombs. The solution does not lie in repression or imposing a moratorium on growth. These would not only consist of an opposition to forces of history that are already in motion, but would continue the state of misery and unrest of the present world. The solutions probably lie in transfer of capital and technology and the rapid industrialization of "underdeveloped" countries to bring a balance between reproduction and production.

Humanitarian actions should not stop with medical aid and food in times of famine. Such humanitarian efforts only prolong the misery of those who suffer and do not lead to a reduction of fertility. The world countries are now interdependent as never before. A worldwide management of resources and technology, of labor management, and a spirit of cooperation is the only viable alternative to world unrest and the potential extinction of the human species. Humanity now is at the end of its tether. It faces, for the first time since the struggle for human survival by our ancestral forms in the savanna of Africa, a challenge to its existence as a life form. Sharing and cooperation saved the day in the dim past of the Pleistocene. It is this very element of human culture that still holds the key to human survival today. The world has become not only a global village, as McLuhan has pictured it, with instant telecommunication, but the inhabitants of the earth also have become a global band who face the same dangers and must exploit the same resources. In the past, natural selection eliminated the individual bands that could not share or cooperate because their chances of survival and passing their genes were minimal. There were many bands, and those that possessed the successful traits were promoted. We cannot at this moment leave our fate to the caprice of natural selection. Our fate is one, and our mistakes can only lead to our global demise. We have amassed a great deal of knowledge. Now we should have the wisdom to use it.

References

Anderson, M. S. 1977. *Europe in the Eighteenth Century 1713/1783*. 2d ed. London: Longman.

FEKRI A. HASSAN

Bartholomew, G. A. and J. B. Birdsell 1953. "Ecology and Protohominids." *American Anthropologist* 55:481-98.
Butzer, K. W. 1971. *Environment and Archaeology*. Chicago: Aldine.
Childe, V. G. 1947. *History*. London: Corbett Press.
Cipolla, C. N. 1970. *The Economic History of World Population*. Baltimore: Penguin Books.
Clark, Colin 1970. *Starvation or Plenty*. New York: Taplinger.
Clark, Grahame 1977. *World Prehistory in New Perspective*. 3d ed. Cambridge: Cambridge University Press.
Coale, Ansley 1974. "The History of Human Population." *Scientific American* 231(3):41-51.
Cohen, M. N. 1977. *The Food Crisis in Prehistory*. New Haven, Conn.: Yale University Press.
Dethier, V. G. and E. Stellar 1961. *Animal Behavior, Its Evolutionary and Neurological Basis*. Englewood Cliffs, N.J.: Prentice-Hall.
DeVore, I. 1971 "The Evolution of Human Society." Pages 297-311 in J. F. and W. S. Dillon, eds., *Man and Beast: Comparative Social Behavior*. Washington, D.C.: Smithsonian Institution Press.
Divale, W. T. 1972. "Systemic Population Control in the Middle and Upper Palaeolithic: Inferences Based on Contemporary Hunter-Gatherers." *World Archaeology* 4(2):222-37.
Engels, F. 1942. *Origin of the Family, Private Property, and the State*. New York: International Publishers.
Fox, J. R. 1967. *Kinship and Marriage: An Anthropological Perspective*. Baltimore: Penguin Books.
Hallowell, A. I. 1956. "The Structural and Functional Dimensions of a Human Existence." *The Quarterly Review of Biology* 31:88-101.
Hardin, Garrett 1964. "Population, Evolution and Birth Control." San Francisco: Freeman.
Hassan, F. A. 1977. "The Origins of Agriculture in Palestine: A Theoretical Model." Pages 559-609 in C. Reed, ed., *Agricultural Origins*. The Hague: Mouton.
_____1979. "Demography and Archaeology." *Annual Reviews of Anthropology* 8:136-60.
_____1979. "The Growth and Regulation of Human Populations." Pages 305-19 in M. Cohen et al., eds., *Biosocial Mechanisms of Population Regulation*. New Haven, Conn.: Yale University Press.
_____1981. *Demographic Archaeology*. New York: Academic Press.
Heidel, A., ed. and trans. 1949. *The Gilgamesh Epic and Old Testament Parallels*. Chicago: University of Chicago Press.
Holloway, R. L., Jr. 1967. "Human Aggression in the Need of a Species-Specific Framework." Pages 29-48 in M. Fried, M. Harris, and R. Murphy, eds., *War: The Anthropology of Armed Conflict and Aggression*. New York: The Natural History Press.
Kenyon, Kathleen 1970. *The Archaeology of the Holy Land* (originally published in 1960). New York: Praeger.

Earth Resources and Population: An Archeological Perspective   215

Kramer, S. N., ed. 1940. "Lamentation over the Destruction of Ur."
    Assyriological Studies, Oriental Institute. Chicago: University of Chicago
    Press.
Lewin, Roger 1981. "Ethiopian Stone Tools Are World's Oldest." *Science*
    211: 806-7.
Lovejoy, C. Owen 1981. "The Origin of Man." *Science* 211:341-50.
Marstrand, P. K., and K. L. R. Pavitt 1973. "The Agricultural Sub-system."
    Pages 56-65 in H. S. D. Cole, C. Freeman, M. Jahoda, and K. L. R. Pavitt,
    eds., *Models of Doom*. New York: Universe Books.
McNeill, W. H., and J. W. Sedlar, eds. 1968. *The Origins of Civilization*.
    London: Oxford University Press.
Mellaart, James 1965. *Earliest Civilizations of the Near East*. New York:
    McGraw Hill.
North, D. C., and R. P. Thomas 1973. *The Rise of the Western World*.
    Cambridge: Cambridge University Press.
Pirenne, Henri 1937. *Economic and Social History of Medieval Europe*
    (originally published 1933). New York: Harcourt Brace & World.
Pressat, Ronald 1970. *Population*. Baltimore: Penguin Books.
Reynolds, V. 1968. "Kinship and the Family in Monkeys, Apes and Man."
    Man. 3:209-23.
Sagan, C., O. B. Toon, and J. B. Pollack. 1979. "Anthropogenic Albedo
    Changes and the Earth's Climate." *Science* 206:1363-68.
Thomlinson, Ralph 1965. *Population Dynamics*. New York: Random House.
Wenke, R. 1980. *Patterns in Prehistory*. London: Oxford University Press.
Wilson, J. A. 1955. "Buto and Hierakonpolis in the Geography of Egypt."
    Journal of Near Eastern Studies 14:209-36.
Wobst, H. Martin 1964. "Boundary Conditions for Paleolithic Social Systems:
    A Simulation Approach." *American Antiquity* 39(2):147-78.
Wright, H. E. 1977. "Environmental Change and the Origin of Agriculture in
    the Old and New World." Pages 281-318 in C. A. Reed, ed., *Origins of
    Agriculture*. The Hague: Mouton.

Commentary by J. Lawrence Angel
Department of Anthropology, Smithsonian Institution

The particular issue I am taking up is the question of population growth
and overpopulation from the Paleolithic hunting and gathering times
through the Neolithic transformation to the present in the area of the
eastern Mediterranean.

On the basis of morphological changes in the pelvic bones caused
by child bearing, you can determine from archeological skeletons the
number of births per woman, which varies pretty much between three

and five, depending upon the length of life of the woman. In Early Bronze Age skeletons from the site of Karatas in Turkey, roughly speaking one-quarter of the mothers were producing half of the next generation of children. Of course, this is fecundity. There is also the question of fertility. If there is an equal chance for all of those children to grow up, then you have a very strong natural selection taking place, which has been continuous from Pleistocene to modern times. Jim Neel mentioned the male aspect of this selection for energy and longevity; I think the female aspect is less strong, but still does exist.

One could conclude from this observation that genes for longevity are being favored. There is an actual genetic trend of which this is probably the tail end, promoting genes for better health and better longevity, both of which would be advantageous in the growth and transmission of culture.

From the early farming period down to modern times in the eastern Mediterranean, population density and female longevity run more or less parallel. In ancient Greece, where female longevity never rose much above thirty-five, in the Neolithic to the Bronze Age, with female longevity in the early thirties, women had about twelve years for childbearing. Under these circumstances, the addition of a half year or a year of life means a considerable potential for population growth. In ancient times only a few women live past menopause. I don't think this has been adequately realized. When you are dealing with early societies or with modern societies where, for reasons of starvation or other restrictions, the length of the childbearing period is reduced, the loss of childbearing years slows population growth.

This is the major point that I wanted to make. The reason why population and also female longevity do not go straight on up involves a very complex series of cultural questions, including, in this case, the breakdown at the end of the Roman Empire in this particular part of the Mediterranean. If you took another area, the timing of variation in population size and female longevity would be a little different. What I am trying to show is that there is a tie-in, perhaps not a very strong one but still an important one, between biological-cultural factors affecting biology and the biological factors feeding back to the cultural possibilities.

DR. HASSAN: Might I comment?

DR. FREDERICKSON: Yes, please, Dr. Hassan.

DR. HASSAN: The problem of population increase in the Neolithic period is a fascinating one and it has a lot of implications for population problems today. There are various ideas. One of them is that people settled after being wanderers, which allowed them to reduce their child-spacing period and thus have more children. There is also a biological argument about the change in diet and its implications for fertility going from a low state to a high state. Also, Professor Angel makes a point about the change in longevity, which has implications for the length of the reproductive period. This is affected by age at menarche, age at menopause, and age at death. It is very difficult to untangle these in the archeological record.

We do know, though, that there was a dramatic increase in population size during the Neolithic of about 0.1 percent per year on average, which, of course, is nothing compared to the 2 percent now. But it is considerable compared with that of the Pleistocene, during which the rate must have been less than 0.001 percent.

Some research I have done indicates that hunter-gatherers as well as agricultural groups in nonindustrial situations—especially in the Pleistocene—had the potential to produce more than they actually did, i.e., they had the potential for high fertility. What we probably have in the Pleistocene, as well as after agriculture, is what we see today in that fertility among humans is often dampened by birth control, abortion, infanticide, or some other means of population control. It seems that the potential for population increase existed throughout the Pleistocene, before agriculture, and that it also existed most probably in the early stage of agriculture. The change that Professor Angel points to implies that population increase was involuntary, a result of women living longer and therefore having the chance to bear more children.

One must also phrase the question of fertility in terms of desired family size. How many children do you really want to have? Because if people have the ability to regulate their numbers and reduce them, an increase through biological mechanisms may also have been dampened. I suggest that the increase resulted from change in the work environment that led to greater use for children.

DR. LASLETT: An early population was very unlikely to survive with 40 percent of the women not having children and only three offspring per woman.

DR. HASSAN: Three is the number of children, but usually between two and three survive. The number that we have varied, given that you have about 40 or 50 percent child mortality. Our empirical data and examples are available and you are welcome to look at them afterwards.

DR. LASLETT: Well, we don't have such good data on most of European history, but where we do, these figures are very dicey. This population looks very unlikely to evolve. Nevertheless, I'm just trying to get it straight in my mind, that we apparently know more about

Neolithic fertility and survival than we do about more recorded European history. This may be true.

Commentary by Theodore Wertime

Department of Anthropology, Smithsonian Institution

I come to this symposium with a different perspective, that of the historian of technology who has been putting together in recent years the story of pyrotechnologic revolution. This is the transformation in man's industrial use of fire, which runs hand-in-glove with the agricultural revolution. Additionally, Mark Cohen and I are looking at all of the six great technologic revolutions to date in a book which we will title *Technology and the Evolution of Human Culture.*

All this is apropos Dr. Hassan's thoughts on the takeoff in human affairs that began during the Neolithic and has come to a focus in our own world. Human population growth is one factor; the expansion in the consumption of energy and materials and in the industrial processing of both is the other. One way to view the exponential growth of mankind and its technologies is to look at the time spans involved in the six great technologic transformations to date.

Nearly 1.5 million years separate the first experiments in stony tools from the beginnings of the joint revolutions in foodstuffs (and urban life) and fire-shaped materials some 20,000 years ago. About 18,000 years separate these two emerging transformations from the serious advent of the mechanistic revolution in the Mediterranean basin during the waning of Greco-Roman civilization 2,000 years ago. To the Industrial Revolution of the eighteenth century was another 1,600 years. But only 200 years elapsed until the beginning of the Edison transformation in electricity, electronics, and the atom which, as the sixth great metamorphosis in technologies, has so shaped the twentieth century.

Apropos Dr. Dubos's presentation, I think we probably grossly overestimate the human capacity to adapt socially and culturally, especially to the future. My comments come out of my efforts to develop some general rules of technologic change. When you examine these six transformations in human life very carefully you find that man, while he was moved by a certain kind of fear and apprehension about the future, never marched forward into the future with flags and arms flying. It was always rump forward, a kind of experimental eclecticism and caution in moving toward the future. We find ourselves very much rump forward today as we confront some disturbing and unbelievable phenomena, including deforestation (half the world's forest gone since 1950), pollution, and inflation. I am very pessimistic about how we are going to get through the no-man's-land of the 1980s and 1990s into the twenty-first century. We have opened a great scientific window upon ourselves in the 1980s. We are in a decade of revelation concerning what we are doing to the earth and life on earth. But we have no sense of how we are going to get to the years 2030 or 2080.

I happen to be not only a historian of technology, but I have spent the past five years in an experiment in developing what I call "the new collecting society." I believe we are in fact moving toward a new kind of collecting society in the twenty-first century. Our technologies are pushing us in this direction, as well as the needs pressing upon us as we find that we really have no substitutes for fossil fuels and for forests. Both the computer and the photovoltaic cell can be instruments in the new collecting society.

We have had crises in the past, but I think we are on the edge of a transformation so different in human affairs, so gross, that nothing that has preceded it prepares us for what is ahead. Dr. Hassan unfortunately was pressed for time in talking about the Neolithic takeoff. That was a fairly total alteration in man's relationship to nature. Man moved from a conservationist ethic about foodstuffs and materials toward an exploitive ethic. It was the beginning of human entropy on a major scale. Human affairs moved from negentropy to entropy. Leslie White is generally right in saying that, to that point, human affairs had generally been negentropic—not energy dispersive but energy collective. They moved into a gross entropic mode, which has accelerated.

Unless we look at the sequence of changes, we just do not have the appreciation of the exponentialism that has been involved since the Neolithic revolution. Here again I support Dr. Hassan's presentation. As for the forces behind this revolution, was it pushed by population drift or by human innovation? The tar baby effect (men being caught up in progressive technologies willy-nilly) is especially notable in the Mediterranean, somewhat less true in China, and much less true in the New World and Africa. We twentieth-century denizens trace our origins to the Mediterranean and the cumulative development of Western civilization launched there.

I would like to cite some of the significant dimensions of the economy of the ancient world. Before I do so I must ask a very basic question as to how, in human culture, a preadaptation becomes an adaptation? Here is where I support what Don Ortner had to say about the relationship between biology and technology.

The shift from preadaptation in technology to an adaptation is from that which we in technologic history would call the toy or plaything to the actual tool. What happens in this process we do not yet really know despite all of the sophistication of the twentieth century. Such shifts were not propelled simply by human inventiveness, nor simply by need. Human invention, as a matter of fact, is the mother of a great deal of necessity. This is why I say that men are launched into great transformations by processes that they do not truly fathom or truly control.

Let me give you a few statistics about Rome. It was a city of concrete and glass. The coming of window glass throughout the Roman Empire was a greater revolution in solar energy than we are witnessing in the United States at the present time. It was prompted mainly by Rome's energy crisis. Rome was going all the way into the Caucasus for firewood. There was a long history of transportation of materials and wood by barge throughout the Mediterranean. But let's look at some of the dimensions of the city of concrete and glass: by the fourth century A.D., 90 percent consisted of four-story concrete tenements with lead water pipes and lead sewage pipes, a city of 700,000 persons. The movement of grain into Rome amounted to 10.5 million bags annually. Arriving at the point of Ostia, the grain was brought into Rome mainly through concrete aqueducts, eleven great aqueducts, averaging 1.5 million cubic meters of water every day. Formerly, Rome relied heavily on cisterns, themselves a preadaptation of underground burial and food-storage vaults. Take any staple of Roman technology, such as lead pipes, and you find an utterly serendipitous adaptation from something quite different. Did anyone every figure out in advance the cost of recreating civilizations out of toys?

Rome paid a price for material growth. At least 60 to 80 million tons of metallurgical slag remained around the Mediterranean periphery at the end of the ancient world. I estimate that between 40 and 60 million acres of forest were destroyed in the process of metals smelting. Now, this is just metallurgy; I haven't said anything about the energy costs of lime, which is probably the most versatile pyrotechnologic material of antiquity and the basis of Roman concrete cement. I haven't mentioned ceramics or glass or household consumption of wood. We know that the Romans were forced by this first great energy crisis, as was China, toward a shift into the forested areas of Europe for most of its pyrotechnologic products. The deforestation of the Mediterranean and China in antiquity was the world's first energy crunch. The deforestation of Europe and North America in the eighteenth and nineteenth centuries was the second. The third is today's exhaustion of fossil fuels and destruction of half the world's forest since the year 1950.

Today the population is probably close to twenty times the world population at the time of Rome. Technology has kept well ahead. If we take world iron and steel production, which amounted in the days of Augustus to about 200,000 tons, the increase is fifty times. In 1970 iron and steel production passed a billion tons for the first time. I might add that two-thirds of that growth has been since the year 1950. But overall, it is a fifty-times increase. It does speak to the fact that man has managed to exploit nature slightly faster than the reproductive capabilities of his own gonads.

This fact has encouraged people like Julian Simon to believe that we can continue with exponential growth. I would argue that we can't. Exponentialism is self-defeating. The book *Limits to Growth* has a logic which is ineluctable. It ends with two phenomena of the twentieth century. One is that of the countervailing effect brought on by human technology—pollution is a countervailing effect and so is energy dispersion. The second effect is that of saturation. Phenomena so overlap that they suddenly overwhelm us. I would leave you with the thought that technology really remains the orphan in our discussions of human cultural adaptation and change. We don't really know how it came into existence, what it really is, or how it changes. Why is one technologic and the other biologic? When you take a look at photovoltaic cells, you see an exact parallel to photosynthesis in the plant. Man is the agency in one case and Nature the agency in the other. But what forces put Nature on an organismic and Man on a mechanistic tract?

Discussion

DR. MASTERS: In the context of Dr. Hassan's essay, I would like to ask a question about something that Jim Neel said about the title of this whole symposium, that unlike the Odyssey of Odysseus, we are not likely to go back again. I would like to know why that is assumed to be the case? My own sense is that if we have to choose between something like returning to a segmentary lineage system, where groups fission and fuse as the Yanomama do, and maintaining huge populations,

FEKRI A. HASSAN

populations in some kind of brave new world, I am for the Yanomama. I would like to go back again.

In other words, in all this debate about whether the northern countries should help to industrialize the countries of the south, maybe Rousseau was right when he suggested that what we should do for an African tribe is to have a gallows at the frontier and hang the first Westerner who tries to enter and hang the first native who tries to leave.

DR. WERTIME: I would merely say that this urge to change the so-called developing world goes back, of course, to the great nineteenth-century dream, which we elevated in the twentieth century. I would not only cite Rousseau's answer to this, I would cite Khomeini's answer to this. I think we are getting the answer, we are getting the feedback today, and the feedback is quite definitive that maybe this equalization, this industrialization, this modernization, this Westernization, this Americanization of the Third World is not going to proceed as we had generally assumed throughout the nineteenth century. We're getting feedback at home also. I work out of Appalachia, and I have gotten quite negative feedback there on many of the ideas that I thought were unassailable.

DR. FREDERICKSON: Jim, it was your rhetorical device. Do you want a shot at it?

DR. NEEL: Well, first, I would ask Dr. Masters if he has spent any time among the Yanomama? I'm not sure he would be quite that enthusiastic. Secondly, as in Thomas Wolfe's novel, which I'm sure you're all familiar with, you can't go home again because the circumstances under which the Yanomama exist are maintained only in a very small part of the world. It isn't there to go back to.

DR. WERTIME: But isn't it also possible, following on what René Dubos said, that major reversals in human affairs have occurred? When we try to look very carefully into the twenty-first century and talk about the "new collecting societies," we may be carrying on such technologies as the computer and the recombinant gene. Nevertheless, in terms of entropy, we will have to go back again. We do not know, at this moment, by what means we are going to maintain the current high technology of our society.

VOICE: It strikes me that going back also involves an enormous reduction in social obligations, the kind of organizational system the Yanomama are already outgrowing. And certainly to go back to any major giving up of our social control mechanisms and a major giving up of our political-economic buffers affecting the ability to move goods around, any major reduction in the commitment we have to produce food in the way we do, you're talking about reducing a population like 90 percent, I mean something enormous.

DR. HASSAN: I don't think, also, that we have the choice. That is probably a false issue.

VOICE: Well, I'm not advocating that.

DR. HASSAN: I'm saying that the issue of going back is probably a false issue because it is not a matter for you to decide whether the nondeveloping countries will choose or not choose an industry, because a lot of them have already decided that and you cannot stop them, so that is not the issue. They have also been exposed to industry and expect certain things, and I think that is the danger.

DR. PASSMORE: There was a hint of that this morning: better red than dead. And I think there is also: better dead than red. Now we're getting: better red Indian than dead. None of these slogans, it seems to me, really advances the matter very much as a slogan. It is pretty clear that we can't go back again. I wish Professor Dubos were right and that tomorrow is another day in the sense that we can always recommence. But I'm afraid that is not true. We have a burden on us. We now have traditions, history, developments, which we can't wipe away overnight and say, clean the slate and tomorrow begin again. It is highly unlikely that any of us has the slightest idea of what is going to happen in the next twenty years.

I think it is probably true that some of the things we're worrying most about now may not be the things that we will be worrying most about by the turn of the century, and that we will have quite a new set of worries. I think some of these problems may look much less serious, and we may be faced with great problems growing out of the technological revolution of the next twenty years, but rather different problems from those that we are now contemplating.

DR. LASLETT: I was waiting until Rousseau came into the conversation. I hope that type of fevered eighteenth-century commentary will not be regarded as evidence. I would also like to add this view, that whatever the condition of Europe before technological change, it was not like the hunter-gatherer or whatever contemporaries there may be in Brazil. There is no point in talking of going back to the tribal conditions described so well by Dr. Neel.

You've got to get an objective story that you're actually concerned with, which is the adaptability of the culture that dominates the material world. We had better get our answers straight and not make the modernization fantasy, which no cultural historian can accept. The transfer from technologically not-very-advanced Europe to the industrial world today is not to be understood in the relationship of a tribe in South America with Americans or British or Germans of today.

DR. FREDERICKSON: How far we need to go back, I suppose, is the question.

DR. LASLETT: Certainly we ought to go back into our own past, not

FEKRI A. HASSAN

the imagined past as inferred from some contemporary preliterate society.

DR. HASSAN: It may not be understood by analogy, but certainly by contrast. If you look at the developing countries and their population explosion, which Western Europe apparently did not undergo, it underscores the differences between the demographic transition in Europe and the demographic change that is happening in most of the world today.

What we, apparently, have seen in Europe is a balance between fertility and mortality because of the transition. What we have today in the nonindustrial nations is a transition overnight, not over 150 or 160 years. They are still nonindustrial economies with industrial technologies, as far as medicine goes, and I think that is what the problem is. You have the family size that, because of the social concept of the family size and the demographic ethos, is still the same, but the demographic conditions are different.

That is despite the difference between high fertility and the reduction of mortality that is happening now in most of the world. So I think by contrast, we can use the transition from hunting and gathering to agriculture and from agriculture to industry, to highlight some of the things that are happening today in most of the world.

DR. BOULDING: I think we are grossly underestimating the unexhausted potential of humans. There are all sorts of scenarios for the future—and some may be pretty awful. But I think you have to remember that it was said in 1866 that you have to choose between a brief splendor and a long continued mediocrity. And as that was being written, we discovered oil and gas in Titusville. The potentials that are on the drawing board now are enormous. I think the carrying capacity of the earth is at least 10 billion, quite comfortably. There is a great deal of evidence for this.

But the ultimate source is the human imagination and the knowledge process and the learning process. Information is not conserved. Entropy does not apply. It is an open system and the second law has been repealed. I think that it is absurdly pessimistic. I am pessimistic, but not to this extent. Evolution is dominant in the human race. Biogenetics is insignificant. Anything the biologists have to say about the human race is insignificant.

DR. FREDERICKSON: In some ways, we ought to end upon those echoes of Condorcet.

DR. MIDGLEY: I am unable to let this pass. I feel I need to point out that the chief dangers come from the human imagination, which is much the most dangerous factor. Who has made this weapons factor? Not history. It is us.

DR. BOULDING: Evolution is punctuated by catastrophe. If it hadn't been for catastrophe, evolution would have stopped two billion years ago. But why doesn't the biological evolution reach an equilibrium? I mean, it certainly ought to, in which all genetic mutations are adverse instead of only 99 percent of them or 999 out of 1,000. And I think the answer is almost certainly, it has been punctuated by catastrophe.

DR. FREDERICKSON: I think one of the features of our culture is that when man stopped having to hunt for his dinner he could talk all night. And I think perhaps it is my responsibility to gavel us down after what I think has been a very pleasant afternoon. Thank you all very much.

8. Food: Past, Present, and Future

NEVIN S. SCRIMSHAW

Director, International Nutrition Program
Massachusetts Institute of Technology

Editor's Summary. Nevin Scrimshaw expresses concern about the increasing dependence on high technology (including commercial fertilizer, irrigation, mechanization and hybrid seeds) and approves of recent attempts to develop crops less dependent on such factors. He emphasizes that the problem of widespread malnutrition involves poor preservation and inequitable distribution of food. He offers suggestions for improved food production and highlights the feedback mechanisms between nutrition and society. Scrimshaw stresses the need for a greater sensitivity to the contemporary realities of global linkages; ultimately what adversely affects the developing countries adversely affects the developed countries.

For all but a tiny fraction of the two million years that humans are said to have existed, they have been hunters and gatherers dependent upon and limited by nature for their food. The history of human civilization includes a steady evolution from procuring food in this way, as do all other animals, to meeting food needs by producing it, processing it, and, most recently with particular reference to vitamins and some amino acids, synthesizing it.

One of the crucial characteristics of both humans and the higher primates from which they evolved is the capacity to live on a wide variety of plant and animal foods. Clearly, man became an omnivore as the result of a rigorous process of natural selection and this gave him an adaptive advantage as a hunter and gatherer. The ability to use a very broad range of substances as foods helps ensure survival in harsh environments. This is dramatically illustrated in the present

227

century (Murdock 1968) by the Australian aborigines, the African Bushmen, and some other tribal societies living in traditional ways in Africa and the jungles of South America (Holmberg 1950; Goldman 1963; Fagundes-Neto et al. 1981).

Australian aborigines, for example, live in a harsh, desert environment and are heavily dependent on annual and seasonal variations in rainfall (Meggitt 1957; Sweeney 1947). Even under favorable conditions, they must exploit a wide variety of food in their environment, including vegetable seeds, yams, root fruits such as onion grass, bush fruits, honey from bees and ants, birds' eggs, insects, as well as a number of small desert marsupials, birds, lizards, leaves and nectar flowers, ant hill clay—to give just a sampling of the extraordinary variety of foods utilized. In good years life is easy; drought years merely increase the area over which they must range for food, and require consumption of some less desirable foods (Berndt and Berndt 1964).

A similar situation is described for the African Bushmen (Lee 1968). Most of their food can be obtained from only 23 of 85 species of edible plants and 54 of 223 local species of animals known and named by the Bushmen. Until they move to another location, they rarely need to utilize more than a radius of six miles surrounding a water source, even though more distant areas are no less rich in plants and game. In dry years, however, they expand both the variety of foods and the area covered.

Varied diets in which no one food plays a dominant role are likely to be highly nutritious because the compositions of the various dietary components complement one another. Ethnographic data on hunter-gatherers indicate that these people usually have adequate supplies of a large variety of foods and suffer from malnutrition much less often than do agriculturalists in the same region (Dubos 1965; Cassidy 1980; Fagundes-Neto 1981).

The ability to use food sources that are progressively less desirable organoleptically and that perhaps require more and more work to obtain is one type of adaptation. As an omnivore, however, man has also had the capacity to adapt to specific environments by depending on a limited number of either animal or vegetable foods. Other hunting and gathering diets differed greatly with habitat zones. The traditional Eskimo diet high in fat and meat represents one extreme of adaptation. The fact that Steffanson (1956) and other Europeans (Freuchen 1961) have maintained good health on such a diet for years at a time argues against any special biological adaptation of the Eskimo (Yesner 1980).

It is the capacity to live almost entirely on a single cereal staple, however, that has been most responsible for the steady numerical expansion of human populations up to the beginning of the present

NEVIN S. SCRIMSHAW

century. In the indigenous populations of Mexico and Guatemala today, tortillas alone, made from whole-ground corn, can supply more than 90 percent of the calories and 80 percent of the protein required, with beans making up most of the remainder. In many Asian populations, rice fills a similar role, as does wheat in parts of Africa and the Middle East. This heavy dependence on a single cereal staple is not only part of the historical record, but is also characteristic of many populations today.

The basis for most of the predominantly vegetarian diets in the world, even when apparently reinforced by religion, has been environmental or economic, and not lack of desire for animal foods. We now find vegetarianism being voluntarily adopted, often at increased cost, by middle- and upper-income populations in the United States and Europe, who are motivated by religious, ethical, or health reasons. The fact that it is entirely possible for wholly vegetarian diets to be nutritious and palatable gives further confirmation of man's adaptive capacity. It is this capacity of man to subsist on the entire range of diets from vegetarian to traditional Eskimo that has made domestication of different kinds of staple successful.

Evolution of Food Procurement. Once *Homo sapiens* became established as a distinct omnivorous species, surviving by hunting and gathering, there is no reason to believe that further *biological* evolution occurred in man's nutrient needs. However, some 10,000 years ago—and apparently independently in several parts of the world—a major *social* change began in the way that his needs were met. This was the beginning of agriculture and of the domestication of animals (Flannery 1965; Heiser 1981). Seeds of wheat and barley and bones of goats dating from 6750 B.C. were discovered in an excavation in Iraq. In deposits after this date in Greece and the Near East, these and other seeds, along with the bones of various domesticated animals, including sheep, goats, and cattle, became progressively more abundant.

In the New World, the earliest Mexican and Peruvian sites with gourd, squash, bean, and chili pepper seeds have been dated between 7000 and 5500 B.C., probably as a result of quite independent domestications. Excavations in Tehuacan, Mexico, show a progressive increase in the size of corn between 5000 B.C. and A.D. 1500 from less than an inch to more than six inches (Niederberger 1979; Heiser 1981). Over this period in Peru, the potato came to play a similar role, since it will grow at much higher elevations than maize (Ugent 1970). The development of agriculture in Southeast Asia is less well known, but archeological evidence from China and Thailand indicates that rice was domesticated about the same time, certainly before 4000 B.C. (Heiser 1981).

As man came to produce more and more of his own food, the adaptations continued to be social, not biological. Societies became less nomadic and required far greater organization and discipline. Seeds had to be planted at certain seasons, protected, harvested, and stored, and animals needed to be cared for. Records were needed and larger population groups became possible. With such aggregations of people, the food supply became vulnerable to unfavorable weather as well as to warfare and civil disruption. Where irrigation projects were developed, a stable social structure was essential. Large concentrations of people also increased the transmission of epidemics of infectious disease, especially the enteric diseases.

The development of agriculture has also frequently had a high environmental cost. Over-cultivation and over-grazing made a desert out of fertile lands in what is now North Africa and currently contribute to the advance of the Sahara desert southward into the Sahel. As populations grew and rotation intervals decreased, slash and burn or shifting cultivation eventually changed the permanent nature of the vegetation, converting forest into savanna. Current massive clearing of the Amazon jungle is not only destroying a fragile ecology, with probably little permanent gain in agricultural production, but it may result in climatic changes as well.

Irrigation has made a major contribution to agricultural productivity, but over-watering can raise the water table to the point where water, drawn to the surface by capillary action, evaporates and produces salination that destroys fertility. Salination helped to undermine some of the earlier civilizations of North Africa and the Middle East and now threatens large irrigated areas of the Asian subcontinent and Middle East as well as parts of China, Australia, and the southwestern United States. Moreover, in parts of India, Australia, Israel, and the southwestern United States, underground water supplies are being used faster than they are replenished from natural sources and, in some cases, it is "fossil" water that is being depleted. Like petroleum, it is nonrenewable.

Erosion is a problem where soil is uncovered for cultivation. The midwestern states of the United States are estimated to be losing from nine to fifteen tons of topsoil per acre per year. Other countries with intensive agriculture are experiencing similar losses of their best agricultural lands (Brown 1981). This progressive thinning of the topsoil layer will lead inevitably to a decline in inherent land productivity.

The Green Revolution—Two Steps Forward and One Back. The latest phase in the long history of man's development of agriculture has been called the Green Revolution. Its achievements are by now familiar to almost everyone—the development of strains of the principal cereals— corn, wheat, and rice—which respond with high yields to fertilizer, pesticides, and adequate water. It has been responsible for greatly increased rice production in Asia and for grain surpluses in India, as well as increased cereal production in many other countries. For the Philippines it has brought self-sufficiency in rice and, in India, a tripling of wheat production between 1965 and 1980.

Green Revolution technology, as developed and applied, has also had some significant limitations. It has been difficult for small farmers to apply because of lack of access to the needed inputs, and it has frequently increased the survival problems of the small farmer and the social inequity in the countries in which it has taken hold (Dumont and Cohen 1980). It has also increased dependence on fossil fuels for the fertilizers, pesticides, and machinery these Green Revolution techniques require. Moreover, by concentrating on cereals, it has led to decreased per capita production and sharp increases in the price of the legumes on which persons with predominantly cereal diets usually must depend for supplementary protein. In Mexico, the Green Revolution produced a dramatic increase in production from 1945 to 1965— mainly by large farmers benefiting from the best lands, government-supported irrigation, and other government subsidies. Thereafter production increases leveled off and Mexico once again needed to import more food each year (Wellhausen 1976).

Only recently has a shift in national policy toward small farmers and a broadening of research attention to their needs, started to reverse this trend. In the meantime, many of Mexico's peasants have been unable to make a living on the land and are abandoning their small plots to find work in the cities or in the burgeoning oil fields. In many of the developing countries that have been the target of Green Revolution research, the small farmers still prefer local varieties for specific qualities that the new varieties lack. For example, in the Puebla area of Mexico, the small farmers prefer tall varieties of corn that give them fodder for their animals as well as grain, instead of the short-stalked varieties thus far developed by the nearby International Center for Wheat and Maize Improvement (CIMMYT).

Only recently has the worldwide network of institutes under the Consultative Group on International Agricultural Research (CGIAR) begun to focus on varieties and agricultural techniques feasible for the small farmer and less dependent on high-energy inputs. These techniques, including multiple cropping, intercropping, and multipurpose crops, hold great promise. However, the record of mixed social consequences of agricultural research that was too narrowly focused is an important lesson for the future.

A similar warning needs to be sounded about the trend to corporate farming in both the United States and developing countries. In general, despite massive inputs, corporate farms are not as productive as those of small farmers, and they often have undesirable social consequences. Where such farms have displaced small farmers and hired them back as seasonal laborers—as in Mexico and many other countries of Latin America and Africa—both self-reliance and tolerable living have been abridged. Conversely, when corporations provide small farmers with extension services, access to improved seeds, fertilizer, pesticides, and credit, plus a fair market for their produce, they can make both a satisfactory profit and a positive contribution to social equity.

By and large, the family farm of a size appropriate to the country and region is the shape of the future, but scarcely in the traditional sense. New techniques of proven value will be accepted by the individual farmer whenever they are available to him and the risk is not too great. Some of these include a steady increase in mechanization (not only for raising plants, but also for producing animals), trickle irrigation to conserve water, biological control of plant pests, multiple cropping, techniques for lessened dependence on chemical fertilizers, and new high-yielding plant varieties less dependent on costly and energy-demanding inputs.

Overall, the evolution of food procurement can be seen as a progressive increase in man's ability to cope with the problems of weather and climate, soil erosion and fertility, seasonal variations, and plant and animal pests, even though some of these developments have at times made him more vulnerable or proved to be inappropriate. An example of the latter is the susceptibility of uniform plant varieties to new plant diseases or to the introduction of pests whose natural enemies have been left behind. The loss of genetic variation through breeding of a limited number of varieties threatens serious future limitations in finding varieties resistant to new pests or suitable for new conditions. Fortunately, germ plasm seed banks have now been established to collect and maintain the thousands of cereal varieties that are still found in nature. These will be an important resource for future breeding programs.

Changes in the Processing and Handling of Food. Concurrently with the accelerating changes in the way in which food is provided, methods of food handling, processing, storage, and distribution have been evolving with increasing rapidity. Food preservation was a major problem for hunters and gatherers. Hunters who killed large animals

NEVIN S. SCRIMSHAW

far from the tribe had little choice but to eat all they could before carrying as much as possible back home. Contemporary hunters in Africa are known to consume several kilograms of meat each at the distant site of a large kill. However, drying, salting, and smoking of meat was developed early. Drying of fruits and vegetables must have become common at about the same time.

To make palatable and safe foods that would not otherwise be edible, processes such as roasting, boiling, and frying were utilized. Fermentation must also have been discovered thousands of years B.C. In historical times, fermentation has been important not only for producing alcoholic beverages, but also for preparing varieties of milk products with better keeping qualities, such as cheese and yogurt. The Indonesians discovered ways of fermenting soy and peanut, respectively, to produce tempeh and ontjom with enhanced nutritional value, flavor, and keeping properties. Other cultures have discovered ways of fermenting fish, vegetables, and other food stuffs, both to preserve them and improve flavor.

The extraction of soy bean curd or tophu is an early Asian process to remove bitter and toxic elements from the whole soy bean, concentrate its protein, and improve its palatability. The process by which potassium cyanide is pressed from bitter cassava in both Africa and the Caribbean is another example. Canning, first described by Nicolas Appert in France in 1810, rapidly became the leading method of food preservation in the industrialized world.

Wherever natural refrigeration was possible in caves, "cold cellars," and ice houses, it was of recognized value for food preservation. Freeze-drying, which results in a food product retaining its original properties to a remarkable degree when rehydrated, was introduced in 1960, but a similar process was used thousands of years ago by the Indians of the Peruvian Andes for the preservation of potatoes. The potatoes were allowed to freeze at night, then were crushed and exposed during the day until, after several days, the desiccated, mashed product could be kept almost indefinitely. Known as "chuña," it is still the staff of life to many Indians in the highlands of Peru and Bolivia.

Other modern means of food preservation (Karel, 1982) include spray- drying, vacuum packing to prevent rancidity of fats and preserve freshness, and the retort pouch, a laminated plastic-aluminum foil substitute for the tin can. Brief, ultra-high temperature sterilization after packaging provides milk, fruit juices, and similar beverages with a fresh flavor, and these products can be stored for several months without refrigeration.

Sterilization can also be achieved by gamma irradiation. When well packaged, some irradiated foods keep almost indefinitely with taste unchanged. Concerns over safety, off-flavors, impaired nutritional value, and cost have thus far blocked the use of this process, originally developed for the military, but gamma irradiation is utilized to reduce sprouting of potatoes and other vegetables, and approval for several other applications is pending in the United States and Europe. Intermediate moisture processing has proved highly acceptable for pet foods because it maintains their fresh, meat-like appearance. The process is entirely safe for human use as well.

Soy protein isolate is the basis for making coffee whiteners, soy milk, and improving the protein content of many processed foods. Soy protein can be spun into fibers and fabricated to simulate ham, chicken, beef, scallops, etc. It can be extruded under pressure much more cheaply into highly acceptable snacks and meat extenders. In Thailand, extruded mung bean protein is an acceptable substitute for meat in many traditional dishes. In many Southeast Asian countries soy milk and soy beverages are filling the role of cow's milk.

We have come to take for granted the increasing ability to preserve food in recent decades. It is a characteristic of today's industrialized world that the means of preserving food for long periods of time have steadily increased, and the variety of processed and packaged foods available is now far greater than ever before.

Biological Responses to Food Shortages. Man is adapted biologically to function well during short periods of food scarcity and is surprisingly resistant to even long periods without food. While the evidence for any special metabolic advantage of racial or ethnic groups is unconvincing, there are metabolic adaptations inherent in the human race as a whole that are important for survival when food supplies are limited. For example, the physical ability of hunters must not be impaired in the periods between successful hunts.

Human adaptation to starvation, short or long, is highly effective (Young and Scrimshaw 1971). In man, glycogen stores in the liver are so limited that, within about four hours after a meal, amino acids begin to be mobilized for the synthesis of glucose in the liver. As fasting continues, fat is broken down to provide fatty acids that can meet skeletal muscle energy needs and ketone bodies that are readily available as an energy source for heart muscle and brain. Because fat provides most of the energy during fasting, the ability to undergo prolonged fasting is heavily influenced by fat stores at the start. As long as fat stores last, loss of lean body mass is minimized.

There is also a metabolic mechanism to delay adverse effects of low protein intake. When individuals are given a protein-free diet with

NEVIN S. SCRIMSHAW

adequate dietary calories, urinary nitrogen excretion drops rapidly and reaches a new plateau within seven or eight days. Some time ago we determined obligatory nitrogen excretion in this way in 100 MIT students (Scrimshaw et al. 1972). A visiting scientist from Taiwan who was at MIT at the time later obtained information from Taiwan University students under closely similar conditions (Huang et al. 1972). The distribution of obligatory urinary nitrogen loss appeared to be significantly lower for the Taiwanese students than for the Caucasians at MIT.

It was tempting to conclude that this difference might represent an adaptation to the predominantly rice diets of Asians over many centuries. However, the differences appeared too small to be of practical significance, and the total obligatory loss of urinary plus fecal nitrogen was the same for both groups. Further doubt was cast upon the significance of this finding when, for some of the same students, urinary losses were still lower when environmental temperatures were higher and sweat losses greater. The drop appeared to compensate for the increase in integumental losses. This is an extremely useful adaptation to tropical climates. These studies also revealed something more of man's capacity to adapt to adverse circumstances. Even subjects consuming a protein-free diet for twenty-one days showed no change in tests of psychomotor coordination, mood, and performance in this period.

Recently, the United Nations University sponspored research studies in Beijing, Taiwan, Thailand, Korea, the Philippines, Mexico, Guatemala, Colombia, Brazil, Chile, Nigeria, Egypt, Bangladesh, India, and Turkey for comparison with those in the United States, Japan, and the Soviet Union (Torun et al. 1981, Rand et al. 1982). For the ten populations for which obligatory nitrogen data are available, no significant differences in obligatory nitrogen loss emerged. Moreover, differences in the amount of protein required for nitrogen balance in all of them seemed totally explainable by variations in digestibility of the protein in the local diets fed experimentally.

Another metabolic mechanism helps adaptation to cold, although it does increase food intake. This is nonshivering thermogenesis, the capacity to increase body heat production upon exposure to cold and reduce it promptly when no longer needed. This ability takes about a month of exposure to cold to develop (Davis and Joy 1962) and is lost in a few weeks when exposure to cold ceases.

Nutritional Requirements of Man. Man differs little from other mammals in the specific nutrients required (Scrimshaw and Young 1976), but he is less limited in ways of meeting them than most other species. There is much still to be learned about the quantitative nature of human nutritional requirements under the wide range of possible physiological and pathological circumstances. Enough is known, however, to evaluate the relative adequacy of diets.

The listings of nutrient requirements in tables 10–12 are intended to demystify the issue of food needs by making it clear that *it is the nutrients that are required and not any specific foods*. However much Americans may view milk as a cultural superfood—at least for children—or Argentinians demand their beef, Asians their rice, Mexican and Central American Indians their corn and beans, northern Europeans their potatoes, Japanese their rice and fish, and so on, none of these foods is nutritionally essential *per se*. It is only the appropriate quantities of the nutrients they supply that are needed. These nutrients can be obtained not only from conventional raw or processed foods, but also from unconventional foods.

Although in the past these nutrients had to be furnished by natural foods, they are already being supplied in synthetic form to some extent. For example, as margarine has replaced butter, synthetic vitamins A and D have been added. Thiamine, riboflavin, niacin, calcium, and iron are added to bread, and iodine to salt. To improve protein value, methionine is added to some soy products and lysine to some cereal products. Fruit juices are commonly fortified with synthetic vitamins C and A. Any essential nutrient can be provided in synthetic form as well as by natural foods and, in the future, more and more nutrients are likely to be supplied from synthetic sources as components of formulated foods.

Consequences of Nutritional Deficiencies. The functional consequences of failure to meet the specific nutrient requirements listed in tables 10-12 have had profound effects on societies. When dietary calorie intake is low, physical activity must be cut back for survival. If this adaptive mechanism and certain metabolic ones of lesser significance are insufficient, wasting—and eventually death—will follow. A large proportion of the populations of developing countries have calorie intakes that limit their capacity for work and especially for social and other discretionary activities essential to well being and community development (UNU 1981).

Even mild to moderate degrees of protein deficit will interfere with the physical growth and development of children and, if the deficiencies are prolonged, they result in impaired resistance and increased morbidity and mortality from infection. The child's cognitive performance may also be affected, often with permanent consequences for subsequent learning and behavior.

NEVIN S. SCRIMSHAW

TABLE 10 Protein, Essential Amino Acid and Fatty Acid Requirements, Functions, and Deficiency Signs

	RDA (per day)	Major Body Functions	Deficiency
Dietary Protein*	0.75 gm/kg body weight	See below	
Indispensable Amino Acids (as components of dietary protein)			
Phenylalanine/ Tyrosine	1,100 mg	Precursors of structural protein, enzymes, antibodies, hormones, metabolically active compounds. Certain amino acids have specific functions:	Deficient protein intake leads to development of kwashiorkor and, coupled with low energy intake, to marasmus
Lysine	800 mg		
Isoleucine	700 mg		
Leucine	1,000 mg	a) Tyrosine is a precursor of epinephrine and thyroxine	
Valine	800 mg	b) Arginine is a precursor of polyamines	
Methionine/Cystine	1,100 mg	c) Methionine is required for methyl group metabolism	
Tryptophan	250 mg	d) Tryptophan is a precursor of serotonin	
Threonine	500 mg		
(Arginine/Histidine)**	not known		
Essential Fatty Acids			
Arachidonic	6,000 mg	Involved in cell membrane structure and function. Precursors of prostaglandins (regulation of gastric function, release of hormones, smooth-muscle activity)	Poor growth Skin lesions
Linoleic	6,000 mg		
Linolenic	6,000 mg		

* FAO/WHO/UNU 1983—For protein of high digestibility, such as milk, meat, egg, and so forth. Must be increased for lower digestibility of mixed diets.
** Essential in the diets of young infants only.

TABLE 11   Vitamin Requirements, Functions, and Deficiency Signs

Vitamins	RDA for Healthy Adult Male (mg.)	Major Body Function	Deficiency
Water-soluble Vitamin B-1 (Thiamine)	1.5	Coenzyme (thiamine pyrophosphate) in reactions involving the removal of carbon dioxide	Beriberi peripheral nerve changes, edema, heart failure
Vitamin B-2 (Riboflavin)	1.8	Constituent of 2 flavin nucleotide coenzymes involved in energy metabolism (FAD and FMN)	Reddened lips, cracks at corner of mouth (cheilosis), lesions of eye
Niacin	20	Constituent of 2 coenzymes involved in oxidation-reduction reactions (NAD and NADP)	Pellagra (skin and gastrointestinal lesions, nervous, mental disorders)
Vitamin B-6 (Pyridoxine)	2	Coenzyme (pyridoxal phosphate) involved in amino-acid metabolism	Irritability, convulsions, muscular twitching, dermatitis near eyes, kidney stones
Pantothenic Acid	5–10	Constituent of coenzyme A, which plays a central role in energy metabolism	Fatigue, sleep disturbances, impaired coordination, nausea (rare in man)
Folacin	4	Coenzyme (reduced form) involved in transfer of single-carbon units in nucleic-acid and amino-acid metabolism	Anemia, gastrointestinal disturbances, diarrhea, red tongue
Vitamin B-12	.003	Coenzyme involved in transfer of single-carbon units in nucleic-acid metabolism	Pernicious anemia, neurological disorders

NEVIN S. SCRIMSHAW

TABLE II (Continued)

Vitamins	RDA for Healthy Adult Male (mg.)	Major Body Function	Deficiency
Water-soluble			
Biotin	Not established	Coenzyme required for fat synthesis, amino-acid metabolism and glycogen (animal-starch) formation	Fatigue, depression, nausea, dermatitis, muscular pains
Choline	Not established	Constituent of phospholipids; precursor of putative neurotransmitter acetylcholine	Not reported in man
Vitamin C	45	Maintains intercellular matrix cartilage, bone and dentine; important in collagen synthesis	Scurvy (degeneration of skin, teeth, blood vessels, epithelial hemorrhages)
Fat-Soluble			
Vitamin A (Retinal)	1	Constituent of rhodopsin (visual pigment); maintenance of epithelial tissues; role in mucopolysaccharide synthesis	Xerophthalmia (keratinization of ocular tissue), night blindness, permanent blindness
Vitamin D	.01	Promotes growth and mineralization of bones, increased absorption of calcium	Rickets (bone deformities in children; osteomalacia in adults)
Vitamin E (Tocopherol)	15	Functions as an anti-oxidant to prevent cell-membrane damage	Possibly anemia
Vitamin K	.03	Important in blood clotting (involved in formation of active prothrombin)	Associated with severe bleeding, internal hemorrhages

TABLE 12   Mineral Requirements, Functions, and Deficiency Signs

Minerals(*)	RDA for Healthy Adult Male (mg.)	Major Body Function	Deficiency
Calcium	800	Bone & tooth formation; blood clotting; nerve transmission	Stunted growth, rickets, osteoporosis, convulsions
Phosphorus	800	Bone & tooth formation Acid-base balance	Weakness, demineralization of bone, loss of calcium
Sulfur	Provided by sulfur amino acids	Constituent of active tissue compounds, cartilage & tendon	Related to intake & deficiency of sulfur amino acids
Potassium	2,500	Acid-base balance; body water balance; nerve function	Muscular weakness, paralysis
Chloride	2,000	Formation of gastric juice; acid-base balance	Muscle cramps, mental apathy, reduced appetite
Sodium	2,500	Acid-base balance; body water balance; nerve function	Muscle cramps, mental apathy, reduced appetite
Magnesium	350	Activates enzymes involved in protein synthesis	Growth failure, behavioral disturbances, weakness spasms
Iron	10	Constituent of hemoglobin & enzymes involved in energy metabolism	Iron-deficiency anemia (weakness, reduced resistance to infection)
Fluorine	2	May be important in maintenance of bone	Higher frequency of tooth decay

NEVIN S. SCRIMSHAW

TABLE 12 *(Continued)*

Minerals(*)	RDA for Healthy Adult Male (mg.)	Major Body Function	Deficiency
Zinc	15	Constituent of enzymes involved in digestion	Growth failure, small sex glands
Copper	2	Constituent of enzymes associated with iron metabolism	Anemia, bone changes (rare in man)
Silicon Vanadium Tin Nickel	Not established	Function unknown (essential in animals)	Not reported in man
Selenium	Not established	Functions in close association with vitamin E	Anemia (rare), cardiomyopathy, tachycardia
Manganese	Not established	Constituent of enzymes involved in fat synthesis	In animals: poor growth, disturbances of nervous system, reproductive abnormalities
Iodine	.14	Constituent of thyroid hormones	Goiter (enlarged thyroid)
Molybdenum	Not established	Constituent of some enzymes	Not reported in man
Chromium	Not established	Involved in glucose and energy metabolism	Impaired ability to metabolize glucose
Cobalt	Required as vitamin B-12	Constituent of Vitamin B-12	Not reported in man
Water	1.5 liters per day	Transport of nutrients; temperature regulation; participates in metabolic reactions	Thirst, dehydration

* In order of concentration in the body

Iron deficiency anemia, which continues to be a major nutritional problem throughout the world, is associated even in subclinical form with impaired physical capacity, reduced cognitive performance, and lowered resistance to infection. It affects the performance of hundreds of millions of individuals in developing countries, and is also the most common nutritional deficiency of industrialized ones. Iron deficiency is heavily influenced by the biological and social environment. For example, the blood loss associated with hookworm and schistosomiasis is an important precipitating cause of iron deficiency anemia. Moreover, availability of iron from predominantly vegetable diets is low.

The most important functional consequence of these nutritional deficiencies in history has been increased susceptibility to infectious disease (Ewbank and Wray 1980). Moreover, the relationship between infection and malnutrition is a synergistic one, with the infections worsening nutritional status and precipitating nutritional diseases (Scrimshaw et al. 1968). It is widely supposed that the advent of modern medicine, including vaccines, antibiotics, and chemotherapy, is responsible for the sharp decline in mortality in this century that has led to the quadrupling of the world's population. It had reached only a billion people by 1840 and is now over four billion. Examination of the historical record, however, shows that the fall in mortality in North America and Europe preceded modern disease-control methods and coincided with improving diets and better environmental conditions. Not only was this true for such scourges as plague and cholera, but for most other infectious disease as well. For example, the fall in mortality in England and Wales from tuberculosis (fig. 24), whooping cough (fig. 25), and measles (fig. 26) can be seen to have preceded effective specific therapeutic or preventive measures (McKeown 1979). Measles is the classic example of a disease with a high mortality in underprivileged developing country populations and essentially none in the more privileged populations of these same countries or of industrialized ones (Gordon 1965). Improved food supplies have thus been a large factor in the drop in mortality that has resulted in the world population explosion.

Food Procurement and Processing in the Future. Man's biological ability to meet nutrient needs from a wide variety of foods and diets has facilitated the recent rapid changes in the way food is procured and processed. If the past is prolog, it must be studied for clues to the future. Historical records show that man has used about 3,000 species of plants for food, and at least 150 have been cultivated. Over the last few hundred years, however, the number used has become smaller. At present, most of the world's population relies on roughly twenty

NEVIN S. SCRIMSHAW

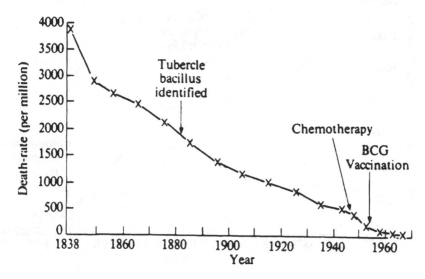

Figure 24. Mean annual death rates from respiratory tuberculosis in England and Wales, standardized to the 1901 population (from McKeown, 1979).

crops (table 13) for food: the cereals—wheat, rice, maize, sorghum, and millet; root crops—potatoes, yams, cassava; legumes, including beans, peas, soy beans, and peanuts; and sugar cane, sugar beets, bananas, and coconuts (Harlan 1976). These twenty crops are the main barrier between man and starvation—not a very large defense. Moreover, apart from fish, only three animals—the pig, the cow, and poultry, including chickens, ducks, and turkeys—account for most of the animal protein consumed. Lambs account for 5 percent and goats, buffaloes, and horses combined, about 3 percent.

To fill future food need, additional plants and animals can and should be better utilized. A wheat-rye hybrid, triticale, was initially over-promoted, but is being improved. It may yet be the first useful new cereal to be domesticated in more than 5,000 years. A National Academy of Sciences (1980) publication described thirty-six under-utilized plant species that could contribute importantly to future food supplies. Plant-breeding improvements of the Green Revolution have thus far been mainly limited to rice, corn, and wheat. There is great potential for the application of conventional plant-breeding techniques to adapting these cereals to the needs of the small farmer and also to improving other common plants such as sorghum, potatoes, and some of the many important legume species (Wortman and Cummings 1978;

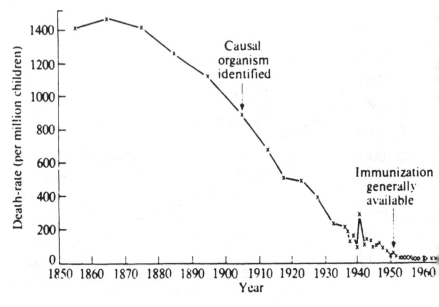

Figure 25. Death rates of children under fifteen in England and Wales dying from whooping cough between 1850 and 1960 (from McKeown 1979).

Frey 1981). These are now receiving more attention from the eight units of CGIAR concerned with plant breeding and from many national agricultural stations.

Thus far, all of these efforts have been limited to conventional plant-breeding techniques. DNA recombinance techniques are only just beginning to be considered for higher plants. They could become the basis for a second Green Revolution. One of the most talked about goals is to enable crops that depend on heavy fertilization to fix atmospheric nitrogen, as do legumes, in order to reduce dependence on chemical fertilizer. Increased efficiency of photosynthesis and greater resistance to insects, disease, salinity, drought, and temperature extremes are reasonable expectations (Flavell 1981; McDaniel 1981).

Animals will not be neglected in future food production. In the near future the most important progress is likely to be more in the manner in which they are raised than in genetic improvement. Already the mechanized feeding of chickens and swine under conditions more resembling an automated factory than a farm has revolutionized the production of poultry, eggs, and pork and kept these foods relatively inexpensive in the United States and Europe.

Modern systems developed for the production of trout have increased the market availability of this fish in North America and Europe despite

NEVIN S. SCRIMSHAW

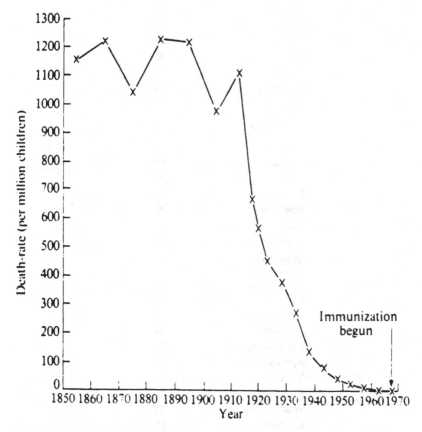

Figure 26. Death rates of children under fifteen in England and Wales dying from measles between 1850 and 1970 (from McKeown 1979).

steady growth in demand. Fish farming will increase steadily in importance as natural yields of fish from oceans and lakes continue to decline as a consequence of overfishing and environmental pollution. Mass production techniques for shrimp, oysters, and mussels are already in wide use.

Fish ponds utilizing agricultural and human wastes for the production of carp and other species have been important in some Asian countries, and the techniques of such fish culture are being continuously improved and extended. Several species can be combined in a pond to utilize fully the organic waste with which the pond is fertilized and the various

TABLE 13   Major World Crops (Millions of Metric Tons)

100 or More

Wheat	360
Rice	320
Maize	300
Potato	300
Barley	170
Sweet Potato	130
Cassava	100

60 or Less

Grapes	60
Soybean	60
Oats	50
Sorghum	50
Sugarcane	50
Millets	45
Banana	35

30 or Less

Sugar Beet	30
Rye	30
Oranges	30
Coconut	30
Cottonseed Oil	25
Apples	20
Yam	20
Peanut	20
Watermelon	20
Cabbage	15
Onion	15
Beans	10
Peas	10
Sunflower Seed	10
Mango	10

food sources and environments the pond offers. In the long term the new genetic techniques are likely to contribute importantly to animal production as well.

Single Cell Protein and Microbiological Biomass Products. Protein from yeasts or bacteria can be produced in virtually unlimited quantities for

NEVIN S. SCRIMSHAW

animal and human consumption without agricultural land, provided that suitable energy substrates are available. Either methane gas or petroleum can be converted to methanol or ethanol to provide chemically pure alcohol substrates. N-alkanes, processed to be free from all carcinogens, can play a similar role. Even "gas oil," a crude fraction of petroleum, can safely serve the purpose if the single cell protein (SCP) produced is solvent extracted afterwards. While such gas- or petroleum-derived substrates are not renewable, the amounts required for this purpose would be relatively small compared with other demands for oil. Carbohydrate sources, such as molasses and sulfite liquors from the paper industry, are renewable and can also serve as substrates.

Although the safety and nutritional value of SCPs for animal feed is well established, there are problems in their use as major protein sources for humans. Because they grow so rapidly, microorganisms have a high ribonucleic acid content. Hence their ingestion in large quantities will increase serum and urinary uric acid levels with the attendant increased risk of gout and uric acid kidney stones. The RNA reduction process seems to release small protein molecules that cause allergic responses in some people. Our research at MIT has demonstrated, however, that both of these limitations can be overcome by simple processing.

Mass production of yeast and bacteria can not only enhance animal and fish production, but can also become a direct part of human food supplies along with filamentous microfungi grown on starch or cellulosic residues. It is noteworthy that food yeast grown on petroluem-derived ethanol produced by Amoco is being used in processed foods in the United States, and that highly palatable forms of filamentous microfungi produced on starch are being test-marketed by Rank-Hovis-McDougal in the United Kingdom. At MIT, we have conducted feeding trials of yeasts grown on ethanol, methanol, and N-alkanes, bacteria grown on ethanol and methanol, and filamentous microfungi grown on starch and wood pulp waste (Scrimshaw and Udall in press). Foods containing SCP as a major protein source can be formulated with texture, flavor, and appearance to be equal in palatability to the foods we commonly consume today. The limiting factor to their mass exploitation is economic, not technological, toxicological, or nutritional.

More economically feasible yet neglected are the potential contributions of microbial biomass product (MBP), in which various organic residues and wastes are treated microbiologically in ways that upgrade their nutritional value for animal feeding and in some cases for human diets (Shacklady 1979). Moreover, MBP production is appropriate for the small farmer individually or collectively. Cellulosic residues from cereal grains in Europe alone amount to nearly 100 million tons. These are of limited value unless processed chemically and microbiologically for use as animal feed. Similarly, millions of tons of fruit, vegetable, and starchy residues that are now discarded could be up-graded for animal feed through fermentation processes.

Asian countries have also long recognized the value of animal and human wastes as fertilizer, but through appropriate technology they can also be used for the production of biogas. The residue is still valuable as fertilizer for fields or fish ponds. Worldwide, the underutilization of biomass produced by man is estimated at 170 billion tons. Future population pressure will demand greater reutilization of such materials, not only to enhance food production, but also to protect and maintain the environment.

Food for the Future—Social Adaptations and Strategies. From the foregoing, a number of conclusions can be reached about future food supplies.

1. The problem of feeding all human kind will not be one of limitations on food production, or at least it will not be the lack of adequate technologies and natural resources for meeting food needs far into the future. The limitations will continue to be the political and social ones of failure to protect the environment and to ensure the equitable distribution of the resources and technical knowledge essential to food production.

2. Given the limitations of current human institutions, the most likely scenario is for food supply to keep up easily with effective demand (i.e., the food that someone is willing to purchase), but fall short of human need for those underprivileged sectors of the population that are unable to acquire sufficient food, whether by raising it, gathering it, bartering for it, or buying it. As long as this continues, part of human kind will continue to be doomed to impaired growth and development, increased morbidity and mortality, and reduced capacity for work and social interaction.

3. The nature of human food supplies will continue to change with more and more technology involved in food production, processing, storage, distribution, and development of wholly new kinds of foods of plant, microbiological, and synthetic origin. The rate of these changes has been accelerating exponentially in this century.

   Animal protein will need to play a somewhat less prominent role in the diets of industrialized countries as rising affluence in today's poorer countries increases global demand for it. Differences in meat

NEVIN S. SCRIMSHAW

consumption as great as 111 kilogram per person in the United States; 6 to 8 in Nigeria and Egypt, respectively; and 1.1 in India cannot persist (Brown 1981). Foods of animal origin will continue to be available, but will need the contributions of SCP and MBP to animal feeding and more efficient animal production systems. Moreover, as world incomes become more equitable in their distribution, animal protein will become more evenly distributed among and within countries.

4. It follows that the world must be viewed as a single ecological unit to be exploited for food wisely in ways that will not jeopardize the survival of future generations. Political leaders, and eventually all peoples, must understand the environmental fragility of the human food chain as population pressures in the environment increase at rates unknown in the evolution of human society and as food production capacity in some areas is deteriorating.

5. Many of the activities of populations, nations, and the international community are neglecting—or having a negative impact on—present and future food supplies and on their equitable distribution. Food, population, energy, and environment are closely interrelated global problems that require the focusing of human resources for their solution. As long as such a large proportion of these resources is devoted to political conflict, present and future, or wasted in a variety of other ways, the future of human kind is threatened, despite a scientific and technological base that gives assurance that all of these problems are manageable. Moreover, the increasing pace of demographic and environmental pressures gives the problems an urgency and global significance unprecedented in human history.

Epilog. It is now generally recognized that most of the famines in nineteenth-century Europe and of the developing countries in the twentieth century have not necessarily been due directly to shortages of food (Sen 1980). Throughout the great Irish potato famine of 1846-47, which resulted in the death of 1.5 million people and the emigration of nearly a million more, shipments of wheat, oats, and barley from Ireland to England continued (Woodham-Smith 1962). The peasants who had lost their potato crops to blight, however, had no way of purchasing other food. In those times nearly one-fourth of the Irish population suffered from hunger even when the potato crop was good.

The Great Bengal Famine of 1943, in which 3.5 million people are estimated to have died, did not affect Calcutta, since food prices were subsidized there (Sen 1980). The peasants of West Bengal had insufficient money to compete with higher prices being offered for food outside their region because of factors associated with World War II. Throughout the famine, the government continued its purchase policy for army use and continued to export food. Hoarding and refugees from Burma also increased demand. As recently as 1974-75, in both West Bengal and former East Bengal, now Bangladesh, people were starving in the countryside for lack of money to purchase food that was abundant in the markets of the towns.

These examples are cited to emphasize once again that it is man rather than nature who bears the responsibility for this kind of human suffering and death. Conversely, the permanent elimination of hunger will require political concern for the food security of common people. Various kinds of entitlement schemes are required by nearly all populations to abolish hunger and malnutrition among the poorest sectors.

In summary, the history of human societies is one of their struggle to develop a social structure that will enable them to cope effectively with their biological and physical environment. A great variety of strategies have been tried, some of them successful and others causing societies to decline or become extinct. Our current *global* society has, or can acquire, the knowledge necessary to solve the pressing problems of food, energy, population, and environmental deterioration. There is no reason for hunger and malnutrition on the earth, either now or in the future, except for our human failure to evolve suitable social, economic, and political structures. The past record of human societies proves no guarantee of success, but it does offer hope that global society will succeed in meeting human food needs.

References

Berndt, R. M., and C. H. Berndt 1964. *The World of the First Australians*. Chicago: University of Chicago Press.

Brown, L. 1981. *Building a Sustainable Society*. New York and London: W. W. Norton & Company.

Cassidy, C. M. 1980. "Nutrition and Health in Agriculturalists and Hunter-Gatherers: A Case Study of Two Prehistoric Populations." Pages 117-15 in N. W. Jerome, R. F. Kand, and G. H. Pelto, eds., *Nutritional Anthropology—Contemporary Approaches to Diet and Culture*. Pleasantville, N. Y.: Redgrave Publishing Company.

Davis, T., and R. Joy 1962. "Natural and Artificial Cold Acclimitization in Man." Pages 286-303 in *Biometeorology: Proceedings of the Second International Bioclimatological Congress, London*. Oxford and London: Pergamon Press, Symposium Publications Division.

Dubos, R. 1965. *Man Adapting*. New Haven, Conn.: Yale University Press.

NEVIN S. SCRIMSHAW

Dumont, R., and N. Cohen 1980. *The Growth of Hunger—A New Politics of Agriculture*. London: Marion Boyars Publishers.

Ewbank, D., and J. Wray 1980. "Population and Public Health." Pages 1504-48 in J. M. Last, ed., *Maxcy-Rosenau Public Health and Preventive Medicine*. 11th ed. New York: Appleton-Century-Crofts.

Fagundes-Neto, V., R. Baruzzi, J. Wehba, W. Silvestrini, M. Morais, and M. Cainelli 1981. "Observations of the Alto Xingu Indians (Central Brazil) with Special Reference to Nutritional Evaluation in Children." *American Journal of Clinical Nutrition* 34:2229-35.

Flannery, K. 1965. "The Ecology of Early Food Production in Mesopotamia." *Science* 147:1247-56.

Flavell, R. 1981. "Needs and Potentials of Molecular Genetic Modification in Plants." Pages 208-22 in K. Rachie and J. Lyman, eds., *Genetic Engineering for Crop Improvement*. New York: The Rockefeller Foundation.

Freuchen, P. 1961. *Book of the Eskimos*. New York: Bramhall House.

Frey, K. 1981. "Capabilities and Limitations of Conventional Plant Breeding." Pages 15-62 in K. Rachie and J. Lyman, eds., *Genetic Engineering for Crop Improvement*. New York: The Rockefeller Foundation.

Goldman, I. 1963. *The Cubeo Indians of the Northwest Amazon*. Urbana: University of Illinois Press.

Gordon, J., A. Jansen, and W. Ascoli 1965. "Measles in Rural Guatemala." *The Journal of Pediatrics* 66:779-876.

Harlan, J. 1976. "The Plants and Animals that Nourish Man." *Scientific American* 235:88-105.

Heiser, C. B., Jr. 1981. *Seed to Civilization: The Story of Food*. 2d ed. San Francisco: W. H. Freeman and Co.

Holmberg, A. 1950. "Nomads of the Long Bow: Siriono of Eastern Bolivia." Washington, D.C.: Smithsonian Institution.

Huang, P., H. Chong, and W. Rand 1972. 'Obligatory Urinary and Fecal Nitrogen Losses in Young Men." *The Journal of Nutrition* 102:1605-13.

Jerome, N., R. Kandel, and G. Pelto, eds. 1980. *Nutritional Anthropology—Contemporary Approaches to Diet and Culture*. Pleasantville, N.Y.: Redgrave Publishing Co.

Karel, M. 1982. "Technological Achievements and Trends in Food Processing, Storing, Transportation and Handling by Consumers." Pages 99-132 in P. Koivistoinen, R. L. Hall, and Y. Malkki, eds., *The Role and Application of Food Science and Technology in Industrialized Countries*. VTT Symposium 18, Espoo, Finland: Technical Research Centre of Finland.

Lathrap, D. 1968. "The 'Hunting' Economies of the Tropical Forest Zone of South America: An Attempt at Historical Perspective." Pages 23-29 in R. Lee and I. DeVore, eds., *Man the Hunter*. Chicago: Aldine/Atherton.

Lee, R. B. 1968. "What Hunters Do for a Living, or, How to Make Out on Scarce Resources." Pages 30-48 in R. Lee and I. DeVore, eds., *Man the Hunter*. Chicago: Aldine Publishing Co.

Lee, R. B., and I. DeVore, eds., 1968. *Man the Hunter*. Chicago: Aldine Publishing Co.

McDaniel, R. 1981. "Plant Genetic Engineering: Possibilities for Organelle Transfer." Pages 185-205 in K. Rachie and J. Lyman, eds., *Genetic Engineering for Crop Improvement*. New York: The Rockefeller Foundation.

McKeown, T. 1979. *The Role of Medicine: Dream, Mirage, or Nemesis?* Princeton, N. J.: Princeton University Press.

Meggitt, M. 1957. "Notes of Vegetable Foods of the Walbiri." *Oceania* 28: 143-45.

Murdock, G. 1968. "The Current Status of the World's Hunting and Gathering Peoples." Pages 13-20 in R. Lee and I. DeVore, eds., *Man the Hunter*. Chicago: Aldine/Atherton.

National Research Council 1975. *Underexploited Plants with Promising Economic Value*. Washington, D.C.: National Academy of Sciences.

Niederberger, C. 1979. "Early Sedentary Ecology in the Basin of Mexico." *Science* 23:131-42.

Rachie, K., and J. Lyman, eds. 1981. *Genetic Engineering for Crop Improvement*. New York: The Rockefeller Foundation.

Rand, W., R. Uauy, and N. Scrimshaw, eds. In Press. "Protein-Energy Requirements of Developing Countries: Results of International Research." *The United Nations University World Hunger Programme Food and Nutrition Bulletin*, Supplement 7.

Scrimshaw, N., M. Hussein, E. Murray, W. Rand, and V. Young 1972. "Protein Requirements of Man: Variation in Obligatory and Fecal Nitrogen Losses in Young Men." *The Journal of Nutrition* 102:1595-604.

Scrimshaw, N., C. Taylor, and J. Gordon 1968. "Interactions of Nutrition and Infection." *World Health Organization Monograph No. 57*. Geneva: World Health Organization.

Scrimshaw, N., and J. Udall. In Press. "The Nutritional Value and Safety of Single-Cell Protein for Human Consumption." *Proceedings of the International Symposium on Single Cell Proteins, 28-30 January 1981, Paris, Association pour la Promotion Industrie-Agriculture.*

Scrimshaw, N., and V. Young 1976. "The Requirements of Human Nutrition." *Scientific American* 235:51-64.

Sen, A. 1980. "Famines." *World Development* 8:614-21.

Shacklady, C., ed. 1979. "Bioconversion of Organic Residues for Rural Communities." *The United Nations University World Hunger Programme Food and Nutrition Bulletin*, Supplement 2.

Stefansson, V. 1956. *The Fat of the Land*. New York: Macmillan and Co.

Sweeney, G. 1947. "Food Supplies of a Desert Tribe." *Oceania* 17:239-99.

Torun, B., V. Young, and W. Rand, eds. 1981. "Protein-Energy Requirements of Developing Countries: Evaluation of New Data." *The United Nations University World Hunger Programme Food and Nutrition Bulletin*, Supplement 5.

Ugent, D. 1970. "The Potato." *Science* 170:1161-65.

United Nations University 1981. "The Uses of Energy and Protein Requirements. Report of a Workshop, 1981." *The United Nations University World Hunger Programme Food and Nutrition Bulletin*, vol. 3:45-53.

NEVIN S. SCRIMSHAW

Wellhausen, E. 1976. "The Agriculture of Mexico." *Scientific American* 235:128-50.

Woodham-Smith, C. 1962. *The Great Hunger: Ireland 1845-49.* London: Hamish Hamilton (New English Library Edition, 1975).

Wortman, S., and R. Cummings, Jr. 1978. *To Feed This World.* Baltimore: Johns Hopkins University Press.

Yesner, D. 1980. "Nutrition and Cultural Evolution Patterns in Prehistory." Pages 84-115 in N. Jerome, R. Kandel, and G. Pelto, eds., *Nutritional Anthropology—Contemporary Approaches to Diet and Culture.* Pleasantville, N. Y.: Redgrave Publishing Co.

Young, V., and N. Scrimshaw 1971. "The Physiology of Starvation." *Scientific American* 225:14-21.

Commentary by Claire Cassidy
Department of Anthropology, The University of Maryland

There are so many points of interest in this paper I am hard put to limit myself to just a few. But true to Indo-European habit, I will pick three.

Beginning at the end of the paper, with the future food section, I understand from Dr. Scrimshaw that his remarks on hydroponics as a farming technique of the future were selectively consumed by the word processor. So I want to get hydroponics back in the record. Hydroponics, or soilless agriculture, has a bright future. It utilizes a few cheap chemicals, recycles water, and requires much less space for production than traditional agriculture. It permits vegetables and fruits to be produced in quantity close to urban centers, which has both nutritional and economic advantages for urbanites. In fact, much of the lettuce used in Washington, D.C., is produced hydroponically within twenty miles.

Second, and in reference to the problem of human dependence upon too few plant crops, I would like to mention a promising new plant now under development. This is the winged bean (Psophocarpus tetragonolobus), a legume of the humid tropics, with nutritional characteristics much like the soybean. This plant is to be developed for home or subsistence production, rather than for export or mass production. Thus, winged bean may be the first plant product of appropriate technology applied to agriculture, and its development is to be contrasted to the development of plants using "Green Revolution" conceptualizations. Time will tell if this results in "one step forward, no steps back."

The sections that I find most exciting in Dr. Scrimshaw's paper are the sections on adaptation. As my third point I would like to address both biological and cultural adaptation by focusing on food scarcity. Dr. Scrimshaw suggests that after humans became established as an omnivorous species, there "is no reason to believe further biological evolution occurred in [human] nutrient needs." I would like to take mild issue with that statement, for I think there is no reason to argue that biological evolution did *not* occur either. Rather, we do not know much about the subject, and I think it deserves study.

There is evidence, for example, that humans are not particularly well adapted to high-grain diets (to which we have been exposed for only perhaps 1 to 2 percent of our total existence as a species). It is possible we are undergoing a process of natural selection with regard to the stresses engendered by dependence upon grain diets. In Dubos's terms, many humans may merely *tolerate* grain diets. Let us speak of dietary adaptation as an equilibrium state (fig. 27). It is a positive state which occurs within a range and describes an optimum with respect to environmental circumstances. Outside the optimal range are toleration ranges, one for deficiency and one for excess. People can survive with either deficiencies or excesses and will not define themselves as sick. Outside the toleration ranges exist disease and finally death.

The locus of equilibrium changes over the healthy life of the

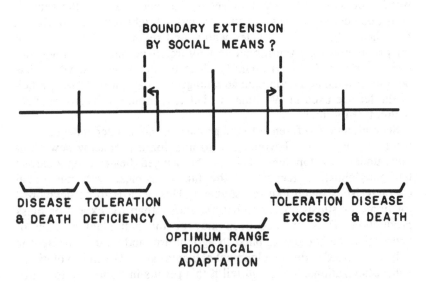

Figure 27. The relationship between biologically and socially defined boundaries for optimal adaptation to diet.

NEVIN S. SCRIMSHAW

individual. Different healthy individuals may also achieve somewhat different equilibria characterized by different levels of energy or nutrient intake within a range that is typical of the human species.

Differences among individuals and populations may be the results of organic evolutionary processes. Take the subject of thriftiness. We are familiar with breeding domestic animals for thriftiness, e.g., chickens and pigs who reach maturity quickly on less feed than their ancestors, and mules that perform as well as horses on less than one-third the feed.

Demole (1969) argues that similar differentials exist for humans, such that it may require as much as 2.5 times as many calories for a constitutionally lean person to gain one kilogram of tissue as for a constitutionally obese person to do the same. Indeed, there may be numerous populational differences in nutrient needs and these deserve to be identified and studied. Several of the nutritional diseases suggest maladaptation to grain diets. Gluten intolerance comes quickly to mind, but others such as dental caries are also linkable. We see clear evidence in both the archeological and ethnographic records that the teeth of hunter-gatherers are less often decayed than those of agricul-turalists.

We may also view protein energy malnutrition (PEM) as an example of maladaptation to grain-based diets; it is a disorder primarily of toddlers weaned onto a high-carbohydrate diet made from grain staples. It is possible to visualize a human with lower protein needs but equivalent growth potentials— perhaps the high frequency of PEM in farming societies actually represents the ongoing construction through natural selection of such a human. That is, the less carbohydrate-tolerant, more protein-needing individuals fail to reproduce. Some of Bill Stini's (1975) work addresses this problem.

But it is equally interesting to travel along the social pathway as Dr. Scrimshaw has done in most of his paper when he discusses social adaptations. To add to his comments I will now turn to a consideration of mechanisms for social adaptations. I would like to ask if, or how, human populations have arranged social mechanisms to deal with nutrient and energy scarcity or excess.

Returning again to the diagram, we may ask if the optimum range is set at birth or if it is subject to plastic response during one's lifetime. If the latter, which is likely, then how do social mechanisms alter the setting? When does toleration transform to adaptation? I shall propose that some societies have social mechanisms which have the effect of transforming toleration into adaptation.

There are several we could discuss. The one I will mention now is what I shall call the *institutionalization of hunger*. By this phrase I mean to include social activities which decrease food access for some segments of the population. It is important primarily in farming populations. I do not have time to detail it adequately, so I ask you to bear with the gaps in my argument.

Hunger or undernutrition is not randomly distributed in populations. It affects primarily toddlers and women of reproductive age. Both these groups have proportionately higher nutrient and energy needs, but it is among these two groups that the most social restrictions on diets occur. These social restrictions and other social habits having to do with food distribution patterns have the *documented* effect of reducing food access for these groups. Thus, we often see in farming populations the apparently paradoxical situation of the food-neediest groups getting the least. How does one explain this?

We may argue that in certain ways the effect of hunger stress on small children is to adapt them to adulthood. They learn details of their social status and how to behave in their social group. They end up smaller—and small-sized people eat less, which is an advantage when food is scarce. Possibly, there is actual physiological adjustment to food scarcity so that those exposed to hunger in childhood are more resistant to hunger in adulthood. If so, then the range of adaptation has been broadened. Finally, a groups' survival is enhanced when those least able to survive on high-grain diets do not reproduce.

Thus, the institutionalization of hunger actually promotes survival, at least in food-scarce environments. If we remember that survival is significant mainly in adults, we understand better why the stress period is focused in toddlers. The economic and social investment in toddlers is much less than in adults.

We can also understand why women of reproductive age are often underfed as well. Rosenberg (1980) has gathered ethnographic evidence to show that practices resulting in the underfeeding of pregnant and lactating women exist where population size is considered adequate or too large. Where increase is desired, reproducing women are fed preferentially.

Another social means of adaptation might be seen in the siesta or even in the development of cooperative work groups. Both these arrangements decrease the amount of energy any one individual must expend and may have developed in response to the lowered work capacity characteristic of marginally underfed adults.

I make these remarks in as value-free a way as possible. I do not want to argue ethical points about the "goodness" of social adaptive mechanisms. It is also perhaps necessary to repeat that "caring for offspring" has different parameters in different societies or environments, as Jane Lancaster has discussed in this symposium. We must learn how people have made and do make adaptations, and do so

NEVIN S. SCRIMSHAW

openmindedly, avoiding the pitfall of reaching premature closure because some mode of adaptation is in conflict with *our* ethical postures.

Our goal is understanding. Its utility for the future is to promote an awareness of human resiliency, and this in itself promotes a certain, though gray-tinged, optimism.

To close, I would simply like to echo Dr. Scrimshaw's remarks at the end of his paper. Indeed, the problem of feeding humans is far less one of food inadequacy than one of dietary inequity. What we clearly need to develop are *social and political mechanisms to distribute food equitably, to protect the land where food is produced, and to protect the farmers—men and women— who produce it.*

References

Cassidy, C. M. 1980. "Benign Neglect and Toddler Malnutrition." Pages 109-39 in L. S. Greene and F. E. Johnston, eds., *Social and Biological Predictors of Nutritional Status, Physical Growth, and Neurological Development.* New York: Academic Press.

Demole, M. J. 1969. "Dietetic Implications of Current Pathophysiological Concepts in Obesity." Pages 396-402 in J. Vague, ed., *Physiopathology of Adipose Tissue.* Amsterdam: Exerpta Medica Foundation.

Rosenberg, E. M. 1980. "Demographic Effects of Sex-Differential Nutrition." Pages 181-204 in N. W. Jerome, R. F. Kandel, and G. H. Pelton, eds., *Nutritional Anthropology—Contemporary Approaches to Diet and Culture.* Pleasantville, N.Y.: Redgrave Publishing Co.

Stini, W. A. 1971. "Evolutionary Implications of Changing Nutritional Patterns in Human Populations." *American Anthropologist* 73:1019-30.

————1975. "Adaptive Strategies of Human Populations Under Nutritional Stress." Pages 19-41 in E. S. Watts et al., *Biosocial Interrelations in Population Adaptation.* The Hague: Mouton.

Discussion

DR. CLAXTON: I wonder, Dr. Scrimshaw, if you would like a couple of minutes of rebuttal time?

DR. SCRIMSHAW: Dr. Cassidy has amplified a number of important points in which we are in obvious agreement. She also highlighted an area of partial disagreement, in that such examples as dental caries and protein energy malnutrition in the agricultural peoples she cites, I view as examples of social maladaptation and not biological maladaptation. It is interesting that she mentioned cereal diets, because there is a recent study in which the traditional Guatemalan corn-and-beans diet was given to Berkeley students, and they were only able to absorb and utilize about 80 percent of the protein, whereas ordinary Guatemalans can use 90 to 92 percent of the protein, which is about the same utilization, or at least the same absorption of protein, that one would get from milk, meat, and eggs.

The other point is that small-size, growth failure is indeed an adaptation which reduces the need for energy, but like the reduced activity to which I referred in somewhat more detail, it has a high price. And for adaptation to protein energy malnutrition by both restricted growth and reduced physical activity, as a restriction in calories, there are consequences.

It is hard to think of children as being immediately responsive to reduction of caloric intakes, but in a recent study in Guatemala, children were given two grams of protein per kilo body weight, which is good and adequate calories. They grew well, playing very actively. Then the calories were cut back by 20 percent. The children continued to grow, but their physical activity was cut way back, and the amount of sleeping increased, which brought them into caloric balance.

The significance of this, I think, goes back to studies of Adolfo Chavas, who observed that the malnourished preschool children in homes were very passive and had very little interaction with other siblings, with adults, or in the environment. But when the supplementary feeding program came in, this changed. Now they were crawling about on the floor, pulling things off the table, getting into the fire, interacting with their environment, and needing attention. This probably is the main effect of malnutrition in this preschool age group, and the effect of supplementary feeding is an increasing of the stimulus that the child gets to social and cognitive development at this critical stage in the life cycle.

DR. BOULDING: There is no such a thing as a world food problem. There are some world problems—with the atmosphere, for instance—but most things we think of as world problems are not. They are quite essentially local problems; that is, Blake said truth is only in minute particulars. Well, one of the greatest errors in thinking is that these problems are global.

Now, this isn't to say that global problems can't be influenced by other localities; they *are*, and even by this little locality here. But I agree absolutely that on the whole, hunger is a problem of poverty, and poverty is quite a local problem, but it can be intervened with on governmental levels. That is certainly true.

There appears to be very little that we can do on a world scale.

NEVIN S. SCRIMSHAW

Something like 0.01 percent of income is redistributed, and it isn't at all clear what the difference in the economy is on poverty. But Japan, for instance, which has the smallest social redistribution, distributes the largest proportion of its national income. These things are very puzzling really if you look at Hong Kong, which I say is an absurdity politically—but I suppose Hong Kong eats fairly well, doesn't it? I don't know. I imagine there is very little malnutrition in Hong Kong, despite the fact that it grows very little. I would be surprised if it produces more than 30 percent of what it eats. What this comes down to is that the scarcest of all commodities is competence, and the great differences among the societies of the world are primarily differences in competence. Now, how you spread competence around I don't know. It has very little to do with ideology. I mean, you've got Cambodia, which is a disaster; you've got Yugoslavia, which is, well, not so bad and you've got Poland, which is on the edge of a cliff; and Rumania, which ought to be; and East Germany, which is doing very well, thank you. The Germans can even get away with being Socialists.

Then you take the other camp and you see the catastrophes of Argentina and Uruguay, and the sort of dulcetlike quality of the British, and the success of the Australians and New Zealand, and then the eager-beaver Germans. I mean, there's just no such thing as a First World, a Second World, or a Third World. Unless we get a sense of the fantastic variety of societies, it seems to me, we will never get anywhere. I have really gotten fed up with the "whole world" approach to things.

The world approach has been an ecological one essentially, and this is spreading information and spreading competence, but how you spread competence I really don't know, and I wish this symposium would give some thought to it because it seems to me crucial.

DR. NEEL: Professor Laslett, in one of his comments yesterday, exploded about the prophets of gloom and doom, and I guess I was scored as one of them. But it wasn't until we had dinner together last night that I realized the essential difference in our approaches. I think in a way he is no less concerned than some of the rest of us about what lies just ahead. But given his long historical perspective—and he is quite capable of speaking for himself if I misspeak—he suggests that a major famine is the one way that man is going to be regulated.

If we allow that as an acceptable solution, then we are pretty close together. I would like to think that by now there is a better way to keep ourselves in balance with the environment than periodically overshooting the mark of our technology and having major dieoffs.

DR. CLAXTON: I sometimes think that if you want a really good thought for the day, it is the fact that the world's most unique extreme population explosion occurred at the same time as the invention of nuclear weapons. Does that tell us anything? Are we determinists?

DR. COHEN: Regarding the whole question of the relationship between maldistribution and contemporary crises as opposing positions, there was an editorial in the *New York Times* probably within the last month suggesting that although we are aware of very large numbers of malnourished people and we report this to ourselves regularly, it is not clear that the percentage of human beings who are thus deprived is any greater than it has ever been. And this editorial was arguing the fact that it is probably less than it's ever been before in human history.

I think there is a real danger in the comments that I hear of confusing what are flaws in human distribution systems and class stratification mechanisms which, I think, in Professor Boulding's words, are local problems in the sense that we can't get at them. I think there is a real danger of confusing some of those with some of the other sense of crises which we have. I think that we would do well to try to sort those two things out.

DR. SCRIMSHAW: I agree with that. Food continues to be produced to meet effective demand. That is food that somebody is willing to pay for, and this simply falls short of human need. And until there are social mechanisms to find ways of closing that gap, we will simply continue to have a portion of the world's population that is poorly fed. One indication of the magnitude of that is in the data for mortality. Another indication comes in the estimate of the extent to which caloric intakes of given populations are below the amounts needed for presumed appropriate activity.

But, of course, as I pointed out, that is very culture-specific. I think the most serious price that we pay for what are certainly localized food problems of societies is this impairment of the physical growth and development of children and the high morbidity and mortality. I do believe that this high mortality is one factor which is impeding the demographic transition in these populations. I also believe that it is having a significant effect on their competence as adults. Not only speaking cognitively and in terms of learning and behavior, but even— and there is very good evidence of this—in their physical capacity.

DR. LASLETT: I don't feel comfortable in being sententious about the future. It seems to me that the only relevance of discussions on the fate of the world for our symposium is how far they enable us to grasp the problem of human adaptation to see whether or not you can cope. But there is no evidence from the future, and the evidence bearing on the present is often from the past.

As we discussed this morning, there are strong indications, given Western tradition and administration, that the necessity of ensuring that there is no starvation goes very deep. A detailed investigation of

NEVIN S. SCRIMSHAW

the relationship of food levels and epidemics, over very long periods of English history, shows that it is very remarkable how very few food crises there have been, how very low the level of casualties from the running out of food there have been, and how strongly developed the ethical imperative of all local administrations always was that food was the ultimate priority.

There were episodes of starvation—not in the history particularly of Britain, but certainly of France—which I think have been drawn out of proportion. The reason they are so conspicuous in the history of the West is because they broke the rules.

What I am insisting on is that there is a tradition of moral responsibility for keeping all members of society alive within the final texture of Western institutions. If the West is in a position to supply the rest of the world with food, I should suggest this is because the West has had that sense of responsibility for individual survival, and the survival of every member of the society. I will end by telling of an episode from a London suburb in 1676, where a pauper was found dead of starvation. The entry in the local record says that he was buried with many tears by the local population, and the name of the overseer of the pauper was written in the record. There was a condemnation against him for his neglect. Now, I am not here insisting upon the moral superiority of Western individuals or Western citizenship. I am drawing attention to the possibility that where Western political techniques have been successful, they have been successful particularly in this sort of direction.   I wanted to make one particular point clear. When I said English people, I meant the inhabitants of the country I was referring to. The failure in Ireland was the type of conspicuous failure of Western political institutions to which it is very important to draw attention. I am not trying to explain away the unfortunate history of the Irish famines. What I am saying is that if you look over the long periods of time of Western history, the exhaustion of food has not been an important population control.

DR. CLAXTON: I don't want to stop this discussion. I want to add to it one essential ingredient we haven't gotten to yet, which is population growth, and it is another subject here. So let us then thank Dr. Scrimshaw.

DR. SCRIMSHAW: Can I make one statement, one sentence?

DR. CLAXTON: All right.

DR. SCRIMSHAW: Although certainly food problems are "local," they are "national" to an extent as never before and they are also "global." There are a lot of factors with which food-short nations must contend that are outside of their boundaries and for which we all must share some responsibility.

9. Food, Energy, and Technology: Perspectives from Developing Countries

EDWARD S. AYENSU

Director, Office of Biological Conservation, Smithsonian Institution

Editor's Summary. Shortages of food and energy, when placed in juxtaposition with rapidly expanding population, create some of the most serious issues confronting human society. Edward S. Ayensu stresses the need to make better use of our knowledge regarding edible foods to improve human diets. Instead of relying on the successful but costly Green Revolution, he proposes greater emphasis on low technology and new foods more compatible with the economic limitations of Third World farming. Also needed are fast-growing woody plants for fuel, along with a social system to make them available to all. Ayensu foresees a major conflict developing between food crops and fuel (alcohol producing) crops which could have a major effect on food supplies in poor countries. Warning poor countries not to adopt industrialized high-energy economies unthinkingly, he suggests that computer technology—because of its small size and low-energy needs—may be of great value to those economies.

It gives me great pleasure to be invited to contribute to the dialogue on the subject of "How Humans Adapt." Having had prior experience in studying the social organization and the behavior of fruit bats in many parts of the world, and how these flying mammals adapt to various situations (Ayensu 1974), I think I am now partially qualified to make impartial observations on how humans adapt. It also seems appropriate at this point in human history to encourage a dialogue on human adaptation because of the rapid emergence of such technological innovations as microprocessors and biotechnology. The latter part of this century has witnessed unprecedented technological innovations that are likely to have profound influence over our lives. Man's ability to adapt successfully to these emerging technologies will undoubtedly make mankind the beneficiary of radically new opportunities for development.

But before I discuss how the developing world can possibly plan an economic quantum jump from the village-based rural societies directly into the computer- based information society of the twenty-first century, I must first discuss the traditional attitudes toward food and energy preferences, the recent scientific influences on new varieties of crops, and the changing habits toward nontraditional foods.

Since the very beginning of human history, man has preoccupied himself with the gathering of food and the acquisition of energy in order to function and survive. As a result of the continuous search for these commodities, methods for obtaining and using food and energy have become closely worked into man's social and economic fabric.

In a significant portion of the world's societies, certain foods have been closely associated with the three major turning points in human life. The Rites of Passage—as these turning points are often called— refer to the time when a baby is born, when he enters puberty (and therefore adulthood), and lastly when he departs from this world.

Since birth implies conception and its attendant period of pregnancy, certain foods are often recommended to the pregnant mother to guard her during this delicate period. In societies where adulthood is marked by religious considerations, such as puberty rites and initiation ceremonies, specific food items are always associated with the occasions. Marriage ceremonies, especially in Indian and other Asian societies, often take on highly elaborate dimensions, with specific food items playing very visible and crucial roles. In southern India married women attending a wedding receive a goodwill offering of coconut, betel leaf (*Piper betle*), and nuts of the *Areca catechu* palm (Ram, personal communication). Even death, the most inescapable and unwelcome part of the *rites de passage*, has certain food items associated with the ceremonies for the departed. Hindus celebrate an annual "death anniversary" of the departed, at which a priest worships the God of Fire, burning twigs of the bo tree (*Ficus religiosa*). Balls of rice with black sesame seeds are symbolically offered to all the ancestors, then fed to a sacred cow representing their forefathers (Ram, personal communication).

Depending on geographical location and availability, man has developed certain food preferences as well as some deep-seated prejudices against certain items, particularly flesh food such as pork, beef, and chicken. For religious reasons, the Jews and Moslems have perpetuated

EDWARD S. AYENSU

the ancient taboo against pork— the pig being considered an unclean animal.

The rejection of pork by very orthodox Jews and Moslems has deep historical roots. But even before the religions of the Jews and Moslems imposed their taboos, the Hindus were practicing a taboo against pork. Regardless of when the prejudice against pork came into being, there are basic underlying reasons for the varying degrees of feeling against pigs in general. Apart from the charge of being unclean, the fact that pork deteriorates rather rapidly under high temperatures has contributed to the establishment of the taboo. For health reasons, and specifically because of the possible contraction of trichinosis, pork has not been accepted in many societies.

To the Hindus, the cow—in contrast to the water buffalo (Hoffpauir 1977)— has always been a sacred organism of high spiritual standing, even above the Brahmins. The cow yields five major products that are important to the daily livelihood of man. These products include cow's milk, milk derivatives, curd, urine, and dung. Milk and milk products, such as clarified butter, are important in the diet, and cow dung is an important construction material as well as an essential energy source for cooking and medicinal healing. On the other hand, most Chinese for centuries have considered cow's milk and cheese to be an abomination!

In certain societies of Africa, Asia, the Far East, and the Pacific Islands, animal milk is considered unfit for human consumption. Various reasons are often assigned to this reaction. However, the prevailing notion—especially in some African and Asian societies—is that they refused to "take on the habit of milking because it was strange to their ways and they did not like it" (Sauer 1952). There are others who object to drinking milk because of the resemblance between milking and urinating.

In a number of East African countries there is a strong taboo against the eating of eggs, especially by girls, because it is believed that eggs cause infertility in women (Simoons 1961). In Somalia many people do not eat eggs and fish because they are snakelike. On the other hand, rotten eggs are a delicacy for the people of Brunei on the coast of Borneo. Maggots and snakes are relished by some Australian aborigines. Snails are regarded as a delicacy by Japanese and by the French. Some Europeans eat putrescent meat, especially during winter, when such rotten meat is referred to as "fowl hung until high."

There is a consistent worldwide prejudice against eating cattle or fowl which have been killed accidentally by road vehicles or by unconventional slaughtering. Although it has been demonstrated that such meat—when cooked—is as nutritious as freshly slaughtered cattle, this cultural prejudice regarding the way an animal dies before it is eaten has been perpetuated throughout human societies.

There are certain foods that were known or believed to cause illness or death. Others were thought to bring about labor difficulties when women were in confinement. Such foods were strictly forbidden from the diets of many societies. Most of these restrictive preferences were handed down by word of mouth from the elders and traditional medical practitioners through generations. Therefore, one cannot find written down anywhere the reasons for such restrictive food selections.

Breaking Down of Taboos. During the last century or so, especially within the context of many African societies, there has been an extraordinary flexibility in the determination of food preferences, along with much relaxation of restrictive practices against certain foods. The Food Science and Applied Nutrition Unit of the University of Ibadan conducted a survey in a Nigerian village to determine how common food restrictions were enforced during pregnancy. The study was geared to finding out which foods were restricted and how those restrictions affect the nutritional requirements during pregnancy. The study concluded that, of the 112 women of child-bearing age (twenty-forty years) who were interviewed, 52 percent did not know about or practice any food restrictions during pregnancy. Of the young husbands interviewed, 64 percent did not know of any food restrictions for their pregnant wives. When the ladies above age fifty-five were interviewed, they all (100 percent) gave long lists of food forbidden to pregnant women.

It seems, therefore, that the traditions and beliefs that make up the large part of food prejudice are dying out simply because of increased contact between peoples of different communities. Most of the changes in the diets of most Africans over the past twenty-five to fifty years have, indeed, been due to this factor.

In 1968 the U. S. Department of Health, Education and Welfare and the Food and Agriculture Organization of the United Nations published the *Food Composition Table for Use in Africa*. It clearly shows the nutritive value of all the foods indigenous to the different regions of the continent. A quick perusal of the book reveals that virtually every available kind of animal and plant is eaten in one form or another by the peoples of Africa. A study of food composition in Latin America showed the same results.

Food Availability. The availability of food is an important contributing factor to the changing food habits of a people. Even if a particular food item acquires the necessary prestige, it will not be eaten unless it is easily available.

In an earlier publication (Ayensu 1976), I noted that many of the bulk staples of Africa are not native to the continent. Crops such as cassava *(Manihot esculenta)*, sweet potato *(Ipomoea batatas)*, cocoyam *(Zanthosoma sagittifolium)*, groundnuts or peanuts *(Arachis hypogaea)*, and maize *(Zea mays)* are all natives of tropical America. Rice *(Oryza sativa)*, which is produced and consumed in large quantities in Africa today, originated in southeast Asia. All these crops have been accepted to the extent that it is very difficult to convince many Africans that they are not indigenous.

In the African context, at least, the factors that seem to transcend and, in many cases, overshadow cultural patterns (as far as food items are concerned) include the ease of cultivation of a crop, the ease of harvesting and, to some extent, the convenience of storage after harvest, and the cost of the commodity at the market.

As I have indicated in a another publication (Ayensu 1978), the United States has also made gradual changes in its food preferences. The industrialization of the farms, beginning at the end of the nineteenth century, changed everything. Various studies reported by the U. S. Department of Agriculture (USDA 1976) have shown that the American public's taste for meat, poultry, and fish increased dramatically between 1900 and 1976. Page and Friend (1978) attributed this change in meat consumption to "change in consumer tastes evidenced by some increase in demand for ground-meat products by fast-food outlets. . . ." Over the same period, the consumption of chicken tripled, with most of the increases taking place after 1950. Here again, the fast-food industry accounted for most of the increase in the use of chicken.

The USDA study further indicates that, apart from butter, the use of dairy products has increased by one-fourth since the beginning of this century. This is due to an increase in the use of cheese and, more recently, is reflected in the popularity of such cheese-containing dishes as pizza.

It is important to point out that changing food habits may sometimes cause nutritional problems. For example, frequent meals of various new food items— such as pizza or other starchy dishes plus a soft drink—may not contain the necessary nutrients needed in a balanced diet. In fact, many people in the developed world, who rely on such foods these days, are liable to be placed in my category labeled "Affluent Kwashiorkor"—kwashiorkor typically occurs when people in poor nations get sufficient calories but inadequate protein in their diet. Aside from the appeal of novelty, newly created and introduced food items often receive quick acceptance, if they have a pleasant and familiar flavor as well as a familiar texture and appearance.

The Green Revolution and Changing Food Habits. A quarter of a century or more ago, the world began to experience a new phase in food research through the establishment of an institute in Mexico with the financial assistance of the Rockefeller Foundation. The institute undertook serious research on breeding wheat strains to become resistant to rust and to produce very high yields. For example, crossbreeding with the Japanese (Norin) short-stemmed varieties led to the production of high-yielding strains (Gaines) that do well in the northwestern United States. A significant factor in the development of the high-yielding, rust-resistant varieties was the application of large quantities of fertilizers and the involvement of irrigation systems. This was also true when the International Rice Research Institute in the Philippines began to breed rice, such as the highly productive IR-8, that would give higher yields and resist fungal and insect attack.

With the improvement in crop yield, as a result of the development of many engineered strains, a number of countries began to experience bigger harvests of rice, corn, soybeans, cassava, wheat, and other crops. In all cases, however, the use of fertilizers, pesticides, and increased irrigation had to be accelerated before the new strains could be introduced.

The selection and concentration of relatively few crops, and the development of monocultures, have changed the food habits of large numbers of people and have, in effect, forced most people to cultivate similar foods. The establishment of the Consultative Group for International Agricultural Research (CGIAR) in 1971 provided the mechanism for concentrating financial resources to support the various international agricultural research centers. The nine international research centers and four associated organizations have specific research responsibilities. The International Centre for the Improvement of Maize and Wheat (CIMMYT) in Mexico is where Norman Borlaug and his associates developed the famous dwarf variety of wheat.

The successes of CIMMYT encouraged the Rockefeller Foundation to join the Ford Foundation in setting up the International Rice Research Institute (IRRI) in the Philippines in 1955. IRRI soon produced high-yielding strains of rice, notably IR-8. The International Institute of Tropical Agriculture (IITA) in Nigeria was established, through the efforts of the Ford Foundation, to concentrate on developing farming systems for maize, rice, sweet potatoes, cassava, yams, cowpea, lima bean, soybean, and pigeon pea. The Rockefeller Foundation also founded the International Centre for Tropical Agriculture (CIAT) in

Colombia to deal with tropical pastures and research on maize, cassava, field beans, and rice. Another Latin American institute, the International Potato Centre (CIP), was established in Peru in 1972 to work on potatoes for both tropical and temperate regions. Also in 1972, a major establishment—the International Crops Research Institute for the Semi-Arid Tropics (ICRISAT)—was founded in India to deal with sorghum, pearl millet, pigeon peas, chick peas, groundnuts, and farming systems.

Two livestock centers, the International Laboratory for Reseach on Animal Diseases (ILRAD) in Kenya and the International Livestock Centre for Africa (ILCA) in Ethiopia, have been established in Africa to deal, respectively, with trypanosomiasis and theileriosis of cattle and with livestock production systems. In Lebanon, the International Centre for Agricultural Research in Dry Areas (ICARDA) was established to deal with farming systems, cereals, food legumes such as broad beans, lentils, chick peas, and forage crops. In 1973 the International Board for Plant Genetic Resources (IBPGR) was created in Italy to promote the conservation of plant genetic material with special reference to cereals. Earlier in 1971 the West Africa Rice Development Association (WARDA) was established in Liberia, with the support of IITA and IRRI, to encourage cooperation in rice research among thirteen countries in the region. Two other organizations within the consultative group are the International Service for National Agricultural Research (ISNAR) in the Netherlands and the International Food Policy Research Institute (IFPRI) in the United States. These organizations are charged with the responsibility of analyzing world food problems.

The establishment of all these institutes has already had an impact on the production of rice, corn, and wheat in many less affluent countries. Asia, for example, has demonstrated man's ability to adapt to new farming technologies in order to help meet her needs. With modern farming techniques at hand— especially those involving high-yielding varieties—Asia has been able to increase the region's food production several-fold during the past decade. The only drawbacks have been the heavy post-harvest food losses and the rapid increases in human population growth.

Tillage Systems and Problems of Soil Loss, Runoff, and Grain Yields. While mechanization of the fields has resulted in major harvests in temperate countries, similar practices have caused great concern in the developing countries. According to FAO estimates, it is likely that about 20 million hectares of land will be cleared mechanically by the next decade to expand agricultural production. Past experiences in the tropics, however, have shown that most of the mechanically cleared lands are laid waste after a few years of planting as a result of soil erosion caused by rain storms.

A study conducted at the International Institute of Tropical Agriculture (IITA) has shown that in traditional farming there was a small amount of runoff of rain (2.6 millimeters per hectare) while there was minimal soil loss (0.01 ton/hectare) and low grain yield of maize (0.5 tons/hectare). The other extreme is the situation whereby Crawler tractors are used. In such a case runoff of rain is very high (250.3 millimeters per hectare), while soil loss is 19.6 tons per hectare, but the grain yield is 1.8 tons per hectare. It is obvious from these studies that much more suitable mechanization should be sought that will extend the utility of the tropical soils.

Nontraditional Foods. The present structure of the international agricultural research network has been designed to promote the Green Revolution; the achievements have been most impressive. However, relatively little research has been carried out on increasing the productivity of tropical crops and animals. Rice and wheat are the two major crops upon which the Green Revolution is based. Wheat is really not suited to tropical agriculture as other crop plants might be.

Another noteworthy factor of the Green Revolution is that the hybrid varieties lack the genetic variability of the "natural" grains from which they are derived and, therefore, their chances of being attacked by pathogens are very high. It is interesting to note that the Philippine farmers, who were given the new strains of rice to grow, decided to abandon them because of several problems. Apart from the increased expenses in growing the new rice—and the high labor involvement in tending to the crop—these high-yielding strains were often considered poor for cultural and other reasons (Dalrymple 1969). Similarly, women who prepare the popular Ghanaian corn meal dough *(kenkey)* have expressed dissatisfaction with the introduction of some of the new hybrid corn, because its dough does not taste the same as the local variants.

In 1975 the U. S. National Academy of Sciences published a report on "Underexploited Tropical Plants with Promising Economic Value." The thirty-six plants described were considered as having significant potential as a source of food in the developing countries. The amaranths *(Amaranthus* species), for example, were among the staple crops in the highlands of Mexico, Guatemala, and Peru during the beginning of the sixteenth century. The Spanish conquistadores suppressed the cultivation of this highly nutritious plant and introduced the less nutritious barley. Amaranths are rich in proteins and contain exceptional quantities of the amino acid lysine. Franklin Martin of USDA

270

and his colleagues (1975) have indicated the nutritious qualities of the leaves, which are used as spinach in Africa. It is important to note that amaranths are among the special plants often referred to as "C4," which use sunlight more effectively than most other plants and which can also germinate and grow in a wide range of climates and soil types.

The winged bean *(Psophocarpus tetragonolobus)*, a native of New Guinea and Southeast Asia, is an important plant whose seeds consist of 37 percent protein, with a considerable amount of the amino acid lysine. The seeds also contain polyunsaturated oil and the roots, edible starch. The leaves are like spinach, and the inflorescence, when cooked, tastes like mushrooms.

Several other plants having similar potential exist in the tropical regions of the world and, even within the United States, there are a number of plants that compare favorably with several common crops. The tepary bean *(Phaseolus acutifolius)* is a crop that has been featured in the diets of the Papago and other Indians of New Mexico and Arizona. Its taste and nutritive value compare favorably with other well-known beans. Being a drought-tolerant crop, the tepary plant seems to do better on dry soils than most other legumes.

There are several food items that are still not exploited because of a lack of promotion in new societies and because of cultural ignorance. In West Africa, for example, the giant land snail *(Achatina marginata)* provides excellent-tasting meat. This *escargot d'Afrique* is nutritious and contains more of the amino acids lysine and arginine than found in chicken eggs. The snails also provide 107 calories, 17.7 grams of protein, 3.5 grams of fat, 132 milligrams of calcium, 110 milligrams of phosphorus, and 4.1 milligrams of iron. With very little effort and at practically no cost, a farmer or housewife can produce 100 to 200 pounds of tasty meat a year with an appropriate number of breeder males and females.

Another example is the least known food from the sea—the microalgae— which the Aztecs used to sprinkle over their food. The spirulina is 75 percent protein, contains all eight essential amino acids, and is rich in vitamins and chlorophyll. It produces twenty times more protein per acre than the well known soybean. In addition to the spirulina, the Chinese and Japanese are making tasty seaweeds very popular the world over. In fact, various cultures that never used seaweeds as part of their diet are now developing a taste for them. On my very first trip to China, I had the opportunity to taste more than two dozen different "sea vegetables" in a period of three weeks. I highly recommend them to those who have not had the opportunity to taste any.

It is necessary that we do not allow the successes of the Green Revolution to drive us away from other potential crops that are available—especially in the areas of the world where populations are increasing at alarming rates. In a sense, it is most unfortunate that we are not taking advantage of the 350,000 potential plant sources and numerous animals for food. As I and others have often stated: throughout the course of human history only 3,000 plant species have been given any consideration as food. Today the concentration is on about thirty species of flowering plants. Nearly half of all human food energy and proteins is derived from only wheat, rice, and corn.

Energy Utilization. Practically all cultures are built around the use of energy. In the less affluent world, firewood and charcoal are relied upon for the bulk of cooking and heating.

It is important that we do not forget that throughout the developing world, human energy and animal power still produce the bulk of energy for agricultural production. Estimates provided by FAO for the years 1974-76 show that human labor provided 66 percent of the power and energy inputs in developing market economies, while draught animals provided 27 percent of the power, and tractors only 7 percent. It is worth noting that tractors accounted for 27 percent of the production power in Latin America, 18 percent in the Near East, 3 percent in the Far East, and 4 percent in Africa. But in the affluent world, the consumer relies on energy in the form of natural gas and electricity to perform similar functions. The cultural lifestyle of the developed world depends on high consumption of energy.

The modernization of agriculture, for example, has meant an increase in consumption of energy for manufacture of fertilizer, mechanical power, and irrigation. These energy sources have all contributed to raise plant yields and helped make labor more productive. FAO records show that the use of energy in agriculture varies greatly among countries and regions. For example, the developing countries as a group used the equivalent of 30.7 million tons of oil in 1972-73. This figure represents only 18 percent of the commercial energy consumed in world agriculture. All the developed countries used 144.2 million tons of oil during the same period. The United States, with only 6 percent of the world's population, consumes about one-third of the world's energy and uses over 20 percent of the world's total production of fertilizer while the whole of Africa uses just about four percent—South Africa using about 50 percent of it.

In general, however, commercial energy production and use in developing countries have been increasing. The 1972-73 FAO figures show that the developing countries used 68 percent of the world's output of fertilizer while the developed countries used only 40 percent. For the use of machinery in agriculture the developing countries used

22 percent as against 57 percent by the industrialized nations. In the area of irrigation the developing countries used 8 percent, while the developed nations increased their irrigation input by only 1 percent. Both the developing and the developed nations used just about 2 percent each of pesticides on their farms.

It is in the areas of processing, transportation, marketing, and the preparation of food that the developed nations use several times more commercial energy than the developing world. For example, the food sector of the United States used 16.5 percent of the country's total energy produced. Farm production accounted for only 18 percent of the food sector's consumption. An example of energy consumption in the United Kingdom around the mid-1970s is most illustrative. The energy used to produce and put a one-kilogramload of white bread, sliced and wrapped, on a retailer's shelf broke down as follows: 19.4 percent to grow the wheat, 12.9 percent to mill the wheat, 16.3 percent to make the bread, and 3.4 percent to deliver the bread to the retailer.

Comparatively, the commercial energy used in food production in the developing countries is mainly directed to actual crop production. Generally, much of the food produced is consumed within a short radius. There is less processing and the so-called noncommercial energy is largely used for processing and preparation.

Competition between Food and Energy. The question of food and energy has been raised in a highly perceptive article by Lester Brown (1980), in which he points out the new competition for the world's cropland— as traditional food items like corn and sugar cane are turned into fuel to meet our energy needs. Another look at the food and fuel equation is provided by David Pimentel et al. (1974), who observed that in the United States today the equivalent of eighty gallons of gasoline is used to produce an acre of corn.

In the face of the world energy crisis and the introduction of modern agricultural technologies, particularly in developing countries, two major facts about energy have to be considered in the light of their effect on changing cultural patterns. The first concerns the availability of energy for food preparation. The second concerns the controversy between food and energy and its effect on culture.

Man has traditionally relied on firewood as his main source of energy. For many decades this source of energy had been taken for granted by at least one- third of the world's population, because of easy access of firewood in areas of human habitation. During the past two decades, however, there has been a growing desperation by the world's poor in their effort to secure firewood from reasonable distances.

In 1980 the U. S. National Academy of Sciences conducted a study on the world firewood situation and selected several fast-growing species that have the potential to do well in various biomes. The study was based on factual analyses of the preferences of societies that rely on firewood as their main source of energy. Among the species selected for consideration was the mesquite *(Prosopis chilensis)*, which is a South American tree. This species is currently turning otherwise useless lands in many arid and semi-arid regions into useful pieces of real estate. In addition, two Australian plants *(Eucalyptus* and *Casuarina)* are equally being propagated in many parts of the world. In most cases these species are doing much better in their new environments than in their natural habitats.

The increase of the world's population, particularly in the poor nations, has contributed to the scarcity of fuelwood. In Nepal, for example, one out of every four persons in a family has to allocate a substantial amount of time to walking great distances to collect firewood for family cooking. In parts of West Africa, people are driven by desperation to steal firewood to cook a simple meal.

It seems, therefore, that various cultures will have to change their age-old habits of free collection of firewood by developing woodlots near areas of human habitation for systematic cropping and replanting. It is obvious that, if such orderly arrangements are made, the social organization of the rural area will have to be adjusted to accommodate the new economic reality. Furthermore, the scarcity of firewood has often rendered this noncash economic activity into a commercial operation that is certainly draining the meager financial resources of the poorest of the poor.

The availability of oil and gas in some of the poor nations of the world has been acknowledged. However, the financial resources needed to develop these reserves are difficult to come by. In some countries where these sources have been developed, the entire orientation has been one of export, with the bulk of the population still pursuing the old energy sources, such as firewood, charcoal, and animal dung, to meet their own needs. Nigeria, for example, is the second largest supplier of oil to the United States, but easily 80 percent of her population still relies on firewood and charcoal to cook their meals.

There are other countries, such as India, that have over the years developed an integrated energy system to meet the needs of people at different levels of the economic ladder. In a recent document presented by the government of India before the United Nations Conference on New and Renewable Sources of Energy in Nairobi in 1981, it was pointed out that a large proportion of the total energy consumed in India was realized through the so-called "non-commercial sources of energy such as firewood, animal dung and agricultural wastes." Also, there was a direct correlation between the level of social and economic development and the consumption of energy. As the standard of living rises, the per capita consumption of energy increases. Figure 28 shows the total energy consumption in India and the share of commercial energy, while table 14 illustrates the consumption of commercial and noncommercial forms of energy, using the coal replacement measure.

Apart from the energy derived from burning agricultural waste and animal dung, substantial energy is derived from draught animal power and human labor. The most important sources of commercial energy

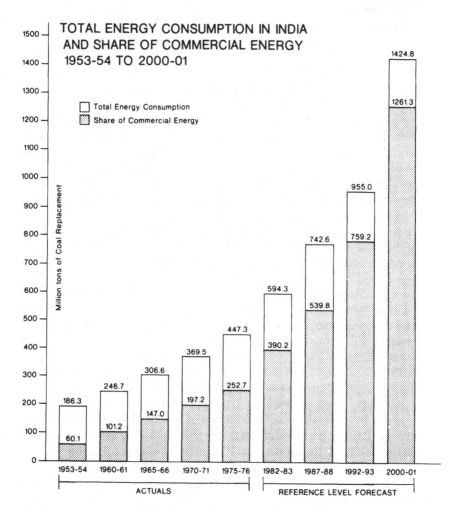

Figure 28. The relationship between total energy consumption and commercial energy use in India from 1953 projected to 2001.

TABLE 14 Consumption of Commercial and Noncommercial Forms of Energy 1953–54 to 1975–76

	(in million tons of coal replacement)				
Sources	*1953–54*	*1960–61*	*1965–66*	*1970–71*	*1975–76*
Total Commercial Energy	60.4	101.2	147.0	197.3	252.7
Total Non-commercial Energy	125.9	194.6	172.2	159.6	145.5
Total	186.3	295.8	319.2	356.9	398.2

are coal, lignite, oil, and hydroelectricity. Nuclear energy provides a minute amount of energy output.

Firewood, however, is the most important noncommercial fuel in India. It is usually obtained from forests and from trees along roadsides, riverbanks, and canals. Other wood is obtained from privately owned plantations and woodlots. It is significant to note that in the past two decades, population pressure on the land has been so dramatic that these hitherto free sources of firewood have now been mostly depleted. It is a fact that the gathering of firewood has by and large ceased to be a noncommercial operation. It is estimated that about two-thirds of the total energy available in rural households, and about 40 percent in urban households, is derived from firewood.

A report by the Working Group on Energy Policy of the Planning Commission (Government of India 1979) has shown that the total consumption of firewood in 1975-76 was estimated to be 133 million tons. Taking into account the uneven information available on the quantity of agricultural waste available as a source of energy, the report estimated that about 302 million tons of this material were used in 1975-76. About 73 million tons of animal dung were used during the same period. Almost all of it was used in households.

Although India is making tremendous strides to obtain energy security by expanding her production of lignite, hydroelectricity, and nuclear power, it is evident that 680 million people are going to rely on the so-called "noncommercial" sources of energy for many decades to come. To be sure, cultural adjustments will take place along the way as the social and economic development of a country enters into various transformations.

As intimated earlier (Brown 1980), the other side of the energy question concerns the new upsurge in the production of agricultural-

EDWARD S. AYENSU

based alcohol fuels. This situation has presented a new conflict between food and fuel needs and the use of agricultural lands. In 1973 the world experienced unprecedented increases in oil prices, which placed the economies of non-oil-producing nations on the verge of economic and, in some cases, social ruin. Brazil and India quickly realized the potential that lies in the production of alcohol fuels from plant- derived materials.

The Brazilian experience is particularly instructive. From 1965 to 1979 consumption of petroleum products grew from 19 to 58 billion liters, at an average rate of 8.1 percent. This meant that the country was importing 85 percent of its oil at an estimated spending rate of 6.5 billion dollars (U. S.) by 1979. Such a situation was obviously unacceptable. In 1975 the Brazilian government took the bull by the horns and launched a national plan to produce energy crops. The program was based on sugar cane, which for the moment is the most efficient alcohol producing crop. At present, sugar cane yields about 3,000 liters of alcohol per hectare or about 75 percent more than corn. Brazil plans to produce 10.7 billion liters of fuel alcohol by 1985. This will require nearly three million hectares of sugar cane, or 10 percent of Brazil's cropland (Solar Energy Intelligence Report 1977). It is obvious that if a substantial number of hectares is converted into energy-producing fields, there will, in the future, be a major competition between the "food fields" and the "energy fields." Such a conflict becomes even more of a problem when projections to convert one-fifth of the country's croplands to produce 20 billion liters of alcohol are being calculated.

Apart from the conversion of sugar cane into alcohol, there are other crops that present immediate problems as far as world food supplies are concerned. In the face of an increasing world population, the pressures of converting large acreages of cropland into fuel lands, and starch crops such as corn, cassava, and sweet sorghum into fuels, are likely to present a very difficult situation. Other developing countries that want to follow the Brazilian example should look hard before they leap.

The Culture of the Poor Farmer. The major transformations in food and energy production and utilization during the past century have affected practically every living culture. However, the degree of impact varies from country to country and from one social class to another. Robert McNamara, former president of the World Bank, described it best during his presentation before the bank's Board of Governors in September 1972:

To many in the affluent world, to be a farmer suggests a life of dignity and decency, free of the irritation and pollution of modern existence: a life close to nature and rich in satisfactions.

That may be what life on the land ought to be. But for hundreds of millions of these subsistence farmers, life is neither satisfying nor decent. Hunger and malnutrition menace their families. Illiteracy forecloses their futures. Disease and death visit their villages too often, stay too long, and return too soon.

Their nation may be developing, but their lives are not. The miracle of the Green Revolution may have arrived, but for the most part, the poor farmer has not been able to participate in it. He simply cannot afford to pay for the irrigation, the pesticide, the fertilizer—or perhaps even for the land itself on which his title may be vulnerable and his tenancy uncertain.

Adaptation of Developing Countries to Technological Innovations. Now the glimmer of hope! During the Industrial Revolution most Third World countries were never involved in the scientific and technological innovations of that period, although some were involved in their own science and technology before the Industrial Revolution (Ayensu 1981). There are several reasons why the Third World countries did not participate and adapt to the changes of the Industrial Revolution and were thus left behind. Without belaboring the obvious, we must first blame the results of *foreign domination*. It seems obvious that there was the *suppression of the inventive spirit* of the peoples of the Third World (Ayensu 1975).

Certainly during the middle of the twentieth century, as the industrial nations began to make rapid strides in technology, there were slight movements in the application of science and technology in some of the least developed nations. But most of these movements took place in the sectors of the economy that were directly linked with the production of raw and partially processed materials for the developed nations.

As we approach the end of the twentieth century—which is only eighteen years away—and as we review the situation in most Third World countries, we see that the gap between them and the developed countries is widening beyond comprehension. If present trends continue, the situation will become so hopeless that during the twenty-first century several of the Third World countries will go under completely if a major effort is not made by them to adapt to the technological innovations of the latter half of this century and to those impending in the twenty-first century. In actual fact, we should not even overlook the application of such innovations as robotics for handling a number of activities, including hazardous types of operations in industrial development. In this connection, I am not referring to the very complicated robots that are available in a number of the United States research laboratories, but rather to a number of very simple robots which handle simple but hazardous operations like painting and spraying, as well as working in very inhospitable environments.

It is my belief that the rapid emergence of the new technologies, such as microcomputers and microprocessors, present the developing nations with a radically new opportunity for development. My optimism is deeply rooted in my observation that village-based rural societies may, in fact, be better predisposed to the kinds of technological innovations that we are witnessing today and the technologies yet to be conceived and turned into reality during the twenty-first century (Ayensu and Trachtman 1981).

I would like to cite two examples to illustrate my point. The first involves communication. Mankind has seen the transformation in communication from the days of the talking drums in African villages to communication satellites covering the globe. In the old days and even today, the talking drums have been used to transmit messages from one village to another. This type of communication is based on a wireless mode. When the developed world began to string telephone wires all over the place to communicate, the village-based rural societies were not included. There are various reasons for this: high cost, inability of the villages to handle the intricate technical details of fixing telephones, and so forth. I believe that the new communication technologies are, in fact, more similar to the talking drum/wireless mode of communication which the African villages were using and, although these new technologies are highly advanced, they are also simple, reasonably priced and, in fact, more compatible with the sociology of most Third World countries. Within the next few years we shall be buying the chip, which has the capacity of storing enormous amounts of information, for less than a dollar.

My second example concerns urbanization. Elsewhere in this book (chap. 13), Lord Briggs makes a lovely presentation about urban life. From my perspective I consider twentieth-century technologies to have contributed substantially to human migration to urban centers. With the beginnings of the Industrial Revolution, people in the developed nations began to move into cities, creating ghettos that have become the hallmarks of New York, Chicago, Liverpool, and so forth. During that period the village-based societies of the Third World continued to stay in the rural areas. With the introduction of science and technology into the Third World economies, villagers began to move into the urban centers, creating ghettos and subjecting these rural inhabitants to a strange way of life that is neither satisfying nor decent.

During the past few years, the introduction of microprocessors and minicomputers has helped many industries to reverse the trend by moving from the urban centers into essentially rural settings. Among the reasons for these changes is the fact that the latest technologies do not require high concentrations and close proximities of people and services. Information is easily transferable over great distances to perform functions without physical involvement. My observation is that the developed nations are learning that it is not necessary for the village setup to be destroyed. In fact, the decentralized societies are more appropriate for the new technologies than the technologies of the Industrial Revolution that brought about the congested cities and towns of the nineteenth and twentieth centuries.

It seems to me that the Third World countries are mostly decentralized at this point, although some of their cities have emulated the bad habits of the developed world. The fact is that most Third World nations never became centralized. The distances between villages are simply ideal for absorbing the technologies of the twenty-first century, rather than for absorbing the current outmoded and cumbersome technologies.

It will therefore be a sad commentary on humanity in general if the developing countries are sold a bill-of-goods by being forced to buy discarded or soon-to-be discarded technologies from the developed world. Are the Third World countries going to be wise enough to take a bold stand by getting involved in the technological innovations of the twenty-first century? Or are they going to be satisfied with the status quo?

I am very much aware that the economics of twentieth-century industrial development place the Third World countries in a double bind. If these countries do not develop, they will certainly be destroyed by poverty. And if they do develop, they are likely to be destroyed by progress. A double bind is always such that it cannot be resolved on its own terms. The way out is to escape from its terms. For the Third World, information technology provides the escape.

Many less developed countries possess traditional economic, cultural, and intellectual assets that are wasted or destroyed by industrial development, but which can be vital natural resources in developing the electronic "cottage industries" of the new information revolution. Traditional knowledge about agriculture and health become information resources with microprocessors, which allow design of decentralized and flexible systems to be adapted to local conditions. To be effective, the new technology requires that each village be provided with information about agricultural practices such as intercropping, land use, soil fertility, disease, and pest-resistance.

Similarly, information about medicinal plants in every locality, acquired by herbalists and traditional healers, can become a basis for developing an effective, decentralized, computer-based health care system. With the introduction of the microprocessors it becomes apparent that it is twentieth-century technology that is primitive. The Third World village represents a sophisticated memory bank, awaiting an adequate information-retrieval system and the technology to use it. As the Third World becomes truly involved in the technological innovations of the twenty-first century, we shall witness the enhancement of the skills of the herbalists and the farmers as they become specialized hunters and gatherers of information, instead of unskilled wage laborers in the urban city centers.

The picture I am trying to portray is that the Third World countries need not feel so badly that they did not participate in the advances of the developed world during the nineteenth and twentieth centuries. In fact, they should turn their disappointment into an appointment to develop the true appropriate technologies of the Information Revolution.

As we have all witnessed over the past quarter of a century, inappropriate technologies sow the seeds of their own failure. For example, the movement of people off land that does not produce enough food into cities that do not provide enough work. The idea of a so-called appropriate technology has been tried as a way to escape this double bind. A lot of the so-called small-scale technologies merely represent an escape back to the nineteenth-century way of doing things. These will certainly not help the developing world to become self-reliant and self-sufficient (Ayensu 1978A).

The science and technology of the twenty-first century may be the true appropriate technologies for the developing societies because the microprocessor, for example, has the advantage of the windmill. It is cheap, simple, decentralized, and it needs almost no energy to operate. But it is a machine that generates the real winds of change.

There is no decree that says Third World countries should always follow the footsteps of the developed world. If the twentieth century has taught us any lessons at all, it is that Third World countries continue to be the recipients of inappropriate technologies that will ensure that they remain underdeveloped forever. The only salvation for the Third World is to escape from the hand-out syndrome of the colonial and neocolonial eras, and move quickly to join the emerging global information society, in which education and economic development are merged into new knowledge industries.

Current indications are that both the developed and the developing worlds are per force going to achieve computer literacy on a broad scale. This will enable all of us to make effective use of cheap, available microprocessors as teachers. There is, therefore, an immediate need for research on an appropriate curriculum designed for Third World cultures. These may have some universal aspects, such as fundamental training in pattern recognition, and an essential orientation toward learning how to learn. The software appropriate to each society must use the relationship between science and culture to make computer education relevant and applicable.

References

Ayensu, E. S. 1974. "Plant and Bat Interactions in West Africa." *Annals of the Missouri Botanical Garden* 61:702-27.
_____1975. "Science and Technology in Black Africa." Pages 306-17 in *World Encyclopedia of Black Peoples*. St. Clair Shores, Mich.: Scholarly Press.
_____1976. "Alternatives for Biological Resources in Africa." *Journal of Washington Academy of Sciences* 66:197-205.
_____1978. "The Farm Goes Industrial." Pages 160-67 in *The Smithsonian Book of Invention*. Washington, D.C.: Smithsonian Exposition Books.
_____1978A. "The Role of Science and Technology in the Economic Development of Ghana." Pages 288-340 in W. Beranek and G. Panis, eds., *Science and Economic Development: A Historical and Comparative Study*. New York: Praeger Publishers.
_____1981. "Biotechnology, Microcomputers, and the Third World." Lecture notes on technological innovations of the twenty-first century.
Ayensu, E. S., and P. Trachtman 1981. "Biotechnology and Microprocessors: The Real Appropriate Technologies for Third World Countries." Paper prepared for the United Nations University in Tokyo.
Brown, L. R. 1980. "Food or Fuel: New Competition for the World's Cropland." *Interciencia* 5:365-72.
Dalrymple, D. 1969. "Technological Change in Agriculture, Effects and Implications for the Developing Nations." *Foreign Agriculture Service*. Washington, D.C.: U.S. Department of Agriculture.
Government of India 1979. Report of the Working Group on Energy Policy. Planning Commission. New Delhi.
_____1981. National Paper for United Nations Conference on New and Renewable Sources of Energy. August 10-21, 1981. Nairobi.
Hoffpauir, R. 1977. "The Indian Milk Buffalo: A Paradox of High Performance and Low Reputation." *Asian Profile* 5:111-34.
Martin, F. W., and R. M. Ruberte 1975. *Edible Leaves of the Tropics*. Mayaguez, Puerto Rico.
McNamara, R. S. 1972. Address to the Board of Governors by the president of the World Bank. September 25, 1972. Washington, D.C.

National Research Council 1975. *Underexploited Tropical Plants with Promising Economic Value.* Washington, D.C.: National Academy of Sciences.
———1980. *Firewood Crops.* Washington, D.C.: National Academy of Sciences.
Page L., and B. Friend 1978. "The Changing United States Diet." *Bio-Science* 28:192-97.
Pimentel, D., et al. 1974. "Food Production and the Energy Crisis." Pages 41-47 in *Use, Conservation and Supply.* Washington, D.C.: American Association for the Advancement of Science.
Sauer, C. O. 1952. *Agricultural Origins and Dispersals.* New York: American Geographical Society.
Simoons, F. J. 1961. *Eat Not This Flesh: Food Avoidances in the Old World.* Madison: University of Wisconsin Press.
Solar Energy Intelligence Report 1979. *Alcohol Production Goals, Programs Explained to U. S. Congress by Brazilian Diplomat.* September 17, 1961. Washington, D. C.
United States Department of Agriculture 1976. "Food Consumption, Prices, Expenditures." *Agricultural Economic Report No. 138.* Washington, D.C.: USDA Economic Research Service.
United States Department of Health, Education and Welfare (Food and Agriculture) 1968. *Food Composition Table for Use in Africa.* Bethesda, Md.

Commentary by Napoleon A. Chagnon
Department of Anthropology, Northwestern University

It is quite clear to me why Dr. Ayensu's paper has the charm and readability of global ethnography, because he apparently very effectively assumed some of the osmotic pressures from Oxford University and did absorb a good deal of the anthropology. That may account for it even more. I was delighted with his allusion to fruit bats and the fact also that he discussed food taboos as one of the things that is significant in understanding human adaptations. I have an anecdote to tell that is as germane to his paper as it is to the comments made by Cavalli-Sforza about sex taboos. The anecdote comes from field research in the Solomon Islands, among the Sivai of Bougainville (Liver 1955).

And as the story goes, the fruit bat was the totem of one of the clans there. The belief was that Fruit Bat Clansmen preserved this totem and would die if they ate a fruit bat. Such beliefs are taken by many anthropologists to reflect some sort of "wisdom of culture" in preserving ecosystems. One day, an aged lady came along with a dead fruit bat in her hand. She decided to cook it and eat it. Oliver, the anthropologist, asked her, "Well, why are you doing that? Aren't you going to die?" She said, "Yes. But I am old and I am going to die soon anyway. I am hungry."

Now, the relationship of this anecdote to some things that have come up in the discussion is that sometimes there are good natural science explanations for taboos, and sometimes there are not. My preference, bias, and prejudice is to look for natural science explanations for taboos. Food taboos would be among them. But not all taboos are explicable in such terms.

I am reminded of a major survey of the food taboo patterns in Tropical Forest South America as some sort of cultural-ecological conservation system of certain agricultural tribes to preserve their ecosystem, an argument with which I do *not* agree.

The point is that I think there are certain almost irrational things that human beings do, or that not everything that humans do is "adaptive." This is germane to the arguments that Cavalli-Sforza is raising about sex taboos as a more effective way of explaining birthspacing in humans. The fact of the matter is that this could be one of those kinds of taboos that is more often believed in by scientists and scholars than some sort of effective and meaningful way to limit or space children.

It turns out that one of the kinds of data most difficult to collect in anthropological field work has to do with copulation, despite what the rest of the academic community thinks of anthropologists peeking through the leaves of the natives' huts, observing their sex habits. The kind of idea that Cavalli-Sforza is suggesting is in fact a testable proposition only if data on copulation are available.

Such a test will require precisely the kind of data that most anthropologists rarely ever get—in fact, I can think of no anthropologist who has meaningful, reliable, and statistically defensible data on the frequency with which the people they study copulate, and how often they violate postpartum taboos.

One additional comment on Dr. Ayensu's paper is that I hope eventually— perhaps not in this paper, but sometime—he will explore the implications of introducing nonindigenous foods in certain developing countries, particularly foods that are replacements for breast milk feeding and the impact that this has in some of these countries on promoting and stimulating rapid population growth and on the deterioration of health in young children.

Commentary by Stuart Marks
Department of Anthropology, St. Andrews Presbyterian College

I wish to comment on some of the subtleties of Dr. Ayensu's paper with some of my own studies of wild animals as food and as symbols in the context of developing societies in Central Africa. Dr. Ayensu's themes are familiar ones to those in anthropology who have shown the close connection between food and culture. One theme I wish to develop is that of identity or ethnicity and the relationship of food and culture to one's well-being. Whereas it is true that people everywhere eat and most of us use energy in the processing of our foods, what we consider palatable, what we eat, and with whom we consume it, are important cultural constraints. What I argue against is the often static, functionalistic, utilitarian mode in which we cast most of what we perceive others to be doing and also against the current industrial world view that defines human needs exclusively in material terms. For it is here that we confront the subtleties and complications of the explanations based upon cultural factors. Let me begin by defining the basis of cultural factors—or what I understand to be the basis of cultural factors—namely, world views or cosmologies.

A world view consists of a set of assumptions and premises which together constitute a particular vision of reality and social order. A world view defines social expectations and problems that need resolution, specifies acceptable methods of problem-solving, and provides a framework for the organization of societal, economic, and political activities. It is based upon a particular scientific and metaphysical tradition and incorporates a particular view of history, progress, and change. More specifically, a world view is a series of values and priorities concerning the goals of social, economic, and political institutions and of the individual and collective needs and responsibilities. It also incorporates other assumptions and attitudes about the role of technology, the scientific method, man's relationship to nature and to other human beings, scarcity and limits, wealth and abundance, boundaries of problems, and time dimensions.

Given this context, I use two examples of wild animals, one as an instance of the malleability of cultural constraints upon food and the other as a symbol of the political and economic dimensions of cultural change. My examples are drawn from formerly subsistence cultivators now in the process of transition in the Luangwa Valley of Zambia. Here the existence of the tse-tse fly precludes livestock. Instead, most of their animal protein is obtained through hunting wild animals.

The Bisa living there refuse to eat the flesh of hippopotamus and zebra. Both of these animals are abundant in their environment, as are the other wild animals which they do hunt and which they do consume. When asked, they provide rational answers as to why they do not consume these animals. These cultural reasons probably are *a posteriori* generalizations and are not what I wish to dwell on here. Both written and verbal accounts demonstrate that they did consume these animals in the past, that in the past they had guilds exclusively for the exploitation of hippos in particular, and that in times of famine they do consume these animals today.

I find the context of these restrictions of interest, just as it was for the Hebrews to prohibit the consumption of pork. With so many tenets of the traditional culture threatened and their autonomy severely eroded and undercut by more powerful social entities, the Bisa refuse to eat these animals because they are a symbol of their identity, their common plight, and their unity against pervasive outside authority and control.

It is not that they do not kill these animals inadvertently and otherwise. They do, and give these animals to representatives of social units to which they have become subordinate. By playing the more blessed role of givers, they maintain some semblance of autonomy and identity when, for the most part, they are on the receiving end of the stick.

It is with the political and economic dimensions of the stick of development and values that my final example is concerned. People in the Luangwa Valley distinguish between three kinds of lions, each of which are named and behave differently. Westerners or northerners see only one. That is the ordinary lion that tourists take pictures of, biologists study, and from which valley people occasionally steal a carcass.

Although northerners do not see these other two lions, they are nonetheless real to local inhabitants. Their discriminations depend upon the context of observable lion behavior and its consequences. Spiritual lions are the mediums through which the spirits of tradition monitor the social and political landscapes, reminding present generations of their past and its continuity. These spiritual lions are rooted in local traditions in reality. They are the lions of identity, of self-respect, and of autonomy. These lions visit villages at night, roar loudly as they move between human dwellings, but generally do people no harm. They are the only cure for the presence of the third lion, the imperial lion of the sorcerer.

The imperial lions appear in times of uncertainty and social crisis. They are the means by which some people seek to influence and to exploit others. They are controlled by the ideas and intents of others and are sent to kill, to maim, and to do harm. These imperial lions are frightful beasts, for they come unexpectedly and their presence disturbs the community until it finds a means to neutralize and dismiss them. These perceptual differences in lion types reiterate in allegorical form, if you do not mind the pun, my "lion" of discourse. Part of our human condition is a yearning for rigidity, for the hard line, and for clear concepts. When we have them, we either must face the fact that some realities elude us or else blind ourselves to their inadequacies. Our own industrial world view not only helps to define the problem, but also becomes part of the problem. We begin to ignore alternative world views. This industrial world view tends to define human needs exclusively in material terms, and to define development in terms of the process of industrialization, and production and consumption of these goods.

Underlying this belief in economic growth is a highly destructive and exploitive attitude toward nature, and a methodology of reductionism that explains the world in terms of isolated components.

The inadequacies of the industrialized world view force us to examine alternatives, some components of which are already emerging. This new view emphasizes the more holistic and integrated vision of the world, of good stewardship toward the earth, anticipatory and preventive (rather than reactive) curative policies and human values that transcend material well-being and that promote sustainable development for developed and developing countries alike.

Although the configuration of the landscape still remains in the realm of the problematic, it must be to the roar of the lion of identity, continuity, and participation rather than the roar of the imperial and ordinary lions to which we respond.

Discussion

DR. HASKINS: I wonder if there is some general discussion following this?

PARTICIPANT: There is one point, I think, that is quite interesting, and that is what one might call nonadaptative learning or mislearning, which I think plays a very considerable role in the modern world. An example might be the people in Africa trying to follow the example of Brazil, when all they have heard about a new idea is that they did such-and-such and they were successful, without any real information about the surrounding circumstances in Brazil or the difference between the two setups or even any information about how the idea has worked in a broader context in Brazil.

I do not know what in practical terms can be done about this, except that I think that people need to be far more cautious in public statements about what they are doing and about the effects of what they are doing, just because this is going to be rapidly circulating around the world and it is not going to have a merely local significance. So we need, I think, a new type of scientific communication that is simply a record of failures, not a record of successes, which you normally get. I cannot really imagine this as a practical thing, but I think it is becoming more and more essential as a mode of communication.

EDWARD S. AYENSU

10. Changing Aspirations for Mental and Physical Health

G. M. CARSTAIRS
Fellow, Woodrow Wilson International Center for Scholars

Editor's Summary. G. M. Carstairs offers some thoughts on the future needs for health care, particularly in developing countries where such care is not keeping pace with demographic changes. There is often an emphasis on high-technology—expensive medical procedures for a few patients when relatively inexpensive procedures would benefit many more people. He proposes greater use of paraprofessionals in health care delivery and a greater stress on preventive medicine through improved nutrition and environmental hygiene. Carstairs also emphasizes that mental illness is not limited to high pressure urban society; it is surprisingly frequent in rural India.

For several decades scientists like Julian Huxley and Theodosius Dobzhansky have been dinning into our heads the fact that today human adaptation depends on the transmission of learned behavior (including very recent learning) much more than on purely genetic change. Thirteen years ago at an earlier Smithsonian symposium on "The Fitness of Man's Environment" (Smithsonian Press 1968), René Dubos stated the limitations of genetic contributions very succinctly when he said: "The genetic endowment of *Homo sapiens* has changed only in minor details since the Stone Age, and there's little chance that it can be significantly, usefully or safely modified in the foreseeable future." This stubborn fact has been highlighted in other essays of the present symposium.

We must therefore rely upon human learning for the exercise of wise choice among the myriads of alternatives presented to us daily in our complex culture. Fortunately, the majority of such choices are relatively trivial, but there are others which may well decide our physical and mental well-being, if not our life expectancy in the years immediately ahead. Before I address the main topic of this paper I should like to pause for a moment to express my astonishment at the long, slow anatomical changes in our central nervous system which have brought learning from new experience—and the transmission of such discoveries—into the repertoire of our patterns of behavior. Some half-million years have passed since *Homo erectus* developed a brain capacity greater than that of all other primates, and at least 120 thousand years since Neanderthal Man could boast of a brain containing as many neural connections (or potential connections) as those of Shakespeare or Einstein.

What I find most extraordinary is that during so many millennia the human brain slowly acquired those surplus millions of unused neural pathways—the biological substrate for a completely new mode of evolutionary change. What prompted man's neural connections to proliferate so far beyond what seemed immediately necessary? And secondly, where was the survival value in this proliferation, which perpetuated the carrying of so much apparent excess baggage for so long? (Presumably, it made possible man's adaptability to new environments.) Even now we are still drawing dividends from this long-stored capital. We know that our brains are equipped to handle many more bits of information and to perform more complex operations than we have yet learned to demand of them [see note following text].

We also know only too well that one part of the human personality, namely our cognitive functioning, appears to have advanced disproportionately in comparison with our emotional, intuitive, aesthetic, moral, and (as some would add) our spiritual modes of experience. Since the beginning of the Industrial Revolution our brains have awakened from a prolonged intellectual torpor to participate—if only as onlookers and beneficiaries—in the increasing torrent of scientific discoveries, soon given practical application in industry. Until about forty years ago such was the exuberance with which these discoveries were welcomed that few people gave any attention to their possible harmful effects but more recently we have come to realize, for the first time, that our planet's natural resources are limited and may be threatened with final depletion. We have also become aware of the consequences of disturbing previously stable ecosystems and of the cost of relying on nuclear power, whose generation leaves us with an ever-accumulating burden of radioactive wastes with half-lives of up to 500 years. This has been—and still is—the era of the Sorcerer's Apprentice. However, I am digressing from my present brief, which

G. M. CARSTAIRS

is to discuss recent changes in human health and health care and to consider some probable future changes in both the rich and the poor countries of the world.

Let me begin by saying that, although I have a high regard for the World Health Organization (WHO)—surely, with all its imperfections, one of the most successful of the UN's international agencies—I think it erred grievously when it declared, in the introductory paragraph of its constitution: "Health is a state of complete physical, mental and social well-being and not merely the absence of disease or infirmity." If this were taken literally how many of us could ever claim to be healthy—and for how long? My late friend Dr. Henry Miller, for years known throughout Britain as the scourge of psychiatrists and advocates of community medicine, once commented on this definition: "My clinical observations lead me to believe that an abounding sensation of positive health usually presages either a cardiac infarction or incipient hypomania."

Health, whether mental or physical, has been defined in many ways. Every society recognizes that its members may sometimes suffer an interruption of their normal activities, accompanied by a varying degree of personal distress, which requires a succoring intervention, but societies differ widely in the range of conditions which they recognize as falling into the category of illness and those which are manifestations of gods, spirits, or black magic. Until recently, death was regarded as the unequivocal manifestation of the end of life—and indeed it is only in some centers of high medical technology that moribund patients are kept for long periods not living, but only partly living, and so confuse the issue. By and large, however, age-standardized death rates still provide a good measure of the differences between countries and between subgroups within countries. Infant mortality rates, in particular, serve as indicators of the level of health care. In London, at the beginning of this century, neonatal death rates were five times higher in the slums of the East End than in prosperous districts like Belgravia. Much greater differences have been reported from populous and poor developing countries as compared with Sweden, Norway, and Denmark, and, in quite recent years, Japan.

It has always been WHO's ambition to bring about major improvements in health care in the Third World. Accordingly, in order to mark the occasion of its twenty-fifth anniversary, WHO in 1973 commissioned a study of basic health services all over the world. This report was duly published (WHO 1973), but instead of providing material for self-congratulation, it revealed some disconcerting facts. Let me give a few quotations from its "Appreciation of the Present Position."

In many countries the health services are not keeping pace with the changing population either in quantity or in quality. It is likely that they are getting worse.

There appears to be widespread dissatisfaction of populations about their health services for various reasons. Such dissatisfaction occurs in the developed as well as in the Third World. The causes can be summarized as a failure to meet the expectation of the populations . . . and a wide gap (which is not closing) in health status between countries and between different groups within countries.

The report also drew attention to "a feeling of helplessness on the part of the consumer who feels (rightly or wrongly) that the health services and the personnel within them are progressing along an uncontrollable path of their own which may be satisfying to the health professions but which is not what is most wanted by the consumer."

These disquieting findings came at an appropriate juncture. WHO had just appointed a new director, Dr. Halfdan Mahler, who had spent eighteen years in India and Africa working on tuberculosis and tropical diseases. He was well aware of the key problem of health care in the developing countries, namely, that in most of these countries— particularly in Africa and Asia—some 80 percent of the people still live widely dispersed in rural areas, but 70 percent of the doctors prefer to work in the cities or better still to find employment in one of the more affluent countries in the West.

Hitherto, both national and WHO planning has been carried out for the most part by Public Health administrators, mostly doctors, who have tended to base their plans on hospitals of different sizes, situated in cities, and Primary Health Centers situated in villages. The personnel were assumed to be trained mental health professionals, including nurses, medical social workers, and pharmacists, working under the direction of doctors. In many countries, the health services are the legacy of former colonial regimes and reflect the thinking of the foreign rulers rather than the felt needs of the peoples themselves. In 1975 WHO published a book entitled *Health by the People* (Newell 1975), which described examples of innovative rural health services in eight developing countries. This helped to stimulate a worldwide reappraisal of the newly independent nations' plans for health care. In particular, it stressed the importance of employing modestly trained medical assistants, such as Assistant Nurse/Midwives and Village Health Workers (reminiscent of China's contemporary campaign to train literally millions of "barefoot doctors") to bring health education and simple health care to the rural population.

India, like many other countries, has been engaged since 1975 in a critical reappraisal of its national policy on health services and on the

G. M. CARSTAIRS

training of health workers. Its latest report, published in 1980, is entitled *Health for All: An Alternative Strategy* (V. Ramalingaswami, director of the Indian Council for Medical Research and J. P. Naik, former secretary of the Council for Social Science Research, eds.). It begins with a categorical statement that it is time that India ceased to base her provisions for health care on a Western model, with emphasis on large, expensively equipped hospitals offering the latest in laboratory investigations and in dramatic medical and surgical treatment for an array of diseases, such as cancer and disorders of the heart, lungs, kidney, or brain. These treatments are remarkable, but they are so costly and so time-consuming that they can be offered to only a fraction of those suffering from these conditions. In many cases they succeed only in prolonging life, in an impaired physical and mental state. In their teeming out-patient departments these hospitals also treat very large numbers of patients whose complaints could be equally well treated by general practitioners in village or neighborhood health centers—and at less cost. The report urges that much greater use should be made of assistant nurses and other trained paramedical staff to bring health education, preventive medicine, and basic health care to the villages.

In India, as in some but not all other developing countries, the national Medical Association has reacted with alarm to such proposals. Doctors fear that medical auxiliaries will soon claim to be doctors themselves and will offer a cheap, substandard type of medical practice. Venezuela was the first country to allay its doctors' alarm by agreeing to introduce its system of "simplified medicine," conducted by paramedicals, only in rural areas where there are no doctors. After some years the doctors decided that even in the cities there might be a place for such medical auxiliaries to attend to common and easily treated illnesses.

Most doctors are so preoccupied with the dramatic advances in therapy brought about by new chemical and antibiotic drugs that they tend to forget that nutrition, environmental hygiene, and the patient's will to recover still play important roles in health care. For example, when Thomas McKeown (1979) claimed that the remarkable reduction in deaths in Britain from pulmonary tuberculosis owed more to improvements in environmental hygiene and in nutrition than to the new potent antituberculosis drugs, many doctors found it hard to believe him. But his figures are incontestable: the standardized death rate for this disease was 2,901 per million in 1848, as compared with 13 per million in 1971, but three-quarters of this decline had already taken place by 1947, when the new antituberculosis drugs first became available. Dr. McKeown does not dispute the accelerated rate of cure of this disease after the introduction of streptomycin, iproniazide, para-amino salicylic acid, and their successors, but he does remind us that we have to look to other social and environmental factors to explain the steady fall in death rates during the previous 100 years.

The illustration also reminds us of a weakness of Western medicine, namely, its "repairman" approach, in which the doctor focuses his attention upon the symptoms and signs of disease in particular organs—in this case, the lungs. Physical examination and laboratory tests first suggest and then confirm the diagnosis, which then indicates the logical treatment. Very little attention is paid to the patient's life history or to his emotional response to the experience of illness and its attendant anxieties. By contrast, most forms of traditional healing are decidedly patient-centered rather than disease-oriented. There are several striking differences between going to see a doctor in the out-patient clinic of either a hospital or a rural Primary Health Center in a country like India and seeking the help of a practitioner of indigenous medicine, such as a shamanistic healer. In all these settings the experience can include long periods of waiting. Indian patients are very tolerant of such delays, which are usually associated with a heightened emotional tension as the moment approaches when the patient—and his accompanying relatives—are face to face with their doctor or traditional healer. If he is the latter, he usually does not have to ask many questions before divining the true, supernatural cause of the patient's illness and instructing him and his family about what must be done in order to rectify the disturbance, which is seen as affecting not only his physical state but also his relations with his community and with the spirit world.

If he is consulting a doctor, the patient is bewildered by having to answer what seem to him to be irrelevant questions about his bodily functions, and to submit to a physical examination whose purpose he cannot understand. He may have to undergo further investigations, such as x-ray, or electrocardiology, or the drawing of blood, which he regards as being bound to weaken him. At the end of the consultation he still probably cannot understand his doctor's diagnosis, even if—as is not always the case—efforts are made to explain it to him in simple terms. Finally, he is dismayed when the doctor fails to give him advice regarding what he should eat or should avoid in order to help his recovery—as every traditional healer would do. Here, doctor and patient have difficulty in communicating meaningfully with each other because each has been tutored in widely differing concepts of disease and therapy.

Cosmopolitan medicine is now in the ascendance both in the

G. M. CARSTAIRS

developing countries and in the West. It receives the lion's share of government support, but does not go unchallenged. In India there are two classical schools of Eastern medicine, *Ayurveda* (meaning the art of living to a ripe age) and *Unani*. *Ayurveda* is followed by Hindu practitioners or *Vaids*. Its teachings were evolved during the centuries just before and after the life of Christ and marked the separation of medical practice from its early preoccupation with healing charms and incantations to counter the influence of evil spirits. *Unani*, practiced by Moslem *Hakims*, was handed down by Persian and Arabic scholars, who had in turn inherited the teachings of Hippocrates. There are also many Indian practitioners of homeopathy in spite of its being a latecomer to the subcontinent.

The Indian census of 1971 reported that there were (in round numbers) 108,000 practitoners of allopathic medicine, 49,000 Vaids, 30,000 homeopaths, 5,000 Hakims, and 35,000 "others." If the last category was meant to include the part-time shamans and others who enlist supernatural powers in the cause of healing, then it is a gross underestimate because these are to be found in a majority of India's 600,000 villages.

Indian tradition idealizes the "true" Vaids and Hakims as learned, holy men, who display a sense of obligation toward each individual patient. Customarily, they did not ask for fees but were rewarded— sometimes very generously—by grateful patients. Indian informants, especially in the villages, often insist that successful healing will only occur when a patient has unquestioning faith in his physician. Today, that degree of confidence is not easily achieved because not only do these traditional doctors expect fees, but they also have largely given up the preparation of the age-old remedies on which their forefathers relied. Research into several thousand prescriptions made out by these practitioners in North and South India has shown that, in both locations, some 80 percent included at least one modern Western drug, one to be obtained from a Medicine Shop (Taylor 1976; Neuman et al. 1971).

If reliance on traditional healers is faltering, it cannot be said that cosmopolitan medicine has altogether succeeded in taking their place. People know that there are new, potent medicines to be had from the pharmacies which have appeared during the last two decades in every town and even in large villages, but they do not know whom to trust to prescribe the correct one.

The Indian achievement in creating a network of Primary Health Centers and subcenters to serve the rural areas is a very considerable one, even though it still falls short of its aims—not least in the centers' being underprovided with even commonly needed medications. If we take the village pharmacies into account, never before has so much effective medicine and informed advice been accessible to so many of the villagers. What is still lacking is confidence and trust.

Earlier this year I spent three weeks in a very spartan little village of farmers, in a semidesert district of Rajasthan, which I have been visiting off and on for thirty-one years. My hosts took me round their fields, showing with pride how their crops had produced a fourfold increase since they had begun to use the new varieties of grain. They had first heard about these new grains through the grapevine of village gossip, and had been given more specific information by traveling agents of the Minstry of Agriculture. Finally, they had begun to listen to early-morning radio broadcasts for farmers, which gave practical advice from week to week about the tasks which had to be carried out if the new grains were to yield their maximum harvest.

I could not help contrasting this great leap forward in the productivity of their fields with another, less happy observation, in 1950, when I had noted that one child out of four born in the previous ten years had died before reaching the age of five. Although a government health clinic has been in existence in a larger village (between six and seven miles away) for a hundred years, the villagers seldom visited it on the grounds that they could not trust its doctors—strangers from far-off cities who did not conceal their disdain for village life and village people. When I repeated this count in 1981, I found the infant death rate to be no better but even fractionally worse. Nor had there been any change in the villagers' reluctance to attend the health clinic. There was no doubt that they cherished their children, especially if they were boys, and I knew that advice on infant health care was included in the regular rural education programs on radio.

Why had one message reached them, while the other had failed to do so? A number of possible reasons come to mind: (1) infant care is women's responsibility; (2) unlike their husbands, most village women only speak the local dialect and cannot understand Hindi, which is used on the radio; (3) in this area, no attempt has yet been made to bring health education to village women in their own language.

For whatever reason or reasons I could not fail to observe that the hygienic condition of the village houses was still at a low level, with an abundance of flies and vermin, and with cattle and goats housed in intimate proximity to the villagers' living space. Water carried from the well was of doubtful purity. Bodily cleanliness was often neglected—by the women in particular, whose observance of *purdah* gave them few opportunities to have a bath. Nevertheless, during my 1981 visit I was interested in getting to know Laxman Singh, a young man in his mid-twenties home on leave from service in the navy. His wife had had a stillborn baby a year earlier, and was now expecting another child. Singh was determined to do all he could to preserve this baby's life. He insisted that his wife come to his village for delivery, although custom usually dictates a young wife's going to her mother's village at this time. He had taken her to the health center for prenatal care and was prepared to take her there should anything go wrong, gladly accepting my offer to drive them in my car should the need arise. When he called me to see the newborn baby, I was impressed to see that he had brought a mosquito net with him from Bombay and had rigged it up over the mother and child to protect them from flies and mosquitoes. This was the first time I had seen a mosquito net in the village.

Laxman Singh was an innovator, both in recognizing where modern medicine could help to protect his wife and child and in having the courage to go against the traditional belief that this was not a man's business. He was rewarded, I am happy to say, by the birth of an exceptionally pretty baby—whom he cherished almost as much as if she had been a boy! Perhaps his example will encourage his fellow villagers to learn how to protect the health of their families, as they have so successfully learned how to improve their crops.

If, in India and other developing countries, modern cosmopolitan medicine has not yet won the confidence of the mass of the people, in the Western world doctors appear to have lost a great deal of their former high esteem. This is partly due to their high earnings but perhaps even more to a change of attitude among their patients. Today people want to be informed about the nature of their illness and about the mode of action and possible side effects of any treatment they are given.

Ivan Illich (1975, 1976) began his celebrated polemic *Medical Nemesis* with the forthright statement: "The medical establishment has become a major threat to health." He went on to describe how the discovery, all in a rush, of more new, potent medicines than had been known throughout man's previous history had swept doctors into a state of *hubris,* or overweening pride. They believed, and so instructed their patients, that soon there would be a medical remedy for every disease, and in the meantime they overlooked the fact that powerful medicines frequently also have powerful side effects. Illich drew attention to the fact that in teaching hospitals, the elite of their kind, no less than 20 percent of patients required treatment to relieve the consequences of modern drug therapy. "Despite good intentions and claims to public service, a military officer with a similar record of performance would be relieved of his command, and a restaurant or amusement center would be closed by the police" (Illich 1976:32).

Above all, Illich deplored the "expropriation of Health," in which doctors act like superior technicians and take over from their patients the responsibility of curing their illness instead of letting the patients be informed participants in their own treatment.

There has been a widespread endorsement, throughout the developed countries, of the need to "de-mystify medicine" so that its scientific knowledge about illness and treatment—and the limitations of such knowledge—can be shared by all those who are involved in medical care, including patients and their relatives.

In the United States, perhaps more than in any other country, the public takes a lively interest in the latest discoveries in medicine, which are reported in the daily press and also in widely read periodicals. Nor is this a purely academic interest. Responding to research findings about the roles of lack of exercise, overeating, and cigarette smoking as causes of disease of the heart and lungs, many people—young and old—engage in daily jogging and even more have reduced their intake of sugar and of saturated fats. A decline in smoking can be seen among the young, but their elders seem to find it difficult to break this long-established habit.

So far, in this paper, I have made no separate mention of mental illness in relation to man's adapting to his changing environment. If it has proved difficult to reach a widely acceptable definition of health in general, this is still more the case with mental health—which so easily becomes a statement of how people ought to behave. Deviance is a departure from social norms, which differ from one culture to another. Although every culture has terms (usually derogatory) denoting outright craziness, they differ as to which forms of behavior should be so regarded, and their differences are still greater in respect to classifying less dramatic departures from normal behavior.

Even Freud, who compelled us to recognize the sometimes excessive influence of irrational (largely unconscious) urges upon our behavior, implied that this was a remediable defect in his summary of the aim of analytic psychotherapy as "Where Id was, let Ego be." At first sight, this reads like an expression of Freud's preference for rationality over impulse, feeling, or wishful thinking, but then one recalls that he depicted the Ego as the intermediary between conflicting urges, reconciling feelings with reason rather than allowing either to exclude the other. This is implicit in his later aphorism, defining the task of analysis as being to enable the analysand "to work and to love"—in other words, to be able to master practical tasks and also to give expression to his feelings. The Greeks went further than this in aiming for a balanced development of the physical *soma* as well as of the cognitive *nous* and the affective *psyche*. This balance is, of course, harder to maintain in situations of uncertainty and threat.

Many people believe that psychiatric illness must be expected to increase in times of rapid social change such as we are experiencing today. This is a time-honored belief. In France, following the social upheavals of the Revolution and the Napoleonic Wars, references was frequently made to the inevitable consequent increase of mental illness. But when Esquirol (1838) studied the statistics of admissions to psychiatric care between 1790 and 1830, he found no evidence of such increase. In London also, as the Industrial Revolution accelerated its development, the changes which came with it were thought to have caused a marked increase in insanity. Indeed, there were even letters in the *Times* (the classic way of expressing public concern) explaining that mental disturbances were the result of damage to the brain which must be caused by the new custom of traveling by train at twenty, thirty or sometimes even forty miles an hour! This in turn was studied by Henry Maudsley (1872), whose paper in the British Medical Journal found no evidence of an increased incidence of insanity.

In retrospect, it seems likely that Esquirol and Maudsley were mistaken in using the numbers of certified psychotics as the criterion for mental stability in their respective societies. They were, however, among the first specialists in psychiatry, the successors of madhouse keepers, and their clientele did not yet embrace the wide range of minor mental disorders or antisocial personalities. In their time, many of such patients would consult general practitioners or specialists in internal medicine and would, in most cases, dwell upon their physical rather than their psychological symptoms. Many others would turn to clergymen or to wise counselors among their family or friends for advice about their problems.

It has been vigorously contended both by social scientists and by renegade psychiatrists, notably Thomas Szasz (1961) and R. D. Laing (1967), that these are not illnesses but reactions to life experiences which call for guidance, not medical attention. (In Shakespeare's *Macbeth,* a witness to Lady Macbeth's tortured sleepwalking endorses this view: "More needs she the divine than the physician".) For some thirty-five years after World War II, however, the popular acceptance of Freudian psychoanalysis as holding the key to the understanding of human behavior led many people—nowhere more so than in the United States—to consult their analysts in preference to any other advisers. Today, this ascendance has declined. There are many contenders for the role of psychotherapist. In part, this can be attributed to the law of supply and demand. We appear to be living in an era of great uncertainty, in which many people feel that not only their own inner stability but also that of their entire culture is gravely threatened. I do not believe that psychiatrists, psychoanalysts, or psychologists can claim exclusive competence in helping such people, but I certainly do believe that we have a legitimate interest in studying the phenomenon and sharing such insights as we may acquire with our patients and fellow inquirers. There is also a heightened awareness of the social factors which contribute to mental illness. There appears to be some recognition that the disintegration of the socializing forces of the nuclear family plays havoc with personality development. This recognition may be the first step in the self-correcting social mechanisms of society.

Mankind has lived through such epochs before, many times over. They have been characterized by many epidemics of unreason, as if people received solace from their anxieties by joining a group that claimed to have found a supernatural means of rescue from their predicament. These temporary abdications of common sense have been well documented; for example, by Charles Mackay, whose late-nineteenth-century classic, *Extraordinary Popular Delusions and the Madness of Crowds,* has recently been reprinted (1980), perhaps because of its relevance to the present times. Mackay tends to ridicule the bizarre, usually short-lived crazes which he describes, although some like the South Sea Bubble or the Tulip Mania brought financial ruin to their followers.

Norman Cohn (1957) in *Pursuit of the Millennium,* on the other hand, not only describes the rise and fall of a long succession of millennial cults, each with its charismatic leader, but also shows elements which they have in common—other than the expectation of 1,000 years of trouble-free existence for the elect. For example, he notes that the leaders were often men from the periphery—if not from another country—as if the element of strangeness made it easier for their followers to believe their unrealistic prophecies. In normal times the teachings of these *prophetae* would not command a large following, but they did do so in times of crisis which occurred, for example, when the old feudal relationship between landlord and serf began to dissolve, leaving the serf free to work for a wage, but also denying him the security of the landlord's support when the harvest failed, or after a succession of famine years, or a visitation of the plague. At such times, large masses of poor peasants found themselves homeless and penniless, with a total lack of security in the future. These formed an audience eager for a magical resolution of their unhappy state. Cohn also shows us the typical life history of such cults, which begin with assertions of the brotherhood of man and the sharing of all worldly

G. M. CARSTAIRS

goods, in which the prophet sets an example. Later, the prophet becomes a divinely inspired (and therefore unchallengeable) leader and lives in regal splendor. When he meets opposition from existing authorities, this is seen as the work of the Anti-Christ, and he now preaches the inevitability of an Armageddon before the elect can enjoy their just reward, and the initially pacific movement usually ends in bloodshed.

Cohn turns to psychoanalysis, and particulary to Melanie Klein's theories about the reemergence under stress of nightmarish fantasies of early childhood, to explain these occurrences. Although this explanation is debatable, the sequence of events is clearly recognizable in such contemporary phenomena as the Ayatollah Khomeini's increasingly bloodthirsty regime in Iran and the gruesome end of the Reverend Jim Jones's People's Temple commune in Guyana on November 18, 1979, when, after the murder of visiting Congressman Leo Ryan, nearly every one of Jones's 900 followers obeyed their leader's command to join him in a mass suicide.

In his book *Journey to Nowhere: A New World Tragedy*, Shiva Naipaul (1981) describes this cult's progress from initial universal benevolence based on collective sharing of worldly goods to its brutal domination by an increasingly paranoid leader. Its tragic career has much in common with those of earlier millennial movements, but Naipaul also shows its affinity with the large numbers of utopian cults which have had their brief heyday in postwar America, especially in California. Mercifully, few have behaved with such disregard for life as the followers of Charles Manson, the Symbionese Army, or the People's Temple. What is remarkable, however, is the apparently unceasing supply of self-surrendering followers attracted by each new guru or prophet of an instant utopia.

This subculture, is of course, not typical of America as a whole. America is currently experiencing a swing toward political and religious conservatism (although conservatism also has its fantasies, such as its unquestioning belief in America's declared role to preserve world peace by maintaining a nuclear overkill even greater than that of Soviet Russia). At this time, however, both Britain and the United States have many alienated citizens, who have suffered during the prolonged depression of the 1970s and who see no prospect of better times ahead. This is also true of the majority of the industrialized societies and of many in the Third World—nor, as we have recently seen in Poland, have even Communist countries been exempt.

Marxists would welcome most of these instances of angry demoralization as the prelude to proletarian revolutions, in accordance with economic determinism. But revolutions are seldom altogether logical, and even more rarely do they remain rational for long. In his recent book, *Fire in the Minds of Men: Origins of the Revolutionary Faith*, James Billington (1980) has shown us the generous intentions of the French revolutionaries who hoped to establish a republic in which *Liberte, Egalite, Fraternite* would truly prevail, and their attempt to perpetuate the joyful excitement of the storming of the Bastille in a succession of spectacular pageants held on the Champs de Mars, culminating on June 8, 1794, with the "Feast of the Supreme Being" (inspired by Robespierre's "Universal Religion of Nature") attended by half a million citizens. But this was too good to last: within a few weeks Robespierre was guillotined, and paranoid suspicions led to the Terror, in which executions of those branded as counterrevolutionaries became almost as numerous as in present-day Teheran.

Billington vividly recreates the sustained nervous excitement of the masses in Paris as their revolution overthrew the king and the ruling aristocracy and challenged the authority of the church. The revolutionaries were utopian in hoping to establish their new society, but unlike their predecessors in the millennial movements they did not rely upon divine intervention to bring about this new world. That was to be their own laborious task.

There must have been great exhilaration, as well as spells of anxiety, in those days of defiance of previous social and religious restraints. One is reminded of the gaiety which attended the student uprising in Paris in 1968, whose demands for reforms in education—and in society in general—were accompanied by an outpouring of rhetoric, by a profusion of witty *graffiti*, and by numerous spontaneous "happenings" which were more theatrical than revolutionary.

In the years immediately following 1798, there were evidences of a sense of incompleteness, a search for an alternative authority or, as Billington puts it, for a "locus of legitimacy" to replace those which had been overthrown. This took some strange forms, such as the attempt to find the new source of authority in Philip D'Orlean's "Club of the Social Contract" or in Nicholas Bonneville's secret society, the "Social Circle," and its public forum, the short-lived "Universal Confederation of the Friends of Truth," both founded in 1790. As the revolution advanced, renewed attempts were made to establish new communal rituals and new symbols to replace those of the Church and the Court. Among the symbols were the monumental Statue of Nature in the Place de la Bastille—and the Guillotine. And among new rituals was a nature-worshipping parody of the Eucharist in which the celebrant addressed a mass assembly: "Friends, this is the body of the Sun which ripens the harvest. This is the body of the bread which the rich owe to the poor. . . ."

But it proved easier to inaugurate new instant traditions than to perpetuate them. Although thousands attended the first celebration, they were soon forgotten and would remain so today but for the diligence of historians. Soon the French Revolution had established itself, together with the American Revolution, as landmarks in human history. They had demonstrated that seemingly impregnable governments *could* be overthrown when they had forfeited the loyalty of their citizens and when a sufficient number of these citizens were ready to risk their lives in order to create a better social order.

Throughout the nineteenth century the revolutionary ideals were kept alive by a large number of societies, both idealistic and fanciful (such as those which believed that the new society should be based on Pythagorean forms: the circle, square, triangle, and Pythagorean prime numbers, especially the numbers five, seven, and seventeen). New societies kept appearing, as their predecessors either lapsed or were suppressed. They included the highly disciplined Order of Illuminists, the Friends of the People, the Society of the Families, the Society of the Seasons, the League of the Outlaws, and many others. As the century advanced and one uprising after another was rudely repressed, the outlawed revolutionaries became increasingly violent. A climax of angry despair was acted out by the Russian nihilists of the 1870s, who resorted to assassination by bombs, sometimes accompanied by self-immolation, as their last resource.

At last, after many false starts and setbacks, the Russian and the Chinese revolutions confirmed that radical changes could be brought about and sustained. Their demonstrations of successful seizure of power have established a new element in man's behavioral repertoire, which has already been successfully reenacted in a number of countries.

Revolutionaries maintain that theirs is the only way to achieve significant social progress, but in this they are surely wrong. Revolutions have been accomplished, but never without a great deal of bloodshed, and, once established, these regimes have shown a striking indifference to individuals' human rights and an intolerance of dissident opinions comparable to that of fascist dictatorships. In other countries, socialist, social-democratic, liberal, and even conservative governments have demonstrated their ability to change their societies for the better without resorting to bloodshed or significantly forfeiting their citizens' civil rights.

The difference between reformers and revolutionaries can be seen as a difference in the degree of urgency with which they view the need to better the lot of the underprivileged. In earlier times, it was possible for large masses of mankind to be underprivileged—and to be unaware of it. An important feature of the new global technology has been the quickening of this awareness. Some thirty years ago President Sukarno of Indonesia said that the movies had opened a window upon the world, through which people in poor countries could look and see how people in rich countries live. "A refrigerator," he said, "can be a revolutionary symbol in a country which has no refrigerators." He was speaking before television had reached his country. Movies had accelerated people's discovery of their poverty. Once they become conscious of being underprivileged, the poor in every culture need little persuasion about the urgency of bettering their lot—and if nothing happens, they become tempted to relieve their pent-up anger through outbursts of violence and destruction.

This too can be learned behavior. In Britain, in early July 1981, a riot broke out in a particularly run-down section of the city of Liverpool. It was fully covered by TV, and at once a series of riots, with looting of shops, broke out in different industrial towns in which there are high rates of unemployment, only to peter out as suddenly as they had begun, although the causes of discontent were still there, unchanged. We deceive ourselves if we fail to realize how strongly people whose eyes have been opened to their deprivation resent it. This applies to people in the Third World no less than to the unemployed in the so-called advanced societies.

I have referred, earlier in this paper, to the extreme difficulty—if not impossibility—of reaching an objective definition of mental health, uncontaminated by culturally determined value judgments. This is equally true of reaching a diagnosis, because psychiatric diagnoses are largely based on what the patient experiences and how he communicates these experiences—here again cultural expectations may well color what he feels and what he tells. We do, however, reach firmer ground when we focus only upon symptoms, because these are recognizable even across the barrier of culture.

A few years ago my colleague, Professor R. L. Kapur of Bangalore, and I carried out a survey of psychiatric symptoms among villagers in a district in South India. We found that 6 percent of our adult respondents reported having four or more such symptoms which were sufficiently troubling to interfere with their normal activities and to have caused them to seek help—usually from a traditional healer. This was an eye-opener, because educated Indians nearly always maintain that village life is peaceful and calm, compared to the stresses of the city.

Among those reporting symptoms there were some with serious disorders, such as epilepsy, schizophrenia, or severe depression, which can be dramatically relieved by modern medication. Already this has been demonstrated by psychiatrists working in Primary Health Centers in collaboration with doctors and paramedical staff, who can quite quickly be trained in the recognition and treatment of these patients. What is equally important, however, is that the patients themselves (and their relatives) have learned that their distressing symptoms can be relieved, allowing them to resume their normal lives. This motivates them to seek out and to persevere with the new drugs.

But these are only a small minority of the patients who reported considerable psychological distress; the majority were passing through a life crisis which found expression in psychosomatic or neurotic complaints. It is only realistic to recognize that anxiety and stress, psychosomatic pains and periods of unhappiness are all part of the human condition—together with occasions for rejoicing and the relief which comes when one has coped successfully with personal difficulties. The contribution of traditional healers, whose treatment involves both patients and their families in a form of group therapy, should not be undervalued. This too is part of each society's biocultural heritage. The very symptoms of neurosis can often be seen not as illness but as what Freud (1930) called the price we pay for civilization, or in Charles Kingsley's (1874) term as "divine discontent"—the spur which keeps us striving to attain a more satisfying life for ourselves, our families, our fellow countrymen, and perhaps, one day, for all our fellow men.

I have dwelt a great deal in this essay on the extremes of discontent and their expression in millennial movements, utopian cults, or revolutionary violence—or in simple unplanned outbursts of violence and anger. I have done so because it seems to me that what we—meaning mankind in general—urgently need to do next is to cultivate a global fellow-feeling which transcends both national and racial boundaries, as our global material culture has already begun to do. Hitherto, our sense of altruism has not been strong enough to motivate this change. Perhaps our sense of self-preservation as a species will succeed where altruism has failed.

NOTE BY DR. CARSTAIRS:
During the symposium I was able to discuss this point with Professor James V. Neel, who suggested that my astonishment over Stone Age man's "surplus millions of unused neural pathways" may have been unwarranted. Later, he was kind enough to elaborate his point of view in a letter: "Let me say that there are no truly rigorous data to refute the suggestion that primitive man, in the jungle, was blessed with a great excess of cerebral power, waiting to be drawn upon as we moved into more complicated times. However, I would be inclined to reject that hypothesis, on two grounds. Firstly, I know of no parallel where

nature has endowed us with a capacity far beyond our needs. This runs counter to everything we now know about the manner in which evolution proceeds. Secondly, and more important, from our own work amongst a number of primitive groups, I have come to feel that so-called primitive man is using his brain power almost as much as we are. I maintain that the calculations that go into sustaining life under primitive conditions are every bit as complicated as those involved in our current culture, the more so since few of us are truly innovative, but are all standing on the shoulders of millions of people who in the past have each made some small contribution to our present complexity. We of the Western world tend to confuse that material complexity with evidences of a sudden burst of unused brain power. I view it as evidence of a human ingenuity seen everywhere, in this case an ingenuity which seems greatly to have outrun our foresight."

I am grateful to Professor Neel for his clarification, and not least for his implication that our neuronal pathways can be employed in many diverse ways including—in rare instances—some which are truly innovative.

References

Billington, J. 1980. *Fire in the Minds of Men*. New York: Basic Books.
Carstairs, G. M., and R. L. Kapur 1976. *The Great Universe of Kota*. Berkeley: California University Press.
Cohn, N. 1957. *Pursuit of the Millennium*. London and New York: Oxford University Press.
Dubos, R. 1968. In *The Fitness of Man's Environment*. Washington, D.C.: Smithsonian Institution Press
Esquirol, J. E. D. 1938. *Des Maladies Mentales*. Paris.
Freud, S. 1930. *Civilisation and Its Discontent*. London: Hogarth Press.
Illich, I. 1975. *Medical Nemesis*. London and New York: Calder and Boyars.
———1976. *Limits of Medicine* (revised edition of Medical Nemesis). London and New York: Calder and Boyars.
Kingsley, C. 1874. *Health and Education*. London: W. Isbister and Co.
Laing, R. D. 1967. *The Politics of Experience and the Bird of Paradise*. Harmondsworth: Penguin Books.
Mackay, C. 1980. *Extraordinary Popular Delusions and the Madness of Crowds*. New York: Harmony Books.
McKeown, T. 1979. *The Role of Medicine*. Princeton, N.J.: Princeton University Press.
Maudsley, H. 1972. "Is Insanity on the Increase?" *British Medical Journal*, No. 576, vol. 1:36–39.
Naipaul, S. 1981. *Journey to Nowhere: A New World Tragedy*. New York: Simon and Schuster.

G. M. CARSTAIRS

Newell, K. W. ed. 1975. *Health by the People*. Geneva: World Health Organization.

Neuman, R. C., J. C. Bhatia, S. Andrews, and A. K. S. Murphy 1971. "Role of the Indigenous Medical Practitioner in India." in *Social Science & Medicine* 5:137-49.

Ramalingaswami, V., and J. P. Naik, eds. 1980. *Health for All: An Alternative Strategy*. New Delhi: Indian Council for Medical Research.

Szasz, T. 1961. *The Myth of Mental Illness*. New York: Harper and Row.

Taylor, C. 1976. "The Place of Indigenous Medical Practitioners in the Modernisation of Health Services." Pages 285-99 in C. Leslie, ed., *Asian Medical Systems*. Berkeley: California University Press.

World Health Organization 1973. *Organizational Study on Methods of Promoting the Development of Basic Health Services*. Geneva: Office Records of WHO, No. 206.

Commentary by Ido de Groot
Faculty of Medicine, University of Cincinnati

The point prevalence of 5 to 7 percent of daily incompetence or psychological disability is a remarkably recurring rate that seems to be just about the rate at which a society can still operate. Consider the presence of people like Mr. Singh. If he had pulled his trick of rigging up a mosquito net in an American village, he would have been automatically banned and put into an asylum or at least a halfway house for semi-acculturation. As Dr. Carstairs points out, Singh is a highly educated man. He was in the Navy and saw the world, or at least a piece of the Indian Ocean. Education is important in the health movement, because for whatever reasons he tried his innovation in the first place—I hope not in order to ward off malaria, filaria, or infant diarrhea—he bought a mosquito net and carted it into the village because it gave him more comfort and pleasure in life. Not being bitten in the butt ought to be a high goal of public health, I would say, regardless of the outcome in terms of specific disease prevention.

There are other lessons in Dr. Carstairs's presentation. When one breaks down the 6 percent prevalence of disorder and discomfort, it appears that the educated, the Brahmins, are relatively better protected against social change—be that good or bad social change we will leave in the middle—than the lower classes. This is a recurring finding also within the United States and certainly within England, where they have a better fix on class than we do. As it turns out, income in this country and in England doesn't differentiate the classes anymore in terms of disease outcome; education does.

To that extent, I think an epidemiologist like Dr. Carstairs, and certainly not myself, has something to contribute to the concept of adaptation. That is, those most educated to the most global view seem to have the greatest resistance in terms of showing disease when social change occurs. They can absorb the change and use it. One of the interesting things about infant mortality, to which Dr. Carstairs also alluded, is that it has gone from approximately 150 per 1,000 live births at the turn of the century in England and here, to approximately 9 or 10 across the population. However, the ratio difference between the upper class and the lower class has not changed materially; that is, the difference between class one and class five.

The differences between haves and have-nots, then, have not changed. That is, we have not broken through the cultural differentiation between those who are capable of acculturating to cultural change or adaptation and those who are not. I would hope that these disease rates will become more useful to the programmatic planners, the city designers, and the like, even economists, in evaluating in advance what we can expect from social change in terms of biological and mental adaptation. The combination of mental and physical adaptation is very manageable. There is no great disaster coming down the pike in this village in India. Although our general feeling is that major social changes will create disasters, they don't.

Commentary by Audrey B. Davis
Department of the History of Science and Technology, Smithsonian Institution

I hope Dr. Carstairs will pardon my looking at his essay as a historian, which probably won't be satisfactory in terms of immediate problems and solutions. I could begin by saying that if you read through the history of medicine going back to antiquity or even not so far back, all of the things that were just reported, as sad as some of them have been, certainly seem like an improvement over the medicine of the previous centuries.

In the first half of his essay Dr. Carstairs spoke of both the abatement and the prevalence, of physical diseases and what he has done is emphasize the limited role of the physician in dealing with these diseases. I think the examples he gave are quite well taken.

There are many ways of looking at medical achievements and, if I could just put them in a little more perspective, even confining my thoughts to the medicine of the eighteenth, nineteenth, and twentieth centuries, we can see the tremendous changes that were made just in those periods, really a fairly short period of time. You would also have to agree that medicine itself is constantly evolving, whether it is in a social or a cultural setting or whether it is a technology in a Western setting, and that there are many forms of healing and medical practices with different objectives constantly being met.

I think you could divide medicine into two forms. The first is the physician or the person who delivers the practice, who has the special skill; the second is medicine as an institution or medicine in the sense of, perhaps, educating people to deal with their own illnesses in many instances, consulting a physician when the time is appropriate.

Therefore, you might say there are two major roles for the physician which I think would apply here: the physician as an environmentalist, in a sense; someone who is advising on ways to improve the community, the people, the way they live, including how to have a child born healthy and remain so. This approach, of course, goes back to Hippocratic medicine. It is not a new idea; physicians have always had this role, regardless of how well they have carried it out, more successfully in some societies than others. I think you could even say that the medicine man and people of that type have carried out this role imaginatively and, therefore, quite successfully.

I was in England just last week and noticed a no-smoking sign that said—the clincher line was—"Most physicians don't smoke." I mean, it is interesting that it was put in these terms, implying the physician is a model for behavior.

The second role for the physician is as a specialist with particular skills. And of course in our Western culture we think of specialists as being very technologically concerned. We think of the physician as the technician. And while this doesn't seem to have much bearing on large groups of people, such as in India, because the types of materials are not available—they don't have the resources, the money, and so forth—I think we certainly should not ignore the fact that technological medicine is still very important, and it is going to be there.

As far as I can see, technology in medicine will continue to grow and evolve. And in fact I would even suggest that when people in other countries, perhaps even in India and China and so forth, do become more aware of medical technology, certain aspects of it, they are probably going to want more of it, too.

The expanding awareness of the body and its functions by larger numbers of people has had an effect on medical research, especially the types of questions posed and the collection of data. At a time when Western medical practices are being questioned as to their effectiveness, and high rates of iatrocentric diseases are reported, it is also being urged that increased scientific evaluation of these medical measures be made before they are applied routinely to patients. Yet public consciousness concerning human experimentation and testing continues to rise and limit the process of medical experimentation. While useful medical procedures are continually being desired by all, the basis on which these are most likely to be provided, i.e., clinical experimentation, is being removed. Medical experimentation is currently being regulated and subjected to more stringent rules, which may limit its effectiveness and potential for discovery. Some of this experimentation is limited, not only by the obvious dangers to the patient-subject, but by the biological and medical astuteness of the patient, whose own knowledge and subsequent behavior enters into the data collected. For instance, in the attempt to measure or record the amount of pain experienced during an illness, or pain perhaps artificially induced, biases sometimes enter into the results. The belief that women and individuals from certain regions are not able to tolerate pain as well has been shown to color the expectations of physicians treating these people.

I have a little more optimistic outlook as to the way technology might be of help to many people, even though it seems so costly and out of reach today. I think we should remember that in this country, when a new device or a method is introduced, it is first used on a few individuals and then, usually within a decade or so, hundreds of thousands of people. How many people are wearing pacemakers today? How many are using kidney dialysis machines?

I can remember, certainly within the decade, when communities set up committees to choose the individuals who would get the use of a kidney dialysis machine, because there were only one or two available to the whole community. Today we have portable units that can be brought into the home and used by the patient directly, perhaps with just a little assistance.

So I think the technological aspect of medicine should not be dismissed too quickly. I think a lot more is going to happen in medical technology, and it will go in directions that will make it available to far more people than we think possible at this moment.

G. M. CARSTAIRS

Aspirations are usually linked to past and present accomplishments. In medicine the achievements of medical technology will continue to enhance the expectations of Western peoples, especially Americans, that more cures for serious and fatal diseases will be found and that more people will be able to live a decade or two longer because of a pacemaker, a daily dose of valium, digitalis, and numerous other drugs, as well as injections to ward off the flu and other viral and bacterial diseases. Others expect to be rescued from a health crisis brought on by an accident or even willful neglect of diet and sound living habits.

Millions of people in India, China, and Africa may not be aware of these effective medical practices and procedures, but when they are (and world communication links will ensure this), they too will demand these treatments, and rightly so. Many acute illnesses are treatable and curable today and there is no sense in denying these successes, whatever the cost and no matter how few individuals, on a relative scale, may benefit from these heroic procedures.

I certainly share Dr. Carstairs's emphasis on the mental and social aspects of living, which seem to affect mental health. People with frustrations will express them in various ways that we may call aberrant behavior or that take on such forms as revolution or cultism. These extremes are expressions not only of frustrations but of the way people are apt to behave in situations in which they find themselves. This clearly is is an area in which medical people could perhaps undertake more study and offer more advice. Clearly, if they aren't interested in these phenomena, it is more than an oversight on their part.

We are learning that some things, like abrupt changes, do not have the great traumatic effect on the human nervous system that we expect them to have, and yet more subtle changes, such as everyday stresses and strains, seem to have a profound effect. I think this point is quite interesting, and it also makes it appear that perhaps we should not only look to the experts for advice along this line, but also within ourselves and at our own behavior to seek ways of coping with and living the lives that we seem to be heading toward.

In summary, I think Dr. Carstairs's message—that many people under stress from all types of living conditions are creating a situation that is forcing us to cultivate our own thinking and feelings—is immensely important. The fact that we can't solve this with technology, not even medical technology, is important, although I would put a little higher emphasis on the technology myself. But I certainly would say that perhaps, where the mind has failed, the heart should take over, which is his final word and is extremely well taken.

Discussion

DR. CLAXTON: Let's ask for questions or comments. Would you please just give your name, so we can identify you. And if you have a question, please say to which member of the panel the question is addressed. If you have a comment, it is most welcome, but as I said a moment earlier, please confine it to about a minute until others have had a chance to speak.

DR. HASSAN: I find myself agreeing with the thrust of the remarks of Professor Carstairs. I however would like to add that medical care, in whatever form, should be considered along with the socioeconomic change. There is no sense in prolonging life if a life is not worth living. And what we see in many countries is that very thing happening; we have more people—which leads to a perpetuation of poverty. The point is that in order to prolong life meaningfully, you need to have good life.

PARTICIPANT: Dr. Carstairs, in your experience in India, have you noticed any linkage between the introduction of nontraditional methods in either physical or mental health delivery and social disturbance, such as the recent upsurge of bandit activity in Rajasthan? It seems to me that many Rajasthanis I've met are looking to medical technology or Western medicine and mental health for their well-being at the very same time that I notice a corresponding upsurge in antisocial activity. I wonder if you have any information on that? Is there any linkage, or is it something I am seeing that is not there?

DR. CARSTAIRS: Perhaps I could say at this point that the failure of Western medical technology came to me as a general practitioner working in the village. I came there from a big hospital working with a big box of medicine, and the real failure was that you knew that you had in your kit extremely good and powerful medicines, but you missed the laboratory support of Western medicine. So I was absolutely compelled to go by the best guess, and the best guess meant picking out the medicines I hoped would be effective in this condition, without being able to verify that that antibiotic was specifically appropriate. On another plane, people are ambivalent about Western medicine because those medicines in the shop are to them extremely dear. So the price of an antibiotic is too high. It is a failure of complete delivery, if you like, and I would agree here that one mustn't denigrate barefoot doctors because the Chinese don't denigrate them. Their barefoot doctor had tetracycline medicine in his kit, because there was quite a wide range of specified conditions for which it was an effective remedy.

The historical dimension reminded me of one of my colleagues in India, Professor Thomas McKeown, who startled us all who were in this mood of terrific complacency about the new wonder drugs. Streptomycin transformed the treatment of tuberculosis, so much so that in a very few months, certainly within one or two years, it started to empty the tuberculosis sanitoriums. One thought that wonder drugs were wonderful, but McKeown said look at the figures. The standardized death rate of 1848 was 2,008 per million. In 1970 it was 13—but he pointed out that 75 percent of that decline appeared before the new drugs were invented, before they were put into action. Why? And then he advanced reasons for believing that improvement of nutrition was the most important thing, and second to that was the improvement of hygiene in the industrial cities in England during the later part of the nineteenth century. So I quite agree that one gets a better sense of proportion when one is reminded by colleagues that there are ongoing processes besides the breakthroughs that the popular press is so fond of announcing.

DR. CLAXTON: Would it be correct, though, to say that both nutrition and hygiene are properly considered as part of modern medicine, certainly public health?

DR. CARSTAIRS: The other speaker was very much on the point by saying you can't separate health provision from socioeconomic advance. It is quite true that nutrition is very important and you need to get socioeconomic advance to be sure to deliver adequate nutrition.

DR. CLAXTON: Those who know nutrition helped to guide the socioeconomic advances.

DR. BECKERMAN: I would just like to ask Professor Hassan to clarify what he said, because I thought he said something I happen to agree with, but which I keep quiet about, because if I suggest to anybody that I had this view, they usually seem shocked and I'm met with silence. I thought he said that one ought to perhaps question the wisdom of policies to increase the longevity of people whose lives were very miserable, and perhaps in some sense, lives that were not worth living. I happen to think he is right about this.

DR. HASSAN: That is what I said. In fact, there is no sense in prolonging life or bringing more lives to this earth when the people whose lives have been extended live under very bad conditions. What is important is the quality of life and not the numbers that we have on this earth. That does not mean that we should let these people die. That would be another unethical position. I think what we have to do is to combine medical care and medical aid with a change in socioeconomic conditions that may allow those that live in a state of dependency on the industrial nations to improve their lives over the long run.

DR. CLAXTON: If you combine the change in the socioeconomic conditions, then you presumably want to provide the systems to bring about that change.

DR. HASSAN: Exactly.

DR. BECKERMAN: If you can do that then there is no problem. I think we can both make people live longer and help them to live better. But as an economist I'm only interested in problems where there are conflicts and choices to be made. But the world being what it is, we can't always eliminate these complex questions, and I'm very interested to note that somebody else is making this point. For example, a well-known institution in this city, the World Bank, publishes the World Development Report. For the last two years it has made a big play of a new indicator of development it has been using, showing the achievements that have been made in the field of increasing longevity and expectation of life, mainly among those parts of the world population where the conditions of life are the most miserable.

If Professor Hassan is right—and I tend to think he is—instead of using this as an indicator of increasing welfare, it should be regarded as an indicator of decreasing welfare. After all, many people would accept some sort of utilitarian objective in the form of maximizing average utility. That is certainly a view that Professor Hassan has put forward and I subscribe to, and which I think follows logically.

Even if you are not an average utilitarian but a total utilitarian, if you think that these people are leading lives with negative utilities and are actually unhappy, then you still have to subscribe to the view that increases in longevity and expectation of life amongst people who are suffering is actually decreasing the world's total satisfaction. Of course this is a potentially dangerous argument. One way to maximize welfare per head is to cut off all the heads and I can see why people are rather shocked by this, but I was fascinated by the fact that Professor Hassan said this and nobody reacted.

DR. CARSTAIRS: I would just like to respond that I keep thinking of phrases by Ivan Ilich, who says that it is only the poor in the Third World who can look forward with confidence to dying in their own homes surrounded by people who care for them, because in the developed countries you know how different it is. He also says that 35 percent of all the expenditure on health care in America in a recent year was spent in order to prolong by a few weeks the existence of rich old people.

DR. DE GROOT: I hear this argument frequently and totally agree with it, but one has to separate the longevity gain into two parts. One is the gain in longevity associated with the reduced infant mortality and early childhood death. The other aspect is the days and years and seconds added after age sixty-five pushing that up; that is only of recent origin.

DR. CARSTAIRS: The expectation of life after sixty-five has hardly changed at all.

DR. DE GROOT: Well, it is beginning to change very dramatically in this country. The thing is that for the first part, the first increase in longevity, medicine can neither take the blame nor the claim. It is totally, nearly completely, out of our hands. The change is related to agriculture, economic factors, and housing. We simply rode to fame on the downward slope, which is a good public health technology, and it is running out of hand now. The other is subject to technological manipulation that we're only beginning to see the effects of right now, specifically in the age group eighty-five and over.

DR. CLAXTON: Our time has expired for this session. The last two comments did, though, fall in a field that I have been acquainted with for quite a number of years. I would just make one observation. The World Bank does not limit, as I understand it, as a test of progress, systems simply to increase life expectancy, which is one of several. They are tending toward using the QLI, the Quality of Life Index, developed by the Overseas Development Organization, which involves several factors including those which would fit within, I believe, Dr. Hassan's concept of an enriched life, literacy, food intake, and some other things. But certainly longevity, I believe, would properly be considered as one of those criteria, because it is primarily related to the reduction in infant-child mortality. May we then consider this session completed? With thanks to Dr. Carstairs and to Dr. Davis and to Dr. de Groot. Thank you very much.

11. Population Growth: Current Issues and Strategies

GEOFFREY McNICOLL
Deputy Director, Center for Policy Studies, The Population Council
MONI NAG
*Senior Associate, Center for Policy Studies, The Population Council,
and Adjunct Professor of Anthropolgy, Columbia University*

Editor's Summary. Geoffrey McNicoll and Moni Nag emphasize that population pressures do have an important and at times adverse impact on human society. Although there has been a decline in fertility in the past two decades, many poor countries will continue to face a critical social problem in absorbing their expected future population increase. The projected rapid growth in numbers in these countries will slow their economic development, impose on them heavy organizational demands if social stability is to be maintained, and in extreme cases constrict what development can hope to achieve for them. Efforts to moderate fertility can of course also be costly, but the experiences of countries that have recently shown rapid fertility declines provide a store of strategic insights for wise policy choice.

Few public issues are as dead as last year's crisis. From a high point some ten to fifteen years ago, intellectual concern about population growth has steadily waned to a position where it falls now somewhere between ocean mining and acid rain. The massive and continuing global transformation wrought by modern demographic change receives the same scant and short-lived attention as a new-found "firewood crisis" or the disappearance of the Peruvian anchovies.

Why is this? One reason is simple boredom—or, related to it, the chronic low-grade depression brought on by repeated economic and ecological alarms. Each such alarm has its day, but life goes on. Sometimes, the anchovies return. In the case of population growth, the highly colored language used by many early publicists of the issue

317

did the field a disservice by joining it to the fast-paced, faddish company of media crises. A population "explosion," for example, fairly describes the phenomenon of contemporary growth seen against past millennia of near stationarity, but is a poor depiction of the shorter-term reality—year-by-year increases of 2 percent or so in aggregate size.

But if boredom were the only reason for neglect, population issues could fade from the spotlight and still continue high on the public agenda of both national governments and international agencies. One purpose of such institutions is precisely to embody long-run social commitments so as to weather transitory shifts in attention and policy focus. More disturbing would be a neglect that is a consequence of shallowness of the commitment itself. For population, such shallowness may well be the case—reflecting either belief in the early attainment of a low-growth-rate demographic regime, drawn from a rosy interpretation of recent trends, or the view that rapid population growth is not after all a serious world problem. The first would seem to be for the most part wishful thinking; the second, if more than political reaction to fit the times, suggests an extraordinary degree of analytical slackness in the intellectual foundation of a major social program.

In this essay we attempt a brief review, from first principles, of the nature and scale of the problem of modern population growth and, on the basis of past experience of individual countries, comment on strategic choices for population policy in the future. We start by assembling the latest United Nations data on the magnitude of current and expected population growth—familiar material to demographers, but increasingly ignored in popular discussions of fertility decline. To interpret properly the significance of this growth would call for exploring its intricate ties to social and economic change, a task far beyond the compass of a brief essay. Here we simply note the major claimed benefits and costs of population growth, and remark on the balance, as we see it, between them. Public policy in this area, of course, should depend not only on this balance but on the costliness and effectiveness of the policy itself, however roughly that must be judged. While the unique conditions of each country require local analysis of policy alternatives, the experience of countries that have rapidly been moving toward demographic modernity is likely to be the main source of policy insight.

Demographic Prospects for the Next Two Decades. Members of the generation born around 1930, who can reasonably hope to outlive the present century, will have witnessed during their lives a trebling of the world's population. From a 1981 total of 4.5 billion, the United

GEOFFREY MCNICOLL and MONI NAG

Nations expects the world to reach 6.1 billion inhabitants in 2000 (in its medium variant assessment made in 1980). The inertia of this growth, built into the age distribution, will carry it forward well into the next century—a population exceeding 9 billion by 2050 would be a plausible forecast.

The regional concentration of population increase is shown in figure 29, based on the UN medium projections for 1980-2000 (United Nations 1981). The width of each rectangle in the figure represents the 1980 population of the region; the height, its expected proportional increase; and the area, its expected absolute increase. The contrast between the two extremes, Africa and Europe, is the most dramatic—each starting from about the same size in 1980, but Africa likely to add thirteen persons for each one added in Europe. More significant, perhaps, is the contrast between the two regional giants, South Asia (here including the Middle East) and East Asia (excluding Japan), each containing about a billion people in 1980. South Asia is still clearly in the high-

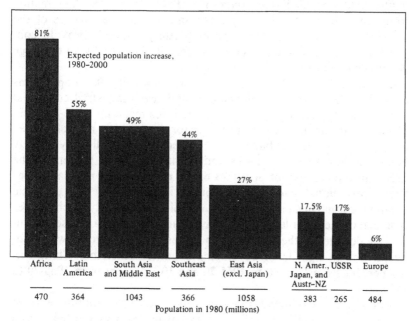

SOURCE: United Nations, 1981

Figure 29. Population in 1980 and projected population increase by 2000 in major regions of the world.

growth category, East Asia not far from the growth rate of the developed world. Dominating this latter contrast is, of course, the differing demographic trends of India and China, which we examine further below.

The UN's most recent assessment of the demographic future does little to buttress the widespread popular belief that population growth is rapidly slowing down, the expansion running out of steam Overall, the world's growth rate seems to have peaked at about 2.0 percent per year in the 1960s and by now has probably dropped to 1.7 percent. The yearly absolute increase in population is continuing to rise, however. In 1980, some 75 million people were added. In the UN's medium projection, the point of inflexion of the trajectory will not be reached until the late 1990s, when the annual increase will be close to 90 million.

The expectation of a substantial slowdown in population growth, however, is based less on trends in natural increase than on recent evidence of fertility decline. There is a temptation to examine country experience here without weighting by population size, so that a significant decline in, say, Singapore or Jamaica tends to offset in the public mind the lack of one in Nigeria or Pakistan. But the data, mainly from the 1970-71 round of censuses and the sample surveys of the mid-1970s (in particular, the World Fertility Survey), provide convincing evidence of some decrease in fertility in virtually all the large developing countries.

The latest UN birthrate data for countries with 1980 populations exceeding 50 million (nine developing and seven industrial countries) are set out in table 15. The sixteen countries make up three-quarters of the world population. The birthrate for the world as a whole, given in the last row of the table, is estimated to have fallen by slightly below 10 percent in the 1960s and slightly above 10 percent in the 1970s, to a level now of about 28 births per thousand population per year. This global average is strongly influenced by the dramatic downward trend in China—somewhat earlier and recently much steeper than that of the other large developing countries—and to a lesser extent also by declines in the industrial countries. The average (unweighted) decline in the eight large developing countries other than China was 4.1 percent from 1955-60 to 1965-70 and 10.5 percent from 1965-70 to 1975-80.

Two significant regional trends in world fertility are not fully evident in table 15: the birthrate declines spreading in East and Southeast Asia and the recent appearance of quite steep declines in Latin America. In the former region, in addition to China, falling birthrates are being recorded in South Korea, Taiwan, the Philippines, Indonesia, Malaysia,

GEOFFREY McNICOLL and MONI NAG

TABLE 15 Birthrates and Birthrate Declines for Countries with 1980 Populations Exceeding 50 Million

Country	1980 Population (million)	Average Annual Birthrate (per 1000) 1955–60	1965–70	1975–80	Change in Birthrate (percent) 1955–60 to 1965–70	1965–70 to 1975–80
Developing Countries						
China	995	37.6	32.4	21.3	−13.8	−34.3
India	684	44.0	42.0	35.3	−4.5	−16.0
Indonesia	148	47.1	43.0	33.6	−8.7	−21.9
Brazil	122	43.1	38.8	33.3	−10.0	−14.2
Bangladesh	88	50.3	49.7	46.8	−1.2	−5.8
Pakistan	87	47.2	46.8	43.1	−0.8	−7.9
Nigeria	77	52.1	51.0	49.8	−2.1	−2.4
Mexico	70	45.7	43.9	38.3	−3.9	−12.8
Vietnam	54	41.9	41.4	40.1	−1.2	−3.1
Industrial Countries						
USSR	265	25.3	17.6	18.3	−30.4	+4.0
USA	223	24.8	18.3	16.3	−26.2	−10.9
Japan	117	18.1	17.8	15.1	−1.7	−15.2
West Germany	61	16.5	16.6	9.8	+0.6	−41.0
Italy	57	18.0	18.3	13.3	+1.7	−27.3
UK	56	16.4	17.6	12.0	+7.3	−31.8
France	54	18.4	17.1	13.8	−7.1	−19.3
World Total						
	4432	36.1	33.1	28.5	−8.3	−13.9

Source: United Nations (1981).

and Thailand; in the latter, most notably in Brazil, Mexico, and Colombia. Identifying the determinants of fertility trends in these two regions is clearly important in making any long-term demographic forecast for the world.

The most visible manifestation of high fertility comes as societies try to absorb successively larger cohorts of young people into their economies. The translation from the economically inert notion of a birthrate to the potentially critical issue of labor force entry is as striking as it is simple. The number of people who will be seeking such entry around the year 2000 is determined essentially by births occurring now—say, over 1980-84.

Figure 30 illustrates the consequences of disparate fertility trends in the four largest countries over the past three decades, under the simplifying assumption that the number of potential labor force entrants in a given year is one fifth of the population then in the 15-19 age group. (The chart therefore ignores such structural changes as the increase in female labor force participation in the United States or the

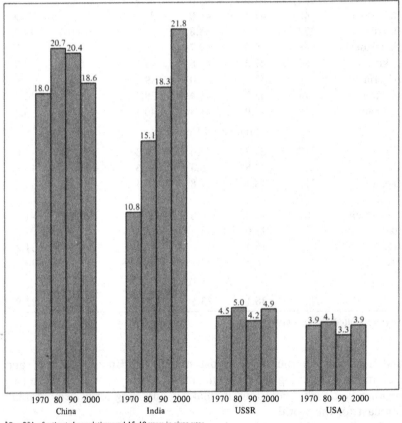

^aOne fifth of estimated population aged 15–19 years in given year

SOURCE: United Nations, *World Population and its Age-Sex Composition by Country, 1950-2000: Demographic Estimation and Projection as Assessed in 1978* (New York: 1980).

Figure 30. Changes in the annual pool of labor-force entrants for four countries from 1970 projected to 2000, in millions.

GEOFFREY MCNICOLL and MONI NAG

effects of changing rates of school enrollment.) It shows, for example, that within twenty years India's planners will have to cope with larger pools of labor force entrants than China's—an extraordinary reversal that will occur abruptly in the 1990s as China's birthrate decline of the 1970s makes itself felt in the economy. India's situation is, of course, the more typical among developing countries.

Corresponding to a present world birthrate of around 28 per thousand is a death rate of about 11 per thousand—or, in more convenient terms, life expectancy at birth of 58 years. Life expectancy in the developing countries has risen from about 35 years in 1930 to some 56 years in 1980 (51 years in the developing world excluding China). There is wide current concern, however, that the era of rapid mortality decline may have at least temporarily come to an end. The optimism that led a 1962 UN report to declare "It may not be too much to hope for that, within a decade or two, the vast majority of the world's people will have an expectation of life at birth 65 years or more" was unjustified, and few would now expect such a state to be achieved in the near future. (See Gwatkin 1980; the UN report referred to is United Nations 1963.)

The estimates of demographic rates we have been relying on above are by no means incontestable. Between the UN's 1978 and 1980 assessments, for example, China's birthrate figure for 1970-75 was raised from 24.0 to 29.5—an increase of more than 20 million in the implied number of births during this period. In a number of other countries, India and Indonesia among them, early returns from the 1980-81 round of censuses show populations larger than had been expected on the basis of previously estimated birth and death rates; so further adjustments in these estimates are likely to be required.

But even with accurate statistics, demographic prediction is a highly fallible art. What is needed to give scientific substance to assumptions about future trends in fertility and mortality is an understanding of the reasons for observed changes in vital rates in the past. For mortality, the reasons for the most part are cut and dried. The effects of improving economic conditions, better hygiene and sanitation, application of new medical knowledge, and expansion of public health services can, in theory, be separated out and quantified by detailed cause-of-death analysis (see, for example, Preston 1976). It is the paucity of cause-of-death data, past and present, rather than any fundamental conceptual conflict that accounts for most of the remaining disagreement. In the case of fertility, however, similar efforts to explain past trends lead immediately into some areas of major controversy. Before venturing there, we take up the second suggested basis for the fading concern with population growth—that on balance it doesn't matter much.

Pros and Cons of Population Growth. What is the balance of costs and benefits associated with population growth? Any answer must be so hedged about with assumptions and conditions that it is not surprising to find continuing argument on the subject. Costs and benefits to whom? Under whose value premises? Over how long a time period? With what resource endowments? In what kind of institutional and cultural setting? Even the comparatively circumscribed issue of the impact of population growth on a nation's socioeconomic development seems to have few questions that both are settled and effectively remain so—not only because answers generally have to be country-specific and contingent on an agreed content of "development," but also because some important technical questions about population-development interrelations are themselves unresolved.

But there is no justification for an agnostic stance here. The questions, both technical and ideological, are clearly amenable to analysis; orders of magnitude for the relationships at issue can nearly always be found; radical outliers in conclusions either dissolve under analytical scrutiny or can be pushed back to extreme, sometimes bizarre, ethical or ideological positions.

We would summarize the broad conclusions of the majority of researchers on the issue as follows: under the conditions existing in most poor countries, rapid population growth slows, sometimes drastically, the absorption of the bulk of the population into the modern, high-productivity economy, while any demographic effects in stimulating innovation or investment call at most for a moderate pace of growth; rapid growth hinders the capacity of poor countries to cope effectively with large changes in their natural or economic environments; and, in extreme cases, it constricts what development can hope to achieve for them. It is not the place of this essay to assemble the mass of evidence supporting this statement, but a brief gloss on each of its three parts is called for.

(1) Rapid population growth imposes obvious costs in needed investment to maintain capital per head. Such investment is at the expense of consumption or of adding to capital per worker and, thus, to labor productivity. At the family level, this effect is compensated by the strong values attaching to children, assuming they are "wanted." Looking just at the economic dimension, the demands of a larger family size provide a strong encouragement to increase family earnings both on the part of parents and by the children themselves. At the aggregate level, however, the net effects on other families—on balance likely to be adverse in most poor-country settings (e.g., through depressed wage rates or over-

GEOFFREY MCNICOLL and MONI NAG

extended infrastructure)—must be added in. Certainly the prospect of continuing large increases in labor force entrants that now confronts so many countries is not counted an economic blessing by the governments affected.

But can such labor force increases act as a stimulus to economic performance, shaking the economy out of an institutional or technological rut? It is often held that population growth induces innovation, stimulates greater private investment, yields scale economies in provision of economic and social infrastructure, and so on. In some cases such benefits may well accrue, but there is nothing to suggest that to capture them requires anything close to the 2-3 percent annual rates of increase common in the contemporary world. (The demographic stimulus, seen by North and Thomas (1970) as contributing to profound economic change in fifteenth-century Europe or by Hicks (1939:302) as a factor behind the Industrial Revolution, was in each case below 1 percent per year.) Moreover, these benefits may in fact be more effectively captured by explicit measures entirely outside the demographic sphere.

Since development has become a self-conscious process, governments have acquired all kinds of direct and indirect means of promoting innovation, raising investment rates, or reaping economies of scale, short-cutting the drawn-out process. A stranger pronatalist economic argument, newly given popular currency by Simon (1981), is that more people mean more geniuses, more spontaneous creation of useful knowledge, as if the poverty of the Third World was not itself the reason for its technological dependency in the face of already preponderant demographic weight.

The significance of a demographic drag on economic performance depends, of course, on the level of that performance. In the dynamic economies of East Asia, with national incomes increasing by 8-10 percent annually, population growth of 2 percent or so is a comparatively minor issue. But most of the rest of the developing world can take no such comfort. Using the World Bank's categories, per capita income growth in the 1970s was estimated at 2.8 percent per year in the middle-income developing countries, and 1.6 percent in the low-income countries (excluding China). In low-income, sub-Saharan Africa, per capita income actually fell over the decade.

(2) In addition to draining off investment to train and equip an expanding labor force, rapid population growth imposes considerable new organizational demands on a society. Social structure is not neutral with respect to scale: doubling population size in a

generation, and city size in sometimes a decade, keeps a country's political and administrative apparatus perpetually off-balance. External shocks, such as sudden shifts in resource prices, which should elicit concerted public and private sector efforts to redirect technology and manpower in accord with the changed market conditions, may become major economic calamities. Effective actions aimed at halting economically or aesthetically damaging processes of ecological degradation (for example, those making for the impending extinction of many plant and animal species), which equally call for organizational competence, may be simply beyond reach.

The connections here are complicated by the likelihood (which we discuss below) that weakness in organizational competence is itself part of the reason for continued rapid population growth. Mutual causation of this kind bedevils the analyst's task in the entire field. Moreover, the connections may be disguised: governments typically respond to social disorganization by deploying the technology of political control, providing at least a veneer of stability. It is plausible that rapid demographic change, by threatening social stability, is a not infrequent contributory factor behind the emergence of military governments in the contemporary Third World.

Popular discussions of "population pressure" often seek to identify direct adverse ill-effects on economic welfare of worsening population-natural resource ratios or to impute a demographic cause for any unhappy ecological outcome. Nearly always such arguments are slippery, however, since they skirt the critical intermediate issue of how a society organizes itself to respond to resource and ecological problems. The current "firewood crisis" in many poor countries, to take an example mentioned earlier, is best seen not as a simple matter of population-induced deforestation but as a result of the failure of societal arrangements that elsewhere manage to maintain a common resource in the face of competing uses and users. The central issue to be explored is then the demographic contribution, if any, to that failure.

(3) In extreme cases, rapid population growth and the resulting absolute size of population constrict what is achievable by economic development in terms of individual wealth and amenity. Countries that have successfully attained the status of industrial economies are able to set about repairing the worst ravages of the industrialization process and creating the amenities that are increasingly in demand by their expanding middle classes. Sheer population density

GEOFFREY MCNICOLL and MONI NAG

seems comparatively unimportant to the outcome of this endeavor—especially since wealth can also buy access to the amenities of the rest of the world. Even if today's brand of international tourism turns out not to be open to the latecomers, simply because of their numbers, amenities less dependent on space and distance can presumably be designed. But it is surely more than mere failure of imagination that we should find it so difficult to foresee the emergence into postindustrial satiety, such as it is, of the massive populations of, say, South Asia or the by-then massive populations of Africa.

If the balance of costs and benefits of population growth is so evident, why is it apparently so easy to call the conclusion into question? Several reasons spring to mind. It may, for one, be a corollary of a belief that, rhetoric aside, there is rather little practical significance to any unweighted summing of numbers of people. To take an extreme case, if a nation's long-run economic policies in reality are designed to benefit a small group within the population, policy achievement is not to be measured by nationwide per capita averages. Less extreme assumptions with certain formal similarities are to be found in neoconservative entitlement theory. (The position has its analogy at the international level. Most countries that have attained a low-mortality, low fertility regime combine economic strength with comparative demographic puniness—Hong Kong, for instance, has the same gross national product as Pakistan. Their natural interest lies in keeping economic weightings in and population weightings out of the international order.)

A different kind of reason for questioning the existence of substantial net costs of population growth is sometimes drawn from dubious attempts at modeling the economic-demographic system. By dint of constant exposure, some analysts may delude themselves into actually believing the assumptions of neoclassical growth theory—an institution-free economic environment in which equilibrium expansion paths pose no maintenance problems and the "steady state" is one of uniform exponential growth of people and product. The disregard of scale effects somehow survives the transition from simple, elegant models in which the stability and efficiency properties of growth paths are explored to attempted depictions of real-world social and economic systems evolving over decades. In one widely publicized simulation exercise (see Simon 1977), alternative fertility trajectories lead to population sizes after two centuries that differ by a factor of 20. Even if we ignore the hubris entailed in 200-year time horizons, such a difference could not fail to reflect utterly different patterns of socioeconomic institutions, with effects extending far beyond levels of fertility. Yet the same model is assumed to characterize the economy in each case, and from it the conclusion is drawn that per capita income is relatively invariant to fertility. A finding more clearly built into the instrument of analysis is hard to imagine.

The relationships between population growth and socioeconomic change, although at first sight a subject ideally suited to formal simulation, have proved highly resistant to compelling modeling. The number of plausible relationships is very large, their empirical basis often cloudy. A complex model can, of course, be set down in short order. The problem is that a few "reasonable" adjustments in its assumptions, adding or deleting a feedback loop or two, can lead it to yield entirely different results. For these particular issues, formal modeling merely papers over our ignorance or exaggerates our biases.

Or, perhaps the simplest explanation, the shrill voices from the 1960s of those foreseeing imminent demographic catastrophe, even though now largely silenced, have damaged public hearing for serious analysis of population problems and policies. Presently predicted global population growth (for the moment admittedly rapid, but sure of eventual halting—with reasonable hope, in a century or so), it seems, can be tolerated and that is all that matters. In this view, possibilities that much lower rates of increase might be attainable, and might yield substantial benefits on the development front and in noneconomic spheres, do not warrant investigation.

Individual and Societal Fertility Interests. We could accept that rapid population growth imposes a whole variety of burdens on a particular society, offset only to a slight degree by its positive features, and yet still conclude that little can or should be done about it. The intractability of population growth to public intervention is perhaps the most important lesson learned over the past several decades of concern with population. It may be that the costs, economic and noneconomic, entailed in any policy action that *would* be effective outweigh the gains to be reaped from its outcome. Such a calculation, quite aside from the intricate distributional judgments it would involve, can only be attempted from a basis of understanding what the feasible options for influencing demographic trends are. With little loss we can restrict a discussion of this to the issue of fertility change.

Over the long haul of human history, societies in a rough sense can be said to organize themselves to support a particular fertility regime appropriate to the resources, technologies, and mortality risks they face. Recruitment of new members, biologically required to be a matter for largely decentralized decision making, is too important a matter to be left wholly up to individual parents. As on virtually any major issue affecting its future welfare, a society intervenes in the demographic behavior of its members to preserve its interests—or the interests of the more powerful groups within it. When an issue such as this must be confronted routinely, routine responses are developed. In other words, some form of control over that particular domain of behavior becomes institutionalized in the society, embodied in the signals conveying social approbation or economic advantage (or their opposites) to individual members. (Realities, of course, do not quite admit of this abstraction of demography from the rest of the social and economic system. But the picture suffices for our purpose.)

What then happens when the situation is suddenly disrupted by new threats and opportunities—new technologies, much larger surviving birth cohorts, a rapidly changing economic and cultural environment? For the society as a whole, the institutional controls that historically governed reproductive behavior have probably become embedded in a broad, more or less coherent sociocultural, legal, and administrative framework and are not readily available for revamping in response to the need for a new demographic regime. So the signals do not change. For individuals and families, as before seeking and acting on their own interests, a variety of demographic outcomes is possible—but only fortuitously an outcome that accords well with the new social interest. In essence, individual and social demographic interests are uncoupled.

Why, it may be asked, did this "uncoupling" not happen historically in the demographic experience of the rich countries? The answer is that it did happen. Community controls over marriage and household establishment broke down under the new economic opportunities and the possibilities for social and demographic mobility they afforded. However, there was no surge in population growth comparable to that in the contemporary Third World because mortality remained high (as it did until well into the present century) and to a lesser extent because emigration was an outlet. Europe's population rose by 7 percent per decade over the nineteenth century, compared to 4.5 percent per decade over 1750-1800 (Durand 1977). The industrial economy, except at the troughs of its business cycle, was well able to absorb the increase it had precipitated.

The individual fertility interests that were formed by the pattern of industrialization, or were coformed with it, eventually worked firmly in the direction of smaller families. One facet of the changes that constitute economic development is a steady rise in the cost of children to parents, far outpacing the characteristically tepid efforts of society to subsidize this cost and thus distribute the burden. Children's income earning opportunities recede; their role as a store of value for their parents' old age or as a hedge against family misfortune is supplanted by new social and economic institutions; educational costs to parents, despite large public expenditures, remain appreciable and become more necessary; the income sacrifice of the time involved in childrearing becomes a large factor in the family economy. More than sufficient reasons can be found in the changing economy and its institutional supports to account for a radical drop in family size.

We do not assume a thoroughgoing material basis for all such changes. Clearly, concomitant cultural developments—particularly in parents' perceptions of children and of the balance of rights and duties between parent and child—have potentially profound import for demographic behavior. Fortunately, there is no need to assign precedence between culture and social organization in this explanatory sketch.

With completed industrialization it would be pleasant to report that there is a "recoupling" of social and individual demographic interests at approximately a replacement level of fertility. There is no evidence for this belief, however. Fertility, just as plausibly, will continue to drop far below replacement, with governments scrambling for ways of subsidizing the cost of children to sustain their national labor force and tax base. Whether their scarcity will again make children more privately valued, thus leading to a fertility rebound, or whether social acceptance of childlessness and one-child families (and voting power accumulating among the elderly and the childless) will further tip the balance toward very low fertility, cannot as yet be seen.

Lessons from Contemporary Experience. While these low-fertility problems are far removed in scale from the current difficulties of rapid population growth in the Third World, there are important analytical similarities. We shall look briefly at a number of contemporary examples of fertility change, all places where substantial declines have occurred, drawing on the same explanatory apparatus. Brazil, on the one hand, and China, on the other, provide in some respects polar cases. The dynamic East Asian countries (principally, South Korea and Taiwan) and Sri Lanka and the Indian state of Kerala are instructive supplementary cases within these extremes. (In a fuller discussion we would, of course, be just as interested in why fertility has *not* declined in certain other places.) The statistics cited for the most part are taken from the first two World Development Reports (World Bank 1980, 1981)—the first of these being used for Taiwan, which thereafter became invisible in UN-system statistics.

Brazil and China. Brazil, the regional giant of Latin America, has shown fairly substantial fertility decline in the last decade. The birthrate

GEOFFREY MCNICOLL and MONI NAG

has apparently fallen by more than a quarter, to a level below 30 per thousand over this period, the decline accelerating in the most recent years. (The latest figures are not yet well confirmed and the 1980 UN data cited in table 15 only partially reflect them.) In terms of Third World averages, Brazil has a relatively high per capita income: close to $2,000 (U.S.). This, for comparison, is equivalent in real terms to the level in Japan as recently as the early 1960s—at which time Japan's birthrate was below 20. It is a highly urbanized country—65 percent of the population lived in urban areas in 1980 (in India, for example, the corresponding figure is 23 percent), and agriculture contributes only some 10 percent of gross domestic product. Manufactured exports have grown rapidly in volume and technological sophistication. Brazil's development strategy has combined vigorous promotion of state and private capitalism, little concern with income distribution, and pronatalist or laissez faire attitudes toward population growth.

While interpretations of demographic trends must still be highly tentative, Brazil seems to be tracking the classic European pattern of transition from high to low vital rates. Death rates declined sharply in the 1930s and by the 1960s life expectancy was approaching 60 years. (Today it is around 62.) On average, there were four or five or so surviving children per family at a time when economic changes were making such a size increasingly incompatible with urban industrial life. For the growing middle class, children competed with newfound consumption alternatives; for the poor, with the daily exigencies of scraping for a living and with the possibilities, although slight, of moving up in the world.

By most assessments of the human costs of economic development, the Brazilian pattern is a high-cost route. Its apparent effect, belatedly, in limiting population growth can be seen in part as a reflection of these costs, albeit in the long run a socially beneficial result. But the future pace of fertility decline is by no means assured, while economic difficulties, particularly problems of labor absorption, loom increasingly large. More interventionist policies in the demographic as well as the economic spheres may become more attractive.

The contrast with China's development strategy in the past three decades could hardly be more extreme. If Brazil's experience is reminiscent of Europe's or America's during the turbulent years of early industrialization, China's experience recalls Europe or Japan in the preindustrial period. In essence, China tried to preserve or recreate in a modernizing economy the kinds of community structure and social controls on its population that characterized these traditional societies. The violent Chinese reforms of the 1950s—the killing or dispossessing of the landlords and powerful lineages and the establishment of collectivized agriculture—paradoxically gave new strength and resilience to the rest of the preexisting social system of family, neighborhood ("team"), and village ("brigade"), a strength evidenced by these groups' successful resistance to the later recurrent excesses of Maoist radicalism.

Just as the eighteenth-century English parish or Japanese village had both an economic interest in and the capacity to determine who married or settled in the community, so the postrevolutionary Chinese village gained a similar stake and role—rendered more effective by covering not just household formation but also marital fertility, and rendered more precise by greatly lowered mortality risks and by the modern technology of birth control. The considerable economic autonomy of brigades and even of teams, including assigned obligations to fund most of their own social services, set up local fertility incentives that in large measure coincided with national government interests in slowing population growth. More direct government pressures, through antinatalist and delayed-marriage campaigns, of course, pushed in the same direction, facilitated by an effective health care system that also provided family planning services. The demographic outcome has been striking indeed: life expectancy rising from below 30 years before World War II to 64 years today and a halving of the birthrate in less than two decades, to a present level of around 20 per thousand.

The same pattern of social and economic organization that has produced the extraordinary drop in mortality and fertility in China has had altogether less spectacular effects on the economy itself. The sustained surge in labor productivity on Taiwan, for example, finds isolated parallels but no broad counterpart on the mainland. In an effort to establish stronger economic incentives, particularly at the individual level, the present Chinese government has begun a shift away from the collective economy toward reprivatization—a shift that, if it proceeds as it seems to be going, could have large, unintended effects on population trends. Both the health system and the local pressures for fertility limitation are intricately dependent on the brigade and team structure, and their course under any weakening of that structure cannot easily be predicted. Selective relaxation of control in the economic but not in the demographic sphere may turn out not to be possible. If so, a consequence of promoting rapid innovation and productivity growth may be, so to speak, to free the fertility genie from the bottle. Concurrent moves in China to introduce draconian economic and administrative measures to achieve one-child families suggest that the government is far from persuaded that a fertility rebound could not occur.

South Korea and Taiwan.   If Brazil and China are the extremes of economic-demographic strategy, is it possible to do better somewhere in-between—combining rapid economic growth with a pattern of social organization that promotes early decline of mortality and fertility and avoids the worst of the human costs of development? If we add to the objectives broad-based political participation, then very likely the problem is overdetermined. Leaving that consideration aside, however, there are a number of countries that seem to have found a successful middle way. Nealy all are in East or Southeast Asia: furthest along are Taiwan and South Korea. Hong Kong and Singapore, the other two members of the so-called Asian "gang of four," are too structurally peculiar for their outstanding success to be relevant here. Malaysia and possibly Thailand would be in the next tier.

South Korea has averaged 7 percent growth in real per capita income over the past twenty years, during which time life expectancy has risen from 54 to 63 years and the birthrate fallen from 43 to 25 per thousand. The corresponding data for Taiwan, for the period 1960-78, are 6.6 percent average yearly per capita income growth, an increase from 64 to 72 in life expectancy, and a drop from 39 to 21 in the birthrate.

Both countries owe their economic success to extraordinary growth of manufactured exports—admittedly achieved in a favorable international economic environment. Domestically, the result was the product of high entrepreneurial capacity, competent and supportive government economic policies, and a well-trained, energetic labor force. (To say this, of course, does not get us far toward *explaining* the performance.) Agricultural productivity has also grown steadily.

The shift to an industrial economy was achieved without generating a high degree of income inequality among households and without emergence of a very large urban-rural wage differential. Although there is no persuasive evidence that an egalitarian development policy necessarily promotes either economic growth or fertility decline, it does seem that the same factors that led to growth with equity in Korea and Taiwan also speeded demographic transistion. These factors included an effective land reform aimed at encouraging small-scale peasant production, supported by a strong agricultural extension program and improvement in rural financial institutions, and rapid expansion of education, health, and family-planning services. The development pattern combined the fostering of vigorous private-sector performance in both industry and agriculture, enlightened provision of social services, and stringent government administrative and political control. Unlike China, discouragement of fertility did not become part of this last sphere, but remained in the second.

Sri Lanka and Kerala. There are various instances of low rates of mortality and fertility being reached without any apparent stimulus from a dynamic economy and without a strong politico-administrative system. Sri Lanka and the Indian state of Kerala are the best known cases—with life expectancies currently 66 and 60 years, birthrates of 28 and 25 per thousand, yet with per capita incomes in the bottom half of the World Bank's "low income" category. (Kerala is one of the poorer states of India.)

No firm agreement exists about how this "modern" demographic regime emerged in these areas. In certain respects, particularly in the wide availability of education and health services, both places are substantially more advanced than their aggregate production perform-ance would suggest. Such services develop in response to demand as well as in anticipation of it, so cannot be straightforwardly posited as contributing factors to fertility decline; but in these places government action seems to have come first.

A different source of explanation is found in the nature of their labor markets—the extensive pattern of labor commuting from rural to urban areas that has developed in Sri Lanka, the organization of agrarian trade unions in Kerala, and in both (partly in consequence) the relative lack of employment opportunities for children. Modern labor relations and the resulting separation between economic and domestic spheres of life have emerged without the accompaniment of high productivity. Finally, an increasingly heard argument in the case of Kerala (and applicable also to Sri Lanka) is that fertility decline is a response to the diminution of agricultural employment and the growing difficulty of finding other work—a situation that widens the gap between aspirations and income and, in turn, makes for a high age at marriage and acceptance of small-family norms. Whatever the reason, Sri Lanka and Kerala have achieved levels of social development and demo-graphic transition somewhat out of line with their extent of economic development—and have done so, in the population case, without substantial government antinatalist pressure.

Economic growth is properly the prime ambition of poor countries, and with little economic success to point to in Sri Lanka or Kerala, it can reasonably be asked: what is to be learned there? The answer hinges on whether or not in these cases the stage has not been set for sustained economic advance. Optimistic observers see the beginnings of such an advance under recent liberalization of Sri Lankan economic policies and the prospect, too, that Kerala over the next decade will move substantially ahead in its economic standing among Indian states. Those who instead see a demographic transition induced by poverty or relative deprivation are less sanguine.

Strategic Options. The conclusions we are led to by these and similar examples of recent demographic history are as follows. First, sustained, rapid economic growth creates conditions that make for fertility decline, largely irrespective of government population policy. Second, the timing and pace of fertility decline in these instances, and the social costs of the overall pattern of economic-demographic development, can vary over a wide range, depending on the particular institutional setting of the society—which in turn is in some measure influenced by government policy. Third, there are combinations of patterns of social organization and designs of governmental programs (notably in education, health, family planning, and rural employment) that foster lowered fertility even without much economic growth. Fourth, in these latter cases, it remains an open question in each instance whether or to what degree poor economic performance is linked to these institutional arrangements favoring low fertility. And fifth, if such a link does exist, political or administrative pressure aimed at speeding a decline in fertility may succeed in narrow terms, but at a considerable cost in foregone economic growth.

General propositions of this sort need detailing for a specific situation if they are to yield an array of feasible options for economic-demographic strategy. Feasibility is constrained in many ways. Existing institutional arrangements are resilient and cannot be arbitrarily altered except at high cost. (For example, the Chinese land reform of 1950-52 was a fundamental restructuring of rural society, but it was achieved at very high cost; the 1958 campaign to establish communes as the dominant rural social unit was abandoned when the evident costs it would entail in overcoming local opposition were judged unacceptable.)

Where there is scope for influencing institutional forms, possible effects on population trends are unlikely to count for much in the choice: both political and economic considerations would inevitably rank ahead of demographic. The feasibility of particular policy directions may change over time as the economy evolves. (For example, massive expansion of urban commuting in poor rural communities—as in Indonesia—made possible by improved transport facilities, erodes the social role of residential groups and thus also the possibility of designing population and development policies that work through those groups.) And, not least, government capacity itself, aside from will, may often be a binding constraint in determining feasibility. (That a government such as Singapore's is able to establish an intricate pattern of antinatalist incentives is of little relevance to the vast majority of Third World governments.)

The desideratum for population policy is that it comprise measures that do not detract from the incentive structure underpinning economic growth, while promoting socially desired demographic ends. A simple division of its content would therefore be into measures that clearly have an economic dimension, and thus call for careful meshing with existing incentive patterns, and measures that, in effect, are divorced from the economy, and so do not raise the possibility of working at cross-purposes.

In the first category, the broad object would be the creation of an institutional framework favoring both economic achievement and demographic restraint—getting the incentives right without waiting for this to happen "naturally." In part, such a framework could be an outcome of government routine activities and development programs, where a shift in emphasis or in design may plausibly have a demographic effect. Fertility levels are empirically linked to factors such as literacy and the status of women, and achievements in these areas earlier rather than later in the development process may have some benefits in demographic terms in addition to their other values. Probably in greater part, however, the institutional framework that generates economic and demographic incentives is rooted in the patterns of social organization that exist in the society: labor market, kin group, village community, local government, and so on. Here the objective would be to seek to modify these arrangements so as to bring home to individuals or families more of the social costs of their demographic decisions.

This "internalization" of demographic costs could be either at the family level or at the level of some larger social grouping that is in a position to exert social pressure on its members. In many preindustrial societies we noted that rural territorial communities had such a stabilizing demographic role; in the contemporary world, however, creating or sustaining that role during economic development is an altogether more problematic task—as the case of China illustrates. Perhaps it is more realistic in most cases today to concede the economic-demographic sovereignty of the family unit. On the kinds of institutional innovation that would act to limit the transfer of demographic costs from the family on to the society at large, no general specifications can be given. The diverse patterns of demographically related social organization found in countries that are reaching or have attained low fertility offer a rich store of insights to draw on. The unique conditions of each setting where policy is to be made, however, demand a strategy founded on thorough local analysis.

The second policy category, divorced from the economy although not entirely resource-free, covers measures aimed principally at promoting social values associated with low fertility or at easing the translation of such values into practice. The former is an intangible goal and one that is not firmly within the grasp of public policy in any sphere. But cloudiness of outcome permits the policies that seek to

GEOFFREY McNICOLL and MONI NAG

direct cultural change to continue to claim efficacy. The latter is where population policy can finally come down to bricks and mortar and management charts.

An emerging socioeconomic setting might clearly support a parental interest in low fertility, but one that parents are for some reason inhibited in pursuing. Contraception may, for example, be culturally disfavored; or the means available may be unsatisfactory. Here is the rationale for government promotion of family planning—as an exercise in legitimation and in ensuring effective, perhaps subsidized distribution of contraceptive services. (It is not, of course, necessarily a rationale for government itself to operate a family-planning program—the direction that action nearly invariably takes, such is the statist orientation of development thinking.)

Demographic Futures. A proper modesty is in order concerning our abilities to predict either economic or demographic futures of nations. In his exemplary review of development experience over 1950-75, David Morawetz (1977) notes that as late as the early 1960s, the most authoritative forecasts of future world economic growth did not foresee such successes of the next two decades as Brazil or the nations of the East Asian rim. Countries that were predicted to "take off" in that period included India, Burma, Egypt, and Ghana. Around 1950, similarly well-informed observers of Japan foresaw a future for that country of Malthusian stagnation or worse. We can have little confidence in making much better calls today.

The reasons for this uncertainty include our considerable ignorance of the origins of social change and the fact that even well-founded predictions can be upset by events, natural or manmade, appearing out of the blue. They also include, however, the degrees of freedom that remain for deliberate choice about the design of social arrangements in any society. Economic development and demographic transition are fundamental, irreversible transformations of these arrangements in which roads not taken are put rapidly out of mind. In retrospect, as a result, the process seems to have an inevitability about it. Failures to accomplish such transformations, which are common enough, similarly seem to have been dealt in the cards. The social scientist, prone to exaggerate the scope of policy choice, in the end gets his comeuppance from the historian, equally, perhaps, prone to diminish it. But the demographic future is indeed in some measure within the domain of social choice, and wise choice here offers potentially large gains in human welfare. We are far from reaching a perfect base of understanding for sound policy action, but that would be a strange reason to turn away from the subject.

References

Durand, J. D. 1977. "Historical Estimates of World Population." *Population and Development Review* 3:253-96.

Gwatkin, D.R. 1980. "Indications of Change in Developing Country Mortality Trends: The End of An Era?" *Population and Development Review* 6:615-44.

Hicks, J. R. 1939. *Value and Capital*. Oxford: Clarendon Press.

Morawetz, D. 1977. *Twenty-five Years of Economic Development 1950 to 1975*. Washington, D.C.: World Bank.

North, D. C., and R. P. Thomas 1973. *The Rise of the Western World: A New Economic History*. Cambridge: Cambridge University Press.

Preston, S. H. 1976. *Mortality Patterns in National Populations: With Special Reference to Recorded Causes of Death*. New York: Academic Press.

Simon, J. L. 1977. *The Economics of Population Growth*. Princeton, N.J.: Princeton University Press.

_____1981. *The Ultimate Resource*. Princeton, N.J.: Princeton University Press.

United Nations 1963. *Population Bulletin of the United Nations No. 6, 1962, with Special Reference to the Situation and Recent Trends of Mortality in the World*. New York.

_____1980. *World Population and Its Age-Sex Composition by Country, 1950-2000: Demographic Estimates and Projections as Assessed in 1978*. New York.

_____1981. *World Population Prospects as Assessed in 1980*. New York.

World Bank 1980. *World Development Report 1980*. Washington, D.C.

_____1981. *World Development Report 1981*. Washington, D.C.

Commentary by J. Lawrence Angel

Department of Anthropology, Smithsonian Institution

I'm going to change the subject slightly, because I am not an expert on the subject of modern demography. I'm going to give you a little bit more of my work in the eastern Mediterranean on ancient demography and make two points, one of which is that the demographic transition in life expectancy from our hunter-gatherer ancestors to our modern industrial era postdates 1900. It is a very, very recent thing, as I think was also brought out earlier. The big increase in population resulting from this demographic change, then, obviously involves the survival of infants who would "normally" have died.

The other change, which is equally difficult to deal with, is that whereas females died before males in all of the prehistoric societies that I know of, they are now outliving them. That, of course, also has an effect on the number born.

It would seem to me then that the solution to the population increase has to be contraception; it has to be family planning. This seemed to me what wasn't really stressed in the essay. I don't know how this is going to be accomplished. We know what the solutions are, but we don't know how they are going to be carried out.

Discussion

DR. CLAXTON: A first question or comment?

DR. PASSMORE: Could I just ask a question—and it is purely for information. You mentioned Kerala and Sri Lanka and suggested that there was no political peculiarity. Of course, Kerala was, at the time, the only communist state in India and Sri Lanka had Dr. Bandaranaike as Prime Minister. Was there no government policy at all arising out of these two special political conditions at that time?

DR. NAG: In both India and Sri Lanka, the government adopted a policy of fertility reduction. Kerala as a part of India, of course, has reduced fertility. Kerala, because of the communist government there, has done certain things, like land reform, which make it atypical.

Sri Lanka, does not have a communist government, and has done better than many other countries. But as to the policy regarding fertility reduction or mortality reduction, I don't think there was much difference between the governments.

DR. LASLETT: I would like to make one comment about this demographic discussion and that has to do with the ineluctable working of demography. It doesn't seem to be generally recognized—at least there is no reference in this paper—that sudden decreases in fertility lead very rapidly to serious aging of the population.

If China succeeds in a one-child per couple policy, I estimate that in the first years of the next century they will have a tiny, inadequate base of active population, in a country whose industrial future is extremely tenuous, trying to support vast masses of elderly.

Whatever we think about the encouragement of a rapid birthrate decline for all of the nonindustrialized countries of the world, we ought to recognize what will happen if those policies are successful and if the countries have a population distribution like England, where 20 to 25 percent are inactive or retired, all of them based on technologically static producers younger than themselves. We can't advocate and succeed in bringing into operation a demographic policy which has to do with fertility without that consequence. The prolonging of life increases the population of the elderly. It is not an increased life expectancy which ages the population in that sense, it is the rapid reduction in fertility.

DR. MCNICOLL: I've no doubt that the one-child policy in China is an extreme case. If the extent of success of that policy came anything close to being an average of one child per family, clearly the policy would have gone far too far, and that would be recognized long before that event.

I see no reason that those costs wouldn't be anticipated in China, as in any other country. So while the aging of the population is certainly a problem, I think in China as elsewhere, it is a much longer-term problem than you suggest. And the rapid decline in fertility in China in the past two decades clearly will have its own aging consequences, even with the 1 percent growth rate that you have there now.

DR. CLAXTON: Could I just add, things aren't perfect. If they go on as they are, most of these developing countries will not succeed in any of their objectives. If they are able to obtain a significant reduction in fertility even down to the level of a two-child family by the year 2000, which is impractical for most of them, nevertheless it will take forty or fifty years before they achieve a population profile like that, say, of England or Sweden or even the United States, which has a considerably higher proportion of small children. During that period, if this is indeed a manageable activity—which of course is questionable in the beginning, but countries are at least trying to do it—they will be able to modify this policy. But the alternative of continuing as they are is for most of them intolerable, if tolerable means to succeed in anything they are trying to achieve, such as providing adequate food, adequate health, education, whatever.

DR. LASLETT: I'm merely drawing attention to the fact that aging is connected with birthrate reduction, which is not normally recognized by those who discuss the problem of demographics.

DR. MCNICOLL: Yes, we certainly accept that.

DR. HASSAN: It seems also if you have high fertility you have a similar problem in terms of dependent's producers, because on the one hand you have the elderly and on the other hand you have the children that you have to support. If this is a short-term problem, then we have to look at the long-term problem, if you want to get the job done.

DR. MASTERS: I think that Peter Laslett's point about the extent to which current trends have already established the age profiles of coming generations is very important. One demographic factor is the correlation between an excess of adolescent males, who do not participate productively in the economic system of their society, and warfare. If we look at everything from the Crusades to major wars throughout the world, we have to realize that the diffusion of conventional weaponry, combined with the existing age distribution in many countries of the Third World, is a recipe for repeated warfare. And future changes in demography are too late to change this.

DR. MCNICOLL: I think that is a more controversial point.

DR. MASTERS: It is intended to be controversial.

DR. CLAXTON: But you can go back one step and refer to the internal instability of countries without getting into the subject of external warfare. The internal instability of the kind which has appeared in Teheran, in El Salvador, and in Guatemala is more urgent in time and is equally destabilizing within a country.

DR. MASTERS: I have to say, as a political scientist, that the most frequent technique for solving internal instability is external aggression, and certainly Iran and Iraq are a perfectly good example.

DR. MCNICOLL: My reading of the evidence leaves the question about the relationship between conflict and population growth very much in doubt in terms of those sorts of conclusions. Clearly, when you have rapid urbanization, that is, growth rates of 5 or 7 percent a year in cities, you get all sorts of social conditions which are conducive to lots of different types of social pathology on account of the pure speed of change. I think regarding the sorts of changes we are talking about—2½ percent per year in population growth, perhaps with somewhat greater growth for the young adult population—any short-term connection between that and warfare and conflict and even, as we put it rather cagily in our paper, between that and totalitarian or authoritarian government, is difficult to demonstrate.

DR. TOULMIN: Could I just put in one more factor here? It does seem to me that migration is associated with warfare. The historical record is that migration tends to turn into warfare when suffering states seek to prevent the migration.

The importance of migration here is that it is one of the ways local problems are prevented from remaining local problems. I think the assumption that we can identify the population problem and treat it is based upon the assumption that sovereign states can in fact operate as closed systems. Now, the existence of uncontrolled migration is one of the ways in which the attempt to operate social states as closed systems breaks down and the migration becomes a force majeure. So I do think that Dr. Masters's point has to be taken into account.

DR. MCNICOLL: But when you look for the evidence of that, then everyone points to the First World War, and I doubt very much whether that is typical.

DR. TOULMIN: I had in mind the Haitian boat people.

DR. MASTERS: And I had in mind things like the French Revolution.

DR. CLAXTON: Dr. Scrimshaw, could I ask a question that is brought up by this population discussion? There is one source of concern that I have with your paper. It is undesirable to leave the idea that the current pressing problems of food, population, and environmental deterioration can be solved without slowing population growth. In talking about population matters, I try to be careful to say that this is only one part of a large problem, which you refer to and which you quoted in your oral statement here, to which I would have added the matter of health.

You referred to food, population, energy, and environment as closely interrelated global problems, and health belongs in there too, because these feed on or contribute to each other. The great difficulty I see in the scientific approach to many of these problems is that scientists tend by training, nature, and practice to deal with a single subject.

All of us together need to try to educate the people who make the policies and run the governments to carry them out—to the degree that governments can have an influence—to understand the interrelations of these matters. This was, I think, a valuable contribution of the Global 2000 study, which was put out recently.

This perspective is still not widely understood or even accepted, but it is a problem which I think I commend the designers of this symposium for bringing up—for bringing together these subjects which have a very distinct interrelationship.

DR. SCRIMSHAW: Obviously, I fully agree with you that health should be added to that list.

12. Demographic and Microstructural History in Relation to Human Adaptation: Reflections on Newly Established Evidence

PETER LASLETT

Fellow of Trinity College, Cambridge

Editor's Summary. The historical perspective involves a much shorter time scale than the time scale of biology. Even so, the changes that have taken place during human history are monumental. Peter Laslett, a historical sociologist, reviews aspects of human adaptation in historical time. He emphasizes that the ultimate success in human adaptation has been industrialization, which has taken place in recent historical time. An important source of data on human adaptation during this period is provided by demographic history, which is the only record having both an unmistakable biological as well as cultural response. Furthermore, it is the only body of historical data which can be brought unequivocally within the realm of the natural, exact sciences. Using the recently recovered demographic record of England, the first society to industrialize, Laslett hypothesizes that industrialization occurred first in northwestern Europe because of the marriage pattern unique to that area; these Europeans married uniquely late and many did not marry at all. Even more important, almost every marriage in that region gave rise to a new, economically independent household. This pattern of marriage does not appear to have been sanctioned by any ethical traditions or norms but rather by what Laslett calls noumenal normative rules or intuitively derived, generally accepted rules of behavior.

A topic as abstract and general as the one we are here to discuss makes the matter of fact historian somewhat uneasy. For one thing it seems to inspire a diffuse anxiety in the commentator, a tendency toward sententious disapproval of the present and gloomy forebodings about the future. These are scarcely the business of the historian, even if he feels impelled to call himself an historical sociologist, which I should have to do. Moreover, unmanageable holdall concepts appear to command the dialogue: expressions like "humanity," "civilization," or "cultures." To which social units, if such they are, almost equally elusive vicissitudes are held to occur: "evolution," "diffusion," "decline," and above all "revolution."

Such processes, we are perpetually reminded, have taken place as human adaptation has proceeded over lapses of time immensely larger than those to which the historian is deliberately confined. This has to be so because biological time, in which adaptation takes place when humans are considered as organisms, has a scale of a quite different order from historical time. The temptation, therefore, to the theorist attempting to relate events and developments which belong to the two dimensions, is to try to elongate historical time by extrapolation, curve fitting, and such operations. And this means conjectural history.

If the question of how humans adapt has to be undertaken by invoking conjectural history, it might be best for the patient recoverer of the knowable past to offer no opinion on that elusive but important topic. For conjectural history almost always requires the use of two expedients which are peculiarly ungrateful to him. It requires in the first place that a sliver of somewhat dubious knowledge shall be spread "like gold to aery thinness beat" over very extended time periods. The tens of thousands of years, for example, which are thought to have been occupied by the hunter-gatherer era before the great "revolution," as a result of which settled agriculture and peasant social organization appeared in the world, have to be understood in terms of scattered and fragmentary archeological evidence difficult to place chronologically, supplemented by such sources as cave paintings. When it comes to early agriculture, things like the Homeric poems and the Judaic scriptures are pressed into use. Now, the use of plastic and literary evidence of this kind, as I have had occasion to insist elsewhere, puts the scholar in the position of looking the wrong way through the telescope. He is faced with the baffling problem of what he should make of the overlapping images, the oblique reflections, the chromatic aberrations in front of him, and of how he should relate them to the archeological findings.[1]

But the second expedient which conjectural history requires is even more disconcerting to the historian because it forces him into deliberate anachronism. The "history" of the hunter-gatherers of Europe turns out to imply the attribution to these remote ancestors the characteristics we can now observe in the extant hunter-gatherers of Africa or Australia. We have to suppose that we see in the paintings on the walls of the caves at Lascaux in France a culture (a form of society in action) whose modes of operation were indeed so close to those of

PETER LASLETT

our contemporary Bushmen or aborigines that their survival was ensured by identical, or very similar, traits of individual and social behavior. Only by such a strategy can we suppose that the transition from hunter-gathering to agriculture was in fact a form of human adaptation. Only by such procedures can we attempt to add historical examples to the biological evidence relevant to the state of human adaptation.

These are somewhat dubious historical procedures—I think you will agree—and there has to be an element of conjectural history even in the discussion of the one particularly well-documented instance of human adaptation which I shall discuss, an example in which biological adaptability can be seen to interplay with cultural. It is, as far as I know, the only such example yet available for *historical* time—the era of written communication and record.

The evidence for this example of adaptation comes from the history of Western society and singles out the mechanism of adaptation which it possesses and has possessed for many centuries. The view will be defended that we are now in a position to make a bold hypothetical inference from known and reliable historical materials as to how the populations of northwestern Europe were able to adapt themselves so aptly to their environment—to climatic, demographic, and biological vicissitudes—that they scored the most conspicuous of all successes as to human adaptation. For in the end, it was the West which gave rise to industrialization, the transfer from underdeveloped or traditional to modern or industrial society. This has indeed transformed the relationship of human populations and polities to the human environment for those peoples to whom it has occurred. The particular file now open to examination and reflection happens to be that of England itself, the first industrial nation. In using the word "modern" and the concept "modernization," I must not be held to imply that I accept all or any of the theory which equates the technological and social history of a country like Britain with the "development" which occurs or which people hope will occur in contemporary undeveloped societies.

Now for the hypothesis, which goes as follows: it was because northwestern Europeans married uniquely late in their lives and were able to vary the age at marriage; because such a relatively large number of them—again a variable proportion—never married at all; and above all because almost every marriage in northwestern Europe over its known history has led to a new social unit (a new household coming into being) that northwestern Europe was the area of the world where industrialization first appeared.[2]

It had been recognized before the end of the eighteenth century by an Englishman, T. R. Malthus (1766-1824), in his *Essay on Population*, that marriage regimes were crucial to human adaptability. At that time it had not become apparent that the technological changes already gathering pace in England were going to transform the relationships between population and subsistence. As an English historian, I have chosen to discuss these issues initially in the terms which Malthus himself used, since he was the first person to work at a model of human adaptability which combined both biology and culture. Or so Charles Darwin thought, and even Karl Marx, much as he disliked the gloomy parson and his dismal science.

Our hypothesis is bold and crude enough to suit any number of versions of conjectural history, you may be disposed to say. Before I go on to expand, expound, qualify, and criticize this position about human adaptation in Europe—or in the European west and northwest—the West of ordinary political discussion[3]—there are some further observations which should be made about the argument to this point. The first is that the adaptive success of the West in attaining industrialization must not be taken as indicating that only such social structures as those of England in early modern times can be supposed capable of effective adaptation, short of industrial transformation.

For the salient fact about all extant human societies—whatever the level of their material culture, technical sophistication, and degree of development—is that they have adapted for survival just as well as we have. Otherwise we should not be contemporaneous inhabitants of the late-twentieth-century globe. It is not the difference between the "primitive" societies of our own day and ourselves which should be our primary interest, or even the differences between them. What we should be after is the differences between them and those social organizations and cultures which once existed in the past, but which failed to adapt well enough to persist to the present. And this is a very difficult question indeed for a historian to investigate. To read into the society which gave rise to the cave paintings at Lascaux in France the features exhibited by contemporary cultures which seem to us to be similar helps in no way to provide information on instances of human adaptative *failure*. Indeed, it has a tendency to attribute to all possible candidates just those features which they may have lacked.

It might be urged that an argument of this kind lacks realism, if only because the units under discussion are so difficult to define precisely. This is true at all points in time and therefore between all points in time. It is impossible to be certain when a society, as it adapts itself to changing circumstances, ceases to be itself and becomes another society. There are many reasons for this, of which our imperfect grasp of how the social world itself is structured as well as how it is related to human biology, is no doubt the most important. There is also the awkward fact that, although the adaptive mechanisms of humans are almost entirely cultural, especially in historical time, they are sometimes biological as well. It is often impossible to make a confident decision as to how much they are the one and how much the other.

Cultural adaptation, moreover, can be tricky. One of its expedients, which happens to be of some significance to our theme, is imitation or *mimesis*, as Arnold Toynbee used to call it. In our own generation, almost every known society other than those which have already industrialized successfully is busy imitating, trying to imitate, or being urged or even forced to imitate the particular collection of social traits which goes to make up industrialization. If any society should fail in this mimetic endeavor, if it should be resistant to development, we shake our heads, however sentimental we may be about the virtues of traditional social arrangements. We prophesy continued and intensifying impoverishment and even starvation, especially in view of the tendency of most societies in this position to be increasing in total numbers. In other cases, we talk of possible or probable extinction, and there are instances enough to justify such forecasts.

Nowadays, in our sententious fashion, we refer to the obliteration of such human associations as genocide and count it as a crime against humanity. But there is nevertheless another possibility of looking at what may happen, one which in no way condones the destruction of any of the world's societies, or the bringing upon its members all the misery of cultural deprivation, but one which illuminates the processes of human adaptation in an unexpected way. We can see it as adaptive mimesis in an extreme form, a people imitating another and hitherto more successful people so completely that they cease to be themselves. They become the still faintly discernible indigenous element in the population of Hawaii today, or the now completely latent Pictish constituent of the people of Scotland, and so on.

Testimony to the physical, the biological persistence of populations in spite of the atrophy of the polity, the institutions, the language, indeed of all the phenomenal elements which once existed as the historical embodiment of the societies which these populations once supported, is to be found in the work of the geneticists. They are perpetually uncovering geographical frontiers in the distribution of shapes of the skull, for example, or of blood groups which are claimed to correspond to the demographic areas occupied by peoples who have disappeared. A whole gradation is evident here, a gradation in demographic or biological survival by means of mimesis.

From the entirely submerged Incas at the one pole, we can proceed by way of the Maori, still very much alive in contemporary New

Zealand but clearly very different from the Maori of yesteryear, to the Japanese, whose mimetic activities have been so successful that in our own decade they have begun to excel the societies which they set out to imitate in order to survive. They excel us in fact in just those attributes I have already singled out as both quintessentially Western and of the greatest significance to the issue of successful adaptation: the industrial arts.

A feature of such mimesis to be observed both in the Japanese and in the Maori—but even more conspicuously in the American Indians— is that some of those elements in their culture which we think of as entirely characteristic were in fact developed in the very earliest days of their contact with the industrial West and developed occasionally perhaps as a defensive mimetic reaction. Only those who interest themselves in the Maori are likely to know that the large and marvelously decorated hall-like structures, which are rightly regarded as one of the great achievements of Maori culture, only appeared after the great sea captain James Cook first initiated continuous contact between these peoples and the West. But every American knows that the Indians had no horses until the European invaders introduced them into the North American continent, after which horse riding and fighting on horseback became one of the hallmarks of the American Indian way of life.

Considerations like these should make it apparent how intriguing, but complicated and puzzling, issues of human adaptation can be when the cultural and biological dimensions of human behavior are brought simultaneously into play. If this essay were to be written out at greater length, other and even more intricate examples could be found. Nevertheless, the fact which faces us is that demographic history provides the only historical record with unmistakable biological reference. Unlike the indications from sources previously used, including the literary and plastic sources already touched upon, it is a record which can be converted unequivocally into numbers and which can for that reason be brought unambiguously within what are usually called the natural, exact sciences.

During the last twenty years or so, we have at last succeeded in deciphering from the parish books drawn up by the clergy of the Christian Church a remarkably reliable demographic history for a century or two in the career of individual communities in some western European countries. But the newly established English demographic study goes very much further, since it covers the whole country, using hundreds of sets of village documents. It is the longest, the most detailed and informative of such national histories, and it has been recovered for nearly 450 years (or fifteen English generations) from

PETER LASLETT

1538 to 1981.[4] The challenge is to try to discover as much as we possibly can from this body of data as to what has been the relationship between the social and biological which did indeed ensure that the English should have been the first to score the ultimate in adaptive human successes: the capacity to industrialize. It has to be read, of course, along with the economic and technological course of English development and in comparison with the more partial indications available for shorter periods for other European countries and for a society decidely outside the European tradition, Japan. It is also necessary to make inferential statements, well supported as historical statements go but much less secure than those for the Wrigley/Schofield era, about medieval England.[5] Here then is the important conjectural element.

A further set of facts has to be set out before we turn our attention to the actual mechanisms of demographic adaptation over historical time. Those of the world's populations whose breakneck pace of growth has caused such justifiable anxiety in the last twenty years are all what the demographer calls high-pressure populations, or they were until the spread of medical knowledge from the developed countries began to reduce their death rates. A high-pressure population is one where a very high birthrate is met by a very high death rate, so that population growth need not be excessive and need have no tendency to cause numbers to outstrip means of subsistence. Birth and death are very frequent, however.

A similar balance in a low-pressure population is one where much smaller birthrates are met by similarly smaller death rates, and both birth and death are infrequent. The population of what we call an advanced society like that of the United States or Britain is one where exceedingly low birth and death rates obtain and where, at least at the moment, there is a tendency for a net loss rather than a gain in numbers to occur, at least over short periods. The great object of recent world demographic policy has been to find ways of ensuring that the high-pressure population regimes of undeveloped countries shall undergo the demographic transition which has occurred in all developed countries and thus become low-pressure regimes.

Though it has been assumed that every one of the world's populations not belonging to a developed country is and always has been a high-pressure population, two further circumstances have now to be reckoned with. One is that the extant hunter-gatherer peoples have low—not high—population pressure. This has been recognized by demographers for some time. The other circumstance, however, is novel and can be said to have been firmly established knowledge for the first time in the work of Wrigley and Schofield already cited. This finding is that, since the sixteenth century, English population has never been a high-pressure population at all, and the indications are that this could have been the case for an indefinite period before the 1500s. One of the implications of all this must not go unnoticed. What I have called the Western mode of controlling the relationship between population and subsistence may well have been in operation for a very long time in England, before even Malthus began to think about it and before the Great Industrial Revolution manifested itself.

We cannot linger on these points, but it should now be apparent why the revolution which conjectural history assigned to the transition between hunter-gatherers and agricultural peoples is potentially of great importance to our topic. It should also be clear that the historical example from English population and social structure raises issues about adaptation and survival which range in time and place very widely indeed. An approach may become possible to examine specifically biological adaptation over recent time.

Until a year or two ago, for example, it looked likely that the English data, along with other information from such unlikely sources as a listing of the Orthodox Christians living in the city of Belgrade in 1773-74, might yield evidence on what seems to be almost an entirely biological or genetic matter: age at menarche in women.[6] It seemed, in fact, as if the fall in that age might be construed as a piece of natural selection going forward in recorded history, since there was a case for its being secular, monotonic, and still in progress. But now it appears that this one possibility of a contribution from history to biological change will be able for the moment to make no further progress than the establishment of the fact of a decline since the late nineteenth century, now leveling out. Other possibilities remain, however. These are the earliest moments in the intellectual appreciation of a fresh dimension in a long-established inquiry.

When Malthus first formulated a theory of how a human population could, or should, adapt to its economic resources through the regulation of marriage, he had no knowledge of the propositions about the demographic history of his country which have been set down above. Nor did he associate regulative marital behavior with the composition of the family group, or even comment on the fact that in Russia, which he actually visited, and the other non-Western countries whose relevant record he consulted, different marriage regulations were associated with different shapes of the familial group. It has been left to an American working in the present decade to demonstrate that the universal early marriage of the great mass of Russian serfs went with large and complex families, the largest and most complex yet known to the historical sociologist.[7]

PETER LASLETT

Malthus was a Christian priest and must have supposed that Christianity was the sanction of the marriage behavior which he believed to be the only human and humane method of population control. To be added to this, of course, was custom and tradition. Rational reckoning on the part of all affected individuals as to the consequences of their action could not be relied upon to ensure that everyone behaved in the indicated fashion. For him, in short, religion and custom mediated between individual decision and its social outcome. This may seem an unsatisfactory proposition, but no one seems to have got much further.

At this point, in fact, we reach the crux of the relationship between the cultural and the biological in the question of how humans adapt. No student of collective comportment of the kind which determines how large the next generation shall be in relation to the present one is likely to conclude that the decision is made on rational grounds by every procreating couple. He is forced like Malthus to suppose that it was the system of social relationships which contains, and always has contained, within it the relevant programmatic instructions and that these instructions must have been resident in the collective mentality of the population concerned.

It is easy to demonstrate this for industrial as well as for preindustrial societies. To recognize that fact we have only to consider the position of our own contemporaries, especially in America, who divorce themselves so easily and have so few children. It is demographically certain that they will spend a high proportion of the remainder of their lives as elderly persons, likely to be wholly dependent at the last on the psychological support of a child—usually a daughter—who is unequivocally and uniquely their own. Yet they behave in such a way as to lower the probability of having such a daughter at all, and so as to run a high risk that if daughters should appear, a close and reliable connection with an individual amongst them will be much less likely than ever it has been before. This is because they may well find themselves in a miscellaneous set of individuals who have undergone a whole intricate history of separation, divorce, remarriage, and consensual unions in which relationships are diffuse and unlikely to give rise to the reliable connection of aging parent with responsible child which has been characteristic of the traditional, Western, monogamous, familial group.

Now this is not rational, calculative behavior, and it patently goes forward in evident ignorance of the implications of the decisions being taken. What is true today seems likely to have been true for all our predecessors and for the the so-called primitive peoples of the present. Historical demographers are not disposed to believe in the adaptive efficacy of individual actions in ensuring the survival and success of whole populations. They have to conclude that such wisdom as does exist and has existed is for the most part collective wisdom. Quite how the fertility decisions of individuals give rise to the adaptive behavior of the group remains problematic, even in the context of our chosen historical example.[8]

For Malthus, no doubt, the solution to this problem was the existence of Christian providence, which watches over humanity in general and individual nations like the British in particular. He was, after all, the near contemporary of Adam Smith, with his notorious confidence that the rational, competitive, economic decisions of individuals constituted the hidden hand which ensures the general economic good. Still, the outstanding characteristic of Western marital and familial behavior—and one which Malthus never ceased to stress—is that it places responsibility for procreative decisions firmly on the individual or the couple who do the procreating. It is they who have to find most of the resources to establish the household, to live independently for as long as they do live, and to maintain any offspring which they engender. Very little is or was shouldered by the family group into which man or wife was born, and practically none of it can or could be shifted from a particular family group to the kinship network, at least in English history. There could be no more effective way of ensuring that adaptation to available resources should be brought home to those whose actions actually determined the size of the group which will succeed them. It also brings the family, its independence and means of subsistence, up against the collectivity, as we shall see, in a way which is often found surprising.

It is tempting, and perhaps justifiable, to see in these circumstances the reasons why the Western pattern was so much more subtle in its responses to the environment than one where marriage could take place without the foundation of a new household. Under such circumstances—under pronouncedly non-Western conditions—couples could and can have children to be cared for and supported within a large coresidential group of associated parents, brothers and sisters, or even more distant relatives. The kin network could be relied upon for subsistence in addition to the family, and universal marriage at a low and invariant age could be a feature of the system. The Russian serf familial pattern is our most useful example of such a polar contrast with the West. I have to admit, however, there are points of inconsistency even here and that the evidence is in a very preliminary condition, especially on the matter of extended kinship.

Once this position is formulated, moreover, and it is as far as I am able to go in suggesting why the Western familial system should have been associated with Western adaptive success, a whole list of further

PETER LASLETT

qualifications and confessions of ignorance suggest themselves. They should convince us that the issue of human adaptation is as yet poorly understood, even in relation to the present case. I have no space here even to write out a list of these points. There are one or two subjects, however, which I would like to put in front of you, in the confidence that the exceptions and the qualifications can be understood as being stated.

The first of these topics is the ever-fascinating one of extramarital procreation. To Malthus, who lived at a time when illegitimacy was particularly high and who died just before it reached a 300-year peak in his country,[9] engendering of individuals outside matrimony was anathema. It defeated the whole system of control, and illegitimacy was the worst form of that vicious behavior which had to be proscribed and which seems to have included contraception itself. For Malthus, in spite of the association of his name with contraceptive practices and devices, especially by the French, never advocated familial limitation and never said anything to imply that it went on in the England of his day. In this, as in other particulars, he has been shown by the most recent research to be remarkably percipient as an observer.[10]

We must add that Malthus had no confidence that the "passion between the sexes" could be expected ever to diminish. He made his disagreement quite plain with those idealists of his generation, such as William Godwin, who saw in the progressive human perfectibility associated with revolutionary radicalism a prospect that procreation would wither away, and individuals live forever. On the points of illegitimacy and contraception, Malthus was quite confident about the preventive check which had to be applied. It was morality, Christian morality, the established ethical system of the English Church, the English State, and English society, for which there was near universal respect and which could be supposed to be universally obeyed if all the forces of State, Church, and opinion could be brought to bear.

Such work as has been done so far on sexual nonconformists, however, has signally failed to bear him out on this point or to confirm any supposition that religious belief or authority could do all the work so often marked out for it in traditional, preindustrial society. Moral rules, as every social scientist knows, are not behavioral rules, and norms may exist not so much to be obeyed as to be used for the purposes of justification of self and disapproval of others. Though most procreation undoubtedly did take place within marriage in the Christian West, it has been reckoned from the English record that "roughly one-fifth of all first conceptions in England may have been extramarital, that the proportion was usually much more like two-fifths and could easily reach three-fifths." (See Laslett et al. 1980, p. 55.) There is also evidence that during the lifetime of Malthus all the girls in a particular village could virtually give up marriage in church as the first step in their procreative career (see the remarkable findings of G. N. Gandy cited in Laslett et al. 1980, p. xiii). Needless to say the priesthood, the moralizers, and the respectable in general were horrified when such facts were brought to their notice, but to little avail. It would seem that the procreative limits established in English society did not depend on morality or on normative regulation to the degree previously supposed.

In fact, one of the conclusions which we have tentatively arrived at in the historical study of familial behavior is that rules of household composition were more consistently carried out than the rules of sexual behavior required of the unmarried. But rules of household composition are not ethical norms. Defiance of them does not bring with it the sanctions of public shame, clerical condemnation, and sententious declarations about a decline in the social fabric. They seem to exist and to have existed without public sanctions and without internal conscientious sanctions either. They were like nothing so much as Thomas Hobbes's natural laws, rules of prudence, of what was advantageous, secure, and desirable to do and which everybody granted to be so. There is no *ethic* of familial composition as far as we can see in English or Western tradition, in spite of the suggested importance of familial composition to Western adaptive ecological success.

Accordingly, we have named such regulations *noumenal normative rules*,[11] noumenal as opposed to phenomenal, and normative because they refer to decisions and behavior which are perceived as matters of choice even if not usually as raising ethical issues in themselves. The possible existence of such regulations is obviously of considerable importance to our general subject, for they may well cover other aspects of the relationship between population and subsistence, and so have a role in human adaptability on its cultural side. If the program metaphor can be accepted, the socially encoded programmatic rates were statements of this character. To identify a regulation as a noumenal normative rule may perhaps be as far as we shall get in explaining the way in which Western populations adapted as they did to environmental conditions. Such an identification tells us nothing, of course, about how such rules arose and why they differed between cultural areas, and specifically between the West and almost everywhere else as to procreation. The relationship of rules of this kind with custom has also to be decided.

Age at marriage itself may rest on a noumenal normative rule, which in the West specifies not a particular age but a limit below which marriage almost never takes place. It may occur nevertheless, in extreme conditions, as it did in the Chesapeake Bay area of the United States in the late seventeenth century, though without any of the notoriety that a "moral collapse" would have brought about. Similarly for proportions marrying, which must rest in the end on a rule which lays it down that independent, responsible household headship, requiring the necessary economic resources, is the appropriate aim for all young persons, but not to be attained by every one of them.

Another and very interesting example is that of breast-feeding which, it has recently been shown,[12] was characteristic of English society in all classes in town and country, in all parts of the country and over the whole of the 400 years we are dealing with. Breast-feeding to the point where weaning is physiologically indicated effectively places a limit on marital fertility, and the suggestion is that this was a reason marital fertility was consistently low in England, from place to place and time to time. There were areas of Europe itself, even as close as Germany or parts of France, where breast-feeding was nothing like as widespread and which even approached high-pressure demographic regimes. This arose because early weaning and artificial feeding led to wives becoming pregnant again sooner and to much higher infant mortality. Though fewer babies were born in England than elsewhere, they consistently survived better. All of which implies that rules of this kind, whatever their nature, must have been culturally specific and programmatic in the way suggested in the mechanism of ecological adaptation. Hence it may be justifiable to classify England as quintessentially Western for our purposes or even to associate English industrial precocity with that fact.

Malthus himself seems to have been so little aware of these possibilities that, like many other anxious commentators on poverty and procreation in preindustrial England, he had no compunction about using the law of the land to enforce the demographically adaptive comportment which he felt was inescapable, if the whole society was not to become impoverished. However, he did not advocate defining illegitimacy as a crime against the state, which had been done at earlier points in English history. Nor did he suggest that improvident marriage should be declared contrary to the law of the land, as it was in southern Germany at the time. It did not occur to him, or to anyone else at any time in the West, to make the form of the family a matter of social compulsion—though a close approach to that position is to be found in serf peasant Russia. In that country, the *mir*—the village community—had the power along with the serf owner to decide whether or when a household should be allowed to divide and a new household spring up. In England, all such decisions were left to the individual prompted by the noumenal normative rule.

What Malthus did advocate, however, was the withdrawal of legally approved social support for those whose improvidence as to marriage and procreation had caused them to become casualties in the system. He solemnly suggested that, after a certain date—long enough ahead to give notice to those who would be affected—the elaborate and highly effective English Poor Law system should be abolished entirely. Maintenance by the collectivity therefore should come to an end for those whose procreative indiscipline threatened the balance between population and resources. This suggestion has been responsible for much of the obloquy loaded on Malthus ever since, especially by radical and progressive critics. Its significance for our subject, however, is for the most part, as in the case of genocide, independent of issues of morality. It demonstrates that Malthus seemed to have been unaware of the historical relationship of family to collectivity in England. It is to this relationship that we now turn our attention.

In a predominantly nuclear family system, little provision can be made for demographic casualties. Widowed persons cannot return to the families of their parents, because the rules of family composition forbid it. Orphans for the same reason have no place. In their final years, the childless or the never-married are likely to be left without company and without support in the most defenseless and dependent part of their adult lives. If the nature of familial "nuclearity" as a whole is such that kinship bonds between families do not carry means of relief, then it will be seen how devoid of all resources the casualties in such a system could be.

Close investigation of these points is not yet complete. It does not confirm that Western nuclear family rules denied support to all casualties, demographic or economic, in England where they seem to have been the most strictly, consistently, and homogeneously obeyed— homogeneously as to region, class, and period. But research in progress seems likely to establish the following:

(1) That casualties due entirely to demographic vicissitudes were considerable even under the low-pressure regime characteristic of England;

(2) That under that regime these casualties, especially orphans and widows with children, could not be spared easily if perpetuation of the population was to be ensured.

(3) That welfare flows along kin lines were minimal.

PETER LASLETT

The last is perhaps the most surprising of these points. It cannot be regarded as unquestionably established, but we have found very little evidence to contradict it. Sons and daughters living away—and they were virtually all living away after marriage—did give psychological and even some economic support to aged parents where they could and where personal association was possible since they lived in the same place. This was a society marked by high levels of population turnover and of extensive, if mostly local, migration so that proximity was not always easy. Brothers might similarly occasionally support sisters. But the flow of resources for the support of the unfortunate seems to have ended at this point. There was no kinship agency beyond the nuclear family which could or would assist that family or individuals in it, should they come to grief.

The outcome could only have been as follows. In a nuclear family system, such as was established in England and over most of the West, much of the responsibility for the casualties rested on the collectivity and little upon the family, however the family is defined. By collectivity we here mean the Church, especially in the Middle Ages, the local community at all times and places, voluntary relief and transfer agencies, but, above all, the State itself with its powers of taxation. In this rather schematic sense we are apt to say that the Welfare State in England is as old as English history. It is neither an innovation of recent generations or of progressive social thought, nor is it an unprecedented extension of the polity into the area previously reserved to the family.

The relevance of what has been hypothesized in the last few paragraphs for the welfare debate of our own day in both our countries must be clearly evident, and the historian has become aware of extraordinarily similar debates taking place at intervals over modern English history. A notable feature of English preindustrial society, which seems not to have been much noticed until recently, was the system of transfer through collective agencies from those parts of the society which were prosperous to those which were not. The latter was appreciably the larger class. The theory with which we are working implies that these transfer arrangements go back historically as far as we can study the English microsocial structure.

We do not yet know enough to say to what extent the flow of welfare was evoked, sustained, or administered by the central government itself, and it has been insisted that the Church and other agencies, many of them informal, were of great importance. Even begging had a part to play, but the role of government in the process fascinates a historical sociologist, if for only one reason. To carry out tasks of this character, government—central and local—had to be efficient, society-wide, informed and responsible, and coordinated as well from locality to locality. To some extent, therefore, though how far we are not yet in a position to say, the polity itself played a regulatory role in maintaining the population as a whole in balance with the environment. We have here one of the origins of the administrative process of the state, an essential accompaniment of industrialization and of the life of high industrial societies like our own.

What prompted those royal and ecclesiastical officials, local administrators, philanthropists establishing relief funds and institutions, and poor law officers in the preindustrial past? What prompted the many intellectuals who wrote about poor relief, its objects and functions, about begging, and about how to cut them both down? What prompted Malthus himself? Not easy questions. I have tried to suggest that there were programmatic principles embedded in collective attitudes, and have given them a title. They are as close as we can yet approach the regulative agency which was finally responsible for the adaptation of the population of the society of the West to its means of subsistence before ever industrialization made its appearance.

This has to be a somewhat disjointed commentary and cannot be said to have a conclusion. There are many matters on which all that has been done is to make suggestive hints. The possible connection, for example, between *mimetic* acculturation and noumenal normative rules, their acceptance or their rejection by the society which has to imitate or to face degradation, has to be left as a hint. You may be disposed to think that a very unhistorical, undevelopmental, undialectic position has been sketched out, since Western attitudes, social rules, and institutions have been handled as if they have been static over time. This is unfortunate, for I most decidedly do not suppose that the elusive reality in question—"Western" or "English" social structure in its reaction to its changing means of subsistence—was the same in the 1300s and 1500s as it became in the 1700s and 1800s. It would be a colossal and an exhausting task, however, to describe the changes which took place. When all was done, it would still tease us with the problem cited earlier as to when it can be said that one of these cultural constructs has really changed or has stayed the same.

If there are those among you, nevertheless—especially the anthropologists—who detect a diffusionist attitude toward culture and change and toward adaptation in particular, I should not complain. Evolutionism in this line of inquiry has had a long run, especially in its catastrophic, revolutionary form. The time has come, I believe, for the interplay between the two poles, diffusion and evolution, to be weighted up and reconsidered.

PETER LASLETT

Notes

1. "The Wrong Way through the Telescope; a Note on Literary Evidence in Sociology and Historical Sociology," *British Journal of Sociology* (1976), concerns itself particularly with the evidence of Homer in relation to the structure of the family group in ancient Greece and of the Old Testament in relation to the political institutions of the Hebrews.

2. The demonstration that the Western, and specifically the English, familial form may have had an intimate connection with industrialization, or early industrialization, is primarily geographical in fact. Factory industry and geometric economic progression appeared in England first and next in contiguous western European areas showing a familial form very close to that of the English, and sometimes identical. Other areas of Europe differed, though not markedly, until the Mediterranean is reached in the south and the Danube in the east (see Laslett, "Family and Household as Work Group and Kin Group, Areas of Traditional Europe Compared," in Wall, Robin and Laslett, eds., *Family Forms in Historic Europe*, 1982). Small differences over many centuries, it is argued, are cumulative in their effects.

   There are other elements in the hypothesis. The Western familial system was at one and the same time accumulatory and exploitatory; it implied high rates of population turnover, especially in the younger age groups in the labor force, indicating labor mobility; neolocal household formation led to a high degree of specialization and division of labor in the handicraft productive process; it also created a considerable demand for consumer durables, and effective purchasing power, since, pace Malthus himself or the view taken of Malthus by many of his followers, the sensitive relationship between population and subsistence prevented population increases from absorbing increases in wealth. The whole case is discussed in preliminary form in an unpublished essay, "The Family and Industrialization" (1977), available at the Cambridge Group. It has been translated into German in *Sozialgeschichte Der Familie in Der Neuzeit Europas*, edited by W. Conze (pp. 147-60), Stuttgart: Klett.

3. The rather awkward disjunction in phraseology between "Western" and "Northwestern" Europe, with the word "European" sometimes being left to cover the area concerned without any further geographical specification, arises because of the terms originally used by John Hajnal in 1965, the first theorist in this line of inquiry. See his famous article, "European Marriage Patterns in Perspective," in Glass and Eversley, eds., *Population in History*. It is of considerable interest and significance that southern as well as eastern Europeans are, or were, outside the "West" for these purposes. This has the somewhat unexpected implication that the Florence of Leonardo and the Italy of Galileo and Vesalius were outside the area due to transform itself in the way which interests us.

4. It is a remarkable irony that this record should in some important respects be less informative for the last 150 years than it is for the first 300, an outcome of the British policy on the accessibility of sensitive statistics. For

the earlier, the much longer and hitherto unavailable part of this story, see *The Population History of England 1541-1871: A Reconstruction*, by E. A. Wrigley and R. S. Schofield (Harvard University Press, 1982). The book contains not simply purportively exact population totals by five-year intervals over the whole of modern English history to the 1870s—indeed back into what is usually called the Middle Ages—but also the following demographic indices for the entire population of the country for each such interval: growth rates; gross and net reproduction rates; expectation of life at birth; distribution of the population by age groups; proportions marrying; proportions emigrating. Cruder indices like birthrates, marriage rates, death rates, and illegitimacy ratios have also been precisely calculated.

Age at marriage, on the other hand, which has been cited above as crucial to the case being argued here, has had to be worked out by a quite independent operation called family reconstitution undertaken on particular villages with exceptional documentation. This statistic is therefore not available for the whole English population but for a collection of some twenty parishes only, and is confirmed by a great deal of other evidence. Measurements of the size and composition of households derive from yet another file, based on all English listings of inhabitants known to be extant from the era before the official census (in England in 1801), supplemented by similar documents or results from countries in western Europe and in other parts of the world. These operations and these collections are those of the Cambridge Group for the History of Population and Social Structure. A bibliography is obtainable by application to 27 Trumpington St., Cambridge CB2 1QA, England.

5. See R. M. Smith of the Cambridge Group, "Fertility, Economy and Household Formation Over Three Centuries," in the *Population and Development Review* (Dec. 1981) and its references, especially Dr. Smith's work on medieval England.

6. See Laslett, "Age at Sexual Maturity in Europe since the Middle Ages," chap. 6, in *Family Life and Illicit Love in Earlier Generations* (1977). The authoritative work on the topic, a classic of biological history, is J. M. Tanner, *A History of the Study of Human Growth* (1981).

7. See Peter Czap of Amherst College, "A Large Family; the Peasant's Greatest Wealth, Serf Household in Mishino, Russia, 1814-1858," forthcoming (1982) in R. Wall, J. Robin, and P. Laslett, eds., *Family Forms in Historic Europe*, with its references and other writings in the course of publication by this scholar and by Andrejs Plakans of the State University of Iowa, a contributor to the same publication.

8. See E. A. Wrigley, "Fertility Strategy for the Individual and the Group," in Charles Tilly, ed., *Historical Studies in Changing Fertility* (1978).

9. In the early years of Queen Victoria, 1855-65; see Laslett, Oosterveen, and Smith, *Bastardy and Its Comparative History*, 1980, pp. 14-17; and especially an article in *Local Population Studies* (Spring 1981).

10. See Christopher Wilson's dissertation on intermediate variables in the study of fertility, an examination of English historical development, sub-

mitted to Trinity College, Cambridge, September 1981, convincingly establishing the fact that the signs of contraception detected by E. A. Wrigley in the 1960s in one particular village cannot have been a widespread pattern. If they had been, of course, the whole marital mechanism discussed in this essay would have been unnecessary.

11. See my discussion in a paper entitled "Typologies of the Family," prepared for a meeting of the International Union for the Scientific Study of Population in Sao Paulo, Brazil, August 1981. Its arguments were repeated and developed at a meeting at Mt. Kisco, New York, a month ago.

12. On all this see the essay by Wilson already cited and especially the references to work done by him with Professor John Knodel at the University of Michigan, as well as Knodel's own earlier work on the topic.

References

Braudel, Fernand 1975. *Capitalism and Material Life.* Vol. I. New York: Harper & Row.

Czap, Peter 1982. "A Large Family, the Peasant's Greatest Wealth: Serf Household in Michino, Russia, 1814-1858." In R. Wall, J. Robin, and P. Laslett, eds., *Family Forms in Historic Europe.* Cambridge: Cambridge University Press.

Godwin, William 1976. *Enquiry Concerning Political Justice.* London and New York: Penguin.

Hajnal, John 1965. "European Marriage Patterns in Perspective." In D. V. Glass and D. C. Eversley, eds., *Population in History.* Chicago: Aldine.

Knodel, John 1977. "Breast Feeding and Population Growth." *Science,* vol. 198.

Laslett, P. 1976. "The Wrong Way through the Telescope: A Note on Literary Evidence in Sociology and Historical Sociology." *British Journal of Sociology* 27:319-42.

_____1977. "Age at Sexual Maturity in Europe." Chap. 6 in *Family Life and Illicit Love in Earlier Generations.* New York: Cambridge University Press.

Laslett, P., K. Oosterveen, and R. Smith 1980. *Bastardy and Its Comparative History.* New York: Cambridge University Press.

Laslett, P. 1981. "Typologies of the Family," paper presented at IUSSP meeting in Sao Paulo, Brazil, August 1981.

Malthus, T. R. 1914. *Essay on Population.* London: J. M. Dent & Sons.

Smith, R. M. 1981. "Fertility, Economy and Household Formation over Three Centuries." *Population and Development Review* 7:595-622. Tanner, J. M. 1981. *A History of the Study of Human Growth.* New York: Cambridge University Press.

Wrigley, E. A. 1978. "Fertility Strategy for the Individual and the Group." Pages 135-54 in Charles Tilly, ed., *Historical Studies in Changing Fertility.* Princeton, N.J.: Princeton University Press.

Wrigley, E.A., and R. S. Schofield 1981. *The Population History of England 1541-1871: A Reconstruction.* Cambridge, Mass.: Harvard University Press.

Commentary by Nathan Reingold
Joseph Henry Papers, Smithsonian Institution

I have no reason to question what Peter Laslett reports about the marriage and family patterns of the English. He may even be correct in ascribing to these phenomena why the English were the first to industrialize, although I am not prepared to accept without qualification his verdict that, and I am here quoting, "the capacity to industrialize is the ultimate in adaptive human successes." Let us, however, skip by that point with the sole observation that industrialization is simply one kind of adaptation, and by no means devoid of pitfalls.

A more interesting problem is what Laslett's position implies for the commonality of economic historians and historians of technology. They are inclined to argue whether the Industrial Revolution was the result of a push or a pull, whether technology or science are exogenous forces, if the revolution goes back to the early sixteenth century or to the years around 1700 or to the last third of the eighteenth century. These no doubt tedious arguments have one characteristic lacking in Laslett's paper. They deal with the events and participants and things of the Industrial Revolution: the factory owners, merchants, skilled and unskilled laborers, machines, raw materials, and finished products. Without a clear linkage to such as these associated with the term "Industrial Revolution," perhaps we are dealing with a chronological coincidence, and not the elemental requirement for that momentous historical transition.

My suspicion is that Laslett has already, or will soon, present such a linkage. Until then, I am skeptical. My skepticism is sharpened by my memory of his wonderful book on the world we lost when industrialization arrived. There he showed clear evidence for differences in height, weight, longevity, and so forth, between classes. No class distinctions occur here. Were wealthy investors and skilled laborers alike in the respects he regards as crucial? The Industrial Revolution did not occur all over England, but in a few specific localities, notably Lancashire, the Midlands, and London (according to Musson and Robinson). Why there—not elsewhere—if the demographic pattern prevailed? Since he finds at least one village with an unusual interest in contraception, can Laslett be sure that the industrialized regions did not differ in some demographic sense from his English norm? What I am driving at is that no theory of industrialization in England will be convincing unless it somehow unequivocally gets

362   PETER LASLETT

the historical drama going—and going right—in those dates and places for which we have so much evidence.

Let me give an example. The coming of steam power is obviously important. Because of the Newcomen engine, the date 1700 (or so) has its adherents. The older accounts credited James Watt's improvements in the engine with transforming the energy basis for manufacturing, yet we now know that the older form of engine continued in wide use, particularly in mining, for decades after Watt. It isn't immediately apparent to me how historical demography explains the origins and development of steam power. That topic is clearly important for any explanation of the origins of the Industrial Revolution.

My remarks, while critical, should not convey the impression of hostility. I am very impressed with his contribution to this symposium. It is a fascinating piece, very readable and provocative. Like all the others submitted to this symposium, it labors under the burden of a great theme not easily amenable to the essayist, no matter how skilled. Laslett is certainly right in his claim for the general significance of the findings associated with him and his colleagues. I am particularly fascinated by his "nuomenal normative rules." There is still a real question here of how they work in concrete situations, how they are inculcated, how they are enforced, where they are likely to be flouted. Perhaps Wrigley and Schofield tell it all. I haven't seen the book. But Laslett himself has given a few marvelous glimpses in his work referred to previously.

As I read his words, I was reminded that Marx somewhere noted that deferred gratification was always a prerequisite for capitalistic growth. Judging by Laslett's figures, the English of those days did not defer gratification so much as limit its form to the nuclear family or the nuclear family coming into being. The treatment of widows and children in the absence of an extended family appears parallel to the deferring of gratification. It does not seem fundamentally too different from Marx. Demographic prudence acts as economic prudence here. That accords well with the jaundiced opinions of the English held by some Scots, Welsh, and Irish. Except that at one point Laslett carefully avoids saying "British."

That raised a little question in my mind. Despite the references to family strategies in other parts of the world, Laslett has a problem even if his demographic explanation for England's first is validated. Belgium was probably the second country to undergo the revolution. Was Belgium the same as England demographically?

DR. LASLETT: Yes.

DR. REINGOLD: If so, why wasn't it first? If not, was its industrialization by a different route? That question applies to all industrializing countries.

Commentary by Mark N. Cohen
Department of Anthropology, State University of New York

I must confess that when, at the beginning of this symposium, Catherine Bateson raised the question of *who is the subject* when we ask how *we* should adapt, I didn't catch its significance. That issue has been discussed several times since then, and the meaning has become gradually clearer. It became clearest to me in reading and rereading Dr. Laslett's paper. I find myself wondering whether we should consider ourselves successful if we succeed in shepherding the human race as a whole into the middle of the twenty-first century or whether we should also be trying to keep the British Empire intact. I offer that comment in jest, of course, but there is a serious note behind it.

One of the lessons of history or, if I may be allowed, prehistory is that the stewardship of cultural evolutionary advance moves around rather rapidly from place to place. It is clear that it has never focused on any one location very long. Civilizations fall even while the human race continues. I wonder to what extent, as we contemplate proposals for the future, we are willing to think about adapting to the twenty-first century while passing the stewardship to some other part of the world? I am not sure we have really faced up to that issue in this group.

Dr. Laslett has—in the written and verbal versions of his essay and in his symposium remarks—thrown down the gauntlet to anthropologists, archeologists, prehistorians, and the like. He has raised questions about data quality and the degree of definition of detail to which we aspire. I must say that I am in some sympathy with his position—up to a point. I used a similar metaphor in a book preface several years ago. I suggested that archeologists, faced with poor resolution of detail, should turn down the power of their microscopes, should stop pretending to quite so much precision of detail as if they were looking under high magnification, and should look to see what patterns appeared of a more general sort over a broader field. So, up to that point, I accept his criticism of our science.

But, on the other hand, if one does turn down the power of one's microscope and broadens the field, there is the chance to see patterns that high magnification does not allow. It is possible to see events in their long-term, historical context and to see comparisons between events in a variety of cultural contexts. This, in turn, greatly adds to our ability both to interpret these events and to test our interpretations. One of the most important such tests is to see whether an explanatory phenomenon really correlates across time and space with the thing it is supposed to explain.

As an interpreter of archeological data, I am concerned to see whether explanations match the scale of the event they are supposed to explain. For example, in the debate about the origins of agriculture, I have been among those archeologists critical of models that describe in great detail why people began to farm in, say, the Valley of Mexico, but somehow fail to notice that the same thing was happening in other places at the same time. Such explanations are incomplete at best. A local factor cannot explain a general phenomenon. At least part of the explanation of any phenomenon has to be a recognition of the common elements it shares with similar phenomena elsewhere.

I can't get the word "skeptical" out with quite the elan that Dr. Laslett does, but I am very skeptical of the kinds of interpretation that historians often provide. I think often they don't have a sense of scale or sequence or range of events that they need to compare when making their explanations. An example appeared yesterday in the discussion. (I don't pretend knowledge about its factual content; I wish only to call attention to the structure of the argument.) Someone described a correlation between various historically recorded social upheavals, including the French Revolution, and the existence of a particular demographic structure, one in which—as I recall—there was an exceptionally large cohort of young men in the population. Whether that is factually defensible I don't know or particularly care, at least for the moment. What I would argue is this: if the pattern is true, anyone who takes on a case such as the French Revolution and analyzes it in largely local, idiosyncratic terms (i.e., terms peculiar to France of the period) without noticing and including this demographic phenomenon in his analysis is being naive in the interpretation of events.

Dr. Laslett interprets the Industrial Revolution in England as a unique event. Certainly he can defend his position that it was unique at least in some respects, but I would argue that the analysis would gain from more careful consideration of the event as one of a larger class of revolutionary transformations. Moreover, if one chooses—as Dr. Laslett has—to emphasize and explain the uniqueness of an event, certain problems in interpretation arise. In particular, it becomes very difficult to use correlation (even at an intuitive level) as a check on the proposed interpretation. There is only one case.

It seems to me that it is, then, incumbent on someone who offers such an interpretation to offer more evidence for—and a more proper test of—his interpretation than Dr. Laslett has done. This might take one of various forms. First, he could show that, *within* the one broad region he is talking about, there is geographic or temporal correlation between the two variables he is discussing. In essence, he could show that the one case is really composed of several separate smaller cases across which a correlation could be tested. Second, he could, and should, explain more fully and convincingly why the two variables are—and must be—related.

Discussion

DR. WOOLF: Thank you very much, Dr. Cohen. That leaves the floor open again to the principal first speaker, if he wishes, or to others.

DR. WIERCINSKI: From the data of, among others, my wife, on the mean age of death in Poland, it appears that it increased with sudden acceleration after the late medieval period. Do you think, sir, that it is possible to induce deliberately on a worldwide scale convenient feedback mechanisms into this demographic explosion, is my first question, and second. . . .

DR. WOOLF: Can we hold the second and try to get at the first question first?

DR. WIERCINSKI: But the second is closely related to the first. You have said that you don't believe in paleodemographic data.

DR. LASLETT: Can I interrupt? I didn't say I didn't believe it. I said I don't know about it, and that I need a much better demonstration than I am familiar with.

DR. WIERCINSKI: I am just trying to provide that. Namely, again, from the investigations of my wife, it appears that there are good measures of quality of life based upon morphological skeletal data, one of which is the length of the longbones. Other barometers of the standard of life are the degree of sexual dimorphism in the longbone length and the variation in the age of death. It appears that all of these traits are considerably correlated with each other, so this correlation proves that there exists a relationship between paleodemographic data, in this case, the mean age of death, with other biological parameters for the standard of life.

Secondly, this paleodemographic data, as is proved by a well-known

PETER LASLETT

Hungarian demographer J. Nemeskeri, oscillates even in Great Britain, in a very sensitive way with the oscillations of the standard of life, because of wars and many other factors which may change the standards.

DR. WOOLF: Let me see if we can sift that observation into a question or two from your statement, and maybe I can assist the panel in addressing it. If I understood you correctly, in the first question you asked about the possibility of a feedback mechanism on a worldwide scale.

DR. WIERCINSKI: Convergent feedback mechanisms.

DR. WOOLF: We hope the feedback mechanisms converge; otherwise, they don't work. The feedback mechanism on a worldwide scale for population control, and this, if I have heard you correctly, is to be tied to certain measures out of the paleodemographic past, that you and your wife have selected to support the possible design of such a mechanism, to wit, the length of longbones, the degree of sexual dimorphism, and the age of death, if I have at least expressed the points you have raised.

DR. WIERCINSKI: Not exactly, and I am sorry. Perhaps I can make myself clear. The question is that the exponential curve of Deevey published in the 1960s is based on estimations comparing the demographic data, so first of all we should believe in the law of demographic data, and secondly, if we have this curve based upon these data, then how do we deal with it.

DR. WOOLF: I think the question has been refined.

DR. LASLETT: Is it wished that I respond to what has been said, or do you want more comment?

DR. WOOLF: I think a response is called for.

DR. LASLETT: Well, can I confess that my ignorance of these arguments about paleodemography doesn't make it easy for me to respond to that very learned and interesting exposition we have just heard from Warsaw. I would simply say this: Do I believe that some archeological discoveries in a particular part of the Near East are indicators of a global demographic movement? The answer is no. The notion of a global change (aside from some type of cosmological event) that could have a worldwide indication is unjustified. One simply cannot take individual events and particular pieces of evidence in a particular part of the world and generalize from that for humanity.

I am not suggesting I know what happened to humanity, but I don't want to pretend I know when I don't. Can I regard this as a map of Europe, with John Bull's other island there, and the Iberian Peninsula there, and Scandinavia there? I am asked on what the industrialization hypothesis depends. I admit everything you say about industrialization.

I would have gone on to say that a prime weakness of this attempt to get at an interest in the problem and an interest in the body of data was based upon the ecologic capacity, it's simply a geographic coincidence. The argument that cumulative capital in relation to ecological vicissitude is typical of a particular area of western Europe is what is contained in it, but I don't intend, of course, to pretend that I can say that it had greater influence on Bradford or the Mill Valleys of Yorkshire than it did in Surrey or Sussex. I am not clearly providing that kind of argument. I know very well you are right in saying every other area of industrialization has to be taken into account, but the area of the Western marriage pattern is specified by Hajnal, and it covers Britain, Ireland, Scandinavia, northern France, and the northern areas of Iberia.

I would like just to return to the nature of the evidence which I have asked the symposium to consider. England has 10,000 villages. Now, if we turned up any archeological evidence from, say, twenty of them or thirty of them, we have got a sample of twenty or thirty out of 10,000 in one country. The effort which has been made is to collect enough material to establish what amounts to a random sample of a whole country, and of course we find enormous variance, not perhaps as much as we find, for example, in the parameters that interest me in the structure of the household. What is interesting about England is, its extraordinary homogeneity, but nevertheless, a piece of evidence from any one of these villages upon which you build a general theory is extremely unlikely to be anything but speculation.

DR. WOOLF: Let me hold you there. There were other hands raised. Perhaps we can broaden the discussion.

DR. CAVALLI-SFORZA: I think we are indebted to Professor Laslett for this beautiful example of British understatement. One of his understatements I fully subscribe to, although I must admit that it is a piece of excellent conjectural history, using the terms that you used . . .

DR. LASLETT: I admit that.

DR. CAVALLI-SFORZA: . . . and that is the importance of the family. One thing I wanted to do was to focus attention on the fact that the French, to whom historical demographers owe some debt, I think, have some interesting information. There was one book that appeared recently that has a fancy title—*The Invention of Facts*—and it is mostly based upon geographical comparisons of the demographics. I will summarize it in two sentences. It shows that there are three types of family structures. One is more like the nuclear one, which belongs to the Franks, who entered into the country in the fifth or sixth century A.D. and is similar to the British type of structure. Then there is the

PETER LASLETT

northwestern type, which is different, and the southern type, which is more an extended family. Everything today in France, including the last political elections, correlates very strongly with the family structure.

This is a geographic correlation where the strength of geographic correlation is not very high, but there are other considerations in this book that make this picture a little stronger than I could make in two sentences. I just think it is very much worth considering. This is an important hypothesis: that the structure of the family has been a very central point in the development of human culture.

DR. HASSAN: I think one of the major problems here is that we are still thinking about current problems in the context of nineteenth-century experience, whereas the problems we are facing today are global. The concept of nations does not make any sense because of the whole world network and nuclear warfare; I think this regional orientation is a problem.

So, it is important to shift from the mentality of the nineteenth century or early twentieth century, when boundaries were meaningful, to a new global mentality, in which anthropology becomes a very important field of inquiry. If we look at the differences between history and anthropology, which Dr. Laslett has pointed out, the pitfalls of history, or what we probably call microhistory, because history doesn't begin with the fifteenth century, unfortunately, human history is more than the last 300, 400, or 500 years, and what happened during these episodes does not reflect the odyssey of human adaptation and the various biological and cultural changes that have happened on the larger scale of human evolution.

The anthropological approach to understanding adaptation is far more inclusive than history, because it looks at the relationships between economy, family, sociology, what have you, without just focusing on an enumeration of events.

DR. WOOLF: I hope you will not think me unkind, but I would suggest you come to a narrow point.

DR. HASSAN: The point is that conjectural history is a misnomer, because all knowledge is basically conjectural. We can say that geology and cosmology are conjectural history and therefore we shouldn't do that. Conjectural history is basically a misnomer, for history—all history—is conjectural, so to speak, and what we have is also an emphasis on information rather than an emphasis on explanation.

DR. LASLETT: I may live in the nineteenth century, and I might belong to a stage of scholarship which is surpassed. All I want to know is what evidence *is* evidence on human adaptation? I don't care what its sources are, but the evidence has to be evidence that demonstrates or contradicts a theory of human adaptation. I call myself a historian with modesty, but I also call myself a historical sociologist, and anybody who knows anything about the development of social sciences in the neck of the woods that I come from would know that the social sciences, particularly anthropology, realizes that it has to have a sense of time, and that no explanation is worth living with.

DR. WOOLF: I would like to insert a small comment and raise a question. Peter Laslett referred to what was happening under Malthus's nose, and the question I would raise is precipitated by a comment from Dr. Cohen about the passing of stewardship. My question is: What are the phenomena of this epoch taking place under our noses, industrially and technologically, and how do they tie to the passing of the stewardship, for example, from the industrialized West to Japan?

13. The Environment of the City

ASA BRIGGS

*Lord Briggs of Lewes, Provost of Worcester College, Oxford, and
Chancellor of the Open University*

Editor's Summary. Asa Briggs argues that the city is a product of human
society and arises from human social nature. Thus it is inappropriate to view
the city as an artificial and hostile environment for human existence. The view
of cities as a natural environment for humans is particularly important since
urbanization is increasing throughout the world, particularly in the Third
World. Briggs emphasizes that we need to view this symbolic center of modern
human existence as a part of our whole experience. Thus if there is something
wrong with the city, there is something wrong with our whole society.

No survey of the human odyssey, however sketchy or cursory, could
leave out the city . . . the city as place of survival, often precarious,
highly vulnerable survival; the city as center of civilization or rather
of richly varied civilizations in time and space; and, not least, the city
as metaphor, with the metaphor itself twisting and turning through the
centuries into old and new shapes.

The etymological fact that in the Western tradition the words "city"
and "civilization" have a common root, along with "citizenship" and
"civility," points to the second of these aspects of the city as an
influence in history . . . to something far more than survival, to the
temple and the theater rather than to walls or shelters, to the creativity
of the individual and of the society, and to the enrichment of human
culture. So also does the haunting preoccupation through the centuries
with the "ideal city," the city of dreams, to which restless and striving
men should aspire; and at this point, of course, fact turns into metaphor,
the metaphor not only of the New Jerusalem but of Babylon.

Yet the first aspect of the city—as place of survival or destruction—has its scaffolding of imagery also . . . Venice under the sea, T. S. Eliot's tumbling of the towers. When Lewis Mumford, a leading twentieth-century surveyor of cities, wrote his second massive book on the city, *The City in History*, twenty-three years after the first, *The Culture of Cities*—with a world war and the atomic bomb in between—it was to the theme of the city as insecure citadel that he returned when pondering characteristically on the relationship between first and last things. "Urban life," he remarked, "spans the historic space between the earliest burial ground for dawn man and the final cemetery, the Necropolis, in which one civilization after another has met its end."

Science fiction takes over in the twentieth century where history ends, often attempting a complete rewriting of history in the process. The city can become nightmare rather than dream as "civilized" relations crack. Yet the technology of the citadel can be strengthened. In one of his short stories, "Caves of Steel," Isaac Asimov envisaged 800 cities on earth with an average population of 10 million. "Each city," he added, "became a semi-autonomous unit economically all but self-sufficient. It could roof itself, gird itself about, burrow itself under. It became a steel cave, a tremendous self-contained cave of steel and concrete." In other science fiction the city returns as symbol after disaster . . . in what I. F. Clarke has called, appropriately, "the Aeneas theme." As the hero in *The Death of Grass* goes out toward a new settlement, he says to the heroine, "There's a lot to do. A city to be built."

The building, adaptation, and transformation of cities has been a major human achievement, providing us in the twentieth century with a whole "prospect of cities," even the newest of them already historically layered. Paradoxically, destruction has often uncovered civilization: thus, the bombing of London in World War II revealed Roman London for the first time. If the twentieth century has changed skylines, the nineteenth century created a whole new city network underground, a network of pipes and sewers, a technological triumph even greater than that of the Romans. We are constantly reviewing our assessments of the achievements of previous ages, not only contrasting present with past but finding in historic cities "similes and analogies for the contemporary architect and urban designer." Napoleon III thought of himself as a new Augustus. Nor is it only autocrats who turn back to the past. As one of the contemporary American architects interviewed last year by Barbarales Diamenstern put it: "When this new wave of architects comes out of the schools, with a sense of caring about context, it seems to me that cities are going to have the concern that you see in a place like Florence . . . some sense of continuity even with changing styles."

It is because both city builders and city dwellers can compare one actual city with another actual city, and not simply with the ideal city, however envisaged, that time scales are as significant as the use of space in judging the appearance of cities. There were more nineteenth-century references to Florence as a particular "place of concern" than there have been twentieth-century references. In Britain's industrial Birmingham, for example, "adventurous orators" in the 1860s would "dwell on the glories of Florence and of the other cities of Italy in the Middle Ages and express the hope that Birmingham too might become the home of noble literature and art."

There was, indeed, a double framework of historical reference in the nineteenth century, with some city reformers and commentators looking back to the city states of the Middle Ages and the Renaissance and some looking further back still to the city in the ancient world, the Greek *polis* and the Roman *civitas*. In each case, there was a strong sense not only of continuity—and of community—but of civic pride. At a time when the actual cities of a new industrial society were generally thought of as problem places, this pride was conspicuous. It was in the United States, not in Britain, that F. C. Howe could write in 1903 that through the city "a new society has been created. Life in all its relations has been altered. A new civilization has been born, a civilization whose identity with the past is one of historic continuity." It was in Communist Poland that historic Warsaw was reconstructed after 1945.

City pride most usually meant not pride in the city but in particular cities, each one was recognized as having an individual identity. Philadelphia was different from Boston or Baltimore or Cincinnati or Chicago; Manchester was different from Liverpool or Birmingham; Warsaw from Cracow; Budapest from Vienna. In Britain, the historian Edward Freeman complained bitterly that some of his contemporaries could not understand how "the tracing out of the features and history" of particular cities could be "as truly a scientific business to one man as the study of the surrounding *flora* and *fauna* is to another." In the attempt to make history "scientific," analysis and imagery could become somewhat confused, as they were when ancient organic metaphors of the city were given new life. Yet the more emphasis is placed on particular cities and the differences between them, the closer we can draw to lost experience. The best nineteenth-century observers recognized (as clearly as Jane Jacobs has done in recent years) that "city processes in real life are too complex to be routine, too particularized for application as abstractions. They are always made up of the interaction of particulars, and there is no substitute for knowing the particulars." This is as true of Vienna or Paris or London as it is of New York.

Yet, though cities as environments have to be treated separately before we can start to generalize about urban structures and styles—and some, like Venice or Kyoto, are visually unique—no city has ever been, in fact, as self-contained in history as in Asimov's short story, least of all Venice. The city has come into existence and developed—sometimes declined—through interdependence both with rural hinterland and within a wider system of cities, linked through trade. The market place has mattered at least as much as both temple or cathedral or fortress and walls. There have, of course, been capital cities which have been above all else centers of power and display, but for every capital city, rival of other capital cities, there have been many cities which above all else have been centers of commerce. As we classify cities or rank them in hierarchies, we can never leave economics out, whether we are concerned with buildings or with ways of life. Indeed, the most fascinating feature of the study of cities is that it must take account of so many subjects which are too often considered separately—economics, demography, geography, ecology, history, sociology, political science, anthropology, architecture, archaeology, to name only some of the most obvious.

No self-contained discipline can cope with the city or with cities. Nor, moreover, are all the disciplines taken together quite enough. Ignoring for a time the ideal city—for this we have to turn to the philosophers—the real city has to be explored before it can be explained. Freeman's historian contemporary, J. R. Green, was once described by Lord Bryce as exploring a strange town and "darting hither and thither through the streets like a dog trying to find a scent." On this side of the Atlantic, Robert Ezra Park, pioneer of the Chicago school of urban sociology, the first academic school of its kind in the world, was fully aware of the need to find scents even in a city which was not on the surface, at least, "strange." In inviting his students to explore Chicago, he always stressed that "the city is not . . . merely a physical mechanism and an artificial construction. It is involved in the vital processes of the people who compose it: it is a product of nature and particularly of "human nature."

There was a similar awareness in Park's British predecessor, sociologist Charles Booth, whose vast survey of London, then the world's largest city, during the last years of the Victorian Age, entailed as much exploration as that then associated with the names of Livingston and Stanley in Africa. "It is in the town and not in the country," Booth wrote, "that *terra incognita* needs to be written on our social maps. In the country the machinery of human life is plainly to be seen and easily recognized. . . . The equipoise on which existing order rests, whether satisfactory or not, is palpable and evident. It is far otherwise with cities, where as to these questions we live in darkness."

In some respects, we still live in darkness almost a hundred years after Booth—despite the boom of the last quarter century in urban studies, specialized and interdisciplinary—although we are perhaps clearer now than Freeman, Park, or Booth were about the influences of the explorer's own attitudes and experience on the selection of facts about the city which he chooses to collect and the images which he seeks to present. The same city means quite different things to different people, even to residents of the city, and in considering impressions we have to distinguish between those of residents and visitors, of privileged and deprived, of reformers and boosters, to note only a few of the relevant categories. If the boom in urban studies has increased our understanding of the city, it is mainly through a sharper realization of the different elements involved in our diverse perceptions, visual and social, of the city. In other words, we have to add psychology to the list of associated disciplines necessary for understanding. As one of the most stimulating recent British writers on the city, Peter Smith, has put it, "Experiencing environment is a creative act. It depends as much upon the subject interpreting the visual array as upon the disposition of objects in space."

During the nineteenth century the collection of facts about the city was one of the most active preoccupations of a new generation of statisticians, some involved in boosting cities, some in problem-solving within the city, and no account of nineteenth-century positivism would be complete without taking stock of it. During the late twentieth century, however, we have focused our attention more on the range of human experience within the city and the perceived pluses and minuses associated with it. Of course, we have left to experts—who were not there in the early nineteenth century—the practical tasks of dealing with the city's pressing problems: surveyors, engineers, traffic analysts, housing managers, leisure controllers, social workers, and above all, planners. There is a gulf between the two kinds of approach, and in recent years the "expertise" of each of the expert groups has been subjected to increasing scrutiny. Meanwhile, city tensions multiply as the volume of writing about the city at every level, not least journalistic, increases. The nineteenth century talked of "the age of the cities." We talk of the "crisis."

I shall end this lecture with four reflections on the contemporary scene, when some observers, like Melvin Webber, have been claiming for more than a decade that we are moving into "the post-city age."

Since we are concerned in the symposium, however, with an odyssey and not with a contemporary scenario or scenarios, I would like to continue, for reasons that I have already given, to range widely over time and space. As Patrick Geddes, Mumford's mentor, put it succinctly in 1905, "A City is more than a place in space, it is a drama in time."

Before we dwell on our current preoccupations, which turn as much on survival as on civilization, it is necessary to acknowledge that in perspective there has seldom been any consensus about the role of the city in human affairs. There has usually been a debate, often crude, occasionally sophisticated, with some crossing of sides. The Christian Bible begins in a garden and ends in a city, and both before and after the Christian Bible, both garden and wilderness have been pitted against town, city, and conurbation. At times, as we have seen, classical literature swayed later generations at least as much as the Bible, both in its portrayal of the urban and of the pastoral. Of course, we quickly move into metaphor here, as we do in Albert Camus's twentieth-century notebooks, where he writes that "as a remedy to life in society, I would suggest the big city. Nowadays, it is the only desert within our means."

The modern debate preceded the Industrial Revolution and was not a by-product of the rise of the industrial city, which was described in one magazine of the late 1830s as "a system of life constructed according to entirely new principles." Go back to the 1770s, the decade of American independence, and you have on the one side William Cowper's unforgettable lines, "God made the country, Man made the town," and, on the other, Dr. Samuel Johnson's almost equally famous rebuke to Boswell, "No, sir, when a man is tired of London, he is tired of life; for there is in London all that life can afford."

In newly independent America, too, the city had its detractors and its defenders. One of the best-known passages in Jefferson's *Notes on Virginia* is that in which he asserts that "the mobs of great cities add just so much to the support of pure government as sores do to the strength of the human body. It is the manners and spirit of a people which preserve a republic in vigor. A degeneracy in these is a cancer which soon eats to the heart of its laws and consitution." The conception of the city as cancer—the organic metaphor gone wrong— was never to disappear thereafter. Indeed, for this reason, biology and physiology should doubtless be added to the list of associated disciplines which have been applied to the study of cities . . . and not merely through imagery, as in Jefferson's case, but through theory, like the theory that city growth depended ultimately on the infusion of healthy tissue from the countryside, on different and older demographic patterns.

The late eighteenth-century debate, which found a place for noise and nuisance as much as for numbers, often looked backwards. Yet it had many new ingredients also. Thus, at the very time that there was talk of "cancer," the new word—and it is difficult to think that it was a new word—"civilization" was coming into use. Related though it was historically to the word "city," the word "civilization" did not come into use until the late eighteenth century. Johnson might sing the praises of London and question the delights of the countryside, yet as late as 1772 when Boswell discovered him preparing the fourth edition of his folio Dictionary, he learned that Johnson would not admit "civilization" as a word, but only civility. "With great deference," Bowell went on, "[he] thought *civilization* from *to civilize* better in the sense opposed to *barbarity* than *civility*."

In a decade of dramatic change, which also saw the introduction of the new word "technology" and of Watt's steam engine, not to speak of the drafting of the American Declaration of Independence and the publication of Adam Smith's *Wealth of Nations*, "civilization" as well as "the city" were already controversial subjects. Indeed, the decade ended in London with urban riot and the open expression of what many Londoners thought of as barbarity in the heart of a great city. Boswell, in talking about the words "civilized" and "civilization," did not add that there was already an alternative vocabulary, itself to become controversial, pivoted on the words "cultivated" and "culture," words derived not from the city but from the countryside.

Some cultivated people then and later were skeptical about or hostile toward "civilization," as were romantic writers like Rousseau and Wordsworth, the former comparing cities with prisons, the latter pointing both to the association of the city with crime—the adjective he used was "dissolute"—and with meaningless bustle, "the same perpetual whirl of trivial objects, melted and reduced to one identity." "Civilization itself is but a mixed good," Coleridge was to write, "if not far more a corrupting influence, the hectic of disease, not the bloom of health."

Ambiguous responses to the idea of "civilization" were equally apparent two decades later, when John Stuart Mill in his brilliant essay on Coleridge, much admired in the late twentieth century, attempted to draw up a balance sheet measuring "how far mankind has gained by civilization." Mill was less interested in centuries-old contrasts between the urban and the pastoral, or in romantic evocations of nature as against culture of the rough against the polished, than in a qualified utilitarian assessment of social and cultural change.

On the credit side, Mill recorded the multiplication of physical comforts; the advancement and diffusion of knowledge; the decay of superstition; the facilities of mutual intercourse; the softening of manners; the decline of war and personal conflict; the progressive limitation of the strong over the weak; (and) the great works accomplished throughout the globe by the cooperation of multitudes." Not all these manifestations of "civilization" explicitly or obviously derived from the city context, although there was a tendency then and later—not least within the Chicago school of urban sociology—to relate all social indicators, socially favorable or unfavorable, to the influence of the city and of urban lifestyles. Certainly when Mill, following Coleridge, identified the items on the debit side, he had the city very much in mind. They include "the creation of artificial wants," "monotony," "narrow mechanical understanding," and "inequality and hopeless poverty"—even though "monotony," at least, and "inequality and hopeless poverty" had often been and were still being associated as much with the countryside as with the city.

Then, as now, it was possible to argue about whether or not the city as such was a causal factor, rather than the society as a whole, and how to weight the different items in the balance sheet. What is curious to note, however, is that there was no specific reference in Mill's balance sheet to the rise of a new kind of city, the industrial city, the advent of which to some extent turned the terms of the argument and in the shadows of which we have lived ever since.

For although there had been cities since the beginnings of recorded history—and earlier—it was only at the time when Mill was writing that it was possible to speak of "the age of great cities." In 1800 there were only twenty-two cities in Europe with a population of more than 100,000 and none in America. By 1850 these numbers had increased to forty-five and eight; there were also four cities in the world with a population of over a million. By 1900 there were 160 cities in the world with a population of more than 100,000 and nineteen with a population of over a million. Significantly, there were as many as twenty-three with a population over 500,000, including new products of the century—often described as "prodigies"—like Chicago in the United States and Melbourne in Australia. The population of London, "the world city," had risen to more than 4 million people. Patrick Geddes called it an octypus or polypus, "a vast irregular growth without previous parallel in the world of life—perhaps likest to the spreading of a great coral reef."

In one turbulent decade, the 1880s, the number of cities of between 40,000 and 70,000 in the United States increased from twenty-one to thirty-five and the number of still bigger cities from twenty-three to thirty-nine, so that the young American scholar A. F. Weber could proclaim in the last year of the century that "the tendency towards concentration or agglomeration is all but universal in the Western world."

Industrial cities constituted only one group of cities in this huge urban expansion, and even the most renowned of them, like Manchester, which was a Mecca for visitors during the 1840s, often became service centers as much as manufacturing concentrations, serving the needs of an adjacent industrial region. It became fashionable, indeed, to classify cities like flora and fauna as well as to deal with them individually or to trace the general processes of urbanization. Yet it was the industrial city which shocked contemporaries into an awareness of the social implications of urbanization. Manchester, where facts were worshipped, became a symbol. In Manchester, wrote the most famous of all nineteenth-century travelers, de Tocqueville, "humanity attains its most complete development and its most brutish; here civilization works its miracles, and civilized man is turned back almost into a savage."

There were four features of the industrial city which received particular attention from critics: deterioration of the environment, social segregation, impersonal human relations, and materialism. The first was obvious enough to the nose as well as to the eye, and it did not need prophets like Ruskin or novelists like Dickens to identify it. It was a business visitor from Rotherham in Yorkshire's West Riding, itself no Athens, who remarked of Manchester as early as 1808, "the town is abominably filthy, the Steam Engine is pestiferous, the Dyehouses noisome and offensive, and the water of the river as black as ink or the Stygian lake." Yet Ruskin, interested as he was in the cities of Switzerland and of Italy, was moved by the experience of the industrial city to probe the relationship between the visual and the social, as well as to indict a whole society and culture, and Dickens in his symbolic picture of Coketown—chapter 5 of *Hard Times,* where the picture drawn is called "the Keynote"—has caught the sense of something more than appearances.

Like de Tocqueville, Dickens places the savage—very much not the noble savage—in the middle of the city. "It was a town of red brick which would have been red if the smoke and ashes had allowed it; but, as matters stood, it was a town of unnatural red and black, like the face of a painted savage." For Dickens, deterioration of the environment and impersonal human relations were two sides of the same question—they are often separated in the twentieth century— and both were related to materialism and monotony:

It was a town of machinery and tall chimneys, out of which interminable serpents of smoke trailed themselves for ever and ever, and never got uncoiled. It had a black canal in it, and a river that ran purple with ill-smelling dye, and vast piles of buildings full of windows where there was a trembling and a rattling all day long, and where the pistons of the steam engine worked monotonously up and down, like the head of an elephant in a state of melancholy madness.

This was highly personal imagery, reminding us that Dickens should always be treated as painter rather than photographer, but what satire could do, statistics could do also, even though *Hard Times* was a satire on statistics. The facts of segregation were obvious enough in the industrial city, and they were made the most of by another of the early critics of Manchester, Friedrich Engels. In the preindustrial city there were social gradations and propinquities. In Manchester, according to Engels—and he was not alone in his analysis—there were hostile classes and socially segregated districts. "He who visits Manchester simply on business or pleasure need never see the slums, mainly because the working-class districts and the middle-class districts are quite distinct. This division is due partly to policy and partly to instinctive and tacit agreement between the two social groups." The word "slum" was another new word, recorded for the first time by the Oxford Dictionary in 1825. Characteristically, it had no ancient roots and emerged from slang. Yet everywhere during the nineteenth century the industrial city became identified with slums as well as with factories.

The processes of segregation are fascinating to trace, whether or not we are dealing, as in nineteenth-century Britain, with segregation by income or, as in the later nineteenth century in the United States, with income and ethnic grouping. Indeed, in the Manchester of Engels, Irish segregation was a particular feature which he dealt with at length, and it was from the vantage point of Melbourne in Australia, a land of rural myth and city fact, that a writer observed in 1886 that "the rich live with the rich and the poor with the poor. The palace and the hovel, except in the imagination of the socialistic romancer, seldom adjut."

The Chicago school, operating in a city which was as much the shock city of its time as Manchester had been half a century earlier, interested itself not only in segregation but in all aspects of urban morphology, in the processes as much as in the structures. "Natural areas," Park was to write, "are the habitats of natural groups. Every typical urban area is likely to contain a characteristic selection of the population of the community as a whole. In great cities the divergence in manner, in standards of living and in general outlook on life in different areas is often astonishing." Park went on to talk of a "sorting out process," and, as memorably as Engels, of "little worlds" in the city, "which touch but do not interpenetrate." Indeed, in a city where there was far more change of land use than in Manchester, he went on to claim that it was only because "social relations are so frequently and so inevitably correlated with spatial relations, and because "physical distances so frequently are, or seem to be, the indexes of social distances, that statistics have any significance whatever for sociology."

Many twentieth-century urban sociologists—and geographers—have tried to place Park—and for that matter Engels—in social perspective. If we wish to see the industrial city itself in perspective, we must not restrict our attention to the four features of it which received most attention from critics or to the generalizations that Lewis Mumford drew out of their and his own criticisms in his description of what he called in 1938 the "insensate" industrial city. It is not true that the new cities of the Industrial Revolution were really "man-heaps, machine warrens, not agents of human association for the promotion of a better life," as Mumford argued both in 1938 and in 1961. Nor is it true that "there were no effective centers in this urban massing: no institutions capable of uniting its members into an active city life; no political organization capable of unifying its common activities."

Can nineteenth-century English history, let alone French or German history, be written in terms of the judgment that "in every quarter, the older principles of aristocratic education and rural culture were replaced by a single-minded devotion to industrial power and pecuniary success, sometimes disguised as democracy?" Finally, it is not true that industrial cities were all the same, variants of Dickens's Coketown, alias Smokeover, alias Mechanicsville, alias Manchester, Leeds, Birmingham, Merseburg, Essen, Elberfeld, Lille, Roubaix, Newark, Pittsburgh, or Youngstown. When night fell it did not fall—and does not fall—on the same urban environment in all these places.

Industrial cities were as varied even in their appearance as preindustrial cities. They had different social structures as well as different appearances. They drew on different heritages from the past, when they had a preindustrial past, and they did not always invent the same history or duplicate the same monuments when they sought to create a heritage for posterity. Many of their buildings, monumental and functional, are worth preserving, and since in some societies they represent the whole of the past, the recent effort to preserve them or to adapt them to new purposes has intensified. They generated more voluntary effort in their own great age of expansion than had ever been generated in cities before, and through the focusing of attention on their problems, which were never minimized by contemporaries, they directed attention for the first time in human history to the full possibilities of social control. They were capable of enunciating civic gospels which combined concern, commitment, and vigor, and their cultural as well as their social life attracts the interest of historians and today can both command respect and evoke nostalgic regret. Perhaps it does not behoove a Smithsonian lecturer to suggest that one of the most misleading of Mumford's judgments was that in the industrial city "sonorous oratory served the double function of stimulant and anaesthetic; exciting the populace and making it oblivious to its actual environment."

If I myself had to generalize in one sentence, I would not be euphoric. I would still fall back on my generalization in *Victorian Cities*, in which I described the growth of Victorian industrial cities in Britain as "a characteristic Victorian achievement, impressive in scale but limited in vision, creating new opportunities but also providing massive problems." I would also want to note the strange coexistence of pride and fear in all the contemporary writing about the industrial city and the continued preoccupation with the creation of an ideal city, not least the idea of a "garden city," marrying town and countryside.

If the pride has until recently been somewhat neglected, the fear continues to dominate historical narrative of the period. "As a stranger passes through the masses of human beings which have been accumulated round the mills (in the industrial north)," wrote one British observer during the early 1840s—and he was an optimist about industrial progress—"he cannot contemplate these crowded hives without feelings of anxiety and apprehension amounting almost to dismay. The population is hourly increasing in breadth and strength. It is an aggregate of masses, our conceptions of which clothe themselves in terms which express something portentous and fearful." The fear could be so great that, as in this case, the observer turned to the upheavals of nature for metaphor, comparing the rise of the masses to "the slow rising and gradual swelling of an ocean," as striking a metaphor as the comparable twentieth-century "winds of change."

Yet the mysterious *terra incognita,* as Booth called it, was not, of course, *terra incognita* to the people who actually lived there. Nor, pace Engels, did most of the people who lived there think of themselves as "masses." The term was originally applied from outside, a new variant in the industrial city of the older term of "mob," associated with the preindustrial city. There were wise city dwellers, who appreciated the dangers of thinking in these terms even at the time. For the most part they were doctors, clergymen, and leaders of voluntary movements, who were prepared to—indeed, expected to— cross urban frontiers into the *terra incognita.* One of them, a Leeds nonconformist minister, warned his congregation in the 1840s against using the term "masses" too easily. "Our judgments are distorted by

ASA BRIGGS

the phrase. We unconsciously glide into a prejudice. We have gained a total without thinking of the parts. It is a heap, but it has strangely become indivisible."

Not all twentieth-century social criticism is so perceptive. We are bound to assess the industrial city, indeed, in the light of our own urban experience in the twentieth century as well as in the light of preindustrial urban experience. It may well be that our cities look more alike than theirs did, that it is we not they who have tampered with the sense of place, that we are more fearful than they were of what we do not experience ourselves within the life of the city, that we are less active in our voluntarism and more disillusioned about our expertise. It is fair to note also that Mumford has criticized the twentieth-century city as sharply as he criticized the nineteenth-century industrial city, dwelling mainly on what he calls "the increasing pathology of the whole mode of life in the metropolis." "The mess is the message."

It is fair, too, to note that Mumford's is not the only view, and that the city generates as much argument now as it did in the 1840s or the 1890s, with first Los Angeles and second Sao Paulo standing out as the shock cities of recent history where all the problems and all the excitements seem to converge. Los Angeles, at least—and it has now passed into a new phase of its history—has always had its passionate defenders, as Manchester had, although it was in neutral Palo Alto, not in Los Angeles, that a conference was held not long ago to compare as shock cities Manchester and Sao Paulo. The English architectural critic Reyner Banham's fascinating *Los Angeles, the Architecture of Four Ecologies*, published ten years ago and concerned with far more than architecture, explains why the twentieth-century city continues to defy consensus. The mobility of Los Angeles can attract or disturb, the visual appearances stimulate or repel, even the weather (apart from smog) seems right or wrong. "Los Angeles has no weather," remarked a journalist, quoted by Banham, in 1969, and the same writer added, "Los Angeles has beautiful sunsets—if man-made."

For the precedents of comments of this kind we have to turn back not to the nineteenth-century industrial city but to the nineteenth-century capital city, particularly London, Paris, and (though it was not the administrative national capital) New York. It was of London, endowed with what Henry James called "general vibration," that Virginia Woolf wrote in her diary in 1918, "they say it's been raining heavily, but such is the civilization of life in London that I really don't know." Twenty years earlier the poet Richard Le Gallienne had proclaimed

London, London, our delight,
Great flower that opens but at night,
Great city of the midnight sun,
Whose day begins when day is done.

The nineteenth-century metropolis inspired more poetry than the industrial city, particularly the poetry of Baudelaire. The great novelist Balzac saw the city of Paris as a stage, the greatest stage in France for human ambition to express itself, while a later novelist, Zola, saw it as a challenge, where individuals and groups were caught in its grip. Baudelaire, however, saw Paris as a place of perpetual change, of fleeting moments and quickened consciousness, beyond good and evil, and he communicated his vision not only to often disturbed contemporaries but to future generations of writers. Thereafter, the old formal distinctions between urban and pastoral acquired new dimensions, as they were to do in the twentieth-century writings of James Joyce, who, as Harry Levin once pointed out, lived in as many cities as the author of the Odyssey, each more polyglot and more metropolitan than the last. Yet his unique vision of the city rested not only on his own experience, but on a rich texture of historical and literary association. We move with him from *The Dubliners* not only to Paris but to *Ulysses*. Feelings of attraction, and of recoil, toward the great metropolitan city, of total absorption and complete disengagement, are all expressed in his work, which is not only of immense imaginative power but of the most subtle complexity—as complex, indeed, as the early- and mid-twentieth-century city itself.

The pictorial artist, too, has expressed something of this complexity in the twentieth century, so that every serious student of cities must visit art galleries as well as archives or newspaper offices. Even in the eighteenth century—in the age of Johnson and Cowper—the poet George Crabbe was advising his readers that

Cities and towns, the various haunts of men
Require the pencil, they defy the pen,

yet it is only in the twentieth century that Western artists, beginning with the Italian futurists, turned enthusiastically to cityscapes. The German expressionists followed, and in New York John Marin anticipated much verbal and visual comment on Manhattan with his paintings of major monuments like the Woolworth Building or Brooklyn Bridge—shaken by a brittle light and seeming to fuse with the sky. They are as much a product of their time as Dutch views of the city in the seventeenth and eighteenth centuries, many of them reaching at the city through green fields and across canals.

There is an apocalyptic element also in some twentieth-century paintings— and photographs—of cities, for the camera too has come into history in the nineteenth and twentieth centuries through films as well as photographs. An appreciation of this apocalyptic element— and it should be added that the camera has brought in a new sense of city beauty also—brings us round by a full circle to what Mumford had to say about megalopolis as necropolis.

I said that I would conclude with four reflections on the contemporary scene, reflections on, not prescriptions for. None of them is original, and none of them focuses on defense or disaster. First, however rapid rates of urbanization were in the nineteenth century, they have been dwarfed by twentieth-century rates. Second, much urban growth in the last half of this century has been in so-called Third World countries. This brings in different approaches to problems and opportunities, some outside our tradition. Third, the position in the West, coexisting at a different stage of development, continues to raise profound questions about the nature of "civilization"—though we seldom now choose to call it such—as well as about survival. Fourth, for various reasons we have a more shaky sense than we used to have of place and what it means, and we find it increasingly difficult to isolate the urban factor either in our analysis or our policymaking.

The reflections must be brief. On the first—more rapid urbanization rates than in earlier centuries—reflection starts in places like Hong Kong—if there is any place quite like Hong Kong—where skyscrapers grow together like trees in a forest and where new towns as big as Manchester was in 1900 can be built within a decade. In 1950 there were 75 cities in the world with populations of more than a million, in 1960, 141, in 1975, 191; the figure for 1985, it has been projected, will be 273. Moreover, as a result of the technogical changes of the twentieth century, particularly those associated with power and trans-portation, there are in many countries huge metropolitan areas, scarcely broken by anything which can be called rural, which may contain 25 million people or more. These are so different in scale and in organization even from nineteenth-century cities, which were, as we have seen, already breaking with tradition that, as Geddes foresaw in the first decade of this century, they seem to consitute a new species.

Second, such cities and bigger concentrations have emerged not only in the so-called "advanced" countries but outside. The proportion of people in "developing countries" increased from 16.5 percent in 1950 to 28.3 percent in 1975. For South Korea, the comparable figures were 18.4 percent and 50.9 percent. *Asia Urbanizing* is the evocative title of a recent collection of essays, yet in Africa, too, and above all in Latin America, urbanization has moved faster than it ever moved in Europe in the nineteenth century. The shock effect has certainly been great, at least to visitors. As Richard Meier put it in *India's Urban Future*, "the restrictions placed by poverty upon urban design always come as a shock to Western visitors," and it is interesting to compare Levi-Strauss with de Tocqueville. Noting—and, as we have seen, it is only one side of the picture—that we are accustomed to associate our highest values, both material and spiritual with urban life, he found in India "the urban phenomenon reduced to its ultimate expression . . . filth, chaos, promiscuity, congestion; ruins, huts, mud, dirt . . . all the things against which we expect urban life to give us organized protection, all the things we hate and guard against at such great cost."

This, too, is one side of the picture for a very different reason. It relates Indian experience entirely to our own views of the city in history to the tradition that stretches back to Greece and Rome. The Asian or African city continues to provoke clashing reactions within and between societies and cultures, not least between generations. Thus, Mills and B. Nak Song in their study of Korean urbanization, a particularly striking phenomenon, insist that it would be much more accurate to refer to Asia's teeming countryside than to its teeming cities and that while all countries that urbanize rapidly have urban problems—of dislocation, adjustment, and serving—urbanization itself is "a normal and desirable concomitant of economic growth." "Korean urbanization," they conclude, "has been a great success story during the third quarter of the twentieth century. The basic reason is that the national government has focused its efforts on the promotion of economic growth instead of on control of urban growth and structure." However different the traditions, the same remark might have been made in nineteenth-century Europe or America.

On the other side, we have José Arthur Rios's comment on the Rio de Janeiro of the late 1960s, which recalls comments on London in the seventeenth and eighteenth centuries before the rise of modern industry: "Rio is the product of a vast maladjustment. Its growth is unparalleled by any other city in the nation. The tendency has been to enslave the whole country to this abnormal growth, like a tumor which drains all the energies of the body. Its high life is fed upon by the misery and backwardness of the rural population."

My third reflection is renewed, if not inspired, by almost every magazine and color supplement. Mumford's contemporary pathology seems to dominate newspaper headlines also, when they deal with crises in urban finance, common to many parts of the world, or with riot, ethnic or social or both. Biology seems to reinforce history. Thus, from a different angle from that of Mumford—though with equal concern—René Dubos has complained that "life in the modern city"—by which he meant the so-called advanced city of the West—"has become a symbol of the fact that man can become adapted to starless skies, treeless avenues, shapeless buildings, tasteless bread, joyless celebrations." Such generalization always invites *riposte*. The inner cities remain problem areas, particularly in Britain, but there have been immense changes in these inner areas—too few in Britain—not all for the worse, since the 1950s. The number of groups of people, concerned not with being adapted but with themselves adapting the urban environment, has greatly increased. There are more trees in the avenues and more people walking on them, more buildings with shape, and not all celebrations are joyless. There is more skepticism about accepted policies and more willingness to look for new answers. The awareness of the visual has been sharpened, although there is still a serious deficiency in visual education.

Fourth, and last, we have a more shaky sense than we used to have of place and what it means, and we find it increasingly difficult to isolate the urban factor in our analysis and in our policymaking. There has always been a tendency to attribute causal influences to the city which would be better attributed to the society and the culture, not to speak of the economy. As Melvin Webber, who has hailed "spatial dispersion" in a new postcity age, "neither crime-in-the-streets, poverty, unemployment, broken families, race riots, drug addiction, mental illness, juvenile delinquency, nor any of the commonly noted 'social pathologies' marking the contemporary city can find its cause or its cure there. We cannot hope to invent local treatments for conditions whose origins are not local in character, nor can we expect territorially defined governments to deal effectively with problems whose causes are unrelated to territory or geography."

It is a salutary remark, yet merely a first point in an argument. If the city cannot deal with all these problems, neither can the nation-state by itself. There are too many interdependencies. Cities have always been in a network, even when nation-states have been at war. Moreover, we cannot afford to leave out the local from our reckoning. Even with spatial dispersion, it is useful to have senses of belonging which extend beyond the private or the national. Indeed, we have to work out new relationships between the local, through our involvement in a particular place, and the global, through our involvement, whether we like it or not, in our whole planet. A touch of pride in where we are as well as where we came from is not out of place, although at the end of our odyssey we might not all agree with Euripides that "the first requisite to happiness is that a man be born in a famous city."

Commentary by Nathan Reingold
Joseph Henry Papers, Smithsonian Institution

A sure sign that Asa Briggs is still a fan of cities is that he questions
Lewis Mumford. Mumford was prone to compare, unfavorably, contemporary cities with a Florence, somehow reminiscent of transcen-
dentalist Concord, Massachusetts. Few people noticed or bothered to
question the bases of Mumford's views.

Briggs could have taken the classic route in his approach to the city:
the comparison of the city with the country, in which the country is
falsely idealized because of the closeness to an uncorrupted nature,
while the city is not merely a human artifact but one inevitably
pernicious of whatever is conceived as *the* true value. A pastoral
country so rhapsodized, of course, is usually a product of timber
clearance, drainage, artificial cultivation, and other forms of human
interference with unimpeded natural processes.

Instead, Briggs adopts a pose natural to him as a very distinguished
historian of human urban life. There is a recital of all of the specialties
required to study the city, with the clear implication that not only can
no one specialty suffice, but that in fact all together they somehow do
not capture everything we need to know about the city. It is reminiscent
of the situation in another discipline. Economists and economic
historians often find that their quantitative labors do not demonstrably
account for 100 percent of the phenomena under study. A residual, as
they say, is left over, and there is some feeling that these residuals
unexplained by theory are extremely important.

Historians wedded to particulars of time and place tend to start with
residuals. What is involved in most discourses about cities is a kind
of implicit theoretical idealization which invariably generates the
residuals. Take one important observation made by Briggs, that we
judge Asian cities by Western standards. Of itself, it means little, but
I read that point—and others made by him—as indicating our inability
to perceive adaptations of cities when these adaptations are unfamiliar
or when they do not accord with a desired conception of the city.

What is involved is that adaptation to the environment is occurring,
but we fail to see it or do not wish to accord it the term adaptation,
which implies to many a degree of approbation. It is a "success" term
and Westerners perhaps overvalue success. When Briggs spoke of

Korean cities, I remember the inevitable Western comments on Cairo as a great overgrown village. By juxtaposing Sao Paolo and Manchester, he makes the same point.

One American student of cities, Sam Bass Warner, Jr., has a piece entitled—I'm quoting, obviously—"If All of the World Were Philadelphia." Philadelphia turns out to be a fine model for the United States, except for New York City. That conurbation, like other great cities, turns out to have peculiar characteristics. Are we faced with the prospect that Philadelphia and Manchester, may pair, but that London, New York, Paris, and Rome are simply not going to fit into the same pigeonhole? If there are some constants influencing adaptation, the city will not do as a unit of analysis. Much as we do not like it, each city, even Calcutta, is a successful example of adaptation. Here Briggs's point on the various disciplines involved underlines his strategy. Successful historians somehow give a sense of a given time and place by presentation of the complexity of particulars. The historian knows that residuals are important, because that is what makes Manchester different from Philadelphia, despite parallelisms one can construct from a given disciplinary perspective.

Beneath textured prose and impressive layers of erudition, Asa Briggs is saying that particulars within each urban environment are important if we want to understand human existence—notice I do not say adaptation, but existence—in cities; that, in fact, a city is many, many environments, each worthy of study by historians and others; that generalizations do not extend one centimeter beyond the particular focus of inquiry.

Dyos on Camberwell reminds one of Philadelphia or New York until one remembers that building contractors usually depart when urban areas become populated, and planners usually know enough to live outside the area of their plans.

By this circuitous route, I am back to Louis Mumford, an estimable, high-minded man with more to his credit than I could recite in the time allotted today. The ethnic slum neighborhoods he scorned are now seen as mechanisms by which the newly arrived adjusted to a new land while preserving a sense of identity. As downtowns decayed, we discovered they provided, among other things, a sense of community. The garden cities of Mumford and his master, Geddes, are not as comfortable to their inhabitants as the old slums unless prudent provision is made for pubs, as is done in Britain. What we need to know is the texture of daily life, not as a residual but as a part of existence valued for itself, whatever our allegiance to discipline and to ideals of human congeniality.

Commentary by Stephen Toulmin
Member of the Committee on Social Thought, University of Chicago

Let me use Asa Briggs's lecture as an excuse to raise some general questions. A well-crafted historical narrative such as he gave us actually spotlights important theoretical issues, and I want to ask some questions that arise, if one sets the story he told us alongside various theories that people have put forward, about why cities grow up in the first place.

In particular, let me compare the theory that my Chicago colleague, William McNeill, puts forward about the rise of the city, as a "population sink" into which the surplus peasantry congregated in order to die from infectious diseases, with the alternative theory that cities arose because of the specialized, centralized activities which could only be carried on in cities—some of them governmental or judicial, some of them industrial, some of them cultural.

When you look, in particular, at the extraordinary rise in the number of vast cities in the less-developed countries at the present time, these questions become questions of some urgency. Living in the United States, one has a sense that the city is becoming *less* necessary as a center of specialized, centralized activities—of kinds that cannot be carried on in the suburbs. Take industry, for instance: the more that industry becomes focused on the second rather than the first law of thermodynamics—the more it becomes information-intensive rather than matter- and energy-intensive—the more dispersed it can be, the less commuting remains an essential part of the industrial way of life. And similarly, if you consider culture: the existence in a city like Chicago, of WFMT and National Public Radio and public television and the rest, means that one can live far out from the central city and still have a rich cultural life that need be punctuated only occasionally by those special occasions when you make a special effort and (say) go to the Lyric Opera.

When you look at the theory of the city as *socially* indispensable, it is really only the governmental and judicial functions which remain. And the danger is, that in the under-developed countries the cities will become examples of McNeill's theory, and that they will indeed become places where the surplus population from the countryside crowds together into a basically unlivable environment.

It was striking to hear Asa Briggs's lecture after Dr. Hassan's. There was a moment in discussion when Hassan turned to us and was inclined to say, "Don't try and help us to avoid *your* mistakes: leave us to make *our own* mistakes, or even to *repeat* your mistakes if we want to.'' And I should be sorry if this was an attitude which we had to sit down under, because the most striking thing that came out of Asa's lecture was the fact that different developing countries have managed the transition to the large modern city in very different ways. Some of the success of South Korea, for instance, has to do with the fact that the Koreans chose to concentrate on electronics, rather than insisting on having large numbers of steel plants, or other nineteenth-century industries, which William Blake himself long ago condemned as ''dark and satanic.''

Discussion

DR. HASKINS: Thank you very much. Is there some discussion of these subjects?

DR. PASSMORE: I would just like to ask a question. I think if you look both at Europe and Australia, a striking feature for the last decade has been the relocation in the cities. It has been accompanied by widespread feeling that there is no point in living on the outskirts of cities. So the actual total population in a metropolitan area, in London, Sydney, and other cities, has tended to diminish. It is much better to live in a small town than in an outer suburb.

On the other hand, I think most people do not feel that they get the rich cultural life by buying a television, with an occasional trip to the village Lyric Opera. And I think the movement into Sydney and a number of other cities that I know—and it has been very remarkable—has been largely provoked by a desire to live in the middle of the range of activities embodied in the theater, films, art galleries. For these, television is no substitute whatever.

I suspect that cities have come to a turning point although that is less obvious in America, because you have certain rather special problems, I think. But the closing down of traffic in the middle of cities, making it available for people to walk around, the building of areas where no traffic is permitted at all, this is occurring in Australian cities and European cities on a very grand scale. These places are becoming strangely medieval. The department store is disappearing. The small shop is coming back as a crucial thing. The city again provides you with the specialties. You go to the supermarket, but if you want to buy something special you go to one of these small shops where the man really knows what he is selling, unlike in a department store. And there is a curious and a very interesting survival of such life. I think actually you can see this in New York to some extent.

DR. REINGOLD: What is happening—and this is relevant to what Steve Toulmin said—is that we are thinking of the city in a certain classical sense as a geographic entity and a political and administrative entity. With the coming of certain forms of transportation and communication, the city is changing and it is becoming something quite different. So that I think what you have in the suburbs is not something that is completely different. You are getting an entity which is, for lack of a better term, a metropolitan entity. And the existence of communications and electronics has provided a kind of a new body or new organism—a new entity. Today the cities are quite different. We are then surprised when we move into a suburb and discover that there have been created certain aspects that we left cities for, so therefore some people will move back into the decaying central city. You are getting a new entity which does not fit the geographic and administrative definitions upon which our discourse is based.

DR. DANIELLI: I think there is an age group phenomenon coming in here as a result of our living to be much older. The situation is changing. I suspect the whole function of cities may be changing as a result of some of these things. I can see that in the past, perhaps the really important thing about cities was the critical mass phenomenon, that is to say, the inner city was the only place where you could get maximum stimulation of the intellect and it was the only place where you could get maximum personal freedom. You put those things together and you have a marvelous situation.

First of all, I think the same thing still applies to young people, who have to set up, as it were, their own personal networks. But then as they want to have children and relax a bit they tend to move into the suburbs or into the country. On the other hand, for people who are optimizing their intellectual lives, I suspect the critical thing now is that you develop an international network, so that no city can in fact fulfill the function that it used to carry out and provide optimization of the intellectual environment. One has to depend, in other words, on a new process which is only just beginning to crystallize. I know, for example, no single city has great satisfactions for me personally any more. On the other hand, my international network of cities is terrific.

DR. BRIGGS: On that particular point, could I just say one thing? It seems to me that if one were just to pick out those two aspects of the historical city, stimulation of the intellect and personal freedom—and they are both very important, for different reasons—they are only two of the classics that one could pick out, and therefore one is in danger, I think, of building an analysis of two of the variables alone.

If you look at cities now you have to relate them not to the hinterlands and to the international dependencies, but you have to relate them to the new conception of communications. Wider areas within each city will have different functions. There I think you are right in saying that the changes in age pattern are perhaps related and could well affect the role of the city. Within the city itself, it is the continuing element of attractiveness. I don't agree with William McNeill's view at all.

DR. CHAGNON: My question is actually apropos the comments made by Lord Briggs. Just now I was struck by the contrast that the McNeill model poses to some of the things that Lord Briggs said, especially this almost poetic allusion to the teeming countryside in certain points of Asia. One thing that has always puzzled me in looking at Latin American cities is, why should they be so attractive? The population density in the countryside is exceedingly small and enormous amounts of tractable, arable, and almost free land exist for the farmer. Despite this, people move into an urban existence which is miserable. The reason probably has very little to do with the stimulation of the intellect. There is an attraction to the city that is quite apart from that. I think the classic example is that of Brasilia, the capital of Brazil, which was set up in the middle of an area that had no countryside to speak of. The only condition in the national competition in the design of the city of Brasilia was that the architects had to build specifically into the city provisions to prevent the development of shanty towns. Before the city was even completed Brasilia was surrounded by them.

DR. BOULDING: Something I think hasn't come up, and that is the distinction between what one might call the urban pull and rural push. The urban city is a result of both of these. Certainly the precondition of the development of the cities is excess food production by the farmer. The urban population is still dependent upon this to a very large extent. In the United States, for example, we went from 90 percent in agriculture at the time of the revolution to four percent today. In the last generation we have pushed 30 million people out of agriculture into the cities. There is a whole first generation of rural people who don't know how to live in the cities, in the Chicagos. They have an awful time when they come off the farm. This is a very universal rule.

I think the trouble is there are two kinds of rural push which are very definite. One is the population push, that is, you have an expansion of the population in the rural areas because of health practices, or whatever, that does not correspond to any change in agricultural productivity.

The other is the rural push that comes from an increase in agricultural productivity, as in the United States for instance. Just the fact that we have doubled the productivity in agriculture in one generation here, that is why we got rid of 30 million people from agriculture and into the cities.

If you look at Sao Paulo, for example, you see the results of more population growth. This is a much more pathological condition than the rural push that comes from agricultural development. I think it is extremely important to look at this.

And, of course, there is also the urban pull involved. After all, civilization is what goes on in cities. This is why we are urbane. Of course, the great achievement of American society has been the almost total destruction of rural life. This is its marvelous achievement. The Iowa farmers now have Ph.D.s. We have developed a superpeasant and this has just been brilliant, an extraordinary development.

14. World Society: The Range of Possible Futures

KENNETH E. BOULDING

Professor Emeritus of Economics, University of Colorado

Editor's Summary. Kenneth Boulding is among those who view the future with optimism. He emphasizes that learning involves uncertainty and that we cannot predict what we will need to know for the future. Despite these uncertainties, Boulding stresses that we should not abandon progress but should assume that creativity and innovation will somehow solve our current and future problems in new but equally effective ways as those which have characterized human success in the past. He notes that human history is punctuated by catastrophe which is likely to continue as an important factor in human adaptation. Boulding's optimistic view may require a larger social rather than individualistic perspective with respect to human adaptation. His optimism arises, in part at least, from his conclusion that solutions to problems are possible because we can conceive of them—that peace is more probable than destruction because it is a much better idea.

All decisions involve choices among images of possible futures. How we get these images, therefore, and how accurate they are, is a crucial problem in what might be called the theory of bad decisions, that is, how do we avoid decisions that we later regret? The testing of images of the future is unfortunately all too easy; all we have to do is wait and observe until the future becomes the past. This is not always helpful, for frequently we do not learn from our disappointments what really went wrong, though sometimes we do learn. If we are to make wise decisions, therefore, we need to have ways of testing our images of the future before they are realized and come to pass. This is not at all easy. Sometimes, indeed, it may be impossible, depending on the nature of the system in which we are operating.

One can identify at least four types of systems in which some statements of the future are possible. I am tempted to name these, by some of their principal examples, as planets, plants, plays, and plagues. The first of these, of which planets are an excellent example, are systems with highly stable parameters in which the record of the past enables us to detect and to measure these parameters and in which, therefore, the future can be predicted with extreme accuracy. Celestial mechanics is the prime and indeed probably the only example of a system of this kind. We can predict eclipses and the movements of the planets with extraordinary accuracy. The reason for this is that the evolution of the solar system has virtually ceased and indeed has left so few traces that we do not even know much about it. Even such questions as the origin of the earth's moon are subject only to speculation.

In the solar system as we know it today, however, the parameters are so stable and the records are so good that once we have caught onto the system, prediction presents no difficulties. The prediction of the movements of the planets is one of the earliest achievements of the human mind. In perhaps a somewhat crude form, it goes back several thousand years, at least, to the Chaldeans. For shepherds on a dark night, indeed, there is little else to do but observe the slow but regular movements of the planets and the other heavenly bodies. This process culminates in Laplace and his equations and, of course, the calculation and discovery of the outer planets—Uranus, Neptune, and Pluto—which were unknown to the ancients.

We practice something like a crude celestial mechanics in daily life when we play tennis and predict where a ball is going to be in the next fraction of a second or when we drive a car and predict where the other cars on the road are going to be as we make a left turn across the stream of traffic. Even in these situations, however, our predictions are sometimes inaccurate because the parameters change. A puff of wind catches the ball and we miss our return shot; some idiot driver accelerates and we have an accident. While there are systems, therefore, which approximate celestial mechanics, there are no other systems which approach it in stability of parameters. The very success of celestial mechanics has perhaps had an unfortunate effect on the other sciences, where it led to a search for stable parameters in systems which do not have them. Mathematical meteorology and mathematical economics and econometrics are perhaps examples analogous to looking at the famous black cat in a dark room that isn't there—in the search for stable parameters of meteorological or economic systems.

KENNETH E. BOULDING

A second type of system, typified by the growth of plants from seeds, animals from fertilized eggs, buildings from blueprints, research from proposals, and investments and planned economies from plans, might be described in the great biologist Waddington's terms as "creodic." They involve growth and development from a "seed," the seed containing know-how, which may not of course be in any sense conscious, and which has the potential for creating a process of growth and development with the cybernetic property that divergence from the "plan" can be corrected. These might also be called genetic systems.

They all involve some kind of genetic know-how in that the seed, the egg, or the blueprint each contains the potential for a process of production. All production starts with genetic know-how. This has to be able to direct energy and information toward the selection, transportation, and transformation of materials into the structures of the product. This process invariably requires space and also takes time. We can think of all processes of production therefore, whether that of a plant from the seed, a chicken from the egg, or a building from its blueprint, as originating in some sort of genetic or know-how potential that is limited by materials, energy, space, and time. These limits may frustrate the genetic potential. The seed will not grow to full maturity without water, the right materials in the soil, energy from the sun, space to grow in, and time to mature. The same is true of the egg and the blueprint.

Once a process like this is under way, our familiarity with similar processes in the past gives us a good deal of confidence in the limits of the future. The potential in a seed, an egg, or a blueprint will either be realized or it will not. It simply will not realize some other potential. A seed will either die or grow up into the plant of which it is a seed and no other. Kittens always grow up into cats, never into dogs. If we have the blueprint of a house before it is built, we have a pretty fair idea of what it will look like when it is built and even how long it will take and how much space it will occupy, what materials and what energy it will require.

One principle that is of importance here, however, is that it is usually very difficult to perceive potential before it begins to be realized. Even a good botanist finds it hard to tell what kind of plant an unknown seed will produce, and all the resources of science cannot detect from a fertilized egg of an unknown species what kind of animal it will produce. It is perhaps easier to tell what kind of a house a blueprint will produce, for in the case of human artifacts we have access to the genetic material which we do not have in the case of biological artifacts. Even in human artifacts as well as in biological, as Burns said, "The best-laid schemes o' mice an' men gang aft a-gley.'' We can never be sure that a plan will be fulfilled. As we move toward complex systems, uncertainty increasingly rules the world.

A third form of system in which projections of the future are possible, though with greater difficulty, might be described as the play or the plot. This bears some resemblance to the plan, but it differs in that the pattern of development is much less specific. In a plan, the outcome is specified in great detail at the very beginning and provision is usually made for the correction of random events which would disturb the realization of the plan. Every good plan has contingencies and it should have in it methods for dealing with unexpected and random events, but it deals with these only to prevent the failure of fulfillment of the original plan. In a play, there is indeed a pattern of development, which we begin to perceive in the first act. What makes a play interesting is not the mechanical following of a previously known plan with all the potential present at the beginning, but rather the creation of new potential under the stimulus of events which could not be predicted. A skilled playgoer, in the first scene, will develop some expectation as to what the play is all about and how it will proceed, but if he could predict exactly what is going to happen, he might as well leave the theater and go home. Even in the case of a very familiar play like *Hamlet*, there may be some pleasure in simply reexperiencing a familiar experience, but much of the pleasure comes from the reinterpretation by the direction and the acting that turns the familiar into the unfamiliar. It is the inability and the failures of prediction that make plays interesting.

The same is true of games, which are a somewhat ritualistic form of drama. It is only uncertainty and the presence of bad decisions that makes games interesting to watch or even to play. If we could predict exactly the outcome of a game before it began, there would not be much point in playing it. In this case, the inability to predict has a positive value. This is why tic-tac-toe—the outcome of which can be predicted even by a child—is seldom played beyond the age at which this prediction becomes possible. One worries whether the computer will not similarly destroy chess, only the unpredictability of which makes it interesting.

Ordinary human interaction, as the soap operas testify, consists very largely of plays. A marriage, a business deal, a political campaign, and a war all have elements of drama, and it is unpredictability that makes us indulge in them. History is largely written by frustrated dramatists. The role that drama plays in religion is also very large. The major religions all seek to interpret the universe in terms of a great drama of good and evil, sin and repentance, heaven and hell,

KENNETH E. BOULDING

death and transfiguration, retribution and forgiveness, justice and mercy, the wages of sin and the promise of salvation. Without profound uncertainty, the power of religion would be broken, so the condemnation of Galileo may have had some excuse!

The fourth type of system in which statements about the future are possible, which for the sake of alliteration I have described as plagues, is perhaps better described by earthquakes and floods, wars and famines, the uncertain horsemen of the Apocalypse. These are systems in which there is some positive probability of an event happening, we cannot tell when. Propositions emerge, therefore, of the form that San Francisco will be destroyed by an earthquake in x years. We do not know what x is, but we are almost sure it is a positive number, simply because of the existence of the San Andreas fault and plate tectonics. In systems of this kind, there are degrees to which the probabilities can be estimated. Thus we can say that everyone now alive will be dead in x years, quite sure that x is not greater than 150. The probability that any particular person will be dead in one year, two years, three years, and so on can be estimated with fair accuracy. The size of a 100-year flood—a flood with an annual probability of 1 percent—is fairly well known from watersheds that have been observed for a long time. These probabilities, of course, change as time goes on and change under the impact of human action. Thus the probability of bubonic plague is much less than it was in the thirteenth century, although we are not quite so sure how much we have reduced the probability of an influenza epidemic. These probabilities change with changes in the parameters of the system, which are usually unpredictable.

Finally, perhaps we should add a fifth category: systems which have irreducible randomness in them and in which, therefore, prediction is impossible, at least in some aspects of the system. We leave aside the question as to whether there is real randomness in the universe, though this is a fascinating matter of speculation. What is important for the human race is whether there is irreducible uncertainty, and certainly there are many systems of this kind. A fair roulette wheel or a lottery would be examples. The existence of gambling certainly suggests that there is a positive demand for uncertainty in situations where uncertainty is the only thing that can give hope. Gambling also represents a demand for inequality, which is worrying to the social philosophers and, while it certainly has a tendency to become addictive and pathological like many other vices, it cannot altogether be ruled out as a part of human nature.

At a more sophisticated level, it is clear that any systems which involve information as an essential element have irreducible uncertainty in them simply because information has to be surprising or it is not information. Learning, therefore, involves uncertainty. We cannot predict what we are going to know in the future, or we would not have to wait, we would know it now—though this does not deny the quest to eliminate, as much as possible, what we don't know that is knowable. This principle can be seen dramatically if we ask ourselves, what would happen if we had a copy of tomorrow's *New York Times*? The answer, of course, is that tomorrow would not happen the way it was printed, for everyone would take measures to guard himself against anything unfavorable and this would profoundly change the future. In systems of this kind, therefore, uncertainty is something that we simply have to learn to live with.

As we look into the future now from 1981, we have to ask how systems in the real world conform to the taxonomy we have developed above. We have to look at three major types of systems of the world, all of which interact with each other—physical systems, biological systems, and social systems. Physical systems, oddly enough, are mainly either planets or plagues. Eclipses we can certainly predict. We cannot, of course, predict the number of satellites the earth is going to have by the year 2000, as we now have political astronomy, which suffers all the unpredictability of social systems. When it comes to such things as earthquakes, eruptions, droughts, floods, and other natural catastrophes, our power of prediction is usually limited to probabilistic statements of our fourth class.

Meteorologists are trying very hard to develop deterministic models of the atmosphere. So far, they have not been very successful. Large physical principles derived from thermodynamics have been invoked in the interest of prophecy, such as the principles of conservation and of increasing entropy. Information, however, which is crucial in the development of both biological and social systems, is not conserved. When I teach, I do not lose the information that the student gains. Both DNA and printing multiply information incessantly. Even thermodynamically, the earth is an open system, receiving energy from the sun and passing it through to outer space, so that the strict second law does not apply, though there are limits, of course, to the throughput of energy. Because of the fact that the social system now interacts quite strongly with both the physical and the biological systems of the earth, there is a profound element of unpredictability even in the earth's physical system.

This does not mean that conservation and entropy at the physical level are irrelevant, but the concepts have to be applied very carefully. The conservation principle expresses itself most generally in what I have called the "bathtub theorem": when more is added to something than is subtracted, it will increase and, if more is subtracted than is added, it will decrease. This principle certainly applies to existing stocks of fossil fuels and ores. A fundamental principle of exhaustible resources is that, if they are used, they will be used up and exhausted. Entropy, I have argued, is a very unfortunate concept, almost like phlogiston, in the sense that it really is negative thermodynamic potential. The second law can be restated in a very general form in terms of potential: if anything happens, it is because of the potential for it happening, and after it has happened, that potential has been used up. This, however, opens up the possibility of the re-creation of potential. This goes on all the time on the earth, even thermodynamically. Water exhausts its gravitational potential by running downhill, and then the throughput of energy from the sun re-creates this potential by evaporating water from the ocean and depositing it in high places as rain and snow. The increase in human knowledge continually re-creates social potential in the form of new discoveries.

At least from the time of Jeremiah there have been Jeremiahs, prophets who might be described as "entropists." The prophets of Israel preached that the Jews were exhausting the resource of Jehovah's patience and good will. W. S. Jevons, the English economist, whose book on *The Coal Question* (1866) was an early "Club of Rome" type of report, predicted the exhaustion of coal on which the then prosperity and development of Britain depended. Ironically enough, almost when he was writing this, oil and gas were discovered at Titusville, Pennsylvania, and a whole new period of expansion began. Nevertheless, his words had a certain prophetic quality about them, when he said, "We have to choose between a brief greatness and a long continued mediocrity."

We might even postulate a sixth type of system in regard to predictability, in which two processes operate in opposite directions, and it is very hard to tell which one will predominate. We might almost call this the "yin-yang system," following the old Chinese image. Perhaps the most striking case of this is the contrast between the exhaustibility of resources on the one hand (clearly a "yin" process) and the discovery of new resources, new processes, new economies in resource use, on the other hand, which is clearly a "yang" process. In the last 150 years, certainly, the yang principle has predominated in regard to natural resources both in terms of energy and materials. Exhaustion has been more than offset by the constant discovery of new sources and new forms of energy, new materials, like aluminum and titanium, and new methods of economizing both energy and materials so that we can get more human satisfaction per unit of input. It could well be, of course, that this period came to an end in 1973 with the formation of OPEC, but we cannot be sure of this.

There are indeed things on the way, perhaps just over the horizon, which suggest that the yang principle is still operating quite powerfully, such as the extraordinary decline in the costs of electricity from solar cells and the great potentialities for the economizing of scarce materials which fibers and lasers have opened up. There are very uncertain but not negligible prospects for nuclear fusion, solar electrolysis, water-producing hydrogen, and even biomass, that is, growing things to burn. The nightmare of an oil-dependent agriculture running out of oil would be somewhat alleviated if a highly efficient tractor could burn farm-produced gasohol, made from crops grown on only 10 percent of the land, with a much greater efficiency than horses. The uncertainties here are very large, but it would be a great mistake to give up the yang principle too soon and subside into the long, drawn-out mediocrity of the yin.

In appraising possible futures, two other dimensions of the total world system have to be borne in mind. One might be called the "quiet-noisy" dimension. It could well be argued that the most important things that happen are the quiet ones that do not get into the information system very much and are therefore apt to be over-looked. These are things like births and deaths, aging, learning, the increase in human knowledge, changes in customs, habits, and val-uations. These often happen quietly and imperceptibly and yet may be major sources of change.

At the other end of the scale, we have the noisy things—wars, revolutions, world catastrophes, plagues, famines, and so on—but these are noticeable and get into the history books and are prominent in the human imagination. Sometimes indeed they are important simply because they are noisy, but their importance usually tends to be overestimated. The music of the spheres is very quiet indeed. The planets rotate around the sun with remarkably little fuss. Plans are noisier. Corn may grow quietly, but kittens and children make a good deal of noise. And the plans of men, if not of mice, are apt to be announced with fanfares. Plots are still noisier. The dramas of politics get onto television, shots are heard around the world, the silent movies succumb to the talkies. Plagues and catastrophes are the noisiest of all. Eruptions, earthquakes, wars, and revolutions are highly notice-able.

Yet the greatest danger in assessing the future is the neglect of the quiet things, though cumulatively these are often much larger than the noisy ones. Evolution is a process of cumulative increase of know-

KENNETH E. BOULDING

how. It is interrupted by catastrophes and these sometimes head it off into a new direction. The role of catastrophe in evolution, indeed, is a very interesting question to which no firm answer seems available. But catastrophe is only the stage for evolution and is not evolution itself, though sometimes it is true that a catastrophe creates a lot of new niches, which are then quietly filled.

Another dimension that we have to look at, in social systems particularly, is that of the nature of human interactions. I have argued (see Boulding references) that these fall into three major categories, described succinctly as threats, exchange, and love. Virtually all human interactions consist of some mixture of these three. A threat system begins when A says to B, "You do something that I want or I'll do something you don't want." What happens then depends on B's reaction, which may include submission, defiance, flight, or counterthreat. Threats help to socialize us as children, they make us pay taxes and support governments, and of course they are overwhelmingly important in the international system of war and peace. Exchange begins when A says to B, "You do something I want and I'll do something you want. Give me something and I'll give you something." Out of this comes trade, money, finance, the whole ongoing evolution of economic life.

The third great class of human interactions, of which "love" (to which one should add hate) is only a shorthand symbol, consists of a vast variety of human interactions all of which, however, revolve around the concept of personal identity. I call this the integrative system. It consists of all the relationships which bring us together or make us fall apart, which create or destroy community. All these interactions are closely involved with images of personal identity. An integrative interaction, indeed, is apt to begin when A says to B, "You do something because of what I am and what you are."

Perhaps the most important aspect of the integrative system is legitimacy, which might almost be defined as "okayness." No person can function in a role unless there is conviction that the role is legitimate. The relation of authority or subordination is impossible unless both parties accept the roles as legitimate. Authority, as Chester Barnard (1938) argued, is granted from below. I have argued, indeed, that the dynamics of legitimacy dominate all other social systems, for without legitimacy, threats have an extremely limited horizon and exchanges cannot be carried through. The bandit can organize a temporary social system; only a legitimated tax collector can organize a larger, continuing one. Casual trade may occasionally take place between enemies; extensive trade can only take place among people who accept, trust, and legitimate each other. When exchange becomes illegitimate, as in the case of slavery or in the case of stocks and bonds in socialist countries, it collapses.

These three systems of human interaction are all present in different degrees in virtually every human relationship, organization, or event. For instance, in the exchange environment, banks or corporations are able to grow mainly because the exchange value of what they sell is greater than what they buy, so they earn profits. Nevertheless, they also have a vital integrative environment. They cannot thrive for long if they are widely regarded as illegitimate. They must maintain internal morale. They cannot survive if they do not have at least some sense of internal community and if what they do is not regarded as generally acceptable. Corporations also exist in something of a threat environment where exchange is monitored by law and where they have to pay taxes and maybe even bribe politically powerful people. They may also be engaged in selling things to military organizations, which operate primarily in the threat system.

Military organizations, though primarily concerned with threat, also operate in the exchange system. A voluntary army has to hire people, though a conscript army relies on threat. They also have to be clothed with legitimacy or they will collapse from sheer lack of support. The organizations that we think of as being primarily in the integrative system, like the family or the church, the foundations or the United Funds, also operate in exchange markets. They have to hire people in the labor market, purchase or rent buildings, furniture, and so on. They also have a certain underlying basis in threat, in the legal system, and the enforcement of contracts. Even the tax system as a threat system is designed to promote charitable contributions. In spite of all this mixture, however, the three systems do have a certain unity and pattern of their own, which it is important to recognize as we look at the future.

Of the three systems, the exchange system is the one that probably has the most stable patterns and comes closest to behaving like the planets. In the United States from about 1950 to 1973, for instance, the proportions of the economy were remarkably stable. Rates of unemployment and inflation were fairly low, at least considering the heterogeneity of the society, income continued to increase at a fairly stable rate, and econometric models worked quite well. We even got the illusion that we could do "fine-tuning" and regulate the economy to produce steady growth and rather stable relative distribution. The exchange system, however, can be subject to quite violent and unforeseen disturbances, as we saw in the Great Depression of the 1930s when unemployment went to 25 percent of the labor force; the national income of the United States in dollar terms almost halved in four years; the relative price structure was severely distorted, with agricultural prices falling much more than manufacture prices and manufacturing wages; when profits disappeared, becoming negative in 1932 and 1933; and so on. We cannot be sure, therefore, that the exchange system will always be well behaved. The 1970s have also seen some disturbing changes, with accelerating rates of inflation now reaching more than 10 percent per year in the United States, 130 percent per year in Israel, interest rates rising correspondingly, and so on. The great hyper-inflations, for instance, of Germany in 1923 and Hungary in 1946, suggest that the exchange system can develop spectacular pathologies in a relatively short time, although in these cases it had to be assisted by a breakdown of the tax system.

The integrative system, even more than the exchange system, seems to be subject to long stabilities punctuated by spectacular changes. The enormous stability of religious institutions is a case in point. There are times, however, when ancient legitimacies suddenly collapse and there are conversions on a large scale, reformations, new religions arising, and so on. These seem to be relatively unpredictable. The creation of evolutionary potential, indeed, in integrative structures is a very mysterious phenomenon, which is hard to identify at the time. It is also very hard to predict when these integrative—especially symbolic—structures lose their power. In systems of this kind we cannot rule out the importance of relatively random events like the rise of a charismatic leader, a Mohammed, a Luther, a Lenin, a Mao Tse-Tung, even a Hitler, who profoundly changes the course of history. These would not succeed, of course, unless the time were right and unless there were a vacant niche. Just because there is a niche does not mean to say that it must be filled. And, of course, it is hard to detect the niches that are not filled.

The threat system has some stable and some very unstable elements. The internal threat system of societies as expressed in the law, in sanctions, and in the tax system are apt to be fairly stable and change slowly. There are occasional examples of spectacular collapse, for instance, like a civil war or revolution, and there are long-run changes in the technology of internal threat that produce things like the income tax, government regulation, and the ever-increasing complexity of law. In the international system, however, which is largely government by threat, instability is very great. There is a certain pattern to war and peace distinguished, for instance, between postwar periods and prewar periods, such as we now seem to be entering under this administration. But just where and when wars will take place is profoundly unpredictable. War and peace do tend to alternate. The systems of stable war are transcended and pass into unstable war or unstable peace, but the alternations are still very irregular. The only stability in such a system, indeed, is stable peace, such as we have developed since about 1815 in Scandinavia and North America.

The reader may now feel that we are looking into the future with a set of telescopes, radar, and time probes that are hardly manageable. Every prediction should be labeled "Believing in this prediction may be injurious to your health." Far and away the most important property of the future is uncertainty, and delusions of certainty, which too much predictive apparatus might easily give, is perhaps the most important single source of bad decisions. When we are fully conscious of the uncertainty of the future, we develop that flexibility which is the key to survival in an uncertain world. We hedge our bets, all our plans are contingent, subject to constant change, we stay liquid and adaptable, we do not make too many commitments, and we keep our mental ears constantly pricked for signs of change in the environment and in the future. Under delusions of certainty we are apt to zero in on catastrophe. With these warnings in mind, therefore, we may venture at a tentative peek through the veil.

In social systems, "planet" systems (unless we are astrologers) are not very significant. Systems like the solar system are extremely rare and indeed one might almost say nonexistent in society, where parameters are subject to constant change. Plans also are suspect unless they have a lot of built-in adaptability. Even though the future of society does involve the realization of potential, we do not really know much about what the seeds are. New mutations are constantly surprising us. Who would have predicted Jesus or Joseph Smith or Karl Marx or Khomeini? The future of the world social system is not merely the realization of past potential. It involves the constant creation of new potential and is much more like a plot than a plan. Still, the plot does have something of a pattern. There are, shall we say, biases and prejudices in it, with a prejudice toward human learning, simply because error is less stable than truth; error can be found out and the truth cannot.

A similar process, it can be argued, takes place in human values, though this is harder to detect. Because human knowledge creates know-how—and know-how is the genetic factor of production that produces human artifacts—the growth of knowledge has irreversible effects on the system. Just because something has always existed does not mean that it will exist in the future. People are not planets. Institutions of ancient legitimacy collapse almost overnight, like slavery or dueling or, in the Socialist countries, stock markets. Nevertheless, there are great stabilities, too. The great religions, the great ideologies, the national states persist in spite of and sometimes because of many

406   KENNETH E. BOULDING

challenges. New patterns are more likely to rise in the future than old ones are to disappear, though their niches may shrink a little.

In terms of these quiet, long-run development processes, there is a good deal of evidence that we are in something of a slowdown. I have sometimes remarked in speeches that the great age of technological change, as far as ordinary human life was concerned, was my grandfather's life—between say about 1860 and 1930. I grew up as a boy in the 1920s, essentially in the modern world—with automobiles, skyscrapers, movies, electric lights, radio, airplanes, and so on. My grandfather as a boy knew nothing of these things, as he grew up in a rather remote part of the west of England. He may not even have seen a steam train until he was a young man.

The situation in regard to economic growth in the world in the last 100 to 150 years can be summed up rather cavalierly by saying that, on the whole, the temperate zones have been doubling their real per capita income roughly every generation, although some parts, like Russia and Eastern Europe, started a little later than others, whereas, on the whole, the tropics have increased their income very little, except for a very small upper class. This is why the world is now divided into the poor tropics (which are not much richer and in some cases even poorer than they were 200 years ago) and the rich temperate zones, with the possible exception of China and the "Southern Cone" of South America. The rich countries are rich because they have gone through 150 years of extraordinary increase in productivity. Most important was the increase in agricultural productivity, which in the United States now enables us to feed the whole population with about 4 percent of the labor force. Two hundred years ago, it took 90 percent. There have been great increases in productivity in manufacturing, some in the provision of health services, to a somewhat lesser extent in education, and probably not at all in government.

We can think of this, if we like, as something of a seed-to-plant process in that the "seed" is the development of the enormous and ever-expanding potential of the scientific revolution and the "plant" is the application of this expanding knowledge and know-how to technology and economic life, which really began about 1860. Why the seed of science did not grow in the tropics is an interesting, but very complex question, to which there are no easy and simple answers. It has something to do with culture and religion, something to do with imperialism and exploitation, something to do with climate and environment. Parts of the tropics started modern development from low levels and had not reached the stage of urban civilization at the time that the scientific revolution first made itself felt. This was not true of Asia or some areas of the American tropics.

A plausible future in this regard is that the rich countries will not get very much richer and that the poor countries will get a little richer, but also not very much richer. The reason for the slowdown in the rich countries (which is very noticeable now in the United States) is primarily that as a society gets richer, its labor force moves increasingly into occupations where increase in productivity is more and more difficult, such as the service trades, government, education, medicine, and so on. Those industries that still have a promise for increased productivity, such as agriculture and manufacturing, are an ever-declining part of the economy. Short of some quite unexpected technical development in terms perhaps of recombinant DNA or something producing fantastically productive new forms of life, it is hard to see where the overall increased productivity in the next generation—and especially the one after that—is going to come from.

In the tropics, of course, the potential for development is there, but the organizational and cultural difficulties seem to be so great (coupled with the fact that many tropical countries are exhibiting considerable population pressures already and virtually all of them are undergoing a population explosion) that optimism is difficult. The gloom reflected, therefore, in the latest report on the year 2000 (Barney 1981) could well be justified, though one should always expect the unexpected. Probably we will find great diversity in the tropics and some parts of it will emulate Japan or even Singapore and move rather rapidly into the modern world. Other parts may continue to find the problem of getting richer virtually insoluble.

A crucial element in the developmental process is, of course, the change in the numbers and composition of the human population itself. Demography has had some ambition to emulate "the planets," and indeed one thing we do know about the future is that everyone who is alive today will either be one year older or dead by this time next year. This is the basis of what is called "cohort analysis." Population can only be predicted, however, if the essential parameters do not change—particularly age specific birth and death rates—or if we carry it to a second order: rates of change in the birth and death rates. Over some periods, these parameters have been moderately stable. In the last 200 years, however, they have been very unstable with sudden and usually unexpected changes in either birth or death rates.

The expansion of European populations all over the world can be traced in no small measure to a sharp decline in infant mortality which took place, especially in Britain, in the mid-eighteenth century and a little later elsewhere. A similar thing happened in the tropics around 1950 with the introduction of DDT and malaria control. A sudden drop in infant mortality is almost inevitably followed by a population

KENNETH E. BOULDING

explosion, as it takes quite a long time to lower fertility correspondingly. Fertility may be mainly related to slow changes in integrative structures, especially in the family—with some connection with exchange structures as to whether people can afford children and even with threat structures when government intervenes.

The so-called "demographic transition" from the high birth and death rates of undeveloped societies to the low birth and death rates of developed societies is distinctly observable in the rich countries. There is no guarantee, however, this will follow in the poor countries, although there are some signs now around the world of decreasing fertility. In demography, however, as in almost all other fields of social life, it is well to prepare for the unexpected. Nobody expected the enormous population explosion after 1950 in the tropics. Nobody expected the "baby boom" from about 1947 to 1960 in the United States or even the trough in fertility that followed it. Sometimes fertility can change very rapidly, up or down. Rumania doubled its birth rate in one year in 1967, thanks to government suppression of abortion and discouragement of birth control.

There are certainly good reasons for Malthusian gloom as we contemplate a country like Bangladesh, which is the size of Iowa, with now nearly 80 million people in it and still growing fairly fast. On the other hand, the relation between population density and riches is by no means sharp. The Netherlands gets away with high densities and even fairly high fertility, and many of the rather low-density countries are among the world's poorest because of the low density of resources. But the Malthusian specter is always there, with its "dismal theorem," so called, that if the only thing that can check the growth of population is misery, the population will grow until it is miserable. It can only be answered by other checks on the growth of population. These, fortunately, are by no means impossible, though they do seem to be difficult. In this field, therefore, the future is fraught with at least uncertain gloom.

Turning now to the possibilty of catastrophes, the overriding danger, of course, is nuclear war. The present international system is very much like San Francisco, which has the San Andreas fault underneath it, which is stable in the short run but cannot be stable in the long run. This is "deterrence," particularly in the variant of "mutually assured destruction," properly named MAD. The proof that deterrence cannot be stable in the long run is very simple: if deterrence were stable it would cease to deter, that is, if the probability of nuclear weapons going off were zero, this would really be the same as not having them. As long as we have them, there is some probability that they will be used, either by accident or by the kind of "domino" scenario which produced World War I from an assassination in Sarajevo.

Just what the probability is of major nuclear war can only be a matter of subjective estimate. My own estimate is that, say three or four years ago, it was on the order of 1 percent per annum, that is, very much like a 100-year flood. Today, I would strongly suspect that it is rising, and it can well be 3, 4, or 5 percent per annum. In the absence of a universe of like events, of course, all estimates of probability are subjective, but we do make them all the time. Certainly it is very easy to write a "scenario," that is, a play in which both the United States and the Soviet Union loose their arsenals at each other with the virtual collapse of each society into roving armed bands, terrorizing the few accidentally spared cities, fighting off the neighbors, with a general collapse of social order.

Here, of course, we are moving into a position in the social system so unfamiliar that nobody knows what it will be like. All we know about for the most part, after all, is the familiar, as the Three Mile Island incident showed so clearly. A nuclear war would be like Three Mile Island on an enormous scale in which nobody would know what buttons to press.

All that deterrence gives us is a short period of time in which it is stable and we have a chance to eliminate the system that must eventually break down into almost total catastrophe. The awful truth about the world is that with the nuclear weapon and the long-range guided missiles, the whole concept of national defense as it has historically been understood has broken down, and the only national defense now is stable peace. Fortunately, this is possible. It must be possible because it exists. Now, indeed, we have a large triangle in the world, stretching roughly from Australia through Japan across the Pacific, across North America, to Scandinavia, consisting of independent nations at stable peace between whom war is completely unplanned and so highly improbable that it never really enters into anybody's calculations.

The most urgent task of the next generation is to achieve stable peace between this present area and the "Socialist camp," as they call themselves. The Socialist camp is so split with division that it can hardly be said to possess stable peace within itself, especially between Russia and China and even perhaps Russia and certain countries of Eastern Europe. But when the choice is between the extremely difficult and the impossible, the extremely difficult has to win out. Stable peace is extremely difficult. National defense in the classical sense is impossible. It can only lead to mutual destruction—and an awakening to this, particularly on the part of the military, is almost a necessary condition for human survival.

If we survive, what then will the world be like? As we move into the area of culture, music, the arts, religion, family, the great tapestry of daily life, prediction becomes more and more difficult, the future more and more uncertain. It is extremely unlikely that the creative potential of the human brain has been exhausted. Cultural, social, and artistic mutations go on all the time. Perhaps the best prophets are the poets. These are the ones who perceive the plot of the great play of human existence, even if perhaps only with the blinding and confusing light of their insight, so we might well end with Tennyson, whose extraordinary vision in "Locksley Hall," well over a hundred years ago, has at least some chance of coming about:

For I dipt into the future, far as human eye could see,
Saw the Vision of the world, and all the wonder that would be;
Saw the heavens fill with commerce, argosies of magic sails,
Pilots of the purple twilight, dropping down with costly bales;
Heard the heavens fill with shouting, and there rain'd a ghastly dew
From the nations' airy navies grappling in the central blue;
Far along the world-wide whisper of the south-wind rushing warm,
With the standards of the people plunging thro' the thunder-storm;
Till the war-drum throbb'd no longer, and the battle-flags were furl'd
In the Parliament of man, the Federation of the world.
There the common sense of most shall hold a fretful realm in awe,
And the kindly earth shall slumber, lapt in universal law.

If the last verse sounds a bit like a wet Sunday in Manchester, let me add a Tennysonian couplet of my own:

We shall learn, and love shall triumph, peace shall reign and art shall thrive,
Every babe will live to murmur, "It is good to be alive."

References

Barnard, Chester I. 1938. *The Functions of the Executive.* Cambridge, Mass.: Harvard University Press.
Barney, G. O. 1981. *The Global 2000 Report to the President.* Vols. 1-3. Washington, D.C.: U.S. Government Printing Office.
Boulding, K. E. 1962. "The Relations of Economic, Political and Social Systems." *Social and Economic Studies* 11:351-62.

_____1974. *Collected Papers*. Vol 4. Boulder: Colorado Associated University Press.

_____1978. *Ecodynamics*. Beverly Hills, Calif.: Sage Publications.

Rolfe, W. J., ed. 1898. *The Poetic and Dramatic Works of Alfred Lord Tennyson*. Cambridge, Mass.: Riverside Press.

Commentary by Wilfred Beckerman
Balliol College, Oxford University

It is very difficult to follow that performance. We have to come down to earth a bit. When I was asked what papers I would feel qualified to comment on, I naturally said Kenneth Boulding's paper. After all, I am an economist, and he . . .

DR. BOULDING: Used to be.

DR. BECKERMAN: Well, he used to be an economist, so I thought I would understand what his paper was all about, but on reading it I became more and more confused, and it looked more and more like poetry than economics. So, I decided that maybe it would be better for me to comment on some other paper, which I am also doing, the paper by Professor Gustafson, which is on a subject I also know nothing about, namely, the interface between theology and philosophy.

Now, those of you who remember exactly what Professor Boulding set out to do in his paper will remember that he begins by talking about four types of systems that will enable us to say something about which systems are predictable and which are not. But he is so inventive that as he goes along he keeps thinking up other systems, and I forget how many he has in the end—about seven or eight or nine. And as I am talking, I have no doubt he has thought up another couple. Anyway, the result is that he is not really prepared to stick his neck out in predicting anything in the end.

He makes one or two fairly safe—what he thinks are fairly safe—predictions about how economies are likely to develop. He thinks, for example, that rich economies probably won't go on growing quite so fast because of the shift out of manufacturing into services and so on. I am not sure how valid that is, but I think most people would subscribe to it.

He then gets on to the question of whether one can make any predictions about the probability of the human race being wiped out as a result of nuclear war. He isn't prepared to stick his neck out on that either—quite rightly. He says he thinks the probability of there

412   KENNETH E. BOULDING

being a nuclear disaster has risen from about 1 percent in any one year a few years ago to probably about 3 or 4 or 5 percent these days, and that seems as good a guess as any, and then he goes on to ask, if there isn't a nuclear disaster, what can we say about the future, and he is not prepared to say much about that, either.

So the first question that occurs to me is, what is the point of all of this elaborate typology of systems if we are not then going to use it in order to say anything at all about the future? If we can't use it to predict the future, I am not quite sure what mileage you get out of this particular classification. It is great fun, and I enjoy reading it, but I am not sure that it does actually help me to say anything about the future.

Secondly, if we agree that we can't predict the evolution of human society in any respects that are really interesting, two supplementary questions then occur to me. First of all, why do human beings have this obsession with trying to predict the future? This has always been a mystery to me. Human beings do seem to have had this obsession over the ages, and people here know much more about this than I. Hence the demand for astrologers and witch doctors and stock market analysts and meteorologists and economists.

I include economists because we are just the modern equivalent of witch doctors in ancient societies. This is because one of the functions that was attributed to witch doctors was simply that they enabled the tribes to reach some decision instead of sitting about arguing inconclusively about whether they should go out hunting or attack the neighboring tribe or something. They had a witch doctor who would go through a lot of mumbo jumbo and look at the entrails of tigers and say, yes, this is a good day to hunt or attack the neighboring tribe. Nowadays you have economists who go through the same sort of mumbo jumbo, look at lots of mathematical symbols and figures and so on, and say, yes, it is a good idea to devalue or have a bigger deficit and so on. This enables society to reach decisions which are no more soundly based—well, they are probably *slightly* more soundly based— than if the advice came from the average man on the street, but not much more, and the difference is about the same as that between the witch doctor and the average man in the tribe.

So, societies do seem to love to have these people who can tell them what is going to happen in the future. I suppose this would correspond to some sort of human passion to feel that one has some control over one's destiny, and isn't completely at the mercy of chance random elements. I don't know. Anyway, there is this human passion for predicting the future, which seems to me great fun, but useless.

People say that we do need predictions for planning purposes, or something like that, but of course that doesn't wash. You can't plan if you can't really predict. And even in a relatively precise area of human activity, such as economics or the pattern of the economy in five years' time or the rate of inflation in two years' time, which are incomparably simpler than the questions that sociologists, political scientists, and so on are talking about, even in those areas, we know perfectly well that we can't predict anything; we have masses of statistical data and people produce beautiful correlations that work perfectly up until last week—and then give completely false predictions of what is going to happen next year.

A second reason why I don't see much point in trying to make these predictions for planning purposes is that the time horizon that is relevant for the sort of questions you are discussing here is mainly far too long to be of the slightest interest to people in power. Politicians really aren't interested in what's going to happen in a hundred years or fifty years. Most of the politicians that I've worked with were more interested in what's going to happen next week or in tomorrow morning's newspapers. One or two exceptional politicians like Mrs. Thatcher may be interested in what is going to happen over the next five years, but she is very exceptional in all sorts of ways.

In my view one main reason why people are so concerned with the future is that it is a form of escapism, and people are concerned with the future because it is an escape from dealing with the problems of the present. Whenever I hear that some politician has decided to organize a task force on the problems of the year 2500 or something, I know he is finding that the problems of getting through the next six months in his own country are becoming exceptionally difficult. I can see it among my own students. Those students who are most passionately concerned with the problems of people living 3,000 miles away in space, or 500 years away in time, are usually those who wouldn't go out of their way to help a blind person cross the road or deal with their own problems of producing an essay for me next week. They prefer to project their problems in space or time—and the longer away the better.

My final question is why, in particular, are we so concerned with the future of the human race? Now, I would have preferred to have dealt with this question after my comments on Gustafson's paper, so I will have to be very brief.

It seems to me that keeping the human race going is not such a big deal anyway, for three reasons. First of all, if we were genuinely concerned with the nonhuman component of the natural world (I don't believe anybody here really is, but I can deal with that in more detail in connection with Dr. Gustafson's paper), it would be better if the

KENNETH E. BOULDING

human race disappeared. Then we wouldn't interfere with all of those animals and fish and trees and all of that sort of thing.

Secondly, if we were concerned with the human race just as it exists now and didn't worry much about posterity, then it seems to me that it wouldn't be very rational behavior to make many sacrifices in the interest of posterity. In other words, if we were maximizing welfare of existing generations, then rationally we should not be prepared to go very far in sacrificing, for example, current consumption of resources for future generations.

Thirdly, even if you want to maximize welfare of mankind over lots of future generations, it seems to me that it is not such a big deal to keep future generations coming into being, because most of them will have a pretty rough time of it. And those generations who don't come into being won't be any worse off as a result. But these matters raise other issues which I think come up in another context.

My main point is that, unlike other species, mankind does have a facility and a capacity to stand back and ask what he is actually maximizing—what is his objective function? So far he has been blindly led by the survival-crazy genes which I have been hearing about. Mankind is in a position to say, "I am not going to be the blind instrument of the survival-crazy genes. I am going to ask myself what exactly I am doing here, what is my objective? Maybe survival isn't the most rational objective for me to pursue." And if mankind stands back and asks himself this question, it seems to me very arguable that he could arrive at the answer that the best thing for him to do is to press the auto- destruct button in this programmed gene-survival robot that mankind apparently is.

DR. WOOLF: Thank you, Dr. Beckerman. I'm amused that the latter part of your statement contradicts the first part, in that you refer to a predictable behavior. Our next commentator is Dr. Wiercinski.

Commentary by Andrzey Wiercinski
Department of Historical Anthropology, University of Warsaw

Perhaps this will be a strong clash between an optimist and a pessimist. First of all, I shall say that the probability that all of us here will be dead within a time span of 200 years is strong. The second prediction is that any star is dissipating energy by radiation in the time elapsed, and it is not coming back to the star.

Dr. Boulding's statement is based upon a new paradigm, which is the systemic approach, and it contains a series of essential truths in regard to the very nature of reality, man, and his possible future. One of them is that any conscious decision of a human individual depends upon his images of the possible future. However, because of deficiencies in information, arbitrariness of imagination, as well as thoughts expressed in abstract denotations, on the one hand, and subconscious motives on the other, man is usually endowed with images of impossible futures which cannot be realized within a given set of circumstances.

Man is very advanced, and acts as if he is endowed with behavioral norms. A behavioral norm may be most simply presented in the following way. Here we have a stimulative pattern, which excites a given receptor. Then through the afferent pathways, the signal comes to the brain, where the information included in the signal is registered and then correlated. So let us symbolize that by the letter R. Then through associative pathways it is related to a center of decision (which may be an emotional center as well) as, for example, related to skeletal muscles. Then in order to perform any behavioral reaction, we must have efferent pathways, which bring this signal about the decision of the selective reaction to the effector, and then the effector may perform a given activity. Of course, this picture may be very complicated by introducing homeostatic activity of consciousness and so on.

I should like to emphasize here that any behavioral norm, which this is, always consists of two endings. One is the cognitive ending, and another ending is the given center of the decision, which may be conscious or subconscious.

But if a conscious and rational decision has to be undertaken, the solutions of the three following cognitive problems must be obtained. First, exploration or identification of facts which are organized systems, since the reality possesses systemic quality. In this case, the explanation involves a simple question, "What is it?" And the reply runs as follows, "This system."

A second problem is classification or ascertainment of the properties, and it is based upon the question, "Which is it?" The reply is, "It is composed of such and such elements, and so belongs to this class or type of systems." And we have a very nice classification in the essay of Dr. Boulding. The third problem is explanation or ascertainment of dependencies. It is the question, "How and why does it act?" And the reply is, "It acts in this way because it is organized as follows."

But the very decision also implies the solution of its own three problems, namely, first, postulation or showing goals: "What to achieve?" The reply? "This system." The second is optimalization, or showing how to achieve it. The reply, "By use of such transformation." And third is realization, or showing the means or resources

KENNETH E. BOULDING

for transformation: "Out of what to achieve it?" The reply, "From such a system."

If we want to get solutions of all of these six problems in regard to mankind's future, we shall find ourselves in a very difficult situation. Why? Because such solutions must be obtained in reference to the biosphere as an organized system, together with all of its interactions with cosmic surroundings. Second, we also need data about man as a species-specific system and as a subsystem of the biosphere. And third, we also need data about the human individual as the most essential subsystem in a net of sociocultural systems.

It should be emphasized that the biosphere, man as a cultural species, and human individuals are by no means equivalent systems. I would like to make clear my view on the nature of the human individual in relation to the statement of Dr. Ortner. He said in his essay that first of all, we are animals. Man is a toolmaker, he uses symbols, and he behaves altruistically, which we correlate with the degree of kinship. However, there also exists in man a specifically human potential which manifests itself in the ability to create ideologies. This is so because we all are endowed with two typically human needs: first, the need for a general world cognition and a world view; and second, the need for a meaning in our life.

These needs appear to be very deep. If the urge for meaning in one's life is not satisfied, a psychoneurosis is inevitable. In general, man differs from animals in that he possesses self-reflecting consciousness. Man is able to act in opposition to his animal needs of self-protection and bioreproduction. Many holy men from different religions have clearly demonstrated that human society was and is governed always by an ideology which supports a given system of values and goals. At the same time, man as a species, as a local sociocultural system, or as an individual, is determined also by the limitations of a given evolutionary channel. We know that within this channel there began to spread about 200 years ago the paradigm of a mechanism upon which a reductionistic science came to be grounded and which made possible, together with other factors, the sudden development of our industrial civilization. Again I shall be less optimistic than Dr. Boulding, because this type of civilization generates threats based upon positive divergent feedbacks.

And finally, I would like to tell you that I am terribly astonished that we don't have here any essay dealing with the comparison between different apocalyptic visions which have been formulated in various religious systems starting from Judaism and Christianity, passing through India, and ending in Ancient Mexico.

Discussion

DR. CAVALLI-SFORZA: As a biologist—Professor Boulding perhaps doesn't like genetics very much—I think a reason for being an optimist is that after all, as Charles Darwin told us, we are machines that made ourselves up and keep ourselves by maximizing our own Darwinian fitness, although he didn't use exactly this word.

Now, fitness, as we all know, is made up of two parts. One is the capacity to have children, but before we can have children we have to survive, so we have to maximize surviving. And because we have been maximized by surviving over so many years, I think we should be able to go through a great deal of crises.

DR. DANIELLI: I agree with much of what Professor Boulding said, but there is one point where I disagree very sharply; namely, that it is improbable that we will become a richer society. The combination of computer science and genetic engineering is opening up totally new domains, and the genetic engineering aspect of this in particular is so rich in potential that it is simply not possible to see the end of the road, except that it will make many things cheaper, more readily available, less polluting, and so forth. So it is likely that the impact of this biological revolution will be just as great as the Industrial Revolution, which was based upon chemistry and physics.

DR. BOULDING: I've changed my mind on this somewhat since I wrote this, largely because of what you say about computers.

DR. TOULMIN: Very quickly, I thought there was a very important point at the end of Kenneth Boulding's paper where he remarked: "In the biological species, the inefficiency is the greatest guarantee of survival." I want to connect this with what seems to me to be the most important traditional error in thinking about political and social affairs in biological terms, which is that people have repeatedly looked at physiology rather than population theory for their analogies. It does seem to me that all of this obsession with prediction is bound up with the modern forms of the organic theory of the state, which assumes that political affairs are run in an oversystemic way, and that what we have to find is how we can establish the pattern of systemic relations in society, which will enable us to discover the salient points we can put in the thermostats and turn things up and down; whereas if you understand Darwinism, you see that the important point is that redundancy is the great source of survival. The important thing, therefore, is to keep resources in the form of the multiplicity of alternative potentials. These alternative possibilities are what we call potential. Probably by seeking to establish too elaborate a system of interrelations within society, you are bound to create a dysfunctional end product which will be overrigid and fail to adapt.

DR. WOOLF: I think that inefficiency is an ambiguous term, and some of the ambiguity might be squeezed out of it by recognizing that "inefficiency" in the sense of not totally adapting, as I believe you are also saying, is a social good from the point of view of the future. But efficiency in using one's organism, where, in fact, a biological revolution is now embedded, is a very desirable situation, and it is the efficiency in using organisms with an absence of total response to the local environment that provides the potential and the resourcefulness.

DR. TOULMIN: But this is the adaptation of Bernard's adaptive system, which is quite different than the adaptation of an historical change.

DR. WIERCINSKI: I am afraid that genetic engineering together with computers, without proper ideology, may lead to the creation of Aldous Huxley's Brave New World.

DR. WOOLF: Other comments? One more, and then we will take a break.

PARTICIPANT: A quick comment. To believe that planning to maximize our future is impossible, without that evidence being in, is not only unscientific, it is counterproductive and self-destructive.

DR. TOULMIN: On the contrary, Wilfred Beckerman produced a wonderful proposition. He said, you can't plan if you can't predict. Now, where do people go from this? A lot of people say, we must pretend that we can predict, so that we can then plan.

PARTICIPANT: Prediction is not the only basis for planning.

DR. TOULMIN: Planning based upon the failure or the illusion that you can predict things which you can't predict is almost—is bound to be disastrous.

PARTICIPANT: Well, Dr. Toulmin, if I hear you and Professor Beckerman correctly, then we are in a know-nothing type of ambience here. If I behave irrationally, I hope somebody will restrain me, and on a global scale, as this gentleman said, rather than this nineteenth-century nation-state philosophy that I have been hearing—on a global scale, man is behaving very irrationally. We can predict that if so many oil tankers are inefficiently operated, we are going to have oil spills, and we can predict certainly that with the plethora of nuclear weapons and the arms race that is going on mankind will not survive. It seems to me that that kind of knowledge is valuable, and everybody in this room depends upon knowledge, I should think, as his way of life.

DR. NEEL: My comments were in the same direction. I think that we have engaged in a bit of academic gamesmanship about the limits of our respective disciplines in terms of their precision for science, but I believe there are some broad generalities that we would agree on surprisingly well, like the limitation to resources, as resources are presently understood, and the need for prudence until we see our way to a new set of resources.

I don't buy genetic engineering as something that is going to get us out of our bind. Bacteria will make insulin and some other compounds very well, but if you are applying genetic engineering to our own species, the last thing I want to see us mess up is the genotype, which we so very poorly understand at the present time.

DR. DANIELLI: I was talking about the economic aspects.

DR. NEEL: Well, there is promise, but there are also strict limitations on that.

DR. BECKERMAN: Is that supposed to be an example of a proposition which would receive more or less universal consent—that there is a genuine problem of resource limitations?

DR. CLAXTON: I just want to make a point that there is a difference which should be drawn between predictions and projections. Perhaps we cannot predict an outcome with certainty and plan on the basis of a particular certainty, but we constantly plan. Everyone plans. Governments plan. Individuals plan on the basis of projections which are, again, based on recognized assumptions. The important thing is to know what your assumptions are, and make your best estimate, and go ahead and do your planning. It is, I think, perhaps missing the point to say we can't plan unless we can predict, because we do plan, based upon the best assumptions.

DR. BECKERMAN: I don't want to produce all of the documentation in the literature but I can provide the references as to why I believe that a notion of any resource limitation on future economic growth or survival of mankind is utterly wrong. If there is any prediction I would subscribe to, it is that the human race will blow itself up in a nuclear war long, long before it runs out of supplies of uranium.

DR. WOOLF: Professor Boulding?

DR. BOULDING: I think I have a right to reply. The reason why I am interested in developing these differences is to widen our agendas, because the whole trouble is, we have concentrated far too much on things like celestial mechanics, and we haven't been enough aware of these other means of developing and improving decisions. That is what I am interested in. I am interested in improving decisions, and if you ask what it is all for, I would say that the most fundamental moral principle is that the inability to realize potential is a pity. I believe the potential of the human race is very far from being exhausted. Now, I

KENNETH E. BOULDING

certainly agree that the human race is only a blip in the universe. I suspect we may produce our evolutionary successor, but if we stop the evolutionary process in this part of the universe, we won't fare too well at the Golden Gates, will we? That is why I added my Tennysonian couplet to my essay, because I believe it is the potential of the human race to say that we shall learn, etc.

I think Tennyson was the greatest writer of the nineteenth century from the point of view of a futurologist. Just read "In Memoriam." It is all there. All we are talking about today is in "In Memoriam." But if we are here for any serious purpose, it is to realize the human potential—and this involves decisions and this means images of the future, and it is better to have better ones than worse ones. We should have more realistic images than unrealistic ones, and all the uncertainty in the world doesn't mean to say we can't make or improve decisions. That is absurd.

You have got to construct the future with a better probability. That is the point, you see. All we have is probability. But, damn it, you can make probabilities better rather than worse. Just the fact that the future is indeterminate, and I agree that it is, doesn't mean we have to give up on it. That is outrageous.

DR. WOOLF: One more final word.

DR. ORTNER: May I make two very brief asides? One is to allude to the well-known principle in mathematics of making a series of successive approximations by which you can achieve some sort of semblance of mathematical truth. I think that is what we are proposing. The other is that I have noted that the highest sense of humor has been exhibited by our economists, which may be an adaptive response to their role as priests of the dismal science.

15. The Future of Democracy

RICHARD J. BARNET

Senior Fellow, Institute for Policy Studies

Editor's Summary. The very tenuous contemporary experiment in decision-making involving broadly based consent by the people of a society (democracy) is threatened in virtually every society in which it is being practiced. Richard Barnet highlights several problems which threaten democracy: fears of scarcity seem to influence policy more than dreams of abundance; conflicts exist over distribution of limited resources, particularly in a slow-growth economy; the size and complexity of modern democratic societies tend to create system overload and a breakdown of a sense of community which is essential to governing by consent. Conditions crucial to the survival of democracy include: a critical and fairly high threshold of economic development; an emphasis on human rights; a better fit between human needs and the distribution of the world's resources; increased literacy; a much more widespread self-sufficiency in agriculture throughout the world; and a system of self-discipline involving a balance between discipline and freedom for the individual.

Human beings are trying to govern a crowded planet with unprecedented organizational problems, and one result is a crisis of democracy everywhere. The modern welfare state grew by developing a moderately egalitarian ideology, but it is under challenge in the industrial West because no one seems to know how to manage growth any more. Although we live amidst abundance undreamed of 200 years ago, the political mood is pessimistic because the processes of growth are not as automatic as we recently imagined them to be. Since we know more than we once did about ecological constraints and resource limits, and hundreds of millions of people once invisible are making increasing demands on those resources, politics in the late twentieth century is more informed by fears of scarcity than dreams of abundance.

All this has profound implications for democracy. In a slow-growth economy conflicts over distribution intensify; there is no surplus with which to buy social peace. Various interest groups have increased their participation in the political process in the last few years, but the result has been social paralysis and disorganization because so many conflicting claims on society are not easily reconciled. Forty years after the defeat of the Axis powers, who once based a bid for world domination on an ideological attack on democracy, popular government is still a rhetorical slogan almost everywhere. Even some of the bloodiest dictatorships pretend to be preparing for the next election. But faith in democracy is flagging.

This essay will consider democracy as a strategy of social survival. We will not ask whether democratic values and institutions are a desirable foundation for late twentieth-century society, but whether they are possible. Is the current organization crisis of the industrial West attributable—as it is becoming fashionable to suggest—to an excess of democracy? Is democracy possible at all in poor countries? Can a state-directed economy devoted to rapid growth and egalitarian goals ever be subject to genuinely popular rule? Given the population explosion, the need for better resource management, and the threat of nuclear extinction, are the survival strategies of the planet to be rooted in democratic principles or, alternatively, do they require us to disenthrall ourselves of an eighteenth-century myth?

Democracy means many things to many people, but there are two ideas which seem basic to the concept of popular government. One is government by the people. Government is legitimated by the consent of the governed, which is given or withheld by some manner of political ritual, such as indirect elections or town meetings. Without a vigorous opposition such rituals merely reinforce authoritarian rule. The other principle is government for the people. A legitimate state cannot be an instrument for the personal enrichment of the ruler, his family, or business associates. However great his power, he is expected to exercise it on behalf of the society as a whole. Egalitarian goals, such as improved income or wealth distribution or the wider distribution of opportunity, are reliable criteria by which to test for democratic health. A society in which a tiny rich minority is able to command more and more of the society's resources and exercise ever more options, while the majority is becoming poor and more restricted in opportunities to participate, is obviously not being governed *for* the people. To be effective, democracy must offer the citizen some significant degree of political participation and economic opportunity on a fair basis.

Let us begin by looking at the crisis of democracy in the industrial

RICHARD J. BARNET

world. In the United States fewer than 20 percent of the electorate voted for the winning candidate in the last presidential election. Voter participation, particularly among the young, has declined precipitously in the last generation. If preelection public opinion polls are to be believed, most voters thought that they had little significant choice in the last election. The Republicans won a major electoral victory, but it reflected as much disillusionment with the incumbent leadership as enthusiasm for the new crop of leaders. Surveys on public attitudes toward authority of all kinds reveal increasing skepticism. The growing apathy and anger about the electoral process is not limited to the United States. There is growing cynicism about elections in France, desperation about preserving democracy in Italy, and in Spain the fledgling democracy was almost destroyed in a *golpe* that may yet succeed.

In the United States perhaps the best index of rising disillusionment with democracy is the formidable growth of an antistate ideology and the extraordinary acquiescence in the extension of corporate power, in effect private government, by those most adversely affected by it. The Republican campaign against "big government" struck a responsive chord in the American people because of two fundamental developments that have occurred over the last generation. One is the rise of the welfare state, an institution with a mandate to interfere directly in the private lives of individual citizens. The federal government has acquired tasks that not so long ago were seen as the exclusive province of the individual himself, his family, or his village or neighborhood. A little less than two generations ago the citizens' encounter with the federal bureaucracy was largely limited to chance encounters with forest rangers in national parks or the county agent from the Department of Agriculture.

Today, large bureaucracies have been created to protect us from poisonous air and dangerous work, to inculcate social values through the school system, to look after us in old age, and to intrude in one way or another into sexual relationships. More than 10 percent of the gross national product is transferred from one group of citizens to another through the mechanism of government, which has become, in the process, a major employer of the middle class. Now that the state is seeking to reduce its obligations in the fields of education, health maintenance, and welfare, millions who belong to the middle class by virtue of their consciousness rather than their pocket books appear to applaud the idea. One of the few clear expressions of popular will, the approval of local referenda to cut tax revenues, such as Proposition 13 in California, resulted in the reduction of services on which the overwhelming majority of citizens depends. Why should this happen?

As government has become more pervasive and intrusive, it has become—so it is widely believed—less subject to popular control. Faceless bureaucracies with their own trajectories defy the traditional democratic steering mechanisms. Anti-statism is so widespread that there are few explicit defenders of the idea of the welfare state in either political party. Expectations of participation in the political process were raised dramatically in the 1960s and 1970s. And indeed the existing democratic structures were significantly broadened. There is far more popular participation in the process of presidential selection than twenty years ago. In 1960 John F. Kennedy entered only seven of the eighteen primaries held that year; major candidates now must enter virtually all of them. The conventions are far more open and representative of the population than in the past. The seniority system in Congress, under which the legislative process was ruled by a few old timers from the South, has been reformed. But despite these significant indices of democratic vitality, feelings of public powerlessness have grown.

In 1975 the Trilateral Commission published three essays (Huntington et al. 1975), one by a Frenchman, one by a Japanese, and one by an American, on "the crisis of democracy." Two common threads run through the essays. One is "system overload." The complexity and size of government cause paralysis in a democratic system because too many people have to be heard before decisions can be executed. There are too many vetoes. Leaders have lost their magical powers, because the public is drowned in a flood of information. No longer a high priest, the politician has become an actor operating in the relentless glare of publicity with a hundred opportunities daily to stub his toe and to cause the crowd to turn on him. Backstage, the professional bureaucrats keep the country running. The chasm separating the democratic fireworks of the electoral arena and the actual decision-making process is huge.

The second thread is the breakdown of community. Every society is becoming what Lester Thurow (1980) calls a "zero-sum society." As government becomes less legitimate, because it falls ever short of the increasing claims, each interest group sees government as a treasure chest to be divided. As profits fall below expectations, corporations demand that the government direct capital to them rather than to labor or to the poor or the aged through transfer payments. Since competition over the division of spoils becomes intense in times of slow growth, powerful interests in a society become perceptibly less permissive about cutting generous slices of a shrinking pie for the less powerful.

Printing money, government debt, and other inflationary policies have been used over the last generation to mask or to moderate political struggles about the fair division of the national product—

between classes and between regions. But these inflationary expedients have, of course, exacerbated the problem. Now that inflation is perceived as a crisis rather than a natural phenomenon, like death and taxes, the response is an austerity program, particularly pronounced in the United States and Britain, but present in almost every other industrial country as well, that rolls back economic opportunity for the majority.

The new economic medicine produces stunning income redistribution effects. In recent years in the United States, one half of the income going to the elderly has been transferred by government from the earnings of younger citizens (Thurow 1980:159). Black unemployment has been twice the rate of white unemployment ever since World War II. Cutting "frills," such as government job-creation programs aimed at increasing minority participation in employment, increases this inequality. As government programs in support of social service, teaching, and related professions are cut back, lower-middle-class unemployment is rising. The lowering of the "safety net" created by the welfare state to protect the unemployed depresses wages in the industrial sector by creating a formidable "reserve army" of the poor.

In the austerity state, the economic security of the citizen, indeed his very legitimacy as a human being, is tied to a job, and thus the competition for a dwindling number of jobs becomes ever fiercer. (It is most improbable that the claim of the "supply-side" economists that increased freedom of corporations to accumulate capital will create more jobs than are destroyed by cutting back the public sector will be borne out by experience.) When freedom to choose is curtailed—save the freedom for a minority to make money and to spend it—the egalitarian myths on which social stability rests are shaken.

In the postcolonial dictatorships of the Third World, similar forces are at work. These societies become tragic caricatures of the developed economies. In many there is not even a theory or pretense at development. In Somoza's Nicaragua, and Amin's Uganda, until recently, and in Duvalier's Haiti today, the country is run as a personal fiefdom or family corporation. A large number of the 170 political entities that send ambassadors to the United Nations are oligarchies. Typically, a few families own most of the land and whatever other wealth there is, while the overwhelming majority of the population goes to bed hungry. In a number of such countries half of the babies typically die from malnutrition-related disease before the age of five. A few of the more "successful" countries of the Third World have imported a particularly brutal form of austerity devised by theorists in the developed world.

The so-called Chilean model is typical. Political opposition is murdered. Perhaps 30,000 Chileans were killed by the government since the coup of September 11, 1973. Almost 100,000 were sent into exile. Opposition parties have been banned and labor militancy outlawed (NACLA 1974). Foreign loans and foreign debt have been used to develop an export industry which employs a fraction of the work force. Unemployment is probably more than 20 percent. The peculiar prosperity is fueled by policies that avowedly favor the rich (Letelier 1976). (The country needs savings and only the rich save.) Scarce foreign exchange is devoted to consumer goods for what appears to be a steadily shrinking class of people who can afford them. The country becomes more and more integrated into a world economy and the gap between rich and poor within the country steadily widens. Argentina, Brazil, and other countries of Latin America are pursuing variations of the Chilean model.

Authoritarianism in the Third World is both a consequence and a cause of political decay. Theorists have observed since Aristotle that democracy requires a certain level of economic development. A small elite cannot maintain control over an impoverished mass by democratic methods. Where political office is the key to maintaining the economic power of a ruling group, it can be safely assumed that economic development will not take place. Encouraging literacy, participation, better income distribution, and other preconditions of a democratic society is widely perceived by those in power to be against their narrow class interests. Even when one set of tyrants is thrown out, the social and economic preconditions for a democratic alternative have been so eroded that one dictator or junta follows another.

In the Socialist world the crisis of democracy is different. For more than three generations an elite in the Soviet Union has defended its dictatorial powers as a temporary expedient, a passing stage on the way to a classless society in which opposition parties are unnecessary because all interests are harmonized. The justifications for concentrating power in an elite deriving its power from the control of a mass political organization have become less and less convincing. No matter how many "counterrevolutionaries" are destroyed, their number is continually replenished and "vigilance," i.e., surrender to party authority and acceptance of official repression, is always required. The Second World War, for a long time a plausible explanation of Russia's political woes, ended almost forty years ago, and people are increasingly impatient, despite impressive economic progress, about the gap between promise and reality. In the Socialist world in contrast to, say, formerly democratic countries in Latin America such as Chile and Uruguay, there has been an increase in freedom and popular participation since the days of Stalin. But the party bureaucracy rules

RICHARD J. BARNET

with an arbitrary hand, and the inevitable consequence is increasing social tension and government repression. Public confidence in the leadership has waned as the system of bureaucratic privilege has become calcified. In China fitful experiments in freedom of expression have been called off, leaving the society with neither a pure Socialist ideology—Mao's moral incentives and efforts at radical mass mobilization are now ridiculed—nor a market-based theory of freedom. Notions of development through foreign investment, harder work for better pay, and social participation through consumerism have all been imported, but the system still seems to be in crisis.

The Costs of Authoritarian Rule. To survey democratic experience in the final quarter of the twentieth century is rather depressing. Functioning democracy is a rarity, and doubts are spreading that democracy can work. In the 1980s neither government by the people nor for the people is regarded as a self-evident proposition. Democracy is no longer an article of faith. The case for it must be argued anew.

Winston Churchill offered what is quite possibly the best defense of democracy when he called it a completely unsatisfactory method of government—except when compared to all the others. The very conditions that make the operation of successful democracy so difficult render new forms of participatory, egalitarian government all the more crucial to survival. To the question, "Can we survive with democracy?" we must, it seems to me, counterpose the question, "Can we survive without it?"

The prospects for humankind are grim unless the species can adapt rapidly enough to head off three disasters which threaten its very existence. The most obvious and immediate threat is nuclear war. A full-scale "nuclear exchange"—to use the curious official euphemism—could wipe out large numbers of human beings immediately, produce hideous mutations in the survivors, and so disorganize the environment and social system as to cripple the adaptive mechanisms of the race.

A war would compound the second crisis, which goes by many names: massive underdevelopment, population explosion, social breakdown in the "poor" countries, global maldistribution of resources. The essence of the second life-threatening crisis for the human species is the prospect of one half or more of humanity on the verge of starvation, without remunerative work, without a valued role in society, deprived of hope for the future or even of a recognized right to life. Massive debt, repression, and suicidal ecological assaults in the name of progress are familiar accompaniments to the standard model of maldevelopment. Why the survival prospects for a divided planet—half human and half subhuman—are not encouraging should be obvious, but they are not, and so I will return below to the links between Third World disaster and the stability of the industrial world.

The third is a spiritual crisis that cripples human efforts to tackle the other two. The political and psychological environment is increasingly hostile to the fundamental rethinking needed for successful reorganization of planetary affairs. The pathological pessimism that goes by the name of realism, the sense of meaninglessness in individual lives, the breakdown of a sense of community, the loss of hope and enthusiasm have the effect of immobilizing constructive efforts to think our way to survival. Our ways of thinking and feeling do not seem conducive to solving our most critical problems. Modern society has developed an elaborate system of incentives to motivate human behavior, but the incentives seem to produce more problem-creating behavior than solutions.

What has all this to do with democracy? The origins of the three crises I have identified are rooted in the loss of legitimacy of previous systems of social and political control. Within the last 400 years or so a number of important political myths have died. Theocracy is dead in most of the world. Even in revolutionary Iran, in which its vestigial hold is strong, the idea of rule by priests who command authority in the name of God is widely rejected by millions who have been touched by modernism. The twentieth-century efforts to resurrect theocracy through Stalinism, with vicars of Marx and holy texts and the like, have not been spectacularly successful.

Feudalism hangs on in various parts of the world, but the radical transformation of agriculture and the disappearance of hundreds of millions of peasants has accelerated the assault on feudal relationships that has been continuing for more than two hundred years of industrial development. With the loss of feudalism goes the internalized sense of place that has perhaps been the most effective means of social control over centuries. In a legitimate hierarchic structure it rarely occurs to one to challenge his place in the hierarchy. As the sun rises in the east and sets in the west, some men are born to be rich and others to be poor. Political, social, and geographic mobility are outside ordinary consciousness. One does not leave family, village, or plot. True, there have been migrations throughout history, but unprecedented migrations on a global scale—displaced farmers from countryside to city, refugees from one country to another in the wake of political crises, and migrant workers on a permanent quest for a job— have helped to undermine the legitimacy of feudal order. Television and radio have expanded the intellectual horizons almost everywhere and have had the effect of exposing masses of humanity to the subversive idea that social and political organization can be different. Expectations rise and greater claims on authority are made.

Aristocracy is also in eclipse. Oligarchs abound, but public-spirited aristocrats are few. The rise of capitalism dislodged the old aristocracy, but business leadership has had difficulty in establishing itself as a new aristocracy because on so many crucial issues, such as environmental protection, worker safety, consumer protection, and so on, it appears to have interests adverse to important segments of the public. The accumulation of private wealth by a corporate elite is an enviable, but not particularly noble, pursuit. It creates political power, but the base of legitimacy for the exercise of the power is shaky. In postcolonial societies, aristocracy takes the form of tribal leadership, but there is a misfit between tribal authority and national sovereignty. Tribal rivalries have made it difficult to build modern nations.

Thus, throughout the world the traditonal roots of legitimacy have been eroded. Secularization and the challenge to family authority have undermined older forms of rule. The rulers of the modern industrial state are unable to carry out the two tasks for which it was created. No one knows either how to defend national territory and such vital "national interests" as access to raw materials with classic military means in the nuclear age or how to "fine-tune" an economy to control the inflation and unemployment that threaten individual security. Max Weber had suggested that the efficiency of the modern state would provide the new basis for legitimacy (Rheinstein 1954). Professional bureaucrats, a new priesthood untouched by special interest, would guide the state along democratic lines. But the resource crises of the 1970s demonstrated that the wonder state was a brief postwar phenomenon. The principal task of rulers today in the industrial West is the restoration of public confidence. Most of the military expenditures are designed not for defense but to provide a psychological lift, to give the illusion and feel of power. The radical surgery of the austerity budgets are also designed to accomplish the same end. The Reagan economic plan is built on a single premise: the redistribution of income to the rich, i.e., "productive," creates a climate in which human energy and creativity will flow in greater abundance. The austerity state seeks legitimacy by offering services and enhanced prospects of enrichment to a minority who, it is assumed, will act as an engine for the whole society and eventually spread benefits for the rest.

A state run ostentatiously for the benefit of a minority is inherently unstable. It must use forms of repression to control those for whom the myth of opportunity has no meaning. In many countries in Latin America, Africa, and Asia, the instruments of rule are death squads, torture chambers, and other forms of mass intimidation. In the United States intrusive methods of data collection and surveillance by computer are being proposed to enforce the cutbacks in welfare and employment benefits. From country to country the state engages in various forms of warfare against the population—from low-level psychological warfare to search-and-destroy missions—all to preserve a shaky social peace. Invariably, legitimacy for internal repression is sought by invoking the magic words "national security." The more unsuccessful the state is in assuring social peace, the more foreign threats—Communist conspirators, greedy sheiks, Japanese carmakers, or whatever—are identified as the source of the problem.

A world of states run by regimes lacking legitimacy is a world at risk. Internal class tensions are controlled by invoking an external threat, but the international tension fostered in order to maintain internal stability erupts periodically in war. As nuclear weaponry continues to spread to the underdeveloped world as well as within the industrialized regions, the risks of nuclear war increase. The most likely way a holocaust will occur is not through the deliberate irrational action of either superpower, but through an unwanted confrontation over a little war in some "strategic" area of the world. Because the superpowers continue to spread vast quantities of lethal hardware to some of the most underdeveloped countries, their power to control the process of escalation is diminished. So-called client states, such as Iraq or Iran, once given billions in military equipment are, as we have seen, capable of acting quite independently.

The "national security state"—austere in everything but a military budget—is a political phenomenon almost everywhere. The world now spends more than $500 billion a year on arms. The militarization of politics not only increases the risk of confrontation but it contributes to the second peril to humankind—economic and social maldevelopment. Scarce investment capital goes for arms instead of education, health, or food. Scarce talent is preempted for the military management sector. The business of the state becomes military planning rather than social or economic development.

The attempt to create an artificial consensus by substituting the idea of national security for political community cannot succeed, because individual security is always threatened in societies where politics has been extinguished. If there are no educational or career prospects for children, no prospects of adequate food or shelter and nothing to look forward to but a perilous life and an early death, security is robbed of all meaning. The national security state cannot compel real loyalty and it can confuse people with promises only for so long. Its survival depends ultimately on being able to exclude from the benefits of the society a large majority of the people. By necessity it must resort to intimidation, pacification, and depoliticization.

The privilege-protecting ideologies for exclusion of the "unproductive" majority have already been developed. There is much talk of "triage," "life-boat ethics" and "basket cases." "Freedom to choose" is celebrated, but the "right to life," it seems, is an exclusively prenatal privilege. The implementation of privilege-protecting strategies increases the perils which most threaten civilization. The crisis of the poor increasingly is visited on the rich. The economic costs of class warfare are escalating. The poor have a way of exacting public money in all sorts of hidden ways, whether it is a $36,000 bill for a year in jail, $100,000 to save a premature baby (far more than decent education of the mother and prenatal care would have cost), or billions in military aid to shore up repressive governments that live in fear of their own citizens. When extraordinary police, custodial, and military expenditures become an ever-escalating cost of maintaining society, the effects are inflationary. This built-in economic instability constitutes the soft underbelly of the austerity state.

An antidemocratic economic strategy has other serious self-defeating consequences as well. Large groups of excluded or exploited people have increasing opportunities to threaten society. The costs of using "throw-away" labor are now being visited on European countries. Mounting racial tension in Britain, France, Germany, and Switzerland is the price of failing to integrate millions of foreign workers into their societies. The United States, which has had a respite in racial violence since the 1960s, is not likely to remain so fortunate if the prospects for excluded minorities continue to decline. (The political origins of dramatically escalating urban crime are not well understood, but it seems, in part at least, to be a consequence of systematic exclusion of certain groups from legitimate participation in society. The costs, social and economic, are staggering and increasing.)

As society becomes ever more complex, the opportunities for "monkey-wrench politics" increase—everything from terrorism, blackmail, and sabotage to blockage of nuclear plants, "sick-ins," and other strategic nuisances that can bring the economic life of great cities to a standstill. The size of the "second economy"—billions change hands under the table and the spectacular growth industry is organized crime—frustrates the efforts of bureaucrats and businessmen alike to bring health to the legitimate economy.

Real growth is impossible in such a situation. The society must pay an increasing tax in guns, private armies, jails, locks, and so on, just to maintain minimum order. Markets do not grow because unemployed, repressed, and rootless masses do not make good customers. A mass-production economy cannot be based on a static or even declining consuming population. The greatest cost of all is the one that is hardest to measure.

In a society under siege, creativity is threatened. The illusion that authoritarian regimes can "get things done" rests on the fallacy that someone in authority knows what to do. The problem is that not even the brightest and most farseeing have the conceptual tools and the experience to deal with the unprecedented problems of the late twentieth century, and those who shoot their way to authority or talk their way to it through demagoguery are typically not the brightest. In a climate of fear, problem-solving is dangerous. Those with the greatest potential for creativity do not take chances; they go into physical or spiritual exile. A society in which leaders spend most of their energy preserving their own power (whether they are candidates who begin campaigning for reelection on inauguration day or corporate executives whose jobs depend upon the next quarterly performance) cannot mobilize the intellectual capacity and moral energy for survival. There is a crippling symbiosis between a demoralized population and its leaders.

The simple democratic notion that government *for* the people is impossible unless it is also *by* the people is inescapably true. Leaders are not and cannot be expected to be altruistic. They have class interests, personal loyalties, prejudices, and peculiar angles of vision that prevent them from conceiving of—much less acting on behalf of—the common good. The limitations of reason and the insatiability of appetite are among the traditional reasons for political checks and balances. Although the institutionalization of checks and balances becomes increasingly difficult as society expands, the need for creative mechanisms for sharing power is greater than ever. The problem is how to create institutions that permit practical self-government without producing a world checkmated by the collision of parochial interests. The world has had a number of successful experiments with democracy for the few, but the principles for renewing democracy in a mass society or implementing it on a global scale have so far eluded us.

Toward a Democratic Vision. If, as I have argued, democracy is the only legitimating principle of government in the historical era into which we are moving—because all the alternatives are even more unsatisfactory—are there principles for adapting democratic ideology to the realities of the contemporary world?

The starting point for a democratic ideology is the definition of the relevant community. Democracy cannot survive in islands of affluence.

RICHARD J. BARNET

The perception of self-interest that led the Roosevelt Administration into World War II was based on a calculation that in a world organized on antidemocratic principles, democracy in America could not survive. It is more true than ever that the antidemocratic trends in the world threaten democracy in every country. Nurturing domestic democratic institutions in an antidemocratic global climate is difficult. For the United States, there are direct internal consequences of the failures of democracy abroad. When this country for national security reasons finds itself compelled to make alliances with regimes whose true nature would offend an alert citizenry, a certain amount of public deception and secrecy is required to carry out the policy. This, of course, undermines the theoretical assumption on which democracy is based: an informed public. Antidemocratic adversaries elicit imitation of their own methods of combat. To fight "totalitarian" regimes it has been deemed necessary to spy on citizens, stifle dissent, deceive the Congress and the public, and engage in a variety of "dirty tricks," all in the name of preserving our "way of life."

If democracy does not expand, it shrinks. That reality has been a source of great confusion. Our foreign policy has been dedicated to spreading capitalism and anticommunism and calling it democracy. Just as the Soviet Union has elaborately confused the fate of worldwide "socialism" and the particular fears, appetites, and prejudices of the men in the Kremlin, as a nation we have asserted—against all evidence—that what is good for America (or more accurately what an administration thinks is good for America) is good for democracy. Our political discourse is so debased that the fact that the forces the United States opposes in most places in the world call themselves democratic, while a majority of the recipients of our most lavish aid do not, is hardly disturbing.

Enormous cynicism has developed about the American commitment to democracy because our methods for spreading it have been so bizarre—military and paramilitary operations, heavy-handed propaganda, attempts to infiltrate, control, or secretly subsidize the democratic institutions of other countries such as newspapers and unions. These efforts do not produce democracy abroad and they weaken democracy at home. The spreading of democracy can come about through example and clear commitment. John Quincy Adams made the point clearly in the early years of the Republic. The world is hungry for a successful political model that guarantees the dignity of human beings. The more Americans made their experiment work, the more its ideas would be adapted and modified to the very different historical and cultural circumstances of other lands. Democracy would spread as a historic idea. It didn't need salesmen, much less executioners, to make people see that it was a good idea: it just need dedicated practitioners.

Today that is more true than ever. Forms of authoritarianism of right and left have brought misery to billions of people. There are no obvious models for imitation anywhere. The Soviet Union is admired nowhere. China is in deep difficulty. The social democracies of Europe are in disarray. There is a demand for greater participation everywhere—Poland, South Africa, Central America—but the institutions to accommodate popular democracy and public order are either undeveloped or nonexistent. The United States has no monopoly on democratic imagination. Quite possibly the next important innovations will be made elsewhere. But we have unique advantages by virtue of history, tradition, and wealth to take the lead in adapting the institutions of the late industrial civilization to the requirements of modern democracy.

The human rights idea is a crucial starting point. In the Orwellian world we inhabit, the insistence upon minimum standards of treatment for human beings everywhere is called "interference in domestic affairs," while the importation of arms to repress populations in revolt is a "blow for freedom." The human rights standard is especially crucial in today's world because the traditional bases for valuing human life have been seriously eroded. Religious faith no longer compels respect for individual souls as it once did. In a world without spiritual limits, everything is permitted. For hundreds of millions, the traditional restraints of family or village life no longer apply. Communities based on reciprocal rights and obligations have been relentlessly destroyed in the quest for "development" and no new communities have been created to take their place. The economic value of human beings, which even in the darkest days of predatory industrialization afforded some protection to workers, has been debased. Approximately one half of the world's population is "redundant," as the British call it. It is not needed to produce the goods the world is able to buy, or it is needed only at certain times and in certain places. Those hundreds of millions who are irrelevant to the productive process—either as consumers or producers—are at great risk. At best they are nuisances to governments, at worst targets to be pacified or eliminated.

In order to arrest this dangerous process, which creates instability for the industrial world as well as for the poor countries—harassed populations migrate and the technological means to exclude them do not exist—principles of protection must be established. We often underestimate the political impact of such measures. The enunciation of positive standards for human rights legitimize and encourage political

RICHARD J. BARNET

struggles for democracy. The Solidarity movement in Poland was launched by an appeal to the rights for workers embodied in the International Labor Organization convention signed by the Polish government.

To be effective in establishing the groundwork for democratic institutions, human rights protection must be universal. Unless every person born anywhere is accepted as having a vested right to a minimal share of world resources by virtue of having been born, there is little hope of building a stable international order. There is no alternative to making human rights truly universal, for there is no middle position that permits dividing humanity into superior races and inferior races, productive individuals and nonproductive individuals, the lucky and the unlucky. The genocidal mind-set that is comfortable with writing off whole peoples and classes threatens the survival of the whole human race because it is subject to no natural limits. Equality of opportunity is the essence of the democratic idea, but it is meaningless if children are handicapped at birth because of the malnutrition of the mother or of themselves or because they belong to a class destined to be forever deprived of the chance to read or write or to go to bed without the pain of hunger.

The human rights idea is essential to the reinvigoration of democracy because without it the processes of compromise which make popular government work are impossible. There is no compromise possible with forces that seek your elimination. Unless there is a consensus on the goal that everyone is entitled to a minimal share—sufficient to maintain life and health—resistance and counterresistance will be so sharp that government will be paralyzed. The familiar cycle of authoritarianism and anarchy will further demoralize societies and cripple their adaptive processes. How to include everyone and what that means in practice are enormously difficult questions. But the institutional framework to move toward these goals will never be built unless there is first a clear ideological commitment.

Another essential element of a democratic consciousness is the necessity to transcend what Karl Polanyi (1947) called the market mentality. The connection between markets and freedom is not well understood. Under certain conditions market mechanisms work well to enhance equality of opportunity; under others they do not. Clearly, a pure market-oriented, antistate ideology—for reasons already discussed—produces undemocratic incentives, behavior, and consequences. The debate about markets, how much state intervention, what kind, and what sort of economic incentives can establish a democratic basis for society, is essential to the reconstruction of democratic theory. But this should not be confused with a larger question: how to promote a sense of common interest that transcends individual self-interest. Since the industrial revolution, modern societies have done what John Stuart Mill accused Jeremy Bentham of doing: "committing the mistake of supposing that the business part of human affairs was the whole of them" (Green 1981:220).

Production has been the goal of modern life, and the dream of personal enrichment and the nightmare of starvation have been the incentives to keep the world working. Whether the exaggerated egoism and preoccupation with self were necessary survival mechanisms in the age of capitalist accumulation is a matter for debate, but they are manifestly not survival mechanisms in a world in which the need is not indiscriminate production, but production for community need and rational distribution. A better fit between basic human needs and the world's supply of goods is essential to head off the life-threatening crises of war and maldevelopment, which cripple developed countries as well as more primitive societies. It would not in itself deal with the third obstacle to adaptation: the spiritual bankruptcy that saps creative vitality. But it is a precondition for dealing with that problem as well.

A fraction of the available human energy is now needed to meet the production needs of the whole human race. Government for the people must therefore be based on alternative ways of establishing links of legitimacy between the individual and society. Redefining work, deepening human relationships, expanding the concept of education, encouraging reflection and self-examination are part of the process of reestablishing a legitimate place for billions of people. Only if people themselves can play the major role in the process of redefinition will democratic institutions evolve.

In developing a democratic ideology for the twenty-first century, it is well to recall the words of Walt Whitman: "Democracy is a great word, whose history, I suppose, remains unwritten because that history has yet to be enacted." In the midst of the authoritarian wave we have been describing, there are powerful democratic cross-currents at work, which are, I believe, the hope of the future. Moves toward the democratization of modern life are occurring in both the industrialized world and in the underdeveloped countries, even as the traditional devices for sharing and legitimizing political power encounter ever greater difficulties.

There is a growing movement for workers' control and self-management of plants in the United States and Europe, East and West. Increased participation of workers in the management of the productive process appears to be a key to improving productivity. Enlightened corporate leaders seem more ready to experiment with greater worker control over the production process itself. One reason is that harder work for more pay no longer motivates workers when the pay disappears in inflation and the worker comes to see himself as a mere extension of a machine. Without security, equity, and dignity, workers produce less and at greater cost. Increasing productivity through participation seems to work best when the fruits of increased productivity are equitably shared. Workplace democracy cannot evolve unless workers are able to share in the management function as they do to a degree in the social democracies of Europe.

The great need in poor countries is to mobilize the skills and energies of the people in the process of development. Neither charismatic leadership nor authoritarian rule works for very long. Only when people mobilize themselves—when they understand the need to store grain properly, dig wells, and so on, and can see that they will benefit from socially constructive community action—does the battle for survival take a favorable turn.

In the industrial countries worker and consumer representation on boards of directors does not represent true participation unless workers are appropriately educated to play an enhanced role in the productive process. Unions and factories should be educating workers at all levels in management techniques, technology choices, and world market conditions needed to convert the workers from human robots into thinking participants.

A crucial component of an education system is the theory and practice of conservation and sound resource management. Without a much deeper consciousness of conservation as a survival requirement, the resource crisis is likely to hasten resource wars and further maldevelopment. Exhortation will not do. People need to have a stake themselves in the resource systems, to understand them, and to make the connections between how they organize their own lives at home and at work and how the planet can be organized for rational resource use. In poor countries this educational process is the heart of any mobilization strategy.

The old idea in democratic theory—that people had to be informed before they could govern themselves—is more true than ever. The gap between the education system and what it should be to make democracy possible in a complex society has never been greater. But here again many innovative ideas and experiments are springing up. The phenomenal success of some underdeveloped countries in mass literacy campaigns—and in catapulting into positions of responsibility men and women who not so long ago were ignorant peasants—suggests that education for democracy is possible if there is a commitment to try.

Nowhere is there greater need for democratization than in agriculture. Much of the extreme poverty and degradation in the world is directly or indirectly related to the maldevelopment of the world food system. Fewer and fewer people are now raising the world's food. Millions of people formerly in subsistence agriculture are being uprooted from their land and are physically separated from their traditional food supply. Land in poor countries, once used for self- provisioning agriculture, is now being used for export crops. Typically, high-technology agriculture—spearheaded usually by the multinational food conglomerates, the fertilizer and seed companies, the farm machinery manufacturers, the grain traders, and the food packagers and proces-sors—moves in and the people who once were subsistence farmers move out, usually to the city. Thus, those who were "self-employed," to use the familiar terminology, enter the job market and the money economy. Separated from their land, they must now buy the food they once raised themselves. Hundreds of millions cannot find work and hence have no money to provide themselves or their families with adequate calories. Even among those who still farm, the cash require-ments for "inputs," especially seeds, fertilizers, and machinery, are now so high that they must sell what they once retained for family consumption. Even the millions who work to provide food for others do not eat adequately themselves.

If the basic human right of survival is to be secured, countries with hungry populations must become more self-sufficient in agriculture. That can happen only with the right technology and the right incentive system. Cash-cropping, sharecropping, and contract farming provide little security or stake for agricultural workers. They must produce for others before they can amass anything for themselves and their families, and they have little incentive to increase production or to decrease waste because they participate so little in the benefits flowing from either. The bare struggle for existence in the countryside and the constant social upheaval caused by the continuing displacement of farm workers are major destabilizing factors in economic and political development in much of the world. Democratization of the food system, by permitting those who farm to make the major decisions about what is to be grown and how it is to be grown and by instituting land reform that will produce greater equity, is a prerequisite for increasing productivity and reducing waste.

A number of studies confirm what common sense would suggest: food security for those who produce food is necessary for a stable agricultural system and there is an optimum size beyond which farms become less productive and more wasteful. (New seed technologies now make small-scale farming more productive than ever, but such innovations have typically led to ever more concentrated control of land.) The essence of democracy is control over the essentials of one's

RICHARD J. BARNET

life. By this definition, the life of peasants is thoroughly undemocratic and insecure, and the rest of industrial civilization, while profiting from the exploitation of the food growers in the short run, is at risk as long as two billion peasants and former peasants are not integrated into a world political and economic order.

In agriculture, democratization means more control by the cultivators over the land, a greater stake in the harvest, and a greater opportunity to bring to bear their own special knowledge and experience of local land conditions. Without such democratization, there is insufficient incentive to produce and to conserve. The struggle for land reform and for democratic control of agriculture, as in Poland, is intensifying around the world.

In the United States great popular movements have served generally as goads to the elected leadership when those authorities have proved incapable of making historic changes required to stabilize and build the society in the face of new challenges. The civil rights movement and the antiwar movement had a profound effect on the national scene. At the local level a variety of citizens' organizations sprang up in the 1960s and 1970s dedicated to direct action on consumer issues, welfare, health care, taxes, and such neighborhood issues as the location of traffic lights. Extraparliamentary organization has been spectacularly effective in West Germany in stopping the nuclear energy program. In Poland grassroots organization transformed for a time at least the governing structure of the nation against enormous odds. The spontaneous outpouring of literally millions of citizens into the streets brought down the heavily armed regime of the Shah of Iran.

It is dangerous to romanticize popular movements. Typically, when they succeed they create democratic space, but they are not easily converted into institutions of self-government. However, the impulse to organize and to take control over critical aspects of public life, because the existing institutions are too corrupt, indifferent, or inefficient, is a powerful political force that can be institutionalized in new and creative ways. Much of the impetus for the antistatist ideology in the United States that is now leading in an undemocratic and antiegalitarian direction comes from widespread disillusionment with big institutions, both public and private. A government bureaucracy can be as unfeeling and as maddening in its relationships with citizens as a corporation. Institutions designated as "public" and "for the people" do not necessarily play that role. But surely in a society as resourceful as the United States there are choices between vesting power in huge, remote public bureaucracies and "private" corporate bureaucracies that are equally inappropriate planners of a democratic society.

The challenge to any democratic society is to balance the social requirement for discipline and the individual need for freedom. The tension between the two can be managed only when people freely discipline themselves. The alternative to the zero-sum society is democratic self-discipline. But it never comes in response to exhortation. How people treat one another, how they treat the environment, whether they honor or dishonor the future depend on the set of incentives that operate in society. In the United States we have erected a complex set of incentives that produce the wrong results from the standpoint of society. We regularly reward waste, sloppiness, and indifference, and make it difficult for individuals or corporations to put human considerations above property considerations. Our tax laws, school systems, and popular culture transmit messages that are inconsistent with the revitalization of democracy. The starting point for applying democratic imagination is the incentive system.

How do we encourage people to discipline themselves in the interests of society? The most powerful incentive is a vision of a just society that is credible because it springs from a popular political movement or party. Although, as I will argue, management functions ought to be decentralized to the greatest possible extent to assure efficiency, encourage participation, offer alternative arenas for experimentation and develop new institutions of checks and balances, the process becomes anarchic unless it takes place within the framework of a unifying vision. The great failure of the political parties in the United States is their abdication of the responsibility to develop such a vision along with the program to bring it to reality.

Democracy is a system for sharing power. As society has become more complex the problems of relating local authority and national and international authority have grown. "Getting the federal government off our backs," now a popular slogan, can be accomplished in a variety of ways. National bureaucracies are essential in a democracy of continental reach that purports to be united. But they do not have to operate as police forces and indeed they did not always do so. Centralized authority can set standards, supply technical assistance, monitor how the incentive systems are actually operating, and act as a redistributive mechanism without undertaking to govern at the local level. (Thus, for example, the government could make it expensive for factories to injure workers, rather than take responsibility itself for plant safety.)

The primary task of the federal government in promoting a democratic revival is to establish the enabling procedures for local authorities to achieve greater economic self-reliance and greater freedom to experiment with the democratic organization of health care, housing pro-

RICHARD J. BARNET

grams, and education within the guidelines of a national program that has been debated in the political process.

The contradiction between the requirements of compromise and sharing, on the one hand, and the need to build stable self-reliant communities on the other, is less than it appears. It is true that there is always a danger that a community once given the power to develop a resource base for itself will become isolationist and will secede, at least spiritually, from the larger community. But the realities of an irreversibly interdependent economy make that difficult in practice. Every great city is now so dependent on the international economy that isolationism is not a real option. The more secure a community is, the more able it is to take a longer view of self-interest and the more willing it will be to subordinate short-term parochial interests to the requirements of a stable global economic system—provided it has a sufficient stake in that order.

As society becomes more complex, the need for self-regulation in various forms becomes ever greater. Centralized hierarchic bureaucracies tend to issue conflicting, confusing, and demoralizing commands, and these produce the wrong political chemistry for building genuine communities. The judgment, skill, and incorruptibility of the regulators is taxed to the breaking point. But a great deal of essential regulation in modern society could be done by those most directly affected. Thus unions could be empowered to play a much more vigorous role in policing health and safety conditions at the plant. They would have more stake in reconciling their own health needs and the health of the firm on which their job depends than an official from Washington would have. Neighborhood organizations could be empowered to play a significantly greater role in consumer education, the delivery of health services, voluntary housing rehabilitation, and neighborhood planning programs. The re-creation of civic space at the local level is essential for the revival of the processes of deliberation. Thus some of the most critical choices of the next generation have to do with alternative resource development and new technology.

No democratic science and technology policy is possible without vigorous local popular involvement in the issues. Whether a city or region develops a coal economy, decides to become a seller of "software," puts its health dollars in the water system, environmental improvement, and prenatal care, rather than into high technology medicine, are critical public choices. If these issues, which affect the quality of public life, cannot be subject to popular deliberation, then the deepening suspicion that nothing of importance can be democratically decided is justified.

The buildup of voluntary institutions with quasi-governmental authority could be the basis of a new democratic consensus in the United States that would unite left and right. More and more people across the political spectrum are impatient with big, intrusive government. Some of the antistate feeling has been deliberately cultivated by the continuing drumbeat of corporate propaganda, which has successfully identified the term "private sector" with big business. The overreaching of the welfare state has certainly helped. (Not long ago the Swedish Parliament voted 259 to 6 to outlaw spanking!)

In the United States we could make much more imaginative use of what de Tocqueville and others have identified as a unique feature of our system: the important role of private, voluntary associations. But the large corporation, while legally fitting into that category, is now clearly a different structure. It is becoming increasingly public in its function and impact by virtue of its size and scope. The ideological assault on big government could turn out to be a healthy democratic corrective, but only if certain ominous trends are resisted. In the wake of the failure of government, a momentous transfer is taking place in the United States. Control over education, police, scientific research, and strategies of resource development is being handed over to global corporations with little or no loyalty to local communities and virtually no institutionalized public accountability. To consign the most crucial public activities to organizations with narrow private goals is to give up on democracy. Democratic practice requires redefining the old distinctions between what is public and what is private to take account of the extraordinary changes in the world economy that have taken place since the present legal and social conventions governing the life of the large corporation were put into place.

Democracy is a system of government that places a high value on individual dignity. In the contemporary world it is difficult to achieve dignity without active participation. Conversely, only individuals with a sense of themselves can act as citizens. For the species, the most critical contribution democratic theory and practice can make is a restoration of a sense of self. Without a proper intimation of individual worth, a human being cannot locate himself in a puzzling universe. He or she cannot understand the meaning of humanness. Unless that meaning is internalized it will not be possible to avoid intraspecies warfare and other planetary disasters. A new sense of planetary identity is crucial for survival, and that means that our ways of thinking about enemies must change. No rational, informed individual could support nuclear war today, but a frightened mob could demand policies that make it inevitable. Only when there is a much higher level of public understanding that the irrational forces impelling us to nuclear war and ecological disaster are a far greater threat than the Russians, Cubans, or whoever (in Moscow: read Americans) will the reorganization of the planet for survival become possible.

A heightened awareness of self and one's own possibilities is necessary before there can be identification with others, particularly those at great distances. Self-awareness is also crucial for the development of effective self-discipline, which is the essential mechanism for organizing a stable society in an increasingly complex world. However, the consciousness needed for survival requires a favorable institutional framework. Have we not come to a point in evolution where only a democratic environment offers the chance to be human?

References

Green, P. 1981. *The Pursuit of Inequality*. New York: Pantheon.
Huntington, S. P., M. P. Craier, and J. Watanuki 1975. *The Crisis of Democracy: Report of the Governability of Democracies to the Trilateral Commission*. New York: New York University Press.
Letelier, O. 1976. *Chile: Economic "Freedom" and Repression*. Washington, D.C.: The Transnational Institute.
North Atlantic Congress on Latin America 1974. Vol. 8, nos. 6 and 8. New York: NACLA.
Polanyi, K. 1947. "Our Obsolete Market Mentality." *Commentary* 3:106-117.
Rheinstein, M., ed. 1954. *Max Weber on Law in Economy and Society*. Cambridge, Mass.: Harvard University Press.
Thurow, L. 1980. *The Zero-Sum Society: Distribution and the Possibilities for Economic Change*. New York: Basic Books.

Commentary by Allen B. Bassing
National Museum of American Art, Smithsonian Institution

As an ethnographer with a specialty in sub-Saharan Africa, I was rather intrigued and daunted by this paper. I don't give a lot of thought to democracy or its future. And after reading it, I was quite ready to change the title to "Is There a Future for Democracy?"

Barnet has a rather somber view of what democracy is and where it is going, and certainly a somewhat pessimistic one. But he responds to the question of a future for democracy with a qualified "yes." The ways of doing it seem to me to harken back to an eighteenth- and nineteenth-century mode. Let the individual come forth. Let the individual do things. Let there be small groups that get together to make decisions. I really question in this day and age whether that is possible any more. We have never had, as far as I can tell, in history—both prehistory, written history, what have you—a pure form of democracy. The closest we have ever come to it is the ancient Greek city-state, or perhaps the town meetings in this country.

What is democracy? In the ideal practice, it is obviously representative government, providing political, social, and economic equality. The problem is, where have we ever had that in our history of mankind? Democracy is an absence or disavowal of hereditary or arbitrary class distinctions and privileges. That, too, has never happened. It is tolerance toward minorities. When has that ever been the case? It is a freedom of expression. That has never existed, either in the media or by individuals. Democracy involves a respect for the essential dignity or worth of the human individual. There should be equal opportunity, which of course has been a very fashionable phrase, but this has never happened. All people have never had the opportunity to develop to their fullest capacity. It just has not taken place, and for very many reasons which I don't think are germane to this particular issue.

But what do we do about all of this? How do we handle it or where do we go from here? I think that what has happened is that we have gotten caught up in phrases, in words, and in definitions. If you think of most other so-called sciences outside of political science, they are looking for new ways of doing things. However, political sciences seem to be quite happy, from my anthropological viewpoint, in keeping with the old terminology. I really question whether we shouldn't be looking for new words, because if we continue to call it democracy—but a new form of democracy—it takes too much of an explanation. People have this old stereotype in their minds. They want to know what you mean when you say democracy today, and that causes all kinds of problems.

Discussion

DR. WOOLF: Thank you very much. May I take advantage of the absence of the second commentator and insert a very brief comment which seems to me to lie embedded in much of what we have been saying, and especially emerges from both the comment of the last speaker and the last paper.

RICHARD J. BARNET

It seems to me on the question of adaptability that the industrial and technological world is undertaking the most rapid action that we can call adaption in human history. A massive shift and relocation of the industrial center, geographically and intellectually, is under way. Having moved to Japan, as we already know, we now see emerging in Southeast Asia, in a very sophisticated cultural matrix, an extraordinary industrial apparatus. This is enhanced by the technology itself, which becomes virtually independent of large numbers of laborers. Automation, what the industry now calls CAD/CAM—computer-assisted design, computer-assisted manufacturing—is becoming extraordinarily efficient. So we are riding a very powerful wave, a tsunami, which is engulfing other forms of political organization and social theory with incredible rapidity. That is adaption with a vengeance, but it is not man-determined, although man-made.

We seem also, it seems to me, to be moving in the year 1982 to a moment of another kind of crisis. If you look at the election events, that is, the political voting process in the world in the year 1982, a concatenation of campaigns is coming to a head, certainly in the voting communities of the Western world and their near equivalents in the Far East. That means that political rhetoric will have to be different from economic fact, because economic fact and technological shift are now both rapid and long-term in their demands. As one hears in economic circles, one must move away from the quarterly report and the annual report as the determinant of investment patterns. It seems to me that social scientists of the kind assembled here, whatever their professions, and even bureaucrats, need to come to grips with two sets of forces which are the most powerful and the most present in our time. These are the force of political behavior, with the quest for an increased democratization of that behavior—incidentally, I believe, enhanced by the potential of the new technology—and the force of real circumstance, redistributing manufacturing processes, thus changing the way we do business in an extraordinary way.

Let me illustrate just one case as a model: the steel industry of the world. At the moment, after some attempt to acquire a steel industry, the Japanese are moving as rapidly as possible to get rid of it, recognizing that this cost-ineffective process, whatever the new technological elements in that process, should be done elsewhere, in lesser societies like the United States or Western Europe.

If one follows the drift of capital, again without ideology, the movement of capital to Southeast Asia is at an enormous pace—to Kuala Lampur, to Malaysia, to Singapore, to Korea, and places of that sort. Perhaps these are instruments that our economists have not shared with us, that they read all the time, and that most of us never see at all. I suggest these are forms of adaptive behavior moving at enormous speed and ought to enter into our discussion and dialogue.

DR. BARNET: Could I make a response to that? I think that what you say is true, that the new technology—particularly the technology of internationalizaton, the data processing, and transportation—has produced this change, and it certainly is an adaptation. But I think what we have not done is look at the consequences of that adaptation, particularly with respect to population. It is very much related to what we were talking about earlier. It goes to the question of whether or not you can plan. If one is talking about a mechanization of agriculture in an area of the world that very rapidly renders irrelevant and landless large numbers of people who are then pulled or pushed into the cities, but the city economy is not nearly developed enough to support more than a fraction of them, I think you have a classic case that we have now all over the world. One could deal with the problem of absorbing people into the cities by a certain amount of anticipation, by a different way of using technology; but I think that, realistically, the decisions about how that technology is going to be used are in fact reinforcing the process, not resisting it. That issue gets to the question of the control of the technology, that is, technology is not controlled in ways that put those considerations as far more of a priority than we do. The technology does not have the effect of solving the problem, but in fact makes it worse.

DR. WOOLF: Well, I would like to respond, but it would be unfair to take advantage.

DR. CLAXTON: Dr. Barnet in his remarks went back to what Dr. Dubos said originally. In his opening address, Dr. Dubos referred to two great negatives, one dealing with nuclear weaponry and the other with the problem of employment or underemployment. He set those aside, and then his address went forward with some of the positives. What you have just been referring to was the problem of the employment of the capacities of mankind. I would like to refer for a moment to the matter of nuclear weaponry, because there are two adaptations involved in this matter which are far beyond what mankind, I believe, has ever done in any other aspect of adaptation. The first is the adaptation of the last forty years or so of a willingness, even a necessity of living under the enormous shadow of potential human destruction. Some of us in this room were alive and active at the time of the bombing of Japan, and others, our children, have grown up with that as a matter of history, but it is a living presence for them. That is one aspect of the adaptation. The other has arisen more recently, because up to the last few months there has been the general agreement of everyone involved in nuclear activities and political relationships that nuclear

RICHARD J. BARNET

weapons could not be a used. The whole concept of armaments on both sides has been to prevent this. In the last few months there have been intimations from this side of the ocean that maybe it is possible to have a nuclear exchange. Here is the small beginning of a concept of the possibility of adaptation to the use of nuclear weapons.

I suggest that these two adaptations may overshadow almost everything else that is going on in the world today. I would like a comment from Professor Barnet.

DR. BARNET: I agree very much with what you say. I think what struck me about prediction of the question of nuclear weapons (I've been looking at it for more than twenty years) is that here is a case where prediction has been almost infallible. The atomic scientists who developed the bomb, Fermi, Oppenheimer, et al., and then the next generation of atomic scientists every few years made predictions about what the next development of the technology would be, and then what the reaction to that would be if the thing was not controlled. They were invariably right.

What also strikes me is that the predictions of those who have continued to assert that security can only be found by increasing these weapons—that there is a technological answer that one has to keep going—have been invariably wrong on the facts. They have based their decisions at the time on arguments about what the other side would or would not do. And my experience is that almost without fail they have been wrong. I guess this suggests to me that the only way we may be able to have some solution to this problem is if the issue of what to do about it becomes, in fact, more democratic—and that seems to be happening. I think what is going on in Europe may be of historic importance. It is much too early to tell, but I suggest that if there is going to be a way out, of averting this particular spiral of nuclear weaponry, it may well be because there was simply a massive opting out of it and a rejection of it by the people who are the potential victims.

One other thing I would say is that one of the crises I alluded to in the paper and that I think is extremely serious, but not very well understood, certainly not by me, is a psychological or spiritual crisis that paralyzes response. Pessimism itself is a central part of it—that there really are no solutions and there is nothing that can be done.

The Gallup poll of three weeks ago showed that in the United States 47 percent of the American people believe that a nuclear war is very likely or quite likely within ten years, and that they personally have no better than a fifty-fifty chance of survival. Most thought their chances much worse than that. This response is quite extraordinary, because it means half the population today believes that because of events well beyond their control, they are not going to have ten years left.

DR. WOOLF: Thank you. Professor Gustafson?

DR. GUSTAFSON: My comment is partly by way of a question, and it pertains to the use of the word "efficiency." Sometimes it gets thrown out in conversation as if there were some thing called efficiency that this term relates to. It seems to me that what we are talking about, when we talk about efficiency, are relations between means and ends. In this context, the issue is not efficiency or inefficiency, but what are the choices of ends towards which our actions or resources are to be used.

DR. TOULMIN: My own observation really comes from this, because I wanted to comment on the way in which Dick actually states the second of his crises, where he talks about half of humanity on the verge of starvation without a valued role in society, or even a recognized right to life, and gives us in this way a package we need to unpack. Perhaps one way we can go ahead is by asking whether things have not got linked together. If you look at the history of human attitudes, there has always been a deep ambiguity, a deep ambivalence towards work. In a way, one of the psychological preconditions for the capitalist industrial transformation was the result of people succeeding, for the time being, in creating a strong positive ethical tone to the idea of productive work. I think that part of what you call the crisis now is the result of the fact that the very success of the ethical policy that was involved in giving this enormous positive attitude toward material production has been so great that we have created a situation in which that ethical valuation is no longer appropriate. I think that the question you pose, namely, how are we to maintain control of the technology, is even worse than the problem of nuclear fusion. I don't believe there is a magnetic bottle within which you can keep technology so as to keep it under control. I think I agree with the implication of what Harry Woolf was saying, namely, that technology has achieved an autonomous status, and that indeed material production would become increasingly a matter of local capital-intensive enterprises. I see no way of stopping this. On the other hand, what we have to do is recognize that not only materially productive work is to be rewarded. I mean, we who work in the universities, God knows, don't deceive ourselves that we are materially productive. We are very thankful that we are rewarded. The bureaucrats who work in government offices in Washington know very well that they aren't being materially productive. They are very thankful to be rewarded; still, there is a great deal of political rhetoric and a great deal of institutional conservatism which seems to me to be built around the assumption that indeed material productivity is what deserves reward, and that people's hope for the future and valued role in society and remuneration are all somehow or other to be seen as bound up with their contribution towards material production. This again is an ideologically neutral observation.

I think what we have to think about much more is work sharing, and all kinds of measures which will break the link as a result of which 40 percent of the population get fully rewarded for their material production and 20 percent of the population are "structurally unemployed" and a lot of people are on the margins in between.

PARTICIPANT: Could I say that I think it is important to correct the impression that the gentleman behind me gave that the use of nuclear weapons some time ago was inconceivable, but suddenly it has become conceivable. The established doctrine of both the Soviet Union and the Peoples Republic of China, and indeed of the political and military figures in the United States, has always been able to conceive of the use of nuclear weapons. Indeed, I think every nuclear power has been able to conceive of the use of nuclear weapons.

Now, segments of the community, particularly the academic segments, find it inconceivable, but it has been standard traditional doctrine within the military and political leadership of all the countries who have nuclear weapons.

DR. WOOLF: Thank you. May I say something about the nuclear weapons issue? While it enters our discussion, and has to in terms of the issues of this symposium, let us confine it to the issue of this symposium and not military policy, which is a very significant topic, but we will stray too far.

DR. MIDGLEY: Isn't that like talking about the crossing by the Titanic without mentioning icebergs?

DR. WOOLF: No, I think not.

DR. WIERCINSKI: I would like to ask one question. What is your attitude or vision of Schumacher in his book, *Small is Beautiful*?

DR. BARNET: Well, I agree with Schumacher's idea, if I read it correctly, that it is precisely the advances in technology which make possible greater decentralization of technology, and therefore the possibility of a wider variety in the use of technology, and more control by the relevant community, as I call it, of that technology. I think he has been terribly misread, to suggest that this means going back. It doesn't mean that, and I am not talking about going back. If I could just refer to the commentator, I think that the technological pool that is available gives us far more choices than in fact we are taking advantage of, and the fact that the choices are being made in the way they are is not, I would argue, a matter of autonomous technology. I think it is a direct result of the political structures within which that technology has been developed and can be changed.

DR. WOOLF: I would like to recognize you, but may I make a short comment on that? The point I was trying to make about the technology story is that it is already an actual practice, for example, to make an automobile in several places in the world, to make the engine in Tokyo, the fenders in Brazil, the main frame, if you like, in Detroit. What that says is that a shift in the nature of employment follows, that one may do his work in different cultures, that the ideological content of these processes will follow lickety-split, including labor regulations, health care for the worker in the several plants in different locations and in different cultural histories around the globe. That suggests to me, at least, a very rapid agglutination of common principles. Now, it may be a leveling down, and hopefully it is a leveling up, but that is not small is beautiful. That is being done on the basis of at least perceived efficiencies, to bring back that word, and also because computer technology allows coordination on a worldwide scale at the speed of a beam of light.

DR. BECKERMAN: The reference to *Small Is Beautiful* reminds me that when I reviewed this book for the *Times* supplement many years ago, I reviewed it under the heading "Small Is Stupid," and nothing I have heard since has persuaded me that I was being unfair to the book. I thought it was a travesty of what economists actually say about the problems of technology, and it seemed to me that the economic problems we are facing—and they are extremely serious today—have very little to do with the sociological implications of technological change. You, Mr. Chairman, threw out a challenge in your comments here when you referred, for example, to the possibility that this enormous shift in the center of gravity around the world toward the Far East might be some sort of an adaptive process that we don't know anything about.

DR. WOOLF: It may be embedded in your secret archives.

DR. BECKERMAN: The reason we haven't revealed it is because we don't know anything about it. I think there is probably wide agreement in the economics profession that the phenomenal changes going on in Japan are rooted in sociological differences between societies, which we don't really understand. In the end, they go back to sociological differences.

DR. WOOLF: Well, I was baiting the hook, and I was hoping that Dr. Laslett would have stayed, because in fact that very issue of the social structure of Japanese families has moved into the social structure of the Japanese factory, and there is an actual linkage. Although I believe at the moment that is about to undergo a very radical change.

DR. BECKERMAN: The point I wanted to make is that we do know something about the problems existing in Western society, which have really nothing much to do with the sociological implications of technological change, work habits, and so on. These are problems that we were worrying about in the 1960s, in the 1970s; we are not worried about them now. We have in Britain the highest unemployment level ever, including the Great Depression of the 1930s, with the obvious social effect that people have seen this year. This is a straightforward, immediate, urgent economic problem, and in principle is a soluble problem. Therefore, as scientists, we ought to be working on just how to get rid of British unemployment. There may be other great problems, such as what we should do about the world's economy in a hundred years' time, and so on, but these aren't soluble as far as we are concerned.

DR. WOOLF: I am suggesting that it is the absence of concern with the powerful undertow that the new technology represents and that will lead to a short-term political solution to unemployment which produces the greater attention.

DR. BECKERMAN: Yes.

DR. BOULDING: That is nonsense. Major social change has been going on for 200 years. I think society is fantastically adaptable to technological change. We chased 30 million people out of agriculture in this country in one generation. We created Harlem, which is all right since Harlem is much better than Mississippi. In 1945 we transferred 30 percent of the American economy from the war industry into the civilian industry without unemployment ever rising above 3 percent. The British unemployment problem is, if I may say so, just the result of idiotic economic policy.

DR. WOOLF: At whose expense, and in what time span?

DR. PASSMORE: Well, it is simply that some of the points you were making, of course, bring out things that were said earlier in the symposium about who counsels "we," and the way in which a crisis has been regarded as a crisis in a particular country or a particular area, but from the point of view of another area it might not be a crisis at all, it might be an opportunity. The movement at the moment of industry to Southeast Asia from the United States is like the movement of heavy industry from the northeast United States to the California type industry, represents possibly a very severe economic problem for the United States. Of course, to Southeast Asia, it is quite the reverse; it represents a remarkable economic opportunity.

DR. WOOLF: And for the world.

DR. PASSMORE: And for the world at large—but then one gets to the second problem: whether the new style of technological innovation, which appears to be obvious at the moment, does really represent a kind of leap that makes these past analogies not really quite applicable. I am not sure about this.

DR. WOOLF: None of us is, but I am suggesting that in spite of Dr. Boulding's admonition.

DR. PASSMORE: People always have a tendency to say, well, we got through that in the past, without realizing that the scale upon which this is happening is increasing, and in the case of some of the new technological innovations, especially in factories, the kind of automation which is now occurring, we can't cope with that. We coped previously by shifting population from agriculture to industry, but the belief that we can shift population in the same way from industry to service industries seems to me perhaps misplaced.

Many problems in the past were solved by large-scale immigration, and that is now largely closed down as a method of solution. One has the same query about the kinds of shift which could possibly occur if factory labor is pretty well abolished.

DR. BOULDLING: It was easier to move Pakistanis to Bradford than to move Bradford to Pakistan. New Bradford is 40 percent Pakistani. We are going to have enormous migrations of South Asians into this country as we are now. Are they going to be good! We are going to have Hong Kongs all over the United States, and it is marvelous.

DR. WOOLF: You are only confirming what I am saying.

PARTICIPANT: I would think that the purpose of this symposium would be ill served if we let the small is stupid comment go by without a reference to the MIT study which showed that from 1959 to 1969 or longer, small businesses were responsible for 70 percent of the innovations in technology. Those businesses, over 500, actually lost in number of employees.

I wonder if Dr. Barnet would agree that to the political crises he alludes to, including nuclear war and underemployment, we should add another crisis, that of overpopulation. Incidentally, I believe David Brauer, in his preface to Paul Erlich's book on the population bomb, was correct in emphasizing the relationship between the underemployment that is taking place and the brutality and terrorism which is growing in this world today.

But I think an even more important consideration for this symposium on human adaptation is that for the past thirty years we have had an institution that seems to me very comparable to the kind of institutions that developed in this country in the eighteenth century, that is, the

RICHARD J. BARNET

United Nations. Shouldn't the social scientists begin thinking now about ways to strengthen that institution? With all of its weaknesses, including the fact that it is not representative of countries democratically, it is one possible hope on a global scale to deal with issues and problems. Don't we need a global institution with all kinds of mandatory police and legal action that is comparable to what was created in 1787 in this country?

DR. WOOLF: Thank you very much. I think we will allow Professor Barnet to say the last word.

DR. BARNET: Just one sentence in response to one of your comments. I would be very reluctant to take the automobile industry as support for the proposition of being smart.

16. Education and Adaptation for the Future

Professor Emeritus in Philosophy and University Fellow in the History of Ideas, Australian National University

Editor's Summary. A social system which relies heavily on decision making by broadly based consensus requires an informed, knowledgeable citizenry. John Passmore expresses a concern over what he sees as a current trend toward present-oriented education emphasizing narrow vocational training. Education for the future ought not be a slot-filling education but a process in which citizens are given a general framework in which to incorporate new facts and ideas. Passmore sees an education for the future as including: emphasis on learning how to learn; developing a capacity for imagination; encouraging and developing a critical spirit in which facts and ideas are challenged and evaluated; stimulating a deep concern and sense of involvement with the future of society.

If John Dewey is right, my whole inquiry is misconceived and ought to be aborted. The present, in his view, is the only proper concern of the educator. The past should interest him only insofar as it is still active in the present; the future should interest him only as the nearly present—that immediate future into which the present ticks away. I shall be suggesting, on the contrary, that the educator should concern himself with a future a good deal more remote than this. And indeed with the past, as providing models for that future.

Why does Dewey think otherwise? An individual, he first of all tells us, can live only in the present. In a certain sense, this is so, self-evidently so. "The present" simply means "the time at which we live." It is scarcely news that this is the only time at which we can live. But just because it is then a tautology, this dictum, thus naively interpreted, is useless for the purpose Dewey has in mind: a stick to beat curricula with. The teacher who wants to teach his children about the past or to prepare them for the future can readily reply: "Of course we can only live in the present; it is in the present that I am talking about the past and preparing for the future." To construct his critique, Dewey needs a quite different sense of "living in the present," in which it just isn't true that we can only live in the present, in which a person can be rebuked, as Dewey is rebuking him, for living in the past or in the future—a rebuke which is quite meaningless if in fact he has no option but to live in the present.

That is the sense in which Paul Valéry could describe a human being as an animal whose "principal home is in the past or in the future." All of us live in the past with our memories and in the future, or at least the imagined future, with our hopes and our aspirations. Amnesia cripples us; we lose not only our memories of the past but those ambitions for the future which had characterized us.

Then what is Dewey talking about when he exhorts us to live in the present? Well, even if we all to some degree live in the future and in the past, there are some people we more particularly describe in this way: "He lives entirely in the past"; "He lives entirely in, or for, the future." Dewey is condemning an educational system in which teachers or pupils fall into these categories.

His fire is particularly directed against past-oriented curricula, even though he quite explicitly condemns educators who attempt to interest the child in the remoter future. That emphasis is not surprising. He was criticizing the traditional curricula, and he could scarcely have complained that such curricula were too future-oriented. That they entirely neglect the future was to be, indeed, the charge leveled against them by Alvin Toffler and his associates. But if we look at Dewey's relatively detailed objections to past-orientation we can easily see just how he could also have applied these same objections to future-orientation.

A past-oriented education, so Dewey attempts to persuade us, develops in those who are subjected to it a particular, and highly objectionable, attitude to the culture of their own time. Culture becomes, for them, not a living force but "an ornament and a solace; a refuge and an asylum." They do all they can to "escape the crudities of the present to live in [the past's] imagined refinements." The culture of their own times they despise: they "make the past a rival to the present and the present a more or less futile imitation of the past" (Dewey 1916:75). Their education, in other words, is an education in escaping, not in living; in fleeing from the world, not in encountering it.

JOHN PASSMORE

How far can these objections be so reshaped as to apply also to a future-oriented education? The contemplation of a remote future, Dewey might have written, "becomes an ornament and a solace; a refuge and an asylum." Those who are encouraged by their education to indulge in it try to "escape from the crudities of the present" in order to "live in the imagined refinements" of the future. They make of the future "a rival to the present," which they dismiss as a "more or less futile" attempt to do in the present what only the future will permit.

Neither Dewey's objections to past-orientation, nor the parallel argument I have constructed for him against future-orientation, are bayoneting straw men. We have all met, I imagine, the sort of person who devotes his life to the study of Shakespeare's dramas, but never goes to see a present-day film or play as inevitably inferior; we all know the sort of person who will not go to see a present-day film or play because in relation to that future described by Trotsky in which the humblest citizen will out-Shakespeare Shakespeare, it cannot be anything better than dross. Or who will accept the contemporary, at best, only when it is a political instrument, propaganda for the future.

Dewey was writing at a time when the emphasis of the historian was likely to be on the glorious heritage of the past and the emphasis of the prophet on the anticipated glories of the future. At the moment, we tend to look at both past and future very differently. When Stephen Dedalus describes history as "a nightmare from which I am trying to awake," he strikes a now familiar note; if we look forward to the world of George Orwell's *1984* or to a world devastated by nuclear warfare, it is scarcely with lively anticipation. But this attitude toward past and future, Dewey might still object, is no less unhealthy. We should forget the horrors of the past as well as its glories, we should keep our minds off merely possible future disasters, as well as off merely possible future glories. Present endeavor is in either case weakened in its resolve by our concern for other times.

In thus exhorting us, Dewey by no means stands alone; there have not been wanting moralists to support him in his demand that all of us, and the teacher *a fortiori*, should concentrate on the present. I shall remind you of a few familiar examples. From Ancient Rome, the poet Horace advises us "to be sensible, to give up all ambitious plans." We should be content, rather, to "seize the passing day, paying as little heed to the future as possible." Then there is the teaching of Jesus, very different in spirit and assumptions from the Epicureanism of Horace, yet issuing in a not dissimilar conclusion: "Take therefore no thought for the morrow for the morrow shall take thought of itself. Sufficient unto the day is the evil thereof." At the beginning of the modern period, Francis Bacon, prophet of science and technology and an exceptionally far-sighted thinker, nevertheless admonishes his readers in similar terms: "Men should pursue things that are just in the present and leave the future to divine providence." Nearer to our own time, this was the recurrent message—providence apart—of the antimoralists of the sixties. "The hipster," Norman Mailer then wrote with approval, "lives in that enormous present which is without past or future, memory or planned intention" (Mailer 1968:271).

If these moralists are right, as I began by suggesting, my whole project is misconceived and ought to be aborted. It is morally wrong to have any but the immediate future in mind when we try to educate; what our pupils should be taught is how to correct the evils of the day, how to act justly here and now, how to live as if every moment were their last. And, as I have already freely admitted, it is not hard to understand why moralists have taken this view. They are warning us against a powerful human tendency to look so much to the future as entirely to ignore present responsibilities. At the level of our family responsibilities, to take an instance, we can be so intent on building a better future for our children that we ignore the love they need here and now. In that very process, we make it impossible for them ever to inherit that better future we have so conscientiously been striving to construct for them. At the personal level, we can easily find ourselves in the situation Pascal described: "If we are always looking forward to being happy, it is inevitable we should never be so." At the social level, the revolutionary ruthlessly destroys and ravages in the name of a glorious future which his very destructiveness makes impossible.

Not only that, not only are there grounds for supposing that it is morally wrong to concentrate our attention on the future, there are practical reasons, too, for setting the future aside. We do not know, cannot know, what the future has in store for us. No doubt, some futurologists are prepared to draw up a relatively detailed blueprint of the future. But when we examine their past efforts at predicting, we can scarcely have confidence in their present efforts. By 1980—we were informed some twenty years ago—the United States would have developed its military resources so far beyond the capacity of the Soviet Union that the gulf would be unbridgeable; the graduate schools, meanwhile, would be so flooded with applicants that they would not be able to find teachers competent and ready to teach in them. These were by no means wild guesses; they were plausible enough extrapolations when they were first promulgated. On the other side, one finds no predictions of such world-shattering events as the Soviet-China breach, the defeat in war of the United States by a minor, ex-colonial power, the rise of Muslim fundamentalism, the oil-powers cartel. All of these, indeed, if tentatively suggested, would have been dismissed as impossible. Predicting the future is clearly a chancy business. How can we possibly educate for adaptation to the future when we do not know what the future is going to be like?

Both moral and practical considerations, then, might lead us to concentrate on a present-oriented education. Pressure for such an education, furthermore, comes from a very different source—intellectually less respectable but politically powerful—from narrow-visioned vocationalists, trading upon that disillusionment with education which is now widely current.

Not so very long ago, even by American standards, a President could roundly proclaim that "the answer to all our problems comes down to a single word: education." Although Lyndon Johnson was a somewhat idiosyncratic President, this was by no means an idiosyncratic statement. An occasional old-style Marxist might have been heard muttering in some dusty corner that fundamental question: "Who shall educate the educators . . . ?" But such eccentric dissidents apart, Lyndon Johnson neither expected nor encountered contradiction. He was heard, indeed, as enunciating the obvious.

If Marx's question was ignored, this is because the last person who is considered in most proposals for educational reforms is the teacher. Two thousand years ago Quintilian pointed out that even those Romans who laid great stress on education were prepared to have their children educated by slaves. Things have not altered all that much. Teachers still have a surprisingly low social status, which they make worse by claiming to be professional educationists, rather than scholars and scientists. Yet without their active and imaginative collaboration, an educational program is but an attempt to write on water.

The century we are in process of enduring—the century which began in 1914—may fairly be described as the century of shattered illusions, as one ideal after another has revealed its darker side or, at best, its limitations. But illusions about the power of education—the illusion that it contains the answer to all our problems and that it is, or could be, the path to liberty, equality, and fraternity—survived even our experience of the best-educated of all societies (Germany) slipping into barbarism. In the 1980s that illusion, if I read the signs aright, at last lies shattered. I doubt whether anyone at all now sees in education a universal panacea; no modern President, certainly, could echo Lyndon Johnson's dictum without arousing derision.

In many respects, this is an excellent thing. The schools cannot possibly do what President Johnson was calling upon them to do. No single institution can properly assume, or ought to have imposed upon it, the burden of solving every social problem, righting every wrong, correcting every injustice supposed or real. The effect of assuming this burden, voluntarily or involuntarily, was to weaken—one hopes not irreparably—the capacity of the schools to do the one thing they can do with at least a limited prospect of success: to teach.

The collapse of utopian aspirations, of social revolutions (peaceful as well as violent) is not uncommonly followed by a period in which the hard-headed men, seizing upon the chaos created by the soft-hearted, take over the system in order to realize their enduring aims. In the present instance, they are in an unusually strong position when they urge that education should prepare its pupils to slot into presently existing jobs. Employers, or the less enlightened amongst them, welcome a proposal which would save them the cost of on-the-job training. The public at large, or so I suggested, has lost sympathy with all three ingredients in the educational process: with the young, after the events of the sixties; with schools, for failing to fulfill impossible tasks; and with teachers, whom they see as self-serving careerists. So it is easy to persuade them that all three—pupils, schools, teachers—should be doing something immediately useful. Fear of unemployment induces the young and their parents readily to acquiesce. As for educators, they so surrendered to the demand for "social relevance" that they are in no position to resist; only the elderly so much as remember the arguments in favor of liberal education. And the elderly are tired.

There are, then, three major objections to an education which sets its sights on the future: (1) the moral objection that the present is our only proper business; (2) the practical objection that we do not know what the future will be like; (3) the vocational objection that the task of the schools is to prepare the young for some quite specific vocation. Now for the counter reply. I shall consider these objections in reverse order.

First, for vocationalism. Let me distinguish two different kinds of vocational education. The first is a narrow training for some highly specific limited task, as the Volkswagen factory trained its workmen to be Volkswagen mechanics, nothing else. The second is a broadly based technical or professional education, of the sort Marx so strongly approved when it was set up in his native Prussia. Only the first sort of vocationalism—let us call it technical training as distinct from technical education—is wholly present-oriented. It prepares the young, purely and simply, for a particular task, limited in its range, in which employment is available, trains them to be not mechanics but—much more narrowly—Volkswagen mechanics. That is what some unenlightened industrialists are now demanding: a form of schooling which would make on-the-job training quite unnecessary, which would treat its pupils as pegs to be fitted into predetermined positions, cultivating no skills, imparting no information, developing no attitudes which are not precisely tailored to such a preparation.

One can understand how training of this sort, not as a supplement to but as the whole of schooling, might be thought appropriate in a society which is in all respects unchanging. One's moral objections to it, that it treats human beings as mere means (factors in production parallel to land, capital, and technology) would persist, but at least there would be no practical objections. Such societies, idealized in Plato's *Republic*, offer just so many forms of employment and the circumstances of their practice is unchanging: no one will discover, later in life, either that his form of employment is obsolescent or that it now needs quite new skills for which he is wholly untrained or new information which he does not possess. He will not suddenly find that he is unemployable. In our rapidly changing society, however, such a contingency is highly likely. An education which concerns itself only with preparation for a presently existing occupation is education for a future scrap-heap. If a schooling which consists solely in vocational training is at odds with a future-oriented education, the advantage lies with the latter.

The objections do not apply, however, to a broadly based technical education. A car mechanic who has learned his mechanics at a technical college (as distinct from being trained, purely and simply, on the job) will in the process have learned, at least in principle, how to service a wide range of mechanical appliances. Even if he must then undertake an on-the-job training course to give him a particularly intimate acquaintance with some particular class of machines, he will not be entirely at the mercy of his employers; he will be prepared to move, to relearn, if he needs to do so. In other words, he will be adaptable to changing circumstances. An education which looks to the future will be an education which anticipates change: a narrowly vocational education is calamitous just because it takes for granted the permanence of the present; a broadly conceived technical education need not do so, if it is designed in a liberal fashion, if it concentrates on the teaching of "open" skills—skills which can be used in a wide variety of circumstances and modified to cope with them—rather than teaching closed capacities, defined as the ability to perform this or that specific task.

At this point, however, we meet the second—practical—objection: that we *cannot* anticipate change because we do not know what the future is going to be like. This objection would have weight against any attempt to transpose a vocational education to the future, to begin from the assumption that in ten years' time there will be a demand for precisely so many people with this or that highly particular skill and to train them now to fill those jobs. That would be wholly stupid. Nobody ten years ago, or even five years ago, could have predicted the present demand for computer operators, word processors, workers with information-bearing chips of one sort or another or predicted, on the other hand, the lack of demand for teachers. In a technological society, future-oriented vocationalism is as short-sighted as present-oriented vocationalism, although the concept of a manpower policy to which education could be subordinated is so attractive to bureaucrats that it persists even after repeated demonstrations of its folly.

But once we drop the idea that a future-oriented education is a future-oriented vocationalism, the practical difficulty vanishes. We do not have to predict the future in detail in order to prepare for it in our schools; all we need predict is that it will not be what we expect it to be. "Expect the unexpected"—that Heraclitean maxim—must be our guiding principle. Education for the future, I said, is education for a changing society. It is education, too, conducted in the recognition that it will not suffice. We are still wedded to the idea, which partly lies behind the recrudescence of vocationalism, that at school we ought to be able to acquire all the skills and all the information we shall need during our lifetime. That concept of education, as a once-in-a-lifetime necessity, has extended the duration of schooling to grotesque dimensions. Young people often spend a third of their expected life, including what ought to be its most imaginative and creative period, enclosed in a schoolroom.

We are beginning to abandon, I hope, the foolish notion that a longer education necessarily means a better education. We may well need to revert to a situation—I cannot now develop this theme—in which even at relatively early stages education is a part-time affair, as both Marx and Matthew Arnold thought it ought to be. But whatever we do now, in later life a large percentage of the population will have to go back to school whether part-time or for longer refresher courses. The problem is to devise an educational system so that this necessity will not lead us into the ridiculous situation Roger Garaudy has envisaged (in which half the population is either teaching the other half or learning how to teach them), but in the full recognition that no once-and-for-all education will wholly suffice, that there is no point in trying to pack more and more into an education to *make* it suffice, as we have been trying to do over the last quarter of a century or so (Garaudy 1970:46-47). To the question how an education which will satisfy this requirement can be designed, both preparing for the future and recognizing that its preparation is inadequate, I shall return. For the moment it is enough to say that such an education is not open to the objection that we cannot predict the future in detail.

What of the moral objections to taking the future into account? In a way, we can dismiss them as exhorting us to do what we cannot possibly do: ignore the future. Every time we wind our watches, we "take thought for the future." Only those who seek to renounce their humanity, who are prepared to abandon wife, family, children, any form of industrial or agricultural life, can ignore the future—so far as that is possible even then. Indeed, one can well argue—as I have argued in *Man's Responsibility for Nature*—that we are now morally called upon, more than ever before, to take account of the future, as we realize to what a degree we can by our present actions make that future humanly unlivable.

Yet it would be wrong, all the same, wholly to dismiss the feeling which lies behind these moral injunctions: we ought not to "live in the future" in a manner or to a degree which leads us to ignore our present responsibilities. Given the practical difficulties in determining what the future will be like even in broadest outline and the moral need not to ignore the present, we can properly conclude that we ought to try to design our educational system so that it enables its graduates to respond both to present and to future needs, opportunities, responsibilities, and obligations. It ought to be a good education whatever the future is like, even when it is an education designed with an eye on the future.

We have been talking about education and the future. Let us look at another word in our title: "adaptation." To survive at all in changing circumstances is to have adapted to them. Those who survive unemployment by accepting welfare payments, narrowing their lives and reconciling themselves to a lower standard of living have shown, in this sense, a capacity for adaptation. If this is what adaptation means, why should we think it either necessary or desirable to educate for it? C. Wright Mills, in his *The Sociological Imagination*, has described the sort of man who "gears his aspiration and his work to the situation he is in." "When he can find no way out," Wright Mills says, "he adapts" (Mills 1967:188). Such an adaptation, he goes on to argue, destroys both the man's capacity to act freely and his power to reason; his adaptation is simply submission. To misquote: "Theirs not to reason why/theirs but to submit and die."

Any of us may find ourselves—although we hope not—in a situation in which we can survive only in this manner: adapting by submitting. Not very long ago, many moralists sought to teach us that this submission was, indeed, our duty. We certainly cannot rule out the probability that our young will grow up in a society in which this stark alternative—submit or die—will be presented to them.

Am I then advocating an education (whether for moral or practical reasons) which will prepare children to adapt in this sense: to learn how to submit? I am most certainly not. That education would suit the vocationalist. "Your job," he is prepared to say, "is to be a plumber. We need plumbers; you will be taught to be a plumber. And you must submit to staying in that job. In the choice of occupation, you are to have no liberty!" This is an attitude which could only be sustained, of course, by a still wider attack on personal liberties. What I am advocating, in contrast, is an education which is centered not on adaptation, in this sense of the word, but on adaptability.

In putting this matter thus, I am drawing on the rich superfluities of the English language to make a distinction, not recognized in our dictionaries, between "adaptativeness" and "adaptability." I should perhaps have called my paper "Education and Adaptability in the Future." Every living organism is adaptive; it has the power of behaving in such a way as to make more probable its survival or—if you prefer—the survival of its genes. Either a lemon tree or a human being can survive in cold times by hardening its skin, at whatever cost to its juiciness.

Adaptability, however, involves more than this. It is the capacity to *flourish* in changing circumstances by the use of information, acquired skills, imaginative enterprise, and to cause to flourish the activities in which one engages, the institutions of which one forms part. I do not have to consider whether it is, or is not, peculiar to human beings. Certainly that is where one finds it in a particularly marked form, related not only to such facts as that we have binocular vision, hands, a developed brain, but in our possession of a culture on which we can draw. That is why it makes sense to talk of education for adaptability. Adaptiveness is something one does not specifically need to educate for. Education, on the contrary, is the obvious way of trying to increase adaptability.

The adaptable person, I should perhaps add, is not necessarily virtuous; criminals, scoundrels, and careerists can be adaptable. A perfectly adaptable person—someone who will flourish under all circumstances at whatever cost to principles—is nobody's ideal type. Poetry knows him as the Vicar of Bray. Adaptability can be misused as much as can any skill, any knowledge. Plato saw this long ago. But in urging that education should try to help children to be adaptable I am not arguing that this is *all* it should do. Single-objective slogans have done enormous damage to education in the past and I am not proposing to compound the damage by adding a new one. One can try

JOHN PASSMORE

to develop in children firm principles which will govern both their adaptability and their use of skills—so that they will not use their chemistry in order to construct bombs or their adaptability in order to lead criminal gangs. But that is not my present theme.

Can adaptability be taught? Yes, it can, even if not by direct exhortation. Furthermore, an education which has an eye on future adaptability—adaptability *in* the future—need by no means be illiberal. On the contrary, it will have to be a liberal education in the proper sense of the word, an education worthy of a free man. An education for adaptability does not quarrel with liberality, but only with an education which is narrowly vocational and, in the long run, totalitarian—a slot-fitting education.

Now to look in more detail at the form of such an education. Only in *relative* detail, of course, but in such a way as to bring out the kind of emphasis which will have to characterize it. First of all, where general principles are known it will emphasize these principles. The main home of general principles is science, including mathematics under that head. The great importance of general principles, for our present purposes, lies in their *adaptability*. A general principle, by definition, is applicable in a great variety of circumstances.

Two objections might, however, be urged against the teaching of general principles. First, that science changes rapidly, that future adaptability cannot, therefore, be secured with its help, since much that is learned at school will turn out to be false. There is exaggeration here, however. For all that my argument has so often referred to rapid change and the resulting obsolescence, I would still demur at the suggestion that change is as absolute as it is sometimes supposed to be. No doubt such standbys of traditional science courses as the laws of Galilean or Newtonian mechanics, Boyle's Law, or the laws of optics have not gone unscathed in the course of scientific progress. The fact remains that in order to understand the problems which beset flights to the moon, we do not need to go beyond Newton to Einstein; it is still possible safely to appeal to Boyle's Law in a great range of circumstances; in designing the vast majority of optical instruments, one need not draw upon any optics except what was already known in the seventeenth century.

True enough, school science ought not to be taught—and often is taught—as if it were indubitable. True enough, no one can hope fully to keep up with science, even in its broadest outlines, on the basis of what was learned at school twenty—let alone fifty—years earlier. Nevertheless, so far as that is possible, a science course ought to be designed so as to facilitate keeping up with science, to facilitate the reading of the better sort of scientific journalism and the capacity to undertake further scientific training when it becomes necessary. As it is, many of our science students leave school with no idea of how to read science for themselves. If they read, however, they will soon discover themselves applying familiar principles, even if in unfamiliar territory.

A second objection is that in the everyday affairs of life, scientific principles will not help people to be adaptable. Even though, it might be admitted, scientific principles are wide-ranging and even though they have technological applications, they are of no use to ordinary persons in the everyday affairs of life; they do nothing to increase adaptability. Unless they happen to be scientists or technologists, they cannot draw upon those principles to confront the problems which daily confront them. When did any of those amongst us who are not scientists or technologists last rummage through our minds for a scientific principle to give us guidance in confronting a practical problem?

To argue thus, however, is to think too narrowly. Without a knowledge of scientific principles, we can have no comprehension of the innumerable technological devices which constitute so large a part of our daily environment. We begin to feel about them as prescientific peoples felt about such natural phenomena as the rising sun—as magical—or come to resent their unintelligibility as an affront. Now that science-based technology has come to be one of the major factors of production, we can have no feeling of being "at home" in society —let alone of being able to cope with its future—without some knowledge of basic scientific principles.

General principles, then, are fundamentally important. To emphasize the importance of principles, however, is not to denigrate facts. There are those, especially Jerome Bruner and his associates, who would like us to free our educational systems from all "fact-laden" subjects, replacing them by subjects which are entirely composed of principles and general concepts—replacing geography and history, for example, by social science as it is commonly taught by a science course which is entirely "structural." Were it possible, indeed, to construct a social science which satisfied Bruner's criteria, which could "provide a set of workable models that make it simpler to analyse the social world in which we live and the condition in which man finds himself," it should clearly constitute at least a considerable segment of any young person's training, just as the natural sciences should. But such "workable models" are far from being at hand. To attempt to introduce the young to nonexistent principles is in fact to introduce them to verbiage and intellectual confusion. The young should be encouraged, to be sure, to think about social structure and social change so far as their limited experience permits, but not to believe that they are (as a result of that thinking) in possession of social science principles which will allow them to face the future with utter confidence. There are no such principles.

Even if principles were available, furthermore, it does not follow that they should be allowed wholly to drive out particularized facts. Learning about the structure of science is all very fine, but it is not the same thing as learning science. To learn chemistry, to learn biology, to learn geology, one has to learn particular facts. Principles, structures, concepts are empty without that reference; even the highest reaches of physics are fact-laden. Of course, the sciences are not just bundles of facts. One of the virtues of science lies in the way in which it interweaves facts and theories, often showing us in the process that what we supposed to be a fact is a fallacy. Nonetheless, science both depends upon and generates information. The serious educational problem is how to select out of the vast body of detailed information that information whch is genuinely *central* and which will enable those who possess it both to understand what has been achieved and to grapple more successfully with what is to come.

This can never be done with complete confidence. As conditions change so does the centrality and the marginality of information. But, to take only one case, at any time or place we need to have at our disposal a spatio-temporal map of the world. Otherwise, we cannot possibly understand the daily flood of information which comes our way or the necessity for action which flows from it. We stand in need of broad but reasonably precise information about the major regions of the world and their own historical understanding of what kind of region they are. It is unfortunately true that history and geography have often been taught in such a way as not to leave the young with this information, so as to enclose them indeed more tightly in parochialism. Under the influence of Herbart and his followers, the entire emphasis was on the local scene; foreigners earned a mention only as wicked creatures who lived beyond, but sometimes threatened, frontiers, or, at best, as picturesque inhabitants of another world. Geography stopped short at national frontiers. Even literature was thus confined. In Germany, Robinson Crusoe was rewritten so that Crusoe could start his adventures in Hamburg rather than Hull.

Those, like myself, fortunate enough to be educated in unimportant countries avoided this kind of nationalism. Indeed, I once set out, on the basis of my own experience, to praise history teaching as helping to break down parochialism, only to discover when I consulted syllabuses that in most countries it carried parochialism to its exceptional limits. Let me nevertheless testify (from my own experience in learning European and American history in a country that was neither European nor American) just to what degree such historical studies can help one to understand the future when it comes. True enough, school history turned out to be not enough; it had to be supplemented when Asia began to come to the fore. Judgments about, say, the final outcome of the Boer War had to be modified. But at least the framework was there into which new data could be fitted. And a permanent possession was the realization that one's own society was atypical and that social and political freedom was a rare phenomenon, inevitably precarious.

Not only individuals but whole nations can suffer in adaptability from the lack of an adequate historical and geographical background. The difficulties which the United States has encountered in its international adaptability in part derive, if I am not mistaken, from its failure to think in historical and geographical terms, its tendency to think of politics conceptually. President Ford was greeted with horror when he remarked that the Russian satellite countries had to be considered individually, that neither their communism nor their character as satellites exhausted their political and social structures. To anyone who thought historically this was the plainest common sense; the past history of these countries and their present geography are highly relevant to their future. No country is simply an exemplification of a general sociopolitical concept.

One not uncommonly hears it suggested nowadays that while there was a time when history could enlarge human adaptability by providing models—as Republican Rome provided a model for modern Republicanism—that time has passed: the present is, and the future will be, so different from the past that there can no longer be anything to learn from it. The point of fracture, at which the past ceased to be relevant, is often put at the destruction of Hiroshima. But this judgment reflects, like the parallel judgment that time outmodes all scientific principles, the degree to which we are dominated by the myth of modernity, by an exaggerated belief in our uniqueness. Nuclear weapons have made a difference but they have not so far transformed the world.

Entering Afghanistan, the Soviet Union, as in its foreign policy generally, fulfilled an ancient tsarist ambition. If we ask why it succeeded when the tsars failed, the answer is not—although this may have had a marginal effect—that modern Russia possesses, as tsarist Russia did not, nuclear weapons. If the Soviet Union's relative power had been solely in conventional weapons, there would have been the same hesitation in resisting it as there was in resisting German and Japanese aggression in the 1930s. A powerful militarist country will always have tactical advantages over countries which genuinely seek peace. Such models from the past are still not irrelevant in understanding our present; they are unlikely to be wholly irrelevant in understanding our future.

Again, the fundamental conflicts of the present day in democratic countries, the conflict between libertarianism and the welfare state, between freedom of contract and combinatory power, are admirably discussed in Dicey's classical work, *Law and Opinion in the Nineteenth Century*. Studying the nineteenth-century debates there described, we are more inclined to agree with de Maistre that "the more things change, the more they remain the same" than with those who stridently proclaim the uniqueness of the present. But neither is right. Some things change, some things remain the same; appearing to change, they sometimes remain unchanged, appearing to remain unchanged, they sometimes change. If the future were wholly different from the present, education would be useless; if it was not going to differ from it except superficially it could be a once-and-for-all affair. In fact, it will almost certainly differ considerably, but not wholly.

The human imagination, one should add, is distinctly limited. In the film *Star Wars*, heroes and villains battle with laser beams, but precisely as if they were swords. Their clothing derives from Roman togas and Japanese samurai armour. That is typical; we *can* only take our models for the future from the past, even if we realize that in its details history never repeats itself. And we have to have models even if we ought not to cling to them tenaciously when they are obviously not working.

Yet if for such reasons I should wish to emphasize the value of information both about the present and the past, information in the form of general principles and particularized information, this is not to deny that, as I have also emphasized, we can learn only a fragment of the information which is not at our disposal and we can never be quite sure in advance what we shall need to know. That is one reason why it is essential that we should learn how to learn, how to acquire information as and when we need it.

At this point, I join hands with the vocationalists in emphasizing the "three Rs," but for a different reason: not, primarily, to prepare the young for jobs here and now—although I do not wholly spurn that function if it is broadly interprted—but, rather, to prepare them to learn. At school, of course, one does not only learn to learn, one learns. Nevertheless, in societies in which the rate of change is extremely rapid, learning to learn assumes a special importance. To be adaptable, one must be able to learn and re-learn, and the three Rs are a path to this learning and re-learning.

Consider arithmetic. Much of what has to be learned in later life is certain to be arithmetical in character. Not to be able to understand statistics, not to be able to rough-check computer calculations, is to be ill-prepared to cope with life either now or in the future. As for reading, that is still a fundamentally important way of acquiring information even if, as time goes on, more such information may be derived by reading computer print-outs than by reading books. Not only is reading an important source of information, reading can help us to acquire skills, to consider alternative futures, to understand how things are going, to enlarge our imagination.

Learning to read, as I understand it, involves a good deal more than being able to transpose written into spoken language; it is fundamentally a capacity to *use* books and other printed sources of information with all that "use" involves. A failure to realize this fact is one source of weakness in traditional schooling, considered as a preparation for the future. That a person can stumble through, with a degree of comprehension, a school reader or a school textbook is no proof that he has learned to read. He has learned to read only when he goes naturally to print in search of entertainment or information. (I was horrified recently when a reviewer wrote of a very elementary book of mine that it could not be used as a text because it was not broken up into subheadings and sub-subheadings. How, knowing only such pap texts, are the young to consult ordinary books for themselves?)

To be adaptable, then, the young need to have some grasp of fundamental arithmetical processes and a firm grasp of how to read. As for writing, one can put the need for that more broadly than did the traditional teacher; the young need to be able to communicate, both orally and in writing, to expostulate, explain, request, demand, object, propose. All this is essential to adaptability. How they are to learn to write clear and succinct prose from teachers who cannot do so themselves is one of the major educational puzzles of our time.

If, however, the three Rs are centrally important in any education for adaptability, the mere acquisition of the skills that entails is not enough. The young have to see the point of reading or—to stretch the three Rs beyond their traditional boundaries—of listening and discussing. Furthermore, they have to learn to read, listen, discuss in a certain spirit, imaginatively and critically. If their reading and listening is uncritical it may tell against, rather than in favor of, their adaptability. All their lives they are going to be subject to a flow of propaganda, misinformation, lies, false suggestions, intellectual confusion; the lesson of much of it will be that they ought to conform, submit, adapt, and adjust, rather than to make use of what they find around them in order to flourish themselves and to help their society to flourish, in the spirit of adaptability. Without imagination, they will not be able to conceive alternatives to what they are told is "inevitable"; without critical abilities, they will not be able to fight against those who would subdue their individuality. The traditional education was weak at these points, both in cultivating the imagination and in developing the power to criticize. One of the best ways of doing both these things is through the teaching of science. But it is often taught, alas, in a way which numbs the critical powers and destroys the imagination, as a series of technical devices, as an array of facts, a set of laws which are presented not as someone's discoveries but as if they had descended from Mount Sinai.

First, for the imagination. I want to begin by making a distinction between imagination and fancy. Fancy is manifested, for example, in daydreaming. For the most part, the mind is then operating in a very conventional fashion; we dream of ourselves as great discoverers or great sportsmen or great lovers or great politicians, as winning the Nobel Prize or the Medal of Honor. Fancy is presumably adaptive. We compensate for our actual submissiveness by dreaming of ourselves as Superman or perhaps as a "strong President" who will save the world as he thunders on, riding a white charger. But it does not help us to be adaptable, to look for real alternatives which will help us to cope more successfully with our circumstances.

By the imagination, as distinct from the fancy, I mean something quite different: imaginativeness, the capacity to think up and to think through alternatives, to see what they involve or—to fall back on the contemporary jargon—to "develop scenarios." This is the central element in the courses of future-directed studies proposed in Toffler's *Learning for Tomorrow*. The pupil is encouraged to think up possible futures ad to consider, as far as can be, what they would be like. Although such courses have their place, they are not enough; the education system, if it is to develop adaptability, needs to be permeated by the cultivation of imaginativeness. So literature and history, which already present us with alternative models of life, can be taught in such a way as to *emphasize* the alternatives they contain within them. It might seem fruitless to discuss such issues, in a sense unanswerable, as "What would have happened if Hitler had not invaded Russia?" or "What difference did it make to the form socialism took in the USSR that it was not an industrialized country?" or "Suppose the United States had not declared its independence, but been content with what came to be called dominion status. How different would it have been?" But such "might have been" questions, realistically discussed and thought through, at once bring out the great importance of the decisions which *were* made and encourage imaginative thinking about possible social alternatives.

Something similar is true in literature with such questions as: "Would

Macbeth have acted as he did without the witches' prophecies?'' or "How far could *Tom Sawyer* be transposed to our own time?'' They at once lead the young into a deeper understanding of the works being discussed and help them imaginatively to appreciate what factors affect decision. Such mathematical techniques as decision theory, the theory of games, and catastrophe theory can be taught in a similar fashion, as aids to imaginative thinking-through. In science one can ask what would happen if certain principles did not hold: if, for example, there were no such thing as gravity or if resemblances between parents and children were determined in a completely random fashion so that (let us say) the chance of two black parents having a white child was equal to the chance of two white parents having such a child. Such questions bring out the role of scientific laws in limiting social choices.

More generally, the pupil is encouraged to think imaginatively wherever teaching centers around *problems* rather than *exercises*, problems where the appropriate rule to be employed has to be thought out rather than, as in an exercise, simply to be applied. One must not exaggerate; not all power belongs to the imagination; skills have to be perfected, habits formed. We do not wish our pilot to have to think out anew his cockpit drill every time he gets into the cabin. But we do wish him to behave imaginatively if he encounters a crisis which is not in the book of rules. And no book of rules is ever complete.

Then there is the critical spirit. This is sometimes opposed to imagination as if it were its antithesis. But in the sense in which I am using the term "imagination"—the sense in which an artist, an engineer, an industrialist, a scientist exercises his imagination (or a farmer or a hunter or a craftsman or a politician)—there is no such antithesis. Imaginatively to think through alternatives is at the same time to consider them critically. Admittedly, some people are better at thinking things up than at fully thinking them through; some people are better at thinking other people's ideas through—perhaps to the point of demonstrating them to be unworkable—than they are at thinking up ideas themselves. But if for that sort of reason the artist and the critic are often different persons, nevertheless an artist who is quite incapable of self-criticism and a critic who cannot see the artist's work as imaginative, who unimaginatively applies standard rules to an unorthodox work, are very poor specimens of their kind. We all know what it is like to read a novel which, as the critics now say, is "self-indulgent," which lacks self-criticism. Many of us know what it is like to be subject to the strictures of an unimaginative critic who will not even look at what we are trying to do and rebukes us for not doing what he expected us to do.

Problems arise in respect to both criticism and imagination because teachers are schoolteachers working in an educational system. When that system is vocation-dominated, they may be discouraged from looking at alternatives, not only by the authorities but even by students themselves. The teacher's task, both students and authorities may agree, is to teach what is known or what, at least, is accepted as known at the particular time. Teachers should, in this view, positively discourage the young from criticizing; they should habituate the young to submit, to adapt to existing institutions, not at all to attempt to transform them.

It has become easier to sympathize with this attitude when, as has now happened in many countries, teachers have been transformed from the unimaginative pedants they used to be at their worst into woolly-minded products of an unfortunate period in the history of our universities. But the objection to such teachers is not—although they sometimes think of themselves in this way—that they teach the young to be critical but, quite the contrary, that they do what they can to teach them to be uncritical, to replace thinking by slogans, the free movement of the mind by enslavement to dogma. They have fixed ideas about what the future *must* be like; there could be no worse education for adaptability. Fortunately, the young nowadays see through them; scepticism rather than fanaticism is the danger to be feared.

There is much more to be said. A person who is prepared to be adaptable, rather than simply to adapt by submitting, will need passion and enthusiasm. A person who does not care deeply about anything will happily adapt to whatever happens. A short time ago I was asked to talk to the boys and girls at what is probably Australia's best-known school. I ended by saying that talents were useless unless they were accompanied by enthusiasm. In the discussion which followed, a boy asked, with the evident approbation of his fellows: "But what is there to be enthusiastic about?" That is the most depressing question I have ever been asked. So long as it represents the mood of many of our schoolchildren, we can have little hope for the future.

Let me sum up thus: to prepare the young for future adaptability, we should proceed in a way which will encourage adaptability in the present as well as in the future. The young need to know the major general principles which human beings have slowly discovered. They need to know the basic facts about the world. But beyond that, and in order that this information should not be too great a burden, they need to learn how to learn. For no information they can now acquire and no skills can wholly suffice for their future.

A great deal depends, also, on the spirit in which their learning is conducted. Unless they learn that learning can be both pleasant and advantageous, they will not use the learning skills their teachers teach them. Unless, too, they learn to think imaginatively and to think critically, those skills will not suffice. And if they come out of school indifferent, uninterested in anything but their own prosperity and security, committed to a policy of every man for himself, adaptation—as distinct from adaptability—is the best we can hope for from them. They may survive, but civilization will not.

NOTE BY DR. PASSMORE: In this paper I bring together, interlace, and develop certain of the themes in my *Man's Responsibility for Nature* and *The Philosophy of Teaching*. In the first of these books I consider the degree to which we have any moral reponsibility in relation to the future. In the second, I work out much more fully pedagogical principles which I can only glance at here, where by the nature of the case the emphasis is on adaptability.

References

Bruner, Jerome 1966. *Toward a Theory of Instruction*. Cambridge, Mass.: Harvard University Press.
Dewey, John 1916. *Democracy and Education*. New York: Macmillan.
Dicey, Albert V. 1976. *Lectures on the Relation between Law and Public Opinion in England during the Nineteenth Century*. 2d ed. (reprint of 1914 edition). New York: AMS Press.
Garaudy, Roger 1970. *The Turning Point of Socialism*. London: Fontana.
Mailer, Norman 1968. *Advertisements for Myself*. London: Panther.
Mills, C. Wright 1967. *The Sociological Imagination*. New York: Oxford University Press.
Passmore, John 1974. *Responsibility for Nature: Ecological Problems and Western Traditions*. New York: Scribner's.
Toffler, Alvin, ed. 1974. *Learning for Tomorrow*. New York: Random House.

Commentary by Marx Wartofsky

Department of Philosophy, Boston University

Passmore's paper is full of insightful, provocative ideas. It also contains several that are wrong-headed and incomplete. In short, it is an interesting paper. I have seven minutes to comment on it, and I want to make four points. By quick calculation that gives me one minute and forty-five seconds per point; so my style is dictated by these temporal facts of life, and I will attempt to be pithy, forthright, and to the point.

Point one: there is a pervasive presupposition that runs through much of Passmore's paper, and it is implied by the title, "Education and Adaptation for the Future." It is that education is a preparation for the future. A child is being prepared for adult life; that is, for work, for social life, for the development and use of one's capacities in a changing future. And there is little to quarrel with here, as far as it goes. My quarrel is with what this leaves out, which I believe is essential to education. What is left out has to do with the present, or if you like, with presentness or immediacy of the concrete experience of one's education, or its palpability. We certainly learn for the future. But we enjoy the present. One of the great defects of education is that it tends to denigrate or ignore the intense, glorious, and magical joy of learning.

The child (as Aristotle already knew and Jerome Bruner took several years to rediscover) is a natural learner. It is the essence of childhood to exult in discovery and to be curious and gratified in the acquisition of competence, or of dexterity, or of language, or of means of expression, or of knowledge about anything and everything, whatever. This is spontaneous, the gift of our animal inheritance, and of our human sociality and inventiveness. And for reasons which have to do with historical forms of society and not with any natural or genetic requirements, this childish affinity for knowledge has been systematically subordinated to domination, to demands of social hierarchy and control. Socialization has come to mean training, preparation for the future has come to mean "slotting," that narrow vocationalism which Passmore eloquently criticizes. Schooling in America (as Gintis, Bowles, Freire, Giroux, and others have pointed out) has been geared to the instrumental, social, and economic functions of providing grist for the industrial mill.

Now there is nothing intrinsically wrong with an education geared to the future needs and purposes of society, but everything depends upon whose needs and what purposes. When education is seen as simply instrumental, as preparation for the future in this restricted sense, then an essential component of its intrinsic worth, what I have called the joy of learning, is drained from it, and the educational system then fails to serve a basic human need, that of the child's self-development in the present. Education then becomes rigamarole.

Let me put this somewhat more dialectically: it is not that instrumental and intrinsic values stand in mutual opposition here, or that the present vies with the future for educational attention. Rather, as Passmore argues very clearly, a liberal education in the present is the best preparation for a changing future. But Passmore argues instrumentally that such a liberal, adaptable, open-ended cognitive and technical competence is best because it is best-suited for an open future.

My argument is somewhat more perverse: that an education ordered to the priorities of an intensive enjoyment of present learning with no eye for the future at all is the richest sort of preparation for just that adaptability which Passmore extols. "No eye for the future" is not quite as perverse as it sounds. The range of traditional subjects, the three Rs, the sciences, history, perhaps above all the arts, together with the liberal and important technological education, are traditional just insofar as they have evolved as the basis for the widest range of possible futures. Thus, the past already provides some eye for the future, in the distillation and preservation of curricular structures which, most generally, have embodied the common consensus about culturally significant contexts of learning. But a traditional curriculum is a dangerous thing, if it is regarded as a limit rather than as a take-off point. The test of the viability of a curriculum is whether it provides the opportunity for free and lively cognitive exploration of alternatives in the present; in short, whether it serves as a basis for the imagination.

Here, I agree with Passmore's emphasis on imagination, that is, that combination of free invention and disciplined construction which envisions alternatives and novel possibilities and which creates new worlds. That is what the arts and literature are about, and so is scientific theory and social theory, and preeminently so is mathematics; and that is what real engineering and design is about when it isn't reduced to formulaic vocationalism.

But here I come to my second point. It concerns freedom as a condition for education. The freedom I have in mind is not the fatuous, "liberal chic" freedom reserved for private "progressive" schools, where the privileged few play out their parents' fantasies about the life of leisure.

The freedom I have in mind concerns the power to make choices which will change the present, and also to have the means available to effect such choices. What I mean here is not education *for* democracy as "future citizens," but education *in* democracy as present students. Let me be clear. Children abhor license. They thrive on disciplined work and on the exercise of responsibility. Why? Because it is a condition for that very achievement which gratifies, satisfies their desire to know, and because structure, order, self-direction have their own rewards in the feelings of autonomous competence.

Such discipline—and here I am agreeing very strongly with the remarks about the relation between imagination and discipline—such discipline, which any baseball-playing, tinkertoy-building, stamp-collecting, or puzzle-solving child exhibits, even to the point of compulsive excess, is a discipline which comes from a sense of self-direction, of enjoyment in gaining expertise.

The freedom here is a practice of concrete purposiveness for which skill and means are recognized as prerequisites, and which, in the achievement of social tasks and projects, also requires a refinement and regard for others which is the root of morality and of sociality.

Passmore touches on this in speaking of adaptability as "the capacity to flourish in changing circumstances by the use of information, acquired skills, imaginative enterprise, and to cause to flourish the activities in which one engages, the institutions of which one forms a part." Agreed. But the flourishing is not simply adaptability *to* changing circumstances, but the ability *to change* circumstances for the sake of one's flourishing. (The emphasis here is therefore on Passmore's second clause in the quotation, namely, the one about "causing to flourish.") Education for the future is, as Passmore says, education for a *changing* society, but I find this too passive a formulation. It should be education for changing society, that is, using the term as an active verb and not as an adjective.

Such an education cannot be deferred until political maturity or voting age is attained. It is then too late. It requires instead the exercise of freedom in the educational experience itself. Here a word in defense of Dewey. Only a word, though a whole page is needed. It was Dewey's point (although not originally his) that the more intelligent an organism, the more it adapts by changing its own environment deliberately to suit its own needs and purposes, instead of adapting to a fixed environment. I can't imagine Dewey to have had quite the limited view of the importance of the future which Passmore attributes to him at the beginning of his paper. Planning, or in Dewey's terminology, *deliberation*, is, as he says, a rehearsal in the imagination of alternative modes of action. How far futureward the projection of such deliberation points is a function of what needs and purposes we envision. As human intelligence and imagination become cosmic in their scope, the earlier forms of millennialism become more than visions of the heaven beyond the earthly present, and are transformed into Utopias for an earthly future with real freedom.

The education for the future which is needed now is therefore an education in the exercise of such real freedom, that is, the filtering down of the virtues of democracy to the schoolroom, as well as to the workplace. End of Point 2.

Point 3 will have to be glossed over briefly. None of the above is feasible simply as educational reform or even as educational revolution.

We are presently in the midst of a crisis in education, a crisis of conflicting desiderata reflecting the schizophrenic social and economic demands of rapidly developing late-industrial and postindustrial societies. Imposed discipline, training for obedience, adaptation by subordination, are too pervasive as features of the working life and the social life of the vast majority of people to require any ideological rhetoric from me to point them out. The public schools—the public school systems in the United States in particular—had this as their explicit original historical premise and purpose. Yet modern industry, technology, and the contemporary service and cognitive industries all require, as conditions of their own viability, just that flexibility and adaptability which Passmore nicely describes.

To put this crudely and in blatantly capitalist terms, cognitive capital is the most valuable and most underused asset in modern economy. The Japanese surprise us when they tap it even slightly under highly controlled conditions in their industries, and we scurry to learn what should have been obvious, namely, that the liberation of human intelligence and inventiveness from the hand of hierarchic vocationalism is the most potent economic force that exists. Thus, it is this exercise of just those capacities, namely, for the enjoyment of learning and the exercise of freedom which I remarked on in Points 1 and 2, that contemporary society needs, above all in economic life, but elsewhere as well.

Now, this conflicts with the older dominative forms of management, of politics, and of education. If there is a general lesson about how humans adapt or will adapt in the near future, it is this. The instinctive or spontaneous capacity for learning and for freedom (and the two, I think, are indissolubly linked) has become a culturally developed requirement, complex in its cognitive and its institutional forms, and impressed with the history of the forms of human cognitive practice. It strains against the older molds, and it will break them. But the present educational system is in a crisis precisely because these two incompatible desiderata—namely, for social control and for freedom—are both strongly operative.

Finally, Point 4—a curricular point: it concerns Passmore's very interesting discussion on the complex relation between learning general principles and learning them in and through the particulars in which they are expressed. It is clear that the sciences and mathematics are major subjects for this sort of education, since they concern just that relation between general principles—laws, theories, axioms—and their particular interpretation. But the arts, and in particular, language and literature, are also appropriate contexts for this sort of learning. Here, form or structure is sensuously imbedded in what we see and hear, what we make or shape; and this, as much in our spoken and written utterances as in our painting or sculpture or music. Needless to say, such realizable and intelligible form is also present to us in all of our industrial and social artifacts as well; in our machines, our buildings, our bridges and automobiles, and also in our social and political structures. Form abounds. General principles are all about us, on every side. The availability of such opportunities for education in general principles, in formal structures, is obvious. The means for such education are not, for the discovery of form, or the creation of new forms is not a matter of simple inductive generalization from many instances. But that such a mode of education is desirable and essential is clear, and I agree wholly with Passmore on that. I would simply propose extending Passmore's insight, and seeing it as a leading idea for all aspects of the curriculum.

Commentary by Elizabeth Torre
School of Social Work, Tulane University

Being placed toward the end of a program, I have discovered, is both an advantage and a disadvantage. I have been constantly revising in my head my remarks during the time that I have been here; I suppose one might say that that is in a sense a sort of adaptation. I am not sure that I shall say anything that has not already been said, but perhaps I may say it in a little different way.

I would first like to make one broad comment about education for an unpredictable future as it has evolved in my thinking these last few days. This is the need to relate what we have learned (through careful scientific study of our biological heritage, human relationships, habitats, and time, and a historical perspective of many ways of adapting) to preventive and remedial interventions in a variety of the applied disciplines. To do so, we may need to work harder at a middle-level theory that allows this movement by continuing to ask and study questions such as to what degree can man purposefully influence the rate of social change? How do motivation and values facilitate or impede change? Is it possible to predict how change in one sector of the human environment might influence desirable or undesirable changes in other sectors?

Specifically addressing Passmore's essay, I was especially stimulated to think about these questions by his contrasting definitions of adaptation and adaptability and by his suggestions for the form an education for adaptability (which he defined as a capacity that allows humans to flourish, not just to survive, in changing circumstances) might take. I would like to place his suggestions about content, method, and the character of an optimal learning environment within a framework of three ways of conceiving of time—not past, present, and future, but developmental time; subject-object interactional time; and historical time. I know these are not mutually exclusive, but I will try to deal with them independently.

My interest in developmental time is that it marks and facilitates both cognitive and socioemotional development, and offers some guidelines for the timing of the learning of basic skills, major general principles, and basic facts about the world. However, as Hans Furth has pointed out in his recent book, *The World of Grownups, Children's Conception of Society*, the guidelines do present us with a chicken and egg question of the relations between children's development of formal operations, as defined in Piagett's scheme, and their development of social understanding.

Furth (p. 70) highlights the difficulty inherent in this question by suggesting that "if formal operations are identified in a narrow sense with expertise in solving and understanding physical problems, societal knowledge comes first and physical knowledge second." If, on the other hand, you focus upon the theoretical attitude as being central to formal operations, "then you could easily accept the fact that a majority of adults reach a mature stage in their societal roles as workers and citizens and yet fail to apply their theoretical capacity to physical problems."

I would like to suggest that an either-or, chicken-egg approach as a guiding principle for curriculum organization ignores the evidence that social development and cognitive development might better be thought of as processes in a relay race toward maturity; that is, both are off at the start of life with one being in the lead at the so-called quarter-mile pole, the other passing it at the half-mile marker, and both needing to finish together.

Practically, this would mean that teaching of the sciences and the humanities be similarly paced, and further attention be given to methods for so doing. Though there are many interesting and important ways to think about such subject-object interactional time, one of which Catherine Bateson pointed out in her commentary to Stephen Toulmin's essay, let me focus on the quality of teacher-student interaction.

Indeed, as Dr. Passmore pointed out, the spirit in which learning is conducted is extremely important in terms of the outcome. If teaching-learning is an interactional exchange between teacher and student, subject and object, both imaginative thinking, which need not take account of other people's viewpoints nor follow logical and factual constraints, and (on the other side) rigorous, logical thinking, which must critically examine and sort out playful images, are more likely to be encouraged. I would like to suggest that the balance of these two leads to a mind-set for expanding and coordinating experience, both congenial and disparate, a capacity for adaptabilty rather than to the accumulating of experience.

Finally, historical time appears to me to be an extremely tricky variable in education for the future. On the one hand, knowledge in both the sciences and the humanities is built on the acceptance, rejection, and reexamination of models from prior times. Also, as individuals, each of us learns not just as children but throughout life. I would like to underscore that learning and teaching are lifelong processes. We learn from experiences with physical forces, adult models, peer groups, social institutions, and social developments.

Yet, as has been said a number of times here, the problem is, no learning can now be for the rest of life, and truth is not, as once believed, ascertainable, waiting out there somewhere just to be discovered. This dilemma, I think, is singularly and perhaps timelessly illustrated by Henry Adams in one of the most famous commentaries written to date, perhaps *the* most famous, on American education, and I would like to quote from Henry Adams as a conclusion to my commentary:

What could become of such a child of the seventeenth and eighteenth centuries when he should wake up to find himself required to play the game of the twentieth? Had he been consulted, would he have cared to play the game at all, holding such cards as he held, and suspecting that the game was to be one of which neither he nor anyone else back to the beginning of time knew the rules or the risks or the stakes? He was not consulted, and was not responsible, but had he been taken into the confidence of his parents, he would certainly have told them to change nothing as far as concerned him.

Then Adams, in his characteristic fashion, continues, "As it happened, he never got to the point of playing the game at all. He lost himself in the study of it, watching the errors of the players."

Actually, letters of students whom Adams taught at Harvard belie these protestations, but I think the warning that in our continuing odyssey we get lost in watching and studying it is worthy of note.

Discussion

DR. THOMPSON: I just wanted to follow on your comment that some of the joy of education comes out of the discipline, because I have really felt that one of the great problems with teaching has been—and I am a teacher—that we have attempted to make it painless. In doing that, we have removed the very fun that it offers. With all of our audio-visual excitements and games and toys, we have taken away the opportunity for instruction in discipline which ultimately can lead to the kind of joy that you are talking about today.

DR. BECKERMAN: Well, not at Oxford. We literally make it as painful as possible.

DR. PASSMORE: Well, it is always a problem when you are talking about education. I find that people complain about what you haven't talked about and what you haven't said and what you have left out, because, of course, the topic is a very broad one, and it has a great many aspects. In this case, I was determined by the fact that I was supposed to talk somewhere within the framework of adaptability, adaptation, and that I had an eye on the future, and that I had in mind a particular class of people that I thought myself to be arguing with, namely, the new vocationalists, who seem to be arising everywhere at the moment, and constantly my emphases were directed particularly in that way.

I did suggest that uncertainty about the future is so great—and is something to be emphasized again and again in these seminars—that education should try to concentrate on things which are valuable in the present as well as in the future. To think of the future as something which is going to determine your educational pattern is a great mistake, so that I don't want to disagree at all about the importance of education being in the present with some of the things that Dewey said, although I think Dewey said so many different things on so many different occasions that it is often extremely difficult to know what his final view on any matter is. You can generally find some quotation which will support any interpretation you care to offer.

But I think Dewey said some very valuable things. He said some other things which he himself felt in the long run did an enormous amount of damage to education, not only in the United States but in my own country. England was for a little while, with its customary insularity, immunized against this, but I think in the long run that Dewey's influence was felt there, and perhaps more intensely than anywhere else. The difficulty was the sort which has just been commented on from the floor. Making education enjoyable in the present was taken to mean that it must not be at any point boring, dull, or unpleasant. If you think in terms of getting children, say, to learn a foreign language, this is terribly exciting at the very first stages, where you learn a lot of words and begin to say things for the first time. Subsequently, however, you tend to get a roundout period in which you settle down into getting greater control of the language, followed by a period in which you can move with greater freedom.

Much the same thing is true in a number of other areas, and it can be true in science. There are facts which you have to build up as a sufficient body of information and principles to go on, and this is less exciting than the initial breakthrough. If you try to make the whole thing exciting, you spend your life doing things like looking at Brownian movements under microscopes, and if it is not as exciting as that and you don't put it in the course, then the effect is, as Dewey himself said, to produce an education which becomes very scattered and disparate. So there are very great problems here.

The problems are elsewhere put in terms of teaching children to care about what they are doing, teaching them not to be careless in the process of doing it, and keeping their schoolwork from turning into a sort of meaningless toil. I regard teaching as the most difficult of all tasks, partly because all things have to be watched simultaneously, and it is particularly difficult, I think, at the secondary school level.

I think if you are introducing children in primary school to lots of things and giving them skills which they can in some degree fairly rapidly master, if they are good primary schools, then such schools can be fairly satisfactory. At the secondary school level, again, in the first year, life is not too difficult, where again you have new disciplines, but the years that follow are years in which many people entirely lose interest in being taught and in education, and I think this period cannot be compensated for by turning secondary schools into places that are simply playgrounds of an elaborate sort or simply child-minding centers, but the problem of teaching in a quite genuine sense, without turning school work into mere toil, is really a very difficult one.

DR. TOULMIN: I wanted to pick up this question of teaching adaptability with an issue that Cathy Bateson raised at the beginning, and that is: Who is doing the adapting? I think one of the problems with the whole notion of child-centered education is that all of the things that you say tend to be interpreted as though you were trying to encourage something in the child or do something to the child. Now, I take it that what you are saying is that you are not seeking to teach children to be adaptable because indeed it may be very difficult to teach children to be adaptable. Whether they are adaptable or not is likely to be a characteristic of their bent, of their natural bent.

I am not saying you can't teach them, but what I understand you to be saying is that indeed the arts that you wish to teach and help children to master are arts that have to be adaptable, that the question is not so much encouraging the child to become a different kind of person, but rather finding a selection of arts that the child is required to master, which themselves will be best suited to the kind of future the child is really going to have, rather than the kind of future the teacher guesses that the child will have.

DR. PASSMORE: I feel like the reference in Plato to a child crying for both, because I would want both. I do believe that you can profoundly discourage children from being imaginative. I have sometimes thought indeed that the Oxford tutorial system is largely designed for this purpose, and that this explains why Oxford has produced so many distinguished civil servants and so few geniuses.

Whether this is true or not, you can very much discourage the child from being critical. I remember having teachers who, if you dared question anything they said, told you that criticism was wicked, and again, if you tried to do an imaginative piece of work, you were immediately suspect, because you might be a mutant in some form.

So that the discouragement was there, but the things that are adaptable in the great range of circumstances, certain skills, like reading, writing, and arithmetic, are adaptable in a great range of circumstances, and the antithesis between teaching children to express themselves and teaching children, let us say, to read and write, seems to be a silly one. In fact, reading and writing are ways in which children express themselves, and if they are incapable of reading and writing effectively, they never will be able to express themselves.

So, the whole of education in my mind is full of false antitheses, and that is why I am like a child crying "both," while realizing the immense difficulty of the task. I taught secondary school for only a short period, but long enough to become conscious of the problems that confront teachers in the classroom.

DR. COHEN: It seems we have an image of education that is somehow supposed to be contributing to the kind of flexibility that we all seem to be hoping for, and it has failed. I wonder whether we ought not to ask whether it is commonly true in human societies that that is not really what education is for. Education isn't really supposed to be doing what you claim Oxford was doing. In other words, what it is really doing is basically making us citizens of the old system, and in some very deep sense it is inhibitory.

JOHN PASSMORE

DR. PASSMORE: I think I would agree; indeed I said that historically speaking the vast majority of educational systems, like the educational systems in many countries now, want to turn out people who will be convergent thinkers, who will accept the customs, the rules, the regulations, the ideas of the society in which we are brought up. In a society which doesn't change at a fast rate, we still might not like this, but there is a certain point. In a society like our own, which is subject to extremely rapid change, this sort of education then is quite disastrous. You see, even in this country I believe the education at the secondary school was thought up as a place of socialization, and this seems to me what went wrong with secondary-school education in the United States.

DR. BOYER: It is absolutely crucial whether we are measuring education against idealizations that have never been accepted. It is hard to find educational purposes stated except perhaps very recently, and then by educational philosophers, not by legislators or pragmatists, as to what founders of educational institutions were trying to do when they created schools. We have an interesting question as to whether we are using yardsticks that no one intended to use.

DR. BOYER: Could we have a final comment?

DR. MIDGLEY: I think there is a lot to be said for this side of the question, but I do feel that I have come across people who were educated at the early progressive schools in England where you weren't ever taught anything or made to do anything you didn't like, and not only didn't they know anything, they weren't capable of learning anything either, and they were also incapable of living with anybody.

Now, I think this is frightfully unfortunate and maddening, but there is this dilemma in which we are constantly living, and I would like to ask you, John, are you, as it were, recommending just jolly good teaching in general? What would you say to this worry that if you want to make somebody adaptable to learning many languages, there is a good case for teaching him only one language first; and similarly, would a saddler not have been better placed to learn to be a tentmaker when that became necessary, then somebody who had never learned saddling. Unless you really get into something, it is extremely hard to get out and go somewhere else.

DR. PASSMORE: That is why somewhat by surprise I never laid down an idea of curriculum, because there are excellent ideas of not teaching any subject you care to mention in school, but on the other hand I know full well that you have to start somewhere, as you say. I agree. I think, for instance, to try to give people six months of a number of languages is quite silly, and once you have learned one language it does help one to learn others.

Again, I don't want formulae like "schools are places where you learn to learn," because obviously you do a lot of learning at school, but still a school is a place where you learn to learn, amongst other things, and one of the great weaknesses of our schools has been that people didn't learn to learn—and not only in our schools. I was once horrified, when I conducted an inquiry into university teaching in Australia, to get a response to some questions from a medical school the remark, "No, we don't teach doctors in any way to consult the literature, we don't have time." It seems to me that nothing could be more important to the doctor when he finished his training than he know where to find the information, and here, the recognition that learning to learn in this sort of sense is quite crucial, it is quite fundamental.

DR. CARSTAIRS: I would just like to put in a word following Mary Midgley's observation. I can readily believe that her experience was an unfortunate one with children exposed to an overlibertarian regime, but I would like to remind her that in 1978 there was published a controlled comparison between extremely liberal elementary schools and more formal ones, and to the disappointment of those enthusiasts for extremely liberal education, those schools came out worse on the comparison, but people usually forget that there was a large sample, and the school that came out best was an extremely liberally conducted school. The suggestion was that it was cheating on its own rules.

DR. BOYER: Any good essay leaves us with greater appetite than when we began, and that truly is the result of this last one. Three issues really came crashing in on me, and they are almost presented in dichotomy. How can one choose between educating to change society and educating to understand and accommodate. These are not adequate options. Practically, there must be a combination of both. No one is going to completely alter, nor should anyone complacently embrace. That is the educator's problem. Nor is it appropriate to put us in the dichotomy of making education painful or joyful. Education toward ends will be meaningful; some of it will be less pleasant than others. I don't find much stimulation in placing the educator's methodology around whether education is joyful or painful. If purposeful, it will be well motivated and some of it will be more pleasant than others.

Finally, I don't see that our choices are vocational or nonvocational. Life is a curious mixture of working and playing and thinking and living. Who in this room is not caught up in the interrelationships of them all? We do our students a disservice to suggest that vocation is somewhat demeaning. The real question is for education to lead towards worthiness.

I was often struck by colleagues on faculties who scornfully rejected students' questions about their own life work, who were devoting night and day to make sure they got tenure. To them, security and vocation were all-consuming. I think educators then have to somehow find that their job is not, as we have heard today, caught up in easy categories, but in the vexing dilemmas whose integration represents life at its best.

17. Ethical Issues in the Human Future

University Professor of Theological Ethics and Member of Committee on Social Thought, University of Chicago

Editor's Summary. James Gustafson explores some of the philosophical roots in Western thought which may contribute to some of the problems confronting human society. Gustafson argues, for example, that anthropocentrism has led to such problems as the rapacious exploitation of our environment and resources. Anthropocentrism, however, has made some important contributions, including the concept that people have intrinsic value and dignity. Associated with this are concepts of justice and rights and the ideological motivation for human achievement. Gustafson argues for a modified and expanded concept of social accountability, with justice involving right relationships between both individuals *and* groups. He states that we must develop a greater emphasis on the common good and that this perspective is inherent in a broader view of human participation in nature in which participation in, rather than mastery over, nature is the proper view.

The thesis of this lecture is that a careful alteration of Western morality and ethical thought is required to take account of important aspects of scientific knowledge and of the impact of technology on relations between man and the rest of nature.

I shall briefly and oversimply describe the strands of morality and ethical thought that need revision and why the revision must be careful. Then I shall state the principal reasons for undertaking an alteration. Finally, I shall describe some of the alterations that are needed.

Anthropocentrism in Morality, Religion, and Ethical Theory. Hans Jonas (1974:6-7) has written, "All traditional ethics are anthropocentric.'' The generalization is too sweeping; much of the "ethics" based on Eastern religions is cosmos-centered, and there are strands in the Western tradition that require modification of Jonas's claim. "Anthropocentrism," nevertheless, is a convenient and sufficiently adequate term around which to develop crucial themes of Western morality and ethics.

Anthropocentrism has been backed by religious beliefs grounded in biblical traditions. John Calvin (1955, vol. I:81-82), for example, wrote, "God himself has shown by the order of Creation that he created all things for man's sake.'' Since man is at the top of the hierarchy of created beings (now that angels are no longer taken seriously), man is also at the top of the hierarchy of values. All other things can properly be put in the service of man.

Without theological backing, a great deal of philosophical ethical theory rests on a similar assumption. Geocentrism (the idea that the earth was the center of the universe) was discarded several centuries ago on scientific grounds, but on religious and moral grounds anthropocentrism continues to be defended. The radical disjuncture between human beings and other animals (a disjuncture crucial to the defense of anthropocentrism) has been challenged by many biological investigations, but the distinctiveness of man continues to be exaggerated for religious, philosophical, and moral reasons.

That anthropocentrism is assumed in Western ethics can be seen by examining some of the principal terms and concepts and their references and usage. *Justice* is one such term. Historically, the most common formal definition of justice is, "To each his or her due." The questions this answers have been varied. What retribution is due a person for his or her violations of law or of morality? What compensation is due a person for labor and effort? What rewards are due a person for achievements beyond the ordinary? What is due a person simply on the basis of fundamental human needs? Is anything due beyond what has been merited when a person has been deprived of capacities through no responsibility of his or her own? Three basic questions are answered when this formal principle of justice is worked out in practice. *What* is due? *To whom* is something due? And, *on the basis of what principle or principles* is it claimed to be due?

In the West we do not have a tradition of asking what might be due to Lake Erie or the atmosphere. To be sure, we ask what is due to certain groups of people, but we do not ask what is due to species of animals. The farmer who uses draft horses might consider that certain things are due them for their beneficial labors, but this is based on their contribution to human ends. Indeed, it initially appears to be absurd to extend the use of the concept of justice to nonhuman forms

JAMES M. GUSTAFSON

of life. Why? Because most applications of justice require that the recipients *deserve* what they are to get. Deserving assumes accountability for actions, and accountability requires capacities that are distinctively developed in the human species.

If it is decided that persons deserve certain things simply on the basis of their need for them, a significant qualification has been made of the merit principle of justice. And there have been persons who believe that needs ought not to be met on the basis of charity and not on the basis of justice, simply because the principle of merit does not apply. Justice assumes the right and capacity to make a claim, either for oneself or groups. Or, as in criminal justice, persons are accountable for doing what they ought not to have done, and thus the community has a claim against them.

It is plausible, with little imagination, to extend a principle of justice based on need to the nonhuman realms of life. The question becomes: Need for what? A romanticism about nature can be grounded by this extension; because a form of life exists it has a claim to flourish. Carcinogenic cells? Ragweed? Amoeba that spread dysentery? As I shall indicate, the ethics needs to be altered in order to account for the interdependence of man and other aspects of nature without slipping into a romantic naturalism.

The concept of *rights* has flourished in Western culture for some centuries, and its application has been extended in recent decades. The concept is usually correlative to justice. A laborer is worthy of compensation; he or she has a right based on the efforts made. Groups have rights. If deprivation of opportunities is based on a classification of persons that ought not to be morally significant (such as race, sex, physical handicaps), their rights have been violated. Arguments continue about what it is that persons have rights to and on what basis the rights are claimed. A right to comprehensive health care is recognized in some societies, and the right implies an obligation for third parties to pay for the care. Rights are the basis for claims, and claims (on the whole) assume some form of deserving. Claims based on needs rather than achievements or on the formal status of being human (as is the right to vote in a democratic society) open the possibilities for extension of the concept of rights.

There is a literature that makes the case for the rights of animals. An implied claim for such rights has been involved in antivivisection movements that have resisted the use of animals for scientific research and in concerns for humane treatment of pets and for endangered species. (It is interesting to note that some investigators still say that the animal is "sacrificed," a term that connotes some recognition of the cost to the animal.) The fact that literature about the rights of animals is highly controversial indicates that the extension of the concept seems odd. Can there be a right where there is no capacity to claim it, either for oneself, or someone representing the interests of another person who is unable to make the claim? Rights to what? To survival? What gradations among forms of life will determine what forms have what claims?

A rights claim is a very strong claim. In practice it is stronger than a claim that something has value to others; in such terms its loss might involve a "sacrifice," but not a violation of a right. Rights claims in Western culture have been traditionally based on a sharp distinction between members of our species and those of other species.

Although it has not received great prominence in recent decades, the concept of the *common good* would appear to modify the individualism that frequently underlies the language of justice and rights. If the common good is only the aggregate of individual goods within a defined community, no modification of individual claims is required. If, however, it requires that a "good of the whole" be considered, restraints on individual claims are morally entailed. Such restraints are often justified in human societies. Military conscription, for example, is justified on the grounds that a threat to the common good of the nation warrants this infringement on normal liberties and the subjecting of persons to threats to life. The issue of what group's common good is to be considered must always be addressed. Is it the good of an ethnic group, or of a nation-state, or of the human community in a present time span, or of the species over future time?

Usually the use of the concept of the common good has been restricted to particular human communities, such as a Greek *polis*, the Christian Church, or a nation-state. Exceptions to this restriction are population geneticists, who are concerned about the well-being and survival of the human species on this planet. The concerns of these persons are still quite anthropocentric; it is difficult to think in manageable terms about the common good of the whole of what we theologians call "the creation." An understanding of the interdependence of man with the rest of nature, however, requires that a more extensive common good be taken into account.

My general point should be clear. These (and other) concepts of traditional Western ethics are based upon anthropocentric assumptions. There is a hierarchy of "being" with humans on top; there is a corresponding hierarchy of value, which backs the anthropocentricism of a great deal of traditional Western morality and ethical theory.

The Values of Anthropocentrism. To push headlong into a rejection of traditional Western morality and ethics, however, would be nothing short of foolhardy. Any alteration must be made with care. Visions of the "good of the whole" give us pause. Plato (1937, vol. 2:645), in *The Laws*, for example, had the Athenian say to the youth:

The ruler of the universe has ordered all things with a view to the excellence and preservation of the whole, and each part, as far as may be, has action and passion appropriate to it. . . . And one of these portions of the universe is thine own, unhappy man, which, however little, contributes to the whole; and you do not seem to be aware that this and every other creation is for the sake of the whole, and in order that the life of the whole may be blessed and that you are created for the sake of the whole, and not the whole for the sake of you.

This vision, which contrasts sharply with John Calvin's, evokes a moral pause. The anthropocentric view has sustained many benefits for both our species and individuals that a more wholistic vision can threaten. I shall briefly remark on a few.

The dignity and distinctive value of the human species that an anthropocentric vision sustains cannot be ignored. The consequences of a rejection of this dignity and value would be morally appalling. This special dignity and value has a biological basis—the development of the human brain. For this no individual, no generation, nor the species as a whole can take credit. It is given and not earned. Nonetheless, there are grounds for the special human claim. Without them there would be no reason to give special protection to human life against threats to it from the forces of the natural world. There would be no reason to protect and preserve the great achievements of human culture any more than the achievements of other animals. If a common good of "nature as a whole" would be best served by the suffering and deaths of human beings or by massive restraints on cultural, and particularly technological, developments, on what grounds could they be negated apart from an assumption of a special value and dignity to human beings?

Anthropocentrism has protected the human species from making the "laws" of nature the sole ground of morality and from assuming that there is a "blueprint" of the whole of the creation of which humanity is a part of no greater value than every other part. To develop ethics that did not care for the special dignity and value of our species would be perilous indeed.

The ideas of justice and rights have preserved many persons from tyranny. Any larger good that is invoked and enforced is defined by persons or groups who have power. Since there are no philosopher-kings disinterested enough and knowledgeable enough to define a common good without taint of particular interests, the threats of tyranny are always present. Without concepts of justice and rights, Thrasymachus would win the day; those in power would have the right to determine what is the common good and to determine at whose cost that good is to be realized. Anthropocentrism has backed the establishment of political and social institutions that involve individuals in the determination of their personal and collective destinies and that protect the weak and the relatively powerless from the powerful. These benefits are not to be modified hastily.

Anthropocentrism has also motivated a great deal of collective and individual human achievement. The assumption, so clearly stated by John Calvin, that all things exist for our benefit has powerfully moved human development. Animals and recombined DNA can be put in the service of man. Clouds can be seeded for rain; fossil fuels consumed, forests cleared and prairie turf plowed, limestone quarried, and so forth. Augustine noted long ago that the order of nature and the order of its utility for man do not perfectly coincide. The tacit or explicit belief that nature exists to serve human beings and that it is right to use it for human interests (including the interests in exposing its "secrets" by scientific investigation) has deeply enriched human life. A prescribed reversion to letting the "natural" ends of nature dictate the restraints on human activity would be foolhardy, indeed. Man would be called to resignation rather than participation and achievement.

Any revision of ethics must preserve the legitimate achievements of the anthropocentric ethics of the present and the past.

A Redescription of the Place of Man in the Universe. Any moral theory rests in part on an implied or explicit description of human life. Moral theories have been developed, for example, on the basis of a description of man as radically free and on the basis of a description of man as deeply determined by inclinations, drives, or desires. What a moral theorist includes in his or her interpretation of human life makes a significant difference in how ethics is explicated.

My proposal is that many aspects of scientific knowledge require that the place of man in the universe be redescribed and that this new description necessarily alters the anthropocentric focus of traditional Western ethics (and religion). For the sake of brevity I shall develop this proposal around three terms: dependence, interdependence, and interaction (or participation). By dependence I mean a necessary reliance by human beings on processes and "forces" that are not the products of human creation and are not under full human control. By interdependence I mean the mutuality of reliance. This can be stated in the simple proposition that what occurs in human activity affects its surroundings, just as what occurs in the surroundings affects human activity. By interaction I mean that intentional human activity involves a participation in these processes and forces and, while they do not have the capacities for intentional response, they nonetheless react. In developing these terms I am attempting to avoid the excesses of a purely organic model of interpretation of how events and actions occur, while taking an organic metaphor more seriously than much of the Western tradition has done.

The redescription of man in the universe around these terms requires extension of the frames of reference in which moral activity is to be interpreted. First there is a "time frame." The traditional Western description of moral agents and acts severely limits the time frame to be considered. Account is usually taken only of specific acts by specific persons. This permits the time frame to be brief: it needs to include only the immediately present conditions for action, the actor's intentions, and the immediate consequences of the act. My proposal is that if man is viewed in relation to what we know about the beginnings of the universe, the development of our solar system, the long processes by which life came to be and our species evolved, and what is now anticipated as the future demise of our universe, the traditional limitation of morality to discrete acts by individuals must be significantly qualified.

Man is dependent chronologically on all these processes that have brought us into being. The existence of our species is brief, in this time frame. From what we now know about the universe, human life is likely to disappear long before the end of the universe. Perhaps our importance is not as great as we would like it to be. As Ernest Troeltsch (Gerrish 1976:117) pointed out vividly to theologians decades ago, "As the beginning was without us, so shall the end be also." Man cannot be the "end" of creation either temporarily or purposively. Our contingent area of life cannot be absolutized in value.

Second, the "space frame" must be extended. Man is not only dependent upon this vast natural history but continues to be interdependent with other persons, with institutions and societies, and with the natural world of which man is a part. A description of our reliance on many relatively minor and relatively vast forces and processes we did not create and cannot fully control qualifies our tendency to absolutize the value and importance of the human. The simple fact that human activities in turn affect many of the forces and processes on which we rely (our interdependence) extends the range of factors that have to be taken into account when human accountability is defined. While a vast "space frame" mitigates human claims to be the center of values, our capacities to affect all with which we are interdependent deepen our accountability. Our capacity to cause wideranging effects extends our moral accountability for our actions and their consequences. The "human good" is dependent upon and interdependent with a wide range of factors. We cannot consider the human good apart from considering a common good that includes the "space" of natural world of which we are part.

A redescription of the relation of human life to other animals requires a third alteration of frame of reference. Mary Midgley (1978), in her sprightly and cogent book, *Beast and Man*, argues against the excessive differentiation between man and beast that has characterized the Western tradition, without reducing the human to other animals. Human dignity, she argues, has traditionally been justified on the basis of a false description of man, one that sets the distinctively human over against biological nature, that sets reason over against inclination, and so forth. Brief quotations from Midgley make the point. "Why," she asks (Midgley 1978:204), "should not our excellence involve our whole nature?" "Our dignity," she argues (Midgley 1978:196), "arises *within* nature, and not against it."

The conception of man as a moral agent necessarily and correctly describes human capacities to form intentions and to govern various powers to fulfill them. But what I call anthropocentrism has led to an exaggerated distinction between "human" capacities and our "whole nature." It has lead to what Midgley describes as a dramatic account of "reason" and "inclination"; these become separate roles in a drama, rather than processes that are in some continuity with each other. A description of man more in continuity with other animal life qualifies the anthropocentrism of traditional Western ethics and religion.

From various sciences that would support these extensions of the frames of reference for understanding man, a further inference can be drawn that affects morality and ethical thought. The universe, our solar system, the natural world of which we are part, human life itself, societies, and culture are processes of *development*. This does not mean that all is flux, that there are no patterns that persist, no laws that control, no natural limits to human capacities. But it does mean that, for example, cultures develop as man interacts with nature, and nature develops in part as a result of the human interventions into it.

It means that nature does not provide a timeless and changeless blueprint for an equally timeless and changeless order of human culture and moral life. While change and development have been more dramatic in human culture and history than they are in nature in the brief time span of human life, the development of nature itself must be taken into account. Recall Augustine's observation that the order of nature does not coincide with the order of its utility to man. But the order of utility (as that is expressed in society and culture) while it develops is always in interaction with the ordering of nature. (Augustine, of course, believed that the order of nature was timeless and changeless.)

Both *possibilities* for human activity and *constraints* upon it occur in the processes of development in nature and culture. What those possibilities and constraints will be in the future, we cannot fully anticipate. Thus ethics of the future must be open to the possibility of extending and revising time-honored principles and values. This is not to say with Epictetus that we should follow events rather than lead them, that we simply adapt morality to the forces of nature, culture, and society. But it is to say that consideration will have to be given to changes in priorities of ends and values and that alterations in traditional morality must be expected. Neither mastery of nature nor conformity to nature are correct. Participation in nature is the proper view.

My basic point is a simple one. By redescribing man in relation to the universe—to the long past span of time and to the anticipated future one, to all the aspects of nature with which human life is interdependent, to animal life, and to the phenomena of development—the traditional anthropocentrism of Western morality is qualified.

Anyone who wants to think seriously about the ethical issues of man's relation to nature does well to read John Passmore's erudite and carefully argued book, *Man's Responsibility for Nature* (1974). Passmore distinguishes several attitudes toward nature that have been developed in the Western tradition: a kind of romantic primitivism; the assertion of man as despot over nature; the notion of stewardship that has deep roots in biblical religions; and cooperation with nature to perfect it. As Passmore analyzes particular problems from his perspective, such as population growth, he correctly comes to the conclusion that difficult choices have to be made. There is "no preestablished harmony" between various ends of human activity, each of which can be reasonably supported in itself (Passmore 1974:135). The contributions of what I have called anthropocentric ethics are worthy of preservation, but public policy requires choices that qualify, for example, a view that individual rights finally shall be supreme. The elegance and simplicity of views of ethics that would develop from my quotation from either Calvin or from Plato does not work out in practice.

In the interests of brevity I shall attempt to develop my own proposals around a few key terms. On particular issues I would come to conclusions very similar to Passmore's.

The redescription of man in the universe that I outlined rules out two extreme perspectives: man as sovereign master of nature (it does not rule out mastery of aspects of nature) and man as simply governed by nature so that the proper human activity is conformity to nature. Despotism and romantic primitivism are ruled out. I propose two terms that need to be construed carefully: consent and participation.

For consent, Samuel Johnson in his Dictionary proposes three meanings. "To be of the same mind, to agree." "To cooperate to the same end." And, "To yield . . . to allow, to admit." The third overlaps with some of his definitions of "resignation": "to yield up" and "to submit without resistance or murmur." I am not proposing an ethics of resignation or an ethics of following events, such as Epictetus over and over counsels. I do, however, propose that consent is appropriate; that human activity and cultural development be receptive both to the limitations and the possibilities that nature requires and provides.

I cannot, of course, affirm that the requirement is to cooperate to the "same end" of nature. In distinction from the classic natural-law tradition in ethics, we cannot be certain that "nature" has a goal or even that various "natures" or species have ends that we can fully know. Nature and culture do not develop in accordance with a predetermined end; species do not have immutable "forms" that provide the normative basis for ethics.

This can be illustrated with reference to human sexuality. Surely propagation of the species is the primary biological end of human sexuality; from this observation came a moral principle in the Christian tradition that sexual intercourse should be engaged in only with the intention of propagation. But sexuality has other "ends"; its "energies" move persons to a whole range of activities. Consequently, the ethics of sexuality, even if derived from its "ends," is more complex than Augustine and many others thought. Yet, even with qualifications, a measure of consenting to nature is required.

Consenting involves acknowledgment of limitations (what we theologians call "consenting to finitude") and the necessity of cooperation with nature. Even physicians cannot *create* health; they can remove impediments to it and medicate so that the body's natural capacities function better. Even the developers of nuclear energy consent to nature; they cooperate with it as they intervene into it and direct its uses.

Consent is the first phase of participation. Cultural activity, including moral activity, begins in response to what is occurring, what has been received from the past, what impinges upon persons and societies. Cultural activity is not creation *ex nihilo*, and moral activity is always within a particular context that the actor did not create. Human activity

JAMES M. GUSTAFSON

is a process of participation in the interdependencies of interpersonal relations, of societies with each other, and of human life with the wider world of nature. I stress again that consenting to what is occurring is not resignation to it; it is a receptivity to possibilities and constraints that exist.

Participation is not *reaction* to overwhelming forces that determine what humans do or avoid doing. Participation involves human agency; it requires the most comprehensive knowledge available of the events and circumstances to which we are responding, carefully developed intentions, careful assessment of the fitting means of action, and as thorough an anticipation as possible of many consequences of human interventions. It requires sophisticated imaging of the reactions of nature to human interventions and of the possible responses of other participants in the drama of interdependence or interaction. The term "participation," rather than "action," is primary because it is implied by the more extensive description of man that I outlined above.

The exercise of human agency is not a matter of discrete acts by "terminal" or autonomous individuals. Agency and action are always in relation: in interrelations with human bodily nature, with the social context in which they exist, the culture of which they are a part, and these in relation to the natural world. When institutions act as agents (for example, developing policies for hospital practice or for use of energy resources), their work involves a receptivity or consent to what is going on and a participation in processes in order to direct them toward justifiable ends. Environmental concerns, for example, did not develop in the last decades because some persons had a deep and reasonable belief, or an ideology, about the environment on the basis of which they achieved political and economic power in order to act in accordance with their convictions. Ecology became a major concern as a result of events and of a new consciousness of the implications of what was occurring in the world. Response to what is occurring is prior to action; human activity is participation in what is occurring.

I have dilated this because the basic image or metaphor we have of human activity has consequences for proposals we make about ethics. "Participation" implies a serious confrontation with both limitations of human control and the possibilities for intervention to give direction to what is occurring in nature, society, and culture. Like the desires of human individuals, so the thrust and directions of a culture and the forces of nature are there; we do not create them, though our actions have significant effects on their courses. By our rational activity we can choose purposes or ends, and by the exercise of various capacities such as "human volition" and technology we can give direction to what is occurring. Since our participation is in "the condition of finitude" (as we theologians call the constraints that are natural to us), there are limits to our capacities to foreknow and to determine all the short- and long-range consequences of human activity.

Since, as Passmore correctly states, there is no preestablished harmony among justifiable ends of human activity, the policies we develop and the actions we take always bear risks, costs, sacrifices for some persons, communities, aspects of culture, and even nature. Since what is useful and beneficial to human life sometimes requires a disruption of natural processes, we cannot revert always to nature to fulfill proper human ends. There is no avoidance of human anxiety in morality; there is no sure way to avoid inconvenience, restraints, pain, suffering, sorrow, and—even in the deepest sense—tragedy, in all circumstances.

To illustrate what follows from this, I return to two of the concepts introduced above: common good and justice. Greater prominence to the concept of a common good is entailed, I believe, by the extension of the frames of reference within which human life is described. Human "good" (or more accurately "goods") is interdependent with the "goods" of nature. Just as what is good for me cannot be determined apart from the relationships, interactions, and common good of my family, so what is good for humans cannot be determined apart from consideration of a "common" good that acknowledges dependence and interdependence with the natural world. Yet that common good defies precise and final definition. The deepest reason for this is that we cannot perceive a single unifying end toward which all things are naturally directed. If Calvin were correct, we could. If Plato were correct, and if we were persuaded that there was a preestablished harmony to the whole that could be known, there would be merit in conforming to the niche that we as individuals have in society or that the human species has in the ordering of nature.

The limitations of ethical prescription for the human venture lie in the fact that, while a common good cannot be defined precisely, we must and do continue to discover patterns of interdependence that require us to take into account not merely the "goods" of other species, and even inanimate objects, but the functions of each in the interdependence of the whole of life. The limits of our knowledge of this common good is one of the best reasons for keeping many of the achievements of the anthropocentric ethics of the West. Yet both restraint on human activities and possibilities for participation that enhance a common good will require (as they always have) the acceptance of risks, costs, and sacrifices of some desires and benefits for human beings for the sake of a large whole of which we are a part.

There is no such thing as moral infallibility, though to resign ourselves to that fact too easily is morally perilous. Only an ethics that is based on a single principle or value—or that is certain about a hierarchy of principles and values so that one is always supreme and overriding—can provide prescriptions for human activity that are elegant and parsimonious. Alterations of Western anthropocentrism and individualism will be required in the face of the claim of a larger common good.

The second term is justice. Justice refers, in part, to right relations between persons. But the application of the term can extend to right relations between groups and, with little imagination, to the right relations between human activity and the rest of the natural world. This was possible when our thoughtful ancestors perceived an order of all things that was changeless and timeless and one of a preestablished harmony of all things. Acts conforming to that natural order of justice were deemed proper.

Our consciousness of development, however, undercuts this happy view; our consciousness of the dynamism of nature (including human nature) does not permit us to define a static, harmonious, just order. Yet all is not in flux; we can perceive relationships that are necessary for the survival, if not the flourishing, of many forms of life. There are relations of things in nature that we must respect, if only because to deny them is perilous for human life. There are relations of things to each other in nature that are disrupted by disease and by human interventions that we can rectify. Like the common good, this extension of the use of the term justice does not provide the precise and absolutely accurate knowledge of an order to which we must conform. But something like a pattern of right relations in nature, and in human activity and cultural relations to nature, is perceived. That pattern requires our respect; it must be taken into account in our activities.

It is not possible to work out implications of the views stated here for decision-making procedures or for particular choices. Since I am a theologian, your suspicions that a certain religious view of the world lies behind this are warranted, but this is neither the time nor the place to develop that. I have attempted to communicate a perspective that requires alterations in Western ethics more than to develop an ethics. That perspective might be summed up in this way: we must see the place of man in the whole creation, and ought to conduct human life accordingly.

References

Calvin, John 1955. Ford Lewis Battles, trans., and John T. McNeill, ed., *Institutes of the Christian Religion.* 2 vols. Philadelphia: Westminster Press.

Gerrish, B. A. 1976. "Ernst Troeltsch and the Possibility of Historical Theology." Pages 100-135 in J. P. Clayton, ed., *Ernest Troeltsch and the Future of Theology*. Cambridge: Cambridge University Press.

Jonas, Hans 1974. *Philosophical Essays: From Ancient Creed to Technological Man*. Englewood Cliffs, N.J.: Prentice-Hall.

Midgley, Mary 1978. *Beast and Man: The Roots of Human Nature*. Ithaca, N.Y.: Cornell University Press.

Passmore, John 1974. *Responsibility for Nature: Ecological Problems and Western Traditions*. New York: Scribner's.

Plato 1937. Benjamin Jowett, trans., *Dialogues of Plato*. 2 vols. New York: Random House.

Commentary by Roy Branson
Kennedy Institute, Georgetown University

Professor Gustafson says that a change of Western morality and ethical thought is required because of increasing scientific knowledge and the impact of technology on man and the rest of nature. Specifically, he emphasizes the need to alter the anthropocentrism of Western thought, rooted, as he says, in the biblical tradition.

It is important to note that Professor Gustafson does not say that he is calling for a rejection or even a fundamental reformation of Western anthropocentrism. Rather, he suggests in his first sentence a "careful alteration." The main point of my response is to indicate just how carefully Gustafson intends that alteration to be.

For example, Gustafson's use of familiar ethical terms indicates that he doesn't advocate a reversal of anthropocentricism. He suggests that the language of the common good is preferable to one of rights, because it better emphasizes the interdependence between the rest of the natural world and humanity. However, Gustafson also says that the limits of our knowledge of this common good is one of the best reasons for keeping many of the achievements of the anthropocentric ethics of the West.

Gustafson makes similar remarks about justice. He says that justice refers not only to right relations among humans, but also to patterns of right relations within nature. The reason those right relations within nature must be respected is "because to deny them is perilous for human life." Gustafson is here suggesting that attention to justice within nature can be defended on anthropocentric grounds.

JAMES M. GUSTAFSON

Rather than science providing new understanding of the sources of ethical behavior or new principles for guiding right action, Gustafson suggests that science has expanded the horizons of Western ethics. We must see man not simply in community, but man within the natural universe.

Man must be receptive to and interrelate with a wider range of forces than Western thought has typically brought into focus. The time frame of Western ethics must be expanded. Ethics cannot look only at discrete acts by individuals, but must consider that man is dependent chronologically, as Gustafson says, on all the processes that have brought us into being. Often, the space frame must be extended. Good must not be found apart from the space of the natural world of which humanity is a part.

Gustafson's careful alterations then are alterations of the scope of Western ethics, not its content. If this reading of Gustafson is correct, there is a sense in which far from taking man out of the center of ethics, Gustafson has extended the horizons of the world within which man remains central.

Professor Gustafson's basic images expand rather than contract man's moral accountability. His suggestion of "participation as an attempt to break out of language" draws our attention beyond right actions of autonomous individuals to the fitting actions of individuals relating to other indivuals. Gustafson has reached for a metaphor that places man's actions within a cosmic setting. Participation, he says, requires sophisticated imaging of the reactions of nature to human intervention and of the possible responses of other participants in the drama of interdependence or interaction. The effect of Gustafson's account, far from undermining the importance of man in the exercise of his ethical responsibility, as he puts it himself, "deepens our accountability."

Gustafson begins his essay by alluding to the biblical traditions and the religious grounds for Western anthropocentrism. When one completes the essay and realizes that he has not proposed a fundamental change in ethics, one can regard his "careful alteration" of Western ethics as a shift of our attention from one biblical strand to another— a movement from metaphors of man entering into specific covenants and contracts found in the Pentateuch to images of man existing within the cosmos found in apocalyptic literature. In apocalyptic images, man does not choose to enter covenants, but discovers that he is a participant in a cosmic drama among superpersonal forces extending throughout the reaches of space and time. Within either biblical tradition, the covenant community or the apocalyptic cosmos, man's faith remains central. So also for Gustafson the central question is not whether trees, streams, and fields have intrinsic worth, but what is to be the destiny of man.

In the last paragraph of his essay, Gustafson reminds us that he is a theologian with a religious world view. If the reading of his essay that I have given is at all accurate, perhaps Gustafson's metaphors for man's place in the universe would encompass not only the apocalyptic tradition's predictions of catastrophe, but also its even more fundamental vision of the sovereignty of a God who is good. But a full exploration of that topic would take a theologian beyond merely the ethical issues in the human future.

Commentary by Wilfred Beckerman
Balliol College, Oxford University

I should like to begin by saying that when I discovered that, in the lottery, I had drawn the number to discuss Professor Gustafson's paper, and I started to look at it, I was a bit alarmed at first to see that it was written by a professor of theology, and theology is a subject that I am afraid I have never had any occasion to study. This would not stop me from commenting on it, because I would not dare have the news get back to my colleagues in Oxford that I had been unable to comment upon a paper simply because I knew nothing about it.

But when I read the paper, I was very glad that I had drawn this number in the lottery, partly because it is written with such beautiful clarity and partly because it struck me as being extremely pertinent to the questions raised in Stephen Toulmin's paper right at the beginning. For example, Stephen's paper asks the question: "Finally, what further adaptive procedures can be devised through which humans can improve the relationships between themselves and their habitat?" Elsewhere, he says that the questions posed for discussion in this symposium rest upon the idea that human beings can arrive at their "proper place in the natural world" and so on. And at many other points Stephen Toulmin's paper talks about this problem of our place in the natural world and the fact that for centuries human beings saw themselves as the rulers of creation and so on. And these are the very questions to which Professor Gustafson's paper is addressed.

A third reason why I was pleased to have a chance to comment on his paper is that, in fact, the issues he raises are familiar to economists, if in a very different sort of guise. Two of the tools of the economists'

JAMES M. GUSTAFSON

trade, which are very closely related, are, first of all, the "Pareto principle," one version of which is that any move is socially optimal if somebody is better off and nobody is worse off. This is one particular principle for optimal social choice. It is a principle which, on the face of it, looks beyond dispute. After all, how can you dissent from the proposition that if somebody is better off and nobody is worse off, society must be better off? In fact, it is not as simple as all that, and it raises various questions, such as by what population is society defined? And this is, of course, one of the issues that comes out of Professor Gustafson's paper.

Perhaps I can illustrate it better by saying what the second economists' concept is. This is the concept of the social-welfare function, which does not necessarily take the form of the Paretian welfare function I have just specified.

A basic principle of orthodox welfare economics is that any self-respecting social-welfare function is in terms of the utilities (or welfares, if you like) of human beings. We do not allow anything else to appear as an argument in the function. We say that social welfare is some function of utility of Mr. X, Mr. Y, and so on. We do not accept the notion that social welfare is the utilities of X, Y, and so on, plus a few fish, trees, and anything else that is not already included as contributing to the utility of X, Y, and other human beings. This limitation on the domain of the function still leaves us scope for lots of lovely discussions about it, such as whether the precise form of the function is that total welfare is some simple sum of individuals' utilities or whether it takes a more complicated form, which would have certain implications for our views on equality, for example.

Nevertheless, all these discusssions are based upon the assumption that we are only talking about some function of utilities accruing to individuals. Fish, plants, stones, and so forth, do not enter into it. And so one of the questions that Professor Gustafson's paper has raised, in fact, is "is this right?" Should we extend the arguments in the function? This is quite important and, in fact, one of my very distinguished colleagues in Oxford (who should have been here rather than I, because he knows all about this) namely, Professor Amartya Sen, has recently been challenging this orthodox view, arguing with some cogency that one ought to put in some other arguments in the social-welfare function.

For example, we could have social welfare as a function of utilities of individuals and certain other things, such as no torture (irrespective of how far that feeds into the utilities of the individuals). Or one could also add kindness to animals or concern for the environment as arguments in the function.

Some people might even ask why not add also plants or other nonhuman components of the natural world; although I do not think many people would push the argument this far. At any rate, this is the sort of issue that does emerge from the question of how widely we extend the domain of the social-welfare function. Is it just man, which is what economists traditionally assumed, or should we extend the function to other components of the natural world?

Another issue that can be looked at in terms of the economist's view is the question of what time period is relevant? When we say, for example, in the context of the basic Pareto principle, that any given action cannot be judged to be optimal however much it makes some people better off, as long as somebody else is worse off, how far in time do we go? Do we mean that one could not judge any action to be optimal even if it improved the position of everybody in Britain today just because there is a chance that somebody in 500 years' time will be worse off as a result of it?

If one accepts that principle, then this is a very strong commitment to the status quo. One would be unable to do anything. Thus the Paretian principle also provides a framework for posing the question of how far one goes *in time* in extending the domain of the social-welfare function, as well as how far one goes in adding to the type of unit that is in the function.

Now, having posed—albeit in very different terms—these fundamental questions, Professor Gustafson then goes on to consider very briefly some of the reasons why one does not, *conventionally,* wish to extend the welfare function beyond mankind. For example, he points out that conventional ethics would not include animals, on the grounds of some theory of justice, such as that justice equates what is due to any being with their having some moral characteristic shared by human beings. Now it is true that Aristotle, who first introduced this notion of distributive justice, did say that distributive justice required that there should be differences between people with respect to their merit, but that people would differ with respect to what they meant by "merit."

I personally do not like the merit criterion, anyway, because I am just 100 percent determinist.

DR. GUSTAFSON: Could I respond to that for a minute, because I think you are misconstruing what I am doing in that section of the paper. I am describing there what I believe to be the traditional uses of it in terms of merit and accountability of justice. And what I am suggesting is that there are some things maybe to be preserved in that, but we have got to go beyond that. It was probably my own lack of clarity. But if anybody detects lack of clarity, it is Oxford professors.

JAMES M. GUSTAFSON

DR. BECKERMAN: You are saying you would not want to include animals on grounds of attaching to them any moral value?

DR. GUSTAFSON: I would want to get at that in terms of the importance of providing necessary conditions for flourishing. And not only for human flourishing, but for flourishing of other aspects of the world of which we are a part. I think you can build a self-interest into that relative to the human; namely, that apart from the appropriate conditions of flourishing of other forms of life, human flourishing might well suffer in some future.

And so I am not prepared to work out a definition or a use of the concept of justice relative to those sorts of things. But I am prepared to say that in terms of my own perception of the long-range responsibility, it is terribly important for us to bequeath the world to subsequent generations with social institutions and cultures, the natural possibility, indeed, with genetic possibilities which do not proscribe certain ranges of possible future development. That is sort of what I wanted to get at.

DR. BECKERMAN: But then you are still limiting it to feedback, somehow or another, to human utilities.

DR. GUSTAFSON: Well, I would do that for these purposes. If I put on my theological hat, I will not.

DR. BECKERMAN: Well, the conclusion I was arriving at was that you do seem to reject claims for including the other things other than the one which you just made.

DR. GUSTAFSON: I reject romantic claims.

DR. BECKERMAN: All claims based on need, for example?

DR. GUSTAFSON: The need of the animal.

DR. BECKERMAN: The need for development? You give examples of needs that you would not regard as justifying giving them rights.

DR. GUSTAFSON: Well, I guess we would have to argue about that.

DR. BECKERMAN: Anyhow, I would be quite happy to go along with your present explanation of your position. The only point I want to make is that, in your paper, I found that, although you seemed to reject some of the arguments that have been put forward for excluding animals and so on on the grounds that they are all biased by their anthropocentric viewpoint and assumptions, you do not spell out the positive grounds for your conclusion that the social-welfare function— my terminology, not yours—should not be confined to human beings. So I am not convinced by your conclusion, although I do sympathize with your reason for rejecting the conventional arguments against it.

Anyway, the conclusion I come to, as an economist, from all of that is that I think that there is a danger of going too far in attaching importance to nonhuman elements in some social welfare function.

For example, there are various movements in Britain that seem to be advocating—or implying—that we should spend the whole of the British national product turning every river in the country into beautiful, clean, swimming pools for fish. And I prefer to give more importance to mankind than to fishkind.

Secondly, I think that one can go too far in thinking there is a conflict between man and the natural world or the environment or anything like that. The real conflicts are between man and man. Those are the dangerous ones, and those are the important, urgent ones.

And thirdly, as an economist, it seems to me that these conflicts are precisely the ones that give rise to current urgent, pressing problems in the political and economic sphere. For example, it is conflicts between man and man that lead to rapid inflation, mass unemployment, and various other acute economic and political frictions. And these are conflicts that, in principle, are relatively soluble.

So that while in my sort of off-duty moments I love playing with these wider problems, such as the domain of the social-welfare function, in my more professional activities, I still think that I am prepared to neglect all arguments in the function other than human beings and to concentrate on the social-welfare function, narrowly defined in this manner.

Discussion

DR. WIERCINSKI: I have the impression that this essay was not written by a theologian. Of course, it is difficult not to agree that morality and ethics grounded on biblical traditions are anthropocentric in practice. However, materialistic ideologies are also anthropcentric.

In theory, at least, Judeo-Christian morality is centered first on God and then on man; man occupies the second place. The God-centered morality includes some solutions of the problems of the relation between God and the cosmos and man within it. In this respect, I would like to mention only the commandment of cultivation of Eden and not of its destruction and a paradisiacal future as prophesized by Isaiah when he speaks about the harmony between the wild animals which do not kill.

Also, the concept of Shekinah in the cabalistic tradition may be mentioned; that is to say, God's presence in the world. Of course, in

Eastern religions, cosmologically oriented, a moral attitude of man versus his natural environment was and is present.

The best example of a rigidly moral attitude on the nature of environment, based upon the principle of purity, was expressed in Zoroastrianism. The second good example is provided by Buddhism in relation to all sentient beings. I agree completely with Professor Gustafson's proposal of widening the area of objects of moral concern. Simply speaking, one possible means of salvation of present-day mankind seems to me to be a strong emphasis on the idea that man should not be only a consumer of the biosphere, but also its coordinator. Let us call man a species which plays the role of the central homeostat for the biosphere.

DR. BOYER: Thank you. Could I just ask you to respond, Professor Gustafson? And I will try to put this a little more bluntly, perhaps. Are we actually saying here that we are now discovering that it is in our self-interest to expand this definition, that it is more out of knowledge than enlightenment? We simply know that we cannot mess around with the other parts as much, and therefore we had better shape up in order to protect ourselves? Is it not the ethnocentricity writ large, based upon some new understandings? Maybe that does violence, but I just thought I would state it rather more pointedly to see whether that is where we are.

DR. GUSTAFSON: Well, Mr. Chairman, you yourself said in talking about education you wanted to avoid dichotomies, and I, too, want to avoid dichotomies. But surely there are perceptions that arise out of events that occur historically which force us not only to expand the range of things we take into account cognitively as pertaining to the human condition, but I think there are at least occasional events that occur which have the effect of simply shaking us out of being curved in on ourselves as particular human communities or as particular individuals.

The human problem is part of the problem of contraction—contraction of vision, contraction of affection, contraction of sensibility. So then what we are talking about are the ways that our vision can be expanded.

The motives for doing that can be very mixed, and I don't really care about that, you see; but I do think that somehow or other we have to conceptualize the world of which we are a part in order to speak more clearly about what the implications of that are.

DR. TOULMIN: I just wanted to follow up this last point, and I think, Jim, you could go further and simply underline the fact that in the general rhetoric of the ecology movement there is a real ambiguity. I mean there are people in the ecology movement who talk about the need to change our attitude toward nature out of a kind of enlightened species interest which, is very much the parallel with the enlightened self-interest of the eighteenth-century moral philosophers and moral theologians.

There are others, I think, who speak about the necessity of doing this out of a genuine piety. Of course we can, as Jim suggests, argue that if we sit down in a cool hour and think long enough, we shall discover that the demands of piety and the demands of the enlightened species interest are going to coincide. That is a mere replay of the eighteenth-century debate about egoism and altruism in a new framework.

But still, I think we need to bear in mind the fact that the demands of piety and the demands of enlightened species interest have at the very least to be stated in differents words, and it is a problematic issue whether they are going to lead to the same conclusions in all future situations.

DR. BOYER: Fair enough.

DR. MIDGLEY: There is something very odd about this notion of the enlightened species interest. Classical utilitarianism did not refer simply to people; it referred to essential beings. And it is thoroughly consistent, with the emphasis which utilitarianism puts on happiness and suffering, to do that.

No reason has ever been given by modern utilitarians who have shifted over to ignoring the rest of creation. They do it because it will take a good deal of trouble to do otherwise. Because while Mill and Bentham thought chiefly about cab horses, we know rather more now about what we are doing to the rest of creation. So, since it would be rather troublesome to look on this as a moral issue, people don't. I think, if I may come back to the word piety, it is not just obvious that these are the only two alternatives.

DR. NEEL: I heard Professor Beckerman on several occasions say that population growth goes too far, but who knows what is far enough? And I think I would like to hear from you what is far enough.

Earlier, when I worried a bit about the growing disproportion between the resources and people, I was told there were plenty of resources; but I presume that you would not challenge the statement that at some point, if the exponential growth of population continues, we will outstrip the finite resources or our ability to deliver those resources. So I think you have to give us the "far enoughs" in your thinking so we can begin to reconcile our biological thinking with that of the economists.

Coming to Jim Gustafson from the ethics side, what we are beginning to see is that each life that exists is a unique solution to a very different set of conditions, some of which may have applicability to ourselves. There are very good reasons for preserving those life forms as long as possible.

DR. BOULDING: I am feeling terrible. I squashed a cockroach earlier this morning.

In the first place evolution has now produced an ecosystem that has human valuations in it that it certainly didn't have several hundred years ago, and this is an important aspect of the ecosystem. As it is, the whole world is an ecosystem, and there is nothing unnatural in it. I mean, that is, anything that is is natural or that is ever going to be.

But all the valuations that we know very much about are human valuations, and I think that we need to bring out what is in the argument of the social-welfare function. I care about the blue whale, but the blue whale doesn't give a damn for me, at least I don't think it does. Environmentalism has applied human valuations to the whole world. It has nothing to do with nature. It has to do with us. This I think is very important.

The other message, it seems to me, is the problem of perspective. Perspective has a certain survival value, obviously. If you perceive that the lion that just is going to eat you is the same size as the one that is a mile away, you're in evolutionary trouble. We have the same problem with morals; the point is that the near tends to be dear. I supported my own children in college, and I haven't supported any children in Szechuan, that is, if you love everybody you don't love anybody very much.

There is a real problem here, and without moral perspective I think the whole moral universe falls apart; there are profound evolutionary reasons for moral perspective.

On the other hand, we've also recognized that just as visual perspective is an illusion—that small person I see over there is the same size as the big one I see here—in a very real sense a moral perspective is also an illusion. But it is a necessary illusion—and I don't really see how we can do without it. To pretend that we can care for the whole universe equally is arrogant. This is more arrogant really than to say the universe is made for us.

DR. GUSTAFSON: Are you drawing that inference from what I wrote in my paper or is that coming from somewhere else, because I don't see it as a necessary inference from what I said in my paper.

DR. BOULDING: It seems to me a possible inference. I don't know. I don't believe in necessary inferences. I never saw any academic who couldn't wiggle his way around it. But I think it is a potential inference, and it is the sort of thing that we have to worry about. I mean we have the cities and all of these energy buffs who seem to think that they have energy theories of value, and it is ridiculous. The only theory of value that makes any sense is to provide for human valuation. Maybe I'm an economist, but that seems to me a very important moral principle.

DR. TOULMIN: Do you doubt that the blue whales care about their young?

DR. BOULDING: They aren't conscious of their being an endangered species, I'm quite sure of that. This is why the human being is so fantastically unique. All of these attempts to derive human behavior from biology, I say, is like determining all about a jet plane by studying the wheelbarrow.

DR. GUSTAFSON: I think that you're responding to some stereotypes.

DR. BOULDING: I haven't finished. The other thing is you haven't taken into account the discounting. You have to take account of it. Discounting is absolutely necessary. The thing that is remote and uncertain just doesn't have the same potency for us, and it should not have.

DR. BOYER: Two responses. The chairman of the conference.

DR. NEEL: I'm sorry, but what I think we are trying to say is that maybe the human being is not so fantastically unique as the view of the last several centuries would have it.

DR. BOULDING: Well, I'm saying it is. I'm saying that with Adam and Eve, evolution went into a totally new deal, and it is just as impossible to be a biological reductionist as it is to be a chemical reductionist.

DR. TOULMIN: You should leave Colorado. It must be getting to you.

DR. BECKERMAN: Well, the question was about the exponential growth of population and the problems that arise when exponential growth and demands on resources exceed the finite supply, and my answer is briefly threefold.

First of all, population will not continue to grow at the same exponential rate. I don't think there are many organisms in human history that have continued the same exponential rate of growth.

Secondly, resources are either finite or they are not. If they are not finite, well, you don't have to worry about them. If they are finite, there is nothing we can do about them, even if we kept on growing at exactly the same level of consumption and standard of living that we have today, we would still use up finite resources some day.

Thirdly, resources are not finite in any sense that makes any sense. If you want further information on this subject, I can give you the references.

DR. BOYER: Mr. Speaker.

DR. GUSTAFSON: Well, I guess I'm most inclined to respond to

Kenneth Boulding, because I really felt that he was responding to a caricature of what I said in my paper. I am as irritated as he is by the sort of romanticism about these sorts of things.

With reference to the discounting and so forth, I don't think life is a zero-sum game. There are some things that grow, but I am firmly convinced that our choices are tragic choices in many instances. To optimally satisfy the interests of certain groups is costly to other groups. These decisions and choices are very hard.

John Passmore makes it very clear in his book, *A Man's Responsibility for Nature*, that there is no ideal order of values—I don't remember his exact words—which are in perfect harmony with each other. It seems to me there is a sense in which our moral choices are exceedingly difficult, in part because of the recognition of the cost to others for those things which we, even justifiably, pursued in our own interest as a group. I would put that, without being romantic about it, even in terms of the animals. There is a marvelous way in which you can say that every time we eat this good roast beef we get fed around here, something has been sacrificed for our interest. I'm not saying that that animal has the value of a human, but nonetheless there are costs.

I should like to make one final comment with reference to Roy Branson's extrapolation from the religious vision. My perception of a deity, which may not interest you at all, is not a deity who guarantees a human good, but that the necessary conditions for all of those things which we can appreciate and fulfill are given to us by those forces beyond our creation. There is, however, no guarantee for the human good in terms of our perceptions of it or even a species good over the long run. Now, that makes me the ultimate pessimist perhaps and rejects part of what Roy Branson reminded us is in the biblical tradition. I don't really worry about that, but it doesn't mean that I'm a pessimist about the human capacities to participate in getting from here to there and from this decade to the next decade.

DR. BOYER: Without being irreverent, I thought about twenty minutes ago of that overly told story of Robert Benchley, who was a student at Harvard and hadn't studied and was confronted with a final examination question in which he was asked to discuss the conflict over offshore fishing rights between the United States and Great Britain. He wrote, "I know nothing about this conflict from the standpoint of the United States, and I know even less from the standpoint of Great Britain. Therefore, I should like to discuss it from the viewpoint of the fish."

Dr. Gustafson's topic genuinely raised, I think, powerful and provocative questions of man in context, and we were asked to consider a paper which proposed the ethics of interdependence. This was heatedly and elegantly discussed and challenged, and I think we have been richly served.

DR. MIDGLEY: Could I have a two-second remark? It is just about what you just said; it was funny and we all laughed. Now, in the eighteenth century when they talked about the rights of man and when anyone mentioned the rights of women, they all laughed—the same sort of sense of something terribly ridiculous. I just want to mention that because it may happen again.

18. Toward a New Understanding of Human Nature: The Limits of Individualism

MARY MIDGLEY

Senior Lecturer in Philosophy, University of Newcastle upon Tyne

Editor's Summary. Mary Midgley warns against excessive individualism while recognizing the obvious need for individual initiative and creativity. She states that the theory of evolution does not imply that people are and ought to be selfish nor does it provide the basis for justifying cut-throat competition. Part of the contemporary problem is that human society now lacks the natural, obvious concepts which have limited human excesses in the past. We are, in a sense, between ethics: the old is unsatisfactory, the new is not yet formulated. Midgley stresses that we are human because we are part of our total environment, thus crass exploitation is self-defeating. The sociobiologists have alerted us to the importance of kinship and Midgley notes that this importance has the potential at least for an ethics which limits our individualism. There will always be tension between serving self and others, but group obligations must be perceived as crucial to both group and individual survival and full social development.

On Living in the World. Mankind cannot bear very much reality. Confronting here the father and mother of all problems, we may deceive ourselves in two ways. We may try to pretend that the problem is smaller than it is, that things have not yet gone badly wrong. This will not get us far. More often we take refuge in fatalism, in concluding that there is nothing that we can do. Obviously, however, there is one part of the problem which is more under our control than the rest. This is our own way of thinking, the general ethos of the West. That does seem to have certain things dangerously wrong with it. Many people, therefore, call for a new ethic. What ought we to do about this?

As John Passmore has very properly pointed out, new ethics cannot be bought like new hats (Passmore 1974:54-6, 110-26). Ethics, being patterns of incredible complexity, have to be evolved, not manufactured, and they always build on what went before. Old ethics do not go in the dustbin. The range of moral insights possible to the human race probably doesn't change much, and all of them go on being needed on occasion. What does change drastically is the emphasis. Quite a small change in emphasis can make an enormous difference to life. In every age, morality has a bias. It is obvious to those who come after, but history shows us how hard even the most astute people find it to detect where the bias of their own age lies. As the Devil points out in the *Screwtape Letters*, the results can be highly ludicrous (Lewis 1942:129):

The game is to have them all running about with fire extinguishers when there is a flood, and all crowding to that side of the boat which is already gunwale under. Thus (says the Devil) we make it fashionable to expose the dangers of enthusiasm at the very moment when they are all becoming worldly and lukewarm. . . . Cruel ages are put on their guard against Sentimentality, feckless and idle ones against Respectability, lecherous ones against Puritanism, and whenever all men are really hastening to be slaves or tyrants, we make Liberalism the prime bogey.

What bias, then, is now misleading us? I shall suggest that it is an unbridled, exaggerated individualism, taken for granted as much by the Left as by the Right—an unrealistic acceptance of competitiveness as central to human nature. People not only *are* selfish and greedy, they hold psychological and philosophical theories which tell them they *ought* to be selfish and greedy. And the defects of those theories have not been fully noticed.

It has been said that Social Darwinism is the unofficial religion of the West, and I think this is true. The official Western religion, Christianity, is well known to be rather demanding and to have its eye on the next world rather than this one. In such situations, other doctrines step in to fill the gap. People want a religion for this world as well. They find it in the worship of individual success. The fast buck itself is not exactly an end, since nobody can eat paper, but it is valued primarily both as a means to this success and as a sign of it.

Contemporary greed is sometimes called *materialism*. This does not seem to be quite right, since people's material needs are limited, and could all be met at quite modest expense. Most of the goods greedily wanted on top of material security are wanted as assurance of personal status. It is widely believed that the theory of evolution proves this kind of motivation to be fundamental and in some sense the law of life. Mystical reverence for such deities as progress, nature, and the life-force is invoked to explain and justify cutthroat competition.

My main business in this paper will be simply to point out that the theory of evolution gives no ground for this kind of fantasy at all. Such a view of the natural motivation of our species is simply a mistake and finds no support in biology. I shall not attack Social Darwinism in the usual way, namely, for reasoning illicitly from facts to values. That attack would concede its view of the facts as correct, which it is not. I shall point out instead how Social Darwinism is wrong on its facts, and consider how to get these very important facts right.

Not all facts are irrelevant to values, only some. This particular set of facts about natural motivation seems relevant enough. *If there were* a social species so extraordinary as to be by its nature entirely egoistic, it would have little choice but to live egoistically, in unmitigated competition, conducting a war of all against all, or controlling it only by bargains made for safety. But it is hard to see how such a creature could ever have become social or capable of reasoning, which requires attention to the views of others. Certainly it could never have entertained—as we are now doing—any criticism of egoism. There is overwhelming evidence that ours is not such a species.

The Bias of Social Contract Ethics. Why, in that case, should anybody ever have thought that it was? Here we reach the sore place, the nub of the problem. Individualism is tied up with much that we rightly value very highly. Ever since the Renaissance, it has been a key project of our culture to free individuals from the pressure of their social background and to enable them to stand alone. Endless devoted efforts have been made to pry each loose from his family, his state, his church, and any other shell to which he might cling, and allow him—indeed force him—to think and act for himself. (Himself, but not always herself—a feature which for some time made the project look more promising than it actually was. Altruism was still expected of women.)

Liberal political theory, from Hobbes onward, called on each citizen to view himself as primarily a distinct, autonomous atom, unlinked to his fellow atoms unless he contracts to join them. The sources of this project, indeed, are far older; it started with the Greeks. Their efforts to encourage individual thought and responsibility gave rise to all that is most distinctive in our civilization. Christianity, with its emphasis on the separate, irreplaceable value of each human soul, also played a key part in the drama, which came into full flower in the Enlightenment. To many people it never looked, until lately, as if we could have too much of that good thing: Individualism.

What has happened now, however, is that we seem to be left with little conceptual ground to stand on when we want to make the opposite kind of point and declare that the world is, after all, in some ways actually one, and that human beings exist only as parts of it. Why, for instance, should a Brazilian farmer *not* cut down the rain forest to raise beef for a few years, moving on when the soil is exhausted and abandoning it to become a desert? Is he not a free agent? Contract-based prudential objections can certainly be found. We can say that the forest is a reservoir of species which may prove invaluable for medicine and other technologies, that the Indians (if provoked too far) will attempt reprisals, and that the effect of forest clearance will probably endanger everyone in the long run, since it would impoverish the whole of Brazil. To this, however, he may simply and consistently reply that, whatever happens, rich Brazilians will always be safer than poor Brazilians, and the point of his present schemes is to make him very rich.

Now, it may be thought that this man has failed to grasp the full meaning of the social contract. And it is true that contractual ethics usually is not expressed in this crude, predatory, ding-dong form. As philosophers display it, it can incorporate all sorts of impersonal, rational safeguards, hypothetical role-exchanges, veils of ignorance, and similar public-spirited devices. But if the atomic individuals who do the original contracting really conceive of themselves as essentially separate, it is hard to see why they should bother about these procedures. The point is not just that, since they are wicked, they are unlikely to consider others, but that, since they are separate, they can have no reason to do so. Our Brazilian might happen to fancy such thoughts, but there is nothing in his essential situation to make them relevant.

Of course, there are further appeals to prudence which can be made. Calculations based on the hope of becoming immensely rich are rather unreliable. A more realistic, enlightened, Hobbesian self-interest might tell him to be less destructive, merely as an insurance policy. But narrowly selfish people tend not to be very imaginative, and often fail to look far ahead. That is one thing which the Industrial Revolution so far has made clear to us, and there is good reason for it. Exclusive self-interest tends by its very nature *not* to be enlightened, because the imagination which has shrunk so far as to exclude consideration for one's neighbors also becomes weakened in its power to foresee

MARY MIDGLEY

future changes. Vice has its martyrs, as well as virtue. A great many aspiring egoists have crashed in attempting the feat which Howard Hughes, in his way, brought off, and his success—such as it was—no doubt involved luck as well as cleverness.

The psychology of this is very important. Hobbes really was mistaken in supposing that people could defer satisfaction indefinitely from prudence, that they were sufficiently patient, dispassionate, timid, and far-sighted to build a harmonious world purely on bargains for self-interest. When other, more direct social motives are weakened, as they are today, human prudence alone turns out quite unequal to the job. On every side now we can see people busily engaged in sawing off the branches on which they (along with many others) are sitting, intent only on getting those branches to market before the price of timber falls. Prudence does not prevent this destruction. It is therefore clear that what did to some extent prevent it in the past was a set of motives quite distinct from prudence and owing nothing to contract. They are, of course, the motives which until recently inhibited the free development of technology. They are motives like conventionality, identification with one's group, the fear of *hubris*, of novelty and excess, loyalty, respect for one's elders, and a general awe at the mysterious otherness of nature.

It would seem a good idea that we should now overhaul this mixed bag of motives carefully, examine them, and sort out what is useful in them from what is not, reexpressing the useful part in terms suitable to our own day. There is plenty of material for this. Anthropologists, showing us how the cumulative nature of culture demands continuity and how the murder of a culture can kill its members, have given us reason to have a far better opinion of conventionality, loyalty, identification with the group, and respect for the elders than we used to. Ecologists have pointed out that on the physical side there was good reason for the fear of *hubris*, and for awe at the mysterious otherness of nature. They make clear that we have only the most superficial understanding of the vast physical systems on which we depend, so that awe—as well as caution in change—is entirely rational. Social psychologists have drawn attention to the complex dependence of human individuals on their background. Ethologists have shown from animal parallels how deep the function of this is likely to be. In psychoanalysis, transactional thinking has broken the grip of Freudian egoism and made it possible to acknowledge human otherness.

In general, whatever reservations anyone may have about particular parts of this development, it must emerge that a whole set of communal aspects of life, which used to be despised and attributed to the corrupting influence of religion, now appear as necessary and understandable in terms of the sciences. They are not just instruments of political oppression but essential conditions of life.

The Dark Side of the Enlightenment. There remains, however, a general difficulty, inhibiting us as Western intellectuals from even considering these motives. Our superegos are very unwilling to allow it. Internalized in each of us is a voice which speaks with the accents of Voltaire and Rousseau, of Mill, Hume, Tom Paine, and Mary Wollstonecraft: a voice which says, "Was it for this that we defied the priests, the fathers, and the kings? Can anything be more important than individual liberty?"

Unluckily, this voice comes now like a fire extinguisher in a flood. It can distract us, but not help us. Of course, there are still tyrants. But what chiefly confronts us today is not an Easter Island row of ossified traditional patriarchs, but a chaotic mob of dollar-snatching cormorants, doing damage of an order undreamed of in previous ages. Even in private life, I am inclined to think that sheer confusion of conflicting claims now makes at least as much misery—if not more— as the confident appeal to traditional authority. But the public issue is the one I must concentrate on. Observing the cormorants, we are in no doubt that we ought to disapprove of them, but our disapproval is forced to be indirect. Our tradition now lacks the natural, obvious concepts by which most of the human race would denounce unbridled human predation. We cannot say—as almost any other culture could— that these people are betraying their ancestors, offending the ghosts, that they are sacrilegious outcasts and matricides, destroying the land which gave them birth. Very likely we are right to throw away this language, but we should not throw away with it the power of expressing certain evident and crucial truths.

Of course, a human being is a distinct individual. But he is also a tiny, integral part of this planet—framed by it, owing everything to it, and adapted to a certain place among its creatures. Each can indeed change his life, but he does not originally invent it. Each receives life in his family (as a petal does in a flower), in his own land (as the flower does on the tree), and in the biosphere (as the tree does in the forest). His environment gives him nearly everything he has, and if, even as an adult, he were deposited with all alien modern conveniences on a planet of Sirius, he would (with all due respect to Carl Sagan) be no more than a shriveled petal.

All this is no derogation of his essential dignity, because dignity is meaningless without a context. The only person who might conceivably exist and make sense on his own is God, and even He apparently

MARY MIDGLEY

prefers not to try it, since He creates the world. And whatever might be true of God, man is no god, but a social being and a part of the fauna of this planet. When the architects of our present ethics were campaigning for individual liberty, this did not need saying. It could safely be taken for granted. Today, with the damage which unrealistic individualism is doing both to the physical life of the planet and to the personal happiness of individuals, it does need saying.

We do not exactly need new concepts, since suitable ones do exist in our culture. (John Passmore is right to insist that such concepts should always be looked for and used when possible, because the idea of an entirely unprecedented moral insight occurring for the first time in this epoch is a fishy one, and people are quite properly suspicious of it.) The concepts exist. But both they and our current moral ideas need adapting in order to show their reality and importance. They have to be seen as taking a major, rather than a minor, part in practical argument. Moral changes are perhaps, above all, changes in what kind of thing people are ashamed of. Till lately, our age was accustomed to classing environmental considerations as marginal, and so treating any emphasis on them as sentimental, emotional, unrealistic and—above all—insufficiently virile. (Some influential libertarians, notably Rousseau and Nietzsche, have been obsessed with virility, a notion which still confuses radical thought.)

The excesses of commercial free enterprise may be repulsive to us, but we are still committed to seeing it as in some way proper and admirable, because it is still a form of freedom. It appears as a monstrous parody of our most sacred ideals. If there is really nothing more important than freedom, must that parody at some level be embraced and accepted? Is there no way to be sure that the destroyer is not an admirable Nietzschean superman?

Distinguishing "Social Darwinism" from the Real Thing. When things look as bad as this, it is usually best to ask: need we have started from here? Having begun to consult the eighteenth-century oracle, perhaps it will help to carry the process further. What has the Enlightenment really got to tell us? It is pleasant to imagine the expression which would rise on the faces of Voltaire, Rousseau, and Kant, if (having recalled them, much against their principles, from the tomb) we explain to them that we find ourselves—two centuries after their deaths—so imprisoned conceptually by their discoveries that we are unable to tackle the problem of adapting them to a new emergency and a different age. They themselves were bold innovators. Insofar as anybody has ever produced a new ethic, they did. They knew that, when the state of the world changes, new ideas must be used. What the ghost of Voltaire requires is that we should be willing to twist the tails of *all* sacred cows, including those from his own herd. The ghost of Rousseau tells us, first and foremost, to understand our own nature and its place in the nature of the universe. As for Kant, he advises us above all to think independently and freely. None of them provides any ready-made conclusions, nor guarantees that the emphasis we need will be the same that was called for in their day. All changes of emphasis in morality are correctives, answering temporary needs. No such change can be final.

Right through the seventeenth and eighteenth centuries, it was reasonable for inquiring people to see their main enemy as Feudalism, the theocratic and monarchical hierarchy in which individuals were paralyzingly embedded. The Industrial Revolution, releasing them to become socially mobile, naturally appeared on the whole as a liberating force. This background explains why Darwin's views, when they appeared, were put to such extraordinary use. The existing intellectual furniture produced a powerful optical illusion, making the doctrine of the survival of the fittest look like the precept "each for himself and the devil take the hindmost." Evolution seemed to endorse egoism and, thereby, unbridled capitalism.

Despite protests from both scientists and philosophers, people still find this interpretation almost irresistible. It accounts for two rather serious confusions today. The first and cruder one is the recent revival of Creationism among educated people and even among some scientists. The project of treating the time scale of the Genesis story literally, as a piece of history, is an amazing one, which serious biblical scholars at least as far back as Origen (A.D. 200) have seen to be unworkable and unnecessary. The reason why people turn to it now seems to be that the only obvious alternative story—evolution—has become linked with a view of human psychology which they rightly think both false and immoral.

The second and more subtle confusion concerns "sociobiology" and the response to it. Liberal and radical people have found this movement very alarming. Many have seen it as no more than a revival of Social Darwinism—another monstrous head on the hydra which they are tired of killing. The essential doctrines of sociobiology do not in fact call for this hostility at all; they are new, inoffensive, and valuable. In fact, they provide the *answer* to Social Darwinism. What does call for alarm is the language, and the unconscious approach to the subject which it betrays.

Sociobiologists often use words like *selfish, altruistic, investment, strategy*, and the like in such a way that, if those words had their everyday meaning, they really would be saying that each must always

MARY MIDGLEY

be only for himself and the devil must take the hindmost. They explain, however, from time to time, that they do not actually mean this. Officially, they are not using the words as the names of motives at all, but, in a quite technical way, purely to describe results. Thus a "selfish" act means one that actually does benefit oneself and so forth. Now the trouble with this is that it is simply not possible to use words of strong everyday import in a private and peculiar sense, when their ordinary sense yields so clear and familiar a meaning, and expect that ordinary sense not to prevail. (This is not, of course, an objection to using everyday language, but an objection to using it carelessly.) And the trick is doubly impossible where that familiar meaning is actually a contentious proposition, already being affirmed and denied in an important controversy.

The whole nature of language prevents such a detachment of meaning from habitual background. Anyone using words like this must mislead his audience. What is still more serious, unless he is exceptionally sharp and sophisticated, he will certainly mislead himself, slipping into the everyday opinions which he supposes are no part of his scientific business. It is clear that this has happened in sociobiology, both from the sporadic—but quite frequent—remarks with a familiar, shocking political sense, which have upset the critics, and also from many unnecessary turns of speech, importing conscious selfish motivation into contexts where it could not possibly have any business, shocking or otherwise. Thus, Edward O. Wilson, discussing what ought to be a plain technical controversy about units of evolutionary selection, suddenly quotes a Finnish athlete who said that he ran for himself rather than for Finland, and calls him "the ultimate individual selectionist" (Wilson 1975:106).

The key to all these surface oddities is the peculiar difficulty which sociobiologists feel about really believing in altruism (or indeed in any sort of motivation not conceived on the model of hunger), even though they describe many other patterns and argue for their existence. Wilson speaks of altruism as "the central problem of sociobiology," asking how such a tendency which "by definition reduces personal fitness, can possibly evolve by natural selection" (Wilson 1975:3). In a couple of sentences he answers this question, but never seems to be convinced by his own solution. The answer is, as he says, kinship. Creatures which act against their own interests in a way which preserves enough of their relatives cause their line to survive. Such lines, therefore, do perfectly well in natural selection.

This answer—first proposed by Darwin and worked out by J. B. S. Haldane— needs to have its details filled in but is, in principle, entirely adequate. It shows plainly that there is no reason at all why creatures need be unmitigatedly selfish and deliberately competitive in order to succeed in evolution. And this ought to be the end of Social Darwinism. But throughout Wilson's book—and others like it—altruism goes on being treated as a raw, worrying, unsolved problem. Its occurrence is admitted, even insisted on, but is always treated as a rather embarrassing and mysterious paradox. Devices are constantly sought to keep it at a minimum and to find hidden egoistic motives, as the real determinants.

Sociobiologists cannot, at a deep level, accept nonselfish motives, because they cannot fit them into the purely commercial pattern of motivation which, at the back of their minds, they accept as the only possible one. They assume, before starting their inquiry, that nobody ever does anything except for gain. Where it has to be admitted that acts don't in any normal sense pay the individual who performs them, some formula has to be found to show that, at a deeper level, they do pay. The first line of defense is to say that they do this by increasing his "inclusive fitness," which is a highly technical property consisting in the prospect that he will have many surviving relatives at some date after his death. This is plainly not an advantage to *him* at all, but it enables the tenacious Social Darwinist imagination to hang on by the skin of its teeth to the notion that he did, after all, act in his own interest, so that each is still really only for himself. Alternatively, as a second line of defense, another entity is found to have the motive, the entity whom the act really does pay directly. And this is the function of the Selfish Gene—a personification ripe for Occam's Razor if ever there was one.

I do not want to say more here about the confused myth-building which at this point has deformed sociobiology. I have dealt with it elsewhere (Midgley 1978:89-103; 1979). Our present business is with the very important truths from which it has distracted attention. I mention "selfish-genery" only in order to lead those put off by it to understand its cultural roots and to see that we all share them. Richard Dawkins did not invent Social Darwinism, nor did the Social Darwinists invent egoism and the social contract. Our culture has for some time been highly individualistic, not just in its commercial systems, but also in some of its most ambitious moral and philosophical beliefs. (Nietzsche has been very influential here.) Since life is many-sided, any such one-sided development is bound at some point to make trouble.

I am not, of course, suggesting that the work of individualism is finished, that tyranny, conformism, and lazy thinking have vanished from the world. Unfortunately, cases of unbalance do not work in this way. It is not necessary for one set of evils to vanish in order for the opposite set to appear. It is quite possible to be at the same time

MARY MIDGLEY

disorderly and tyrannical, conventional and destructive. In fact, people find these feats quite easy.

Facts and Values. Now it may seem that I am bringing two quite different kinds of individualism together here and that, even if liberal thought is making a mistake, it is an entirely different mistake from any made by Social Darwinism. All Darwinist inquiry, it may be said, is empirical, aimed at establishing facts about human nature, and its mistakes are therefore empirical mistakes. Social contract thinking, by contrast, may seem to be purely philosophical (prescribing ways in which we *ought* to think if we want a just world) and explicitly refusing to use any concept of human nature at all.

I answer that these two aspects of the problem are indeed distinct and must not be confused, but they cannot be unrelated. The philosophical inquiry makes no sense alone (see Midgley 1978: chap. 9; 1981: chap. 1). The assumption that it needs no reference to human nature flows only from taking for granted a convenient view of that subject. One cannot sensibly ask how people ought to be related to one another without making some assumptions about the kind of beings that they are. And even the idea that nothing can matter to them *except* their relation to each other—that their relation to animals and plants and the rest of the physical universe is morally neutral—is a factual assumption about their nature and a very surprising one.

When words like *right*, *justice*, and *duties* are defined in such a way that they cannot in principle apply to any claims except those of another human being, the point made is either a trivial verbal one or a startling moral one, which could only be explained and defended by an appropriate supporting theory of human nature, one which showed how everything else really did not concern us. Philosophers already admit that definitions of such words which—like some in Hobbes and Freud—reduce all claims to egoist terms depend on a peculiar and empirically false view of human psychology.

A similar dependence is present in theories like Kant's—that we cannot have direct duties to animals. Words like *duty*, *right*, and *justice* have in ordinary speech quite a wide meaning. Accordingly, phrases like "no rights" or "no duties" do not sound like mere verbal corrections, but seem to convey general license and absolution, to mean "it doesn't matter how you treat these beings." This is not self-evident at all. Indeed such views are usually not meant to settle the moral question out of hand. Philosophers often add that they are simply clearing it out of the way because it does not fall under a convenient heading for their present business. (See Kant, 1781, p. 239, on Duty; Hume, 1751, sect. 152, and Rawls, 1971, p. 512, on Justice;

and McCloskey, 1965, on Rights. All add that we are *not* absolved from real responsibility on these issues.) But the way in which the use of such terms has become narrowed in the social contract tradition makes it natural to feel that rationality commits us morally to detachment and unconcern about everything nonhuman. A moment's thought will show that this cannot be right. The mere fact that a previous age did *not* interest itself in something cannot show that that thing does not matter. And since people always narrow their terminology to suit their interests, the withdrawal of terms like *duty* from this area cannot possibly bind us not to expand them again.

First, people today do not commonly accept Kant's intellectualist view of human beings as creatures whose essence is their reason, related only contingently to their bodies and feelings. That view is what produced the denial of duties to nonrational beings. And second, no previous age was confronted by the current situation where (as President Reagan's official scientific advisers informed him on taking office) one-fifth of the species on the planet may well be extinct by the end of the century (Joyce 1981:197).

When we ask whether such things can concern us, we are asking whether we are the kind of beings whom such things concern. This is the kind of question which Kant asked and answered about departed spirits. He said that these were no concern of ours, that we had no business with them, because nothing that we did could have any effect on them. That is not an argument which can be used about the environment. Other arguments might be brought forward, but they could not just be ones drawn from the fact that a certain moral tradition has been systematically dumb on the subject. They would have to concern the nature of human motivation, and would therefore be at least partly factual. The way in which motives work is a matter to which empirical evidence is relevant. The reasons why social contract ethics took a restricted view of motives was not theoretical but political. It needed to assert and emphasize the importance of individual human interests against governments and family systems which were treating them as expendable and religions which failed to do them justice. There is nothing in this to show that these individual interests are the only values which we shall ever have to consider, or that they will even make sense if deprived of their proper context.

The Plurality of Values. It should be noticed that I take for granted here an irreducible plurality of values. Human life, I flatly remark, does not have only one good thing in it. Besides individual liberty there are—to take a hasty random sample—such things as affection,

MARY MIDGLEY

generosity, laughter, celebration, homecoming, shared work and play, old people and children, cats and cucumber plants, frogs and fresh water, skies, seas, and mountains. Though we must often choose between these many good things, there is no need at all to set up a general competition between them, with the assumption that one will turn out fundamental and make all the rest unnecessary. In particular, liberty could never take this place, since it is essentially the negation of a negation, the removal of obstacles to our doing the things we want to do. Someone who no longer wanted anything but liberty would not be in a position to make a free choice.

Psychological theorizing has had a remarkable tendency to distrust this plurality, to look always behind it for a single aim such as pleasure, power, self-preservation. It does this partly because it is called in to search for a mechanism to resolve conflicts, partly also from a notion that simplicity is always more scientific. For many purposes the reductions are indeed both useful and instructive. But it is quite wrong to go beyond these limited uses and suppose that the more general aim must always be the real one. We are so framed as to want and need many things which are highly specific. You cannot turn a musician into a painter, or make a mother accept another child instead of her own, by convincing them that what they really wanted was an abstraction which may as readily be found elsewhere. And these strong, specific feelings do not come just from culture, since they sometimes develop in direct opposition to the local culture, and very often in forms which that culture sharply rejects. The attempt to reduce plurality of motivation to abstract unity shrinks the essential self to a wizened old nut, a bare intellectual center of choice, unattached to particular people and things, and equally capable—if its one abstract need is met—of living anywhere. This is the sort of thinking which produces ideas like that of transplanting the human race to a different planet. We may not make much sense here, but we would make none anywhere else.

Reduction of our many motives to one simply does not fit the phenomena. Concrete human aims are often bad, mistaken, and conflicting, but they are not vacuous and infinitely interchangeable. Neither egoism nor the pleasure principle nor any other single explanation can swallow the lot. Even selfish people commonly have particular affections, things out in the world which they mind about. Those who lose all interest in these tend to become incapable of living at all. Howard Hughes died alone. Certainly there are ascetic doctrines which advise us to "give up the world" for certain supreme ideals. But these do so after admitting the full range of values which the world contains, and showing reason why certain conflicts demand this solution. They do not rule out values leading outside the self as unreal from the start.

It is not possible to find order in the plurality of motives by reduction. In resolving conflicts, we do indeed look for underlying principles, and detect general needs which can have varying concrete expressions. We find a structure which will guide us and, where it does not guide us, we assume that we have not fully understood it. We accept the picture of a rough balance of needs, arbitrated with difficulty, showing some underlying shape, but never enough to resolve all conflicts, an ordered, but never fully organized plurality, of motives and of values. That, as we normally suppose, is both the problem of life and the only solution it is ever likely to get (Midgley 1979: chap. 11, 13; 1981: chap. 1, 6, 9).

If we now turn from ordinary life to the theory of evolution as it actually is (and not as Social Darwinism misrepresented it), we find that this is exactly the kind of constitution which natural selection can be expected to produce in a social mammal. Once a species becomes social at all, its continued prosperity does *not* depend only on traits of behavior likely to produce an individual's own survival, but also, and quite as much, on those favoring the survival of kin and group. I cannot put the point better than Edward O. Wilson does, in one of the many passages where he speaks as Jekyll and not as Hyde. In such creatures, he says: "Love joins hate; aggression, fear; expansiveness, withdrawal; and so on, in blends designed not to promote the happiness of the individual, but to favour the maximum transmission of the controlling genes" (Wilson 1975:4). (The word "designed" should not put off the sensitive reader here. As I have explained, it is not intended literally, as if the genes were engineers, but means merely "adapted" or even "likely.")

The ambivalences stem from counteracting pressures on the units of natural selection . . . what is good for the individual can be destructive for the family; what preserves the family can be harsh on both the individual and the tribe to which its family belongs; what promotes the tribe can weaken the family and destroy the individual, and so on upward through the permutations of levels of organization (Wilson 1975:4). [It follows that] the theory of group selection . . . predicts ambivalence as a way of life in social creatures (Wilson 1975:129; cf. 563).

This diagnosis—which is the real message both of sociobiology and of ethology—has nothing to do with the crude Social Darwinist view: the bleak, superficial conflict between serving self and serving others, which the self is somehow bound to win. It shows why we must expect

MARY MIDGLEY

the exact contrary, namely, a most complex emotional constitution open to very varied and subtle social claims, fitting each individual to bear his part in a tremendous orchestra which existed before he was born and over which he can never have more than a limited control. It shows both why this adaptation is inevitable and—what is really new—why it must always be incomplete, why we are not perfectly shaped by nature for our place in society.

The clash between the interests of the individual and those of his various groups is real. What natural selection produces is not a perfect machine for a single social function, but a workable compromise between many possible ones. Conflicts are inevitable. It is important that other animals have them too. They too are not simple machines perfectly adapted for a harmonious social life. Systematic study of animal behavior has now made clear that they too endure conflicts and frustrations, suffer from them, and attempt various solutions. Where *Homo sapiens* is peculiar is not in his basic problems, but in his use of the intellect to solve them. An intelligent creature perfectly adapted to its society would not have free will and morality as we do, because it would not need them. (The proper revision of Genesis, as regards human beings, seems to be that the Lord wanted free servants, and therefore so devised evolution that it was bound to produce them. This is, of course, too anthropocentric to be the whole story, but as far as it goes, it makes sense.)

We do right, then, to use our intellect and indeed to respect it. But the intellect itself does not produce premises. For practical thinking, these must be drawn from our emotional constitution, from our possible wishes, desires, fears, and aims. If it were true that each of us was incapable of caring for anyone but himself, we would not be able to recognize a duty to others. But it is not true.

In the same way, if it were true that each of us was naturally indifferent to everything outside human life, we could perhaps have no duty to the rest of the biosphere other than a prudential one for the sake of our fellow humans. We would then regard it at all times simply as a scratching post, a device for giving human beings satisfactory experiences. But we do not so regard it. Is it even true (as Passmore suggests) that we can accept that we have no need to worry about what happens to it after (say) the death of our grandchildren or whatever human beings are the limit of our human concern (Passmore 1974:91)? This suggestion is a consistent expression of recent, exclusively humanistic thinking. I find it very interesting because it does not convince me, and it seems a fair test for such thinking as a whole. People's love for their native land (and for other lands), their admiration for other species, their response to seas and mountains simply does not take that form. It is direct. It does not see the nonhuman simply as a means to human ends, but values it precisely because it is not such a means. In spite of Kant, we take these things to be ends in themselves. This only sounds strange against a certain limited and somewhat arbitrary conception of how love and respect work.

Suppose we could be sure that our present tendencies to wreck the biosphere would not take effect until after the death of certain chosen human beings to whom we limit our concern, would that make us feel fully justified in accepting those tendencies? It seems to me that most of us would find scarcely any comfort in this assurance and would still view those tendencies with guilt and horror. That guilt and horror are certainly not trivial, artificial products of our culture: first, because people in almost any culture would share them, and, second, because our own culture has for the last two centuries done everything it could to get rid of such feelings. They are real scruples, as genuine as any that we can feel, appropriate to the occasion and fundamental to human experience. The relation of a part to the vast whole in which it arises can never be morally insignificant.

I want to end this discussion with a note of salute and thanks for Passmore's book. It is one of the best, most serious explorations that we have of this fearfully difficult topic. I have found it endlessly helpful. This paper arises out of an attempt to throw light on one of his most worrying problems: the relation between what he calls *conserving* the environment for human use and *preserving* it for its own sake. Passmore sees plenty of arguments for the first, but is much less confident about the second.

In practice, this doubt might possibly not make much difference. Passmore's very generous conception of *human use* and enlightened, long view of human interests might so work that nearly everything which we have any hope of saving might appear on both lists anyway. But it does make a difference in the way we conceive the enterprise. Where pains are taken to make people ashamed of their natural and direct motives, their indirect ones tend to become confused and unreliable. Prudence, I have suggested, cannot be effectively extended to form an all-purpose policy. What prudence requires is commonly done from motives other than prudence. Those motives must therefore be taken seriously. (This is why utilitarianism, especially in its current highly abstract and exclusively humanistic form, is often unhelpful and misleading.) And a prudence which extends to the whole human race has already abandoned the notion of people as essentially discrete social atoms. Had they been such atoms, it would indeed have been impossible to find reason why they should consider any other being for its own sake, whether inside or outside their own species. But they are not such atoms, and that problem does not arise.

MARY MIDGLEY

References

Hume, David 1751. *An Enquiry Concerning the Principles of Morals.* LaSalle, Ill.: Open Court Pub. Co. (1947).

Joyce, Christopher 1981. "Reagan Warned about Deserts and Extinct Species." *New Scientist* 89 (Jan. 22, 1981).

Kant, Immanuel 1781. Louis Infield, trans., *Lectures on Ethics.* London: Methuen, 1930.

Lewis, C. S. 1942. *The Screwtape Letters.* London: Union.

McCloskey, H. J. 1965. "Rights." *Philosophical Quarterly* 15:115-27.

Midgley, M. 1978. *Beast and Man: The Roots of Human Nature.* Ithaca, N.Y.: Cornell University Press.

_____1979. "Gene-juggling." *Philosophy* 54:439-58.

_____1981. *Heart and Mind: The Varieties of Moral Experience.* New York: St. Martin's Press.

Passmore, J. 1974. *Man's Responsibility for Nature.* London: Duckworth.

Rawls, J. 1971. *A Theory of Justice.* Cambridge, Mass.: Harvard University Press.

Wilson, Edward O. 1975. *Sociobiology—The New Synthesis.* Cambridge, Mass.: Harvard University Press.

Commentary by Paula Thompson
Director of Education, Brooklyn Botanic Garden

Hugh Iltis in his essay "Can One Love a Plastic Tree?" recounted the Hans Christian Andersen fairy tale "The Emperor's Nightingale." Those of you who are familiar with this tale know that the nightingale the emperor usually had at his side had flown away, so some of the people around the court concocted a mechanical one and had it move and make some of the sounds that were like a nightingale. Of course, the emperor was not at all satisfied with what came out of this mechanical nightingale and indeed he became very ill, with death at his bedside. However, at the last moment the real nightingale flew back in the window and all was well and the emperor was happy.

Indeed, it is not possible to love a plastic tree. It is an old moral, but an essential one that we understand. Iltis goes on to ask if the love of a living tree is taught by culture or is it a product of genetic evolution. Is it something in the genes? Is this call to love plants and to love things green something deeply within us?

As a very young child I grew up in New York City and we all know that if we have a vision of New York, it is of concrete towers and ventless windows. Now, where I grew up it wasn't quite like that, but there was a kind of restriction on my experiencing the green environment. My father was a road salesman, and when he came home on weekends the last thing he cared to do was to take us out to the country. So those were very unusual times when I got to actually be out in the forest experiencing nature. And yet when I did go out there I felt something very special.

I had been through the mill of getting a degree in what strangely turned out to be indoor botany. There I was for four years, conducting research on the top three centimeters of the seven-day-old wheat leaf tip, and I wonder about the relevancy of that. And then for nine years after that, something akin to the lives of cats, perhaps one a year, I taught in a college environment, and now I find myself at the Brooklyn Botanic Garden in the city where I first felt my kinship with plants.

Martin Kreeger has noted that genuine, unspoiled nature is routinely only offered to the rich, those people who rent planes and fly off to Alaska and get to climb where no one else has been. But our peculiar brand of social justice gives to the poor a tiny city park with concrete pathways and but a subway ride from the tenements. Now what is very special about the Brooklyn Botanic Gardens is that it is fifty acres of greenness in the middle of the chaos and horror of New York City.

Dr. Dubos had something very interesting to tell me about my new home in New York and that is that he looks upon the Brooklyn Botanic Garden as an example of one of these places where people thought globally and acted locally. It once was a city dump, which through the vision of our founder more than seventy years ago, became a garden. What was it that was in the genes of that man that helped him have the vision to create this very special environment that I now have the pleasure to enjoy? This pleasure, of course, doesn't just mean what is green in Brooklyn. It is a very special environment—brown, black, white, and green alike—that I have gotten to be a part of.

Daniel McKinsky has said that it is time for men and women to commit themselves to a contemplative study of nature, however hard that may be for us to begin. As Midgley notes, it's a difficult ground to stand on, to declare that the world is, after all, in some ways actually one and that human beings exist only as a part of it.

Commentary by Marx Wartofsky
Department of Philosophy, Boston University

I was told recently by a philosophical colleague born and raised in India that his favorite reading as a child was Kipling's *Jungle Stories*.

I was tempted to ask how a self-respecting Indian could appreciate the work of such a bloody colonialist as Kipling, but I held back my question because I too grew up loving these stories. Riki-Tiki-Tavi and Shere Khan were more than mongoose and tiger to me. They were animals that exhibited wit, cunning, dignity, and with whom Mowgli was able to communicate. They represented an animal nature humanized, just as Mowgli represented human nature naturalized.

It is no small virtue of Mary Midgley's book *Beast and Man* to have cunningly explored the human bestiary to discover how we express our dependence on animal nature as a resource for our self-understanding, and how we transform and relate to this nature as a changing mirror of this human self-understanding.

In her paper "On the Limits of Individualism," as in her book, Professor Midgley argues in defense of a benign interpretation of the latest and most systematic attempt to biologize human nature—sociobiology. She does this in the context of a discussion of human options of a socioeconomic, political, and environmental sort, and ultimately of a moral sort.

Thus Midgley joins her critique of egoistic individualism with a defense of sociobiology against the charges of Social Darwinism. The gist of her argument, as I understand it, is that egoistic individualism has outlived its viability and now becomes a threat to future survival of human well-being; that it is not a genetically determined dominant human trait; that sociobiology argues in effect against the Social Darwinist view that it is, and provides an alternative genetic desideratum, altruism, as the pattern (if not the motive) of genetic selection.

No one has done a better job of deanthropomorphizing sociobiological lingo than has Mary Midgley. She argues that Wilson's sociobiology has yet to clean up its act, at least its speech act, and cleanse itself of mistaken and misleading metaphorical usages which still preserve the presuppositions of a culturally inherited, selfish individualism, as the old Adam which needs to be overcome.

Finally, the argument proceeds to an environmentalism which projects human valuation of nature beyond the more narrow projections of future concern, which Passmore proposes.

In the brief time allotted for this comment I will focus on two points. First, what I regard as Midgley's overly generous and mistaken interpretation of sociobiology. Here I will argue on Midgley's own grounds that sociobiology is a redundant and empty explanation and that its genetic reductionism of culture (still preserved in Lumsden and Wilson's latest work, which set out to correct the earlier biological reduction of culture) is not simply a matter of redundant or careless use of language or metaphor. Second, I will address the question of an enlightened environmentalism, or what I will call natural piety, in order to extend Midgley's argument and perhaps add my two cents to the discussion.

Egoistic individualism—what C. B. Macpherson has brilliantly described as possessive individualism in political theory—is a historical phenomenon. Midgley sees this and puts it in its place as such. She associates it with the theoretical formulations of Voltaire, Rousseau, and Nietzsche and with the contract-theory tradition of individualism from Hobbes to Rawls. But the rise of this individualism is not simply an ideological or conceptual phenomenon, as Midgley also recognizes. It is the emergent social praxis of nascent urbanism and capitalism, both with respect to the emphasis on individual liberty as freedom from constraint, i.e., as negative liberty, suited to an entrepreneurial bourgeoisie chafing under the communalism of a church- and state-dominated economy; and it is the abstract individualism of the emerging class of wage laborers, each of whom is now free to shift for himself or herself in the free market and to sell his or her *proprium*—the ability to work—in an act of free contract or exchange. So far, *no genes*.

Selfishness is older than capitalism, of course. Moreover, individuality, the celebration of selfhood, of freedom to become oneself as an autonomous, responsible person, is also older than individualism as a human value. And it needs to be distinguished from the egoistic or rapacious individualism of the dog-eat-dog or lion-eat-lamb sort that Midgley decries.

Still, the law of fang and claw, the struggle for survival, as a competition for scarce resources, the relation of predator and prey which underlies Darwinism and its Social Darwinist interpretation, maps this individualism and competitiveness into nature and into the very mechanisms of natural selection with which evolutionary theory has been identified. In this sense, once Darwinism melds with genetics in neo-Darwinism, egoistic individualism seems to find its sources in the biological determination of these traits. They got mapped into the genes. It was earlier observed by Marx in a letter to Engels (and since then has become commonplace in the discussion) that Darwinism had transferred the morality of rapacious capitalism to nature, reading into the jungle the values and practices of the economic jungle. In short, biology was infected with social and historical metaphor from the start. It wasn't that we read the jungle into society as Social Darwinism did, but rather whether the very formulations of Darwinian biology had already used that metaphor, borrowed (as Darwin had borrowed it) from Malthus and others (although to be fair, Darwin had presented alternative models of symbiosis and of other forms of relation as well, and not only that of rapacious competition).

Midgley argues that sociobiology does not make this Social Darwinist mistake, but in fact opposes it with evidence, not of a gene for selfishness but rather an enlightenedly "selfish" gene, whose self-reproducing benefit lies in programming altruistic behavior traits which will serve to preserve it through generations.

What Midgley argues, however, is *not* that either selfishness or altruism is preset as a genetic trait, but that the genetic or sociobiologic argument yields a polyvalent, underdetermined basis of dispositions, needs, biological desiderata, among which the evolved intelligence then chooses, building morality as epigenetic choice and eschewing Wilson's earlier reduction of morality to biology.

But if the genetic program is, in effect, indifferent in determining cultural or moral choice and if it leaves intact the plasticity of human nature in this regard, then the claim for the very existence of some genetic basis for cultural or moral traits must be based upon independent evidence, and can't be read back into nature from either phenotype or from cultural or social practice or from ideology at large. Such evidence must be at a level of biochemical or biostatistical argument, where precisely such cultural traits as selfishness, altruism, or whatnot have no purchase because they are uninterpretable at this level. They are redundant.

But perhaps Midgley wants to argue that the genetic level of species determination provides the constraints—that is, the limits—within which such post-genetic options or choices become available and beyond which human choice is meaningless. This is Midgley's view, I believe. But then it is vacuous for (a) we can't read these constraints off the genes. We have no gene reader which tells us what to project from the DNA-RNA structure. We can only read back from phenotypic variation and from human cultural variation what *may be* a genetic determinant and (b) we can't know in advance what the range of possible cultural and historical choices may be, for this would be to limit human variability to already given or predictable modes of variation; and this, one may show, is the historicist or inductivist fallacy—the one in which generals are always fighting the last war.

But if the sociobiological account is operationally redundant on Midgley's own grounds, and there are no independent grounds upon which the mapping from culture to gene or from gene to culture can take place, then the claim for the relevance of sociobiology is vacuous. We then have to ask why it is made, whether its redundancy is indifferent, as Midgley says, or whether it bears fruit at variance with Midgley's own account of morality as the exercise of intelligent choice among the options of emotional or practical life. (This is just the converse of Hume's view, namely, that "once reason has laid out all the alternatives, 'tis sentiment that decides. . . ." Though I can't argue it here, I believe the sociobiological account, whether of selfish, self-preserving genes or of altruistic genes serving phenotypical behavior, remains Social Darwinist in the requisite sense: society mirrors, with whatever vacuous plasticity may be allowed, the constraints and determinations of genetic structure. Biology remains destiny even if the new fatalism is clothed in biostatistical, stochastic garb instead of in the raiments of an older causal determinism.

On my second point: human recognition of dependence on nature is as old as religion. Worship of the sun, of water, of trees, of the vine, of hilly places, or salt, or blood, or bread, is enshrined in the objectified deities who embody these features or represent them, and in the sacraments which celebrate them. What humans worship is the source of their being. Totemism, animal worship, even animal sacrifice are expressions of a reverence for the beneficence of animal life and awe in the face of its dangers and opportunities. Attitudes of natural piety, extending to cosmic reverence, to the oceanic sense, to feelings of being "at One with the One," are not new environmentalist creeds, but are as old as humankind—or almost.

Feurbach explored the foundations of religion in this sense, and he saw a dual source. First, *Homo homine deus est*: man adores his own species nature as one of infinite possibility beyond individual attainment and man also adores external nature as the source, the eternal source of his being. Spinoza was ready to say *Deus sive natura*—either God or nature—in characterizing substance or self-created being. Nature, then, extends as far into the future, as human beings recognize their dependency on it. But in religion this extends to infinity. Of course, every generation's infinity is conceived in the sensuous or conceptual forms of that generation's most advanced ideas, and innovation in science and art and morality is just that enlargement of a given infinity.

Individualism is the infinite world of the meanly limited ego. It is a highly circumscribed infinity, imbued with the properties of limited historical class interest, and content with the margin of its own death as marking the margins of the world. The biocultural odyssey of humans adapting is an odyssey of historical and conceptual transcendence of such limits, expressed as much in the concrete institutional forms of social, economic, and moral life as in the more abstract representations and projections of the imaginative vision.

Midgley is right about seeing beyond *mere* conservation of nature to a sense of the whole, to the preservation of a nature with which our symbiosis is a necessity. But such an understanding is an understanding of the infinite potentialities of a self-transcending human nature, and how such a nature comes to value the condition of its own existence. Species consciousness—perhaps fragmented by the limitations of class or nationalism or kinship—nevertheless emerges and inscribes itself on a nature and a future which is as large as the species' own conception of its value and its possibilities. If the limited individualist consciousness has had its day, it still impinges upon and restricts human action and human society. It is neither written into our genes nor into our stars and here Midgley is right. What replaces it, however, must be a social individuality, where the enhancement of each is the enhancement of all.

Discussion

DR. BOYER: I would ask Dr. Midgley to comment first.

DR. MIDGLEY: Thank you. I liked the second part of that very much. You supply examples of the concepts which we have because they are traditional, and I think it is very important to say that we have them and I like that, so I don't think I want to say anything more about it.

About my not being savage enough about sociobiology, it is a general principle with me to keep controversy at a minimum and to take what I can get out of it and not say too much about the rest. But I think we really do still have an issue here. I think that the seriously bad thing is the suggestion that sociobiologists are doing the psychology of motive when in fact they profess not to be doing it. It would not perhaps matter if they could keep clear of motives, just talk about cells. But when you use this language of selfishness and so on you appear to have said something quite substantial about what underlies all human motives, namely the one motive of selfishness.

I think it is really necessary to say that that inconsistency ought to be held in their favor. That is, when a chap says on one page, "I don't really mean this," and there on the next page he callously uses the words, I think one should take him at his best because it saves time.

But still, if you take what they certainly do constantly say when they say they don't mean it, this kind of analogy to Freudian theory that instead of sex underlying everything what underlies everything is selfishness, then indeed what they say is objectionable.

You are chiefly worried, it seems to me, about the remaining determinism, which is something far more general. Am I right that even if we take them on their good days as not saying anything about selfish motives, but simply saying that whatever our motives may be—and they may be very complex—our genes are part of the causal system and our emotional constitution was in a way bound to those needs, the root of the offense is still there?

It is difficult to reconcile determinism with free will because causal explanations in the sciences are constantly developed without thought of free will. In a scientific age this is particularly troublesome, but it is just as bad a problem in the social sciences as it is in the biological ones. If the social sciences want to be taken seriously as sciences, they make a great deal of the fact that they can predict, that their predictions are serious—and that seems to be just as threatening.

I don't see that one can do anything about free will by saying, "I want to regard all the causes of human conduct as social and not as biological," and thus get rid of determinism, because determinism is a presumption of all science. I don't suppose we ought to go into the metaphysics of this now. What I would do in a sentence is to say that determinism is never fatalism. That is, the doctrine is never a compelling force. It ought always to be a modest view of knowledge, simply the view that one would be able to predict if one had the data.

As one never does have all of the data for human conduct, this is not very threatening and it does not say, then, that something outside the person is a compelling demon. I think that the language of sociobiology is troublesome here again, that these genes are personified. Again, they say they don't mean it, but they keep saying it. So that certainly it does sound as though I am compelled to be here now doing what I'm doing by a nasty little creature which wished to transmit itself and therefore gave me curiosity, the habit of speech, and so forth. Now that is poppycock, and they say that it is poppycock in their thinking. Fatalism is indeed a terrible thing and one is quite right to complain of it. I don't think fatalism is involved with determinism in general, so I don't think it's involved in biology, but if I did think it was, I can't see how the threat to the social sciences would be any less.

DR. WARTOFSKY: If I can just make a brief answer, I don't think that determinism of that sort is a condition for any science, especially since it is a metaphysical or a methodological question of the sort which you have given, namely, that you are compelled to be here, or of the sort that says that determinism is identical or ought to be identified with predictability. That is a whole other thing.

I think that the upshot, especially of the Lumsden-Wilson book, is that it really doesn't matter what you do. You can do anything you damn please and you have all the freedom you like to do that. In the

MARY MIDGLEY

long run it won't matter because the genes are going to have their way. Now they may have their way by accommodating themselves to some of the funny things you do, so that there is some feedback which is a kind of revival of an alternative version of that, not Lamarckian but sort of post-Lamarckian, namely, that there is phenotypical guidance of genetic mutation; in effect, that it is not simply random selection.

But that is a variation, which is why I argued that it is vacuous, because if on your account and on that account the genes do no more than provide us with a polyvalent set of alternatives among which we are able to choose, and the plasticity is there, then the explanatory function of this genetic base simply doesn't exist.

DR. DANIELLI: I really think that this is beating a dead horse. What Wilson and his colleagues talk about as altruism has very little to do with altruism in human beings, and there are much better ways of handling the biological aspect of human altruism than are used by Wilson.

The problem, as I see it, is not whether what Wilson has done is as applicable to human conduct as he maintains, but why Wilson and a few others have gone to such extraordinary lengths to try, as it were, to establish the dominance of this particular approach to life. I really think that Professor Gustafson's review of Wilson's book is much more illuminating about the situation than it is to consider the relevance of what Wilson did to human affairs.

DR. MIDGLEY: That doesn't tell us much because we don't know what Professor Gustafson said.

DR. DANIELLI: Well, he doesn't seem to be here.

DR. BOYER: Yes, he is. He's just behind you.

DR. GUSTAFSON: May I speak, Mr. Chairman?

DR. BOYER: Yes, please.

DR. GUSTAFSON: What I did was take the human nature book of Wilson and show that in the course of the development of the argument, he asked and provided an answer to every question that a systematic theologian has ever asked in the Western tradition. And so I interpreted it as a secular version of a systematic theology. I won't go into further detail on that, but that was the point that Dr. Danielli was picking up.

DR. MIDGLEY: I'm sorry, but I do think it would help if we could not have a fight with Wilson, but stick to the issues. I would just as soon the whole set of books had never come out. It is very easy to criticize them, and may I just ask that we have people put things in their own words and don't get into Wilson at all?

DR. PASSMORE: I think there is one feeling that many of us eighteenth-century figures have, and that is that we still regard individual liberty as tremendously important, and when we hear a phrase like enhancement of the one is enhancement of the whole we begin to shudder because we know what the implications of this tend to be.

I myself, however, would want to say that the mistake that was made was the supposition of the social aspects of this. In fact, the only way that you can become free is through voluntary associations from the family onward. Totalitarian societies regularly atomize people by trying to destroy every voluntary organization, everything that stands between the individual and the state. The limit of voluntary organizations is, in a sense, a crucial thing. If you have them, the liberty of individuals can almost be taken for granted. Without them we cannot survive.

Now, if people are left free, many of them will do things that we wish they wouldn't. It's a bit like what I said about imagination earlier. There is no humanly valuable sort of activity that I can think of that cannot in a sense sometimes be conducted in a manner that leads to things being done and consequences being brought about that we don't like.

Utopianism is precisely the doctrine that we can have everything that we want at once. We have never had it and we have to recognize, I think more fully than we know, what sort of costs are inherent in the existence of voluntary organizations. While we are doing that, we have to set against them the alternative paths for social organization, because costs are part of whatever form of social organization we have. But this might be regarded as a pessimistic view.

DR. WIERCINSKI: I would like only to mention that there exists a strange coincidence between the accelerated rate of biological individual variation and the trend toward democracy and the ideologies behind it. Of course, perhaps it is only coincidence, but I would like to mention it.

DR. TOULMIN: I do think that your question illustrates one of the reasons why some of us are unhappy about the general systems theory, because indeed you can lay out relationships which are significant in our understanding of social affairs in the kind of diagram that the general systems theory people offer us. But all the lines that join the boxes in general systems theory look exactly the same, even though, as we know very well, some of them represent causal interactions, some of them represent states of affairs which people take into account in deciding how to act, so that they provide people with reasons for acting.

For many of the things that Mary was saying it seems to me this distinction is crucial. I had a fascinating set of encounters in Chicago recently with Israel Goldiamond, who is a hard-line radical behaviorist

MARY MIDGLEY

who nevertheless insists that when he talks about conditioning within his behaviorism it doesn't matter whether the factors that are responsible for behavior change operate causally or whether they operate by serving as reasons.

As far as he is concerned, he is only interested in how people behave differently in consequence of uncertain changes in external conditions. For his purposes it maybe doesn't matter, but for the philosophical purpose nothing could be more important.

DR. WIERCINSKI: I'm sorry, it's not like that. First of all, when we make this scheme, all of the arrows denote that we write some words which say what kind of relationship is mentioned. And, secondly, as a systemist I would say always that any effect has two causes. One is input and the latter one is the specific activity on the system, and there are two questions which must be answered, one about the input from outside and another about how the reactivity of the system evolved.

DR. MIDGLEY: Perhaps I will say a word about this. I think behaviorism must bear a lot of the blame for the fact that people don't make the distinction Steven just mentioned. The whole terminology of conditioning is meant to abstract from the difference between doing a thing because you are kicked, as it were, downstairs and going downstairs because you have decided to.

Now, of course, from some rather remote perspective, if you are merely talking about the physical movements of objects, this doesn't matter. But if you are talking about human life, it matters very much. I have every sympathy with people in the humanities and the social sciences who go off the handle when they find this kind of distinction ignored.

I think behaviorist psychology and various other branches of the social sciences have aspired to look like physics and have talked in this way, causing a great rift between science and the arts which is utterly disastrous, and I want to make this kind of concession extremely fully. I am myself from the humanities. I spent a lot of time thinking about reasons. I think they are perfectly genuine parts of the world and nothing could matter more than whether somebody does something because he is forced to or because he thought he wanted to. But, you see, I don't think these two kinds of explanations compete. I think one must have room in one's thinking for both.

So for instance, I'm sorry, there are fifty sorts of examples. May I take—a genius. We find some exceptionally and extraordinarily gifted people in the world. Now it is quite habitual to think that this came in their birth. When you talk about genes you at once get unhappy because this mythology has been invented of these looking like the kinds of beings which are puppetmasters pulling us about on strings. But one doesn't think in that way if one says what was different about Shakespeare. One doesn't say he went to a good school and read some good books and met some interesting people. One cannot avoid saying he was different, for a start. Now, it is possible to be dualist and say they sent down a special kind of soul, but no serious religious thought compels one to think like that and I don't think it is very intelligible to do so. It seems clear to me that in the meat inside his head there was something a bit different from what there is in the rest of us. This is not in any way setting him up as a puppet who has a puppetmaster. Causes should not be anthropomorphized and thought of as mythical beings, powers, and fates which force us to do things. They should be thought of as the conditions of what happens.

I think this genius business is rather good when discussing determinism, because it is a kind of case where you cannot make the mistake of saying somebody made him be like that and it is not very easy to make the mistake of saying it could have been predicted. You see, they set all of the psychologists to work and they do all kinds of tests and he has psychotherapy and they come up with a report which says next week he will start a play called "Othello," and it will run as follows and they ought to be able to write it down.

Now that is poppycock and it is the case of the humanities to keep shouting that that is poppycock, and if the psychologists say we can't do it yet, they should be sent to do some homework or be kicked downstairs.

Now the other angle, you see, which one still doesn't wish to deny, I think, is that inside the meat inside the head something is happening when he does all this. He has got his hormones and he has his food and if you interfered with that lot, it couldn't be done. Knock him out very severely, feed him badly and the like, and you won't get your play or you will get a slightly different play.

That is to say the thing could be easily damaged from outside. Well, if that is so, then there was something physical there which could have been damaged. I am not being reductive, I am not putting this on any level except this sort of understanding of the play, which is a proper understanding. But I think its existence is quite compatible with it. I will bring in another related example, which is that of sex. We wouldn't get far, I guess, without a lot of hormones and the like, which we did not ourselves will and invent out of free existential choice. Had we not had them we couldn't have done it. That isn't to say that the explanation in terms of the hormones is a complete explanation or anything of that kind, but it does help if you want to understand what is going on.

MARY MIDGLEY

Now Freud, a clever fellow, guessed that there were hormones before anyone found them, and it wasn't odd that he could do this. And may I enter a point that was about this, that one couldn't reason as the sociobiologists do to this kind of cause unless one had the detailed biochemical apparatus, the observation of the genes. Of course, one can't get that and that's absolutely right. But because they talk about that sort of observation, they make asses of themselves. The kind of data which we do have and which have forced people to think in this way comes from the observation of children. Children do many things without having been conditioned. Children do both things which people are pleased with—play and the like, show affection—and things which they thoroughly object to and have been trying to stop spontaneously, all the time. Anyone who lives with children cannot but be struck by this.

Now if our response to children is caused by society, one has to do a most elaborate performance of saying, oh, but you were conditioning that subconsciously, your expectations, you see. Well, it can so happen that one knows what one's expectations were. One's first child was of a quite different type.

DR. WIERCINSKI: I would insist upon a presentation of individual liberty and democracy, even with Social Darwinism behind them.

DR. BOYER: This has been a long day and a rich session. I think I would do a service now by communicating that we are deeply in debt to those who have made major presentations and I am struck that there has been an important theme that has stuck with us throughout— with only the wish perhaps that now we could return again to the discussion of education, because the issues that were raised could now be thought of much more carefully. As we have looked at our deliberations about the nature of the individual and about the issues of the individual in context, our view of that, it seems to me, is all important when we now return to the question of educating for the future.

Indeed, as one who spends more time in education than perhaps with some of the other issues that were raised, I think we are impoverished in our schools and in our colleges precisely because we haven't engaged in the weighty matters that have been presented here. How do we view the individual? How does the individual respond to context?

I was impressed that two speakers took as their themes the limits of individualism and the ethics of interdependence, which, in some very important ways, represent interesting views toward a kind of converging theme. Now, how do those of us who care about, both formally and informally, transmitting whatever we consider to be culture and, more important, freeing the same kind of creativity that Professor Passmore mentions, how do we take this new kind of discussion we have had and convert it into anything like a reasonable educational procedure for the future? I think the schools and colleges have an enormously important obligation, to say nothing of the informal teachers in our culture, mass communication included.

On behalf of all of us I wish very much to thank Professor John Passmore for his presentation and Professor James Gustafson and also Professor Mary Midgley for what I think to be an absolutely first-rate series of papers that have brought us together to think carefully about essential issues.

Epilog:

Adaptability: A Curse or a Technique of Survival?

WILTON S. DILLON

Director, Office of Smithsonian Symposia and Seminars, Smithsonian Institution

During the 1973 Smithsonian Symposium, "The Nature of Scientific Discovery," honoring Copernicus on his 500th birthday, Professor John U. Nef, founder of the Committee on Social Thought at the University of Chicago, told the following story[1] that deserves repeating in the context of "How Humans Adapt." (I add this to what we take away from this extraordinary pooling of human talent as we try to adapt to a future celebrated by René Dubos.)

Years ago, Gertrude Stein came to the University of Chicago. According to a report—I wasn't there—she went to dinner with the president of the University, Robert Hutchins, and there was a long discussion. As she was going down the stairs leaving with Alice B. Toklas behind her, Alice B. Toklas turned back and loudly said for Hutchins and others to hear, "Gertrude has said things tonight that will take her years to understand."

The point of remembering this charming anecdote is to praise the occasional virtue of *misunderstanding* language or to realize that the same words evoke different responses like Rorschach ink blots. Words reflect a *Rashomon*-like variation in how we perceive reality according to the values we bring with us into a symposium. For example, Professor Kenneth Boulding, as already noted by Donald Ortner in the Preface, had no place in his rich repertoire of language for the concept "adaptation." It implied spineless, medusan adjustment to morally unacceptable conditions, something like being a "good soldier," suppressing dissent. As a prophetic social critic, pacifist, and philosopher of political economics, Boulding's discomfort remains, for me, one of the major legacies of the exercise.

"I don't want to adapt. I came to this country to avoid adapting to life in England," said Boulding the last morning of the symposium in the Regents Room of the Smithsonian Castle . . . built with money given the United States by the Englishman, James Smithson. A fellow Englishman of an earlier century had given Boulding and others a forum for piquant assertions. Boulding was carrying out that great British cultural civilizing mission of "putting the cat among the pigeons." He wanted to disturb, to challenge, to question—and adaptation meant compromise. The biological underpinnings of the symposium were less familiar to an economist.

Coincidentally, in the same room a few weeks later I convened some advisors to help plan the eighth international symposium to honor George Orwell by producing a book, *Closed and Open Societies: 1984*. Still under the influence of how humans adapt, I asked whether scholars have any insights and data about the speed with which a society might adapt to military rule, as witness the case of Poland in the wake of the Solidarity movement. We also had discussed the sources of motivation of the El Salvador rebels to fight. Two of my consultants spoke immediately of "the curse of adaptability," and advised me that any symposium on closed and open societies could not neglect the adaptability factor as a key to closing up a society or to maintaining long-term social, political, or military controls over a population. I thought that Professor Boulding had marched smartly right back to the table. We were recycling his views.

Abbott Gleason, historian, an Orwell buff and then secretary of the Kennan Institute for Advanced Russian Studies at the Woodrow Wilson International Center for Scholars, had to leave the meeting. His exit line: "Remember, the curse of adaptability."

Such cries for resistance to tyranny or arbitrary authority must have been heard throughout human history, in preliterate, preindustrial, and postindustrial societies. Yet individual or collective freedom may be a new concept in human affairs. I wonder what clues could be found in the legacies of these societies about the timeless debate over techniques of survival. Was survival consciously debated or was it a suicidal game of following the leader or a deity, as in Masada or Jonestown? The Ik people of Uganda, in Colin Turnbull's *The Mountain People*,[2] survived their transplantation by penalizing the young and the old for consuming scarce food. Outside observers can analyze this as adaptation and see patterns not beheld by people caught up in their respective food chains, thus indifferent to tribal council agenda ques-

WILTON S. DILLON

tions like "Eat now, starve later." At what point does one adapt to cannibalism or dare to violate an incest taboo? Is it a luxury of modern humankind that we possess weapons for total self-destruction, designed by our humanoid opposable thumbs, and use human brainpower and moral persuasion to prevent their use? Jonathan Schell's *The Fate of the Earth*,[3] on nuclear war, picks up where René Dubos's untimely exit left off, but does not ask Dubos's other implied question, "Can we survive the malaise of unemployed youth?"

Those of us who still have a little relief from constant hunting and gathering for food—like the playful otter after some energizing protein—can use our leisure to sort out what was said or neglected at the symposium. We need to ask ourselves whether adaptability is a cure or a technique of survival.

Such a heavy agenda can be digested only with some lightheartedness. I believe that one can still be responsible and try to get through each five-minute segment of our lives in "the most pleasant manner possible." Remembering what Alice B. Toklas pleasantly said about Gertrude Stein, indeed, it may takes years for us to understand what we said or ought to have said.

Notes

1. See Owen Gingerich, ed., *The Nature of Scientific Discovery* (Smithsonian Institution Press, Washington, D.C., 1975), p. 501.
2. Simon and Schuster, New York, 1972.
3. See the *New Yorker* magazine, February 1, 8, and 15, 1982.

Index

Armageddon, 301
Asimov, Isaac, 372
Atomic age, 30
Australopithecines, 34
Authoritarianism, 428, 429, 434, 436
Automation, 447
Axes, iron, 18
Ayurveda Hindu medicine, 295
Aztecs, 271

Bacteria, as food, 247
Balkan peninsula, 124
Bandkeramik people, 111
Bands, tribal, 72
Barefoot doctors, 292, 312
Bean, 229
Beaumarchais, Pierre, 183
Becker, Gary, 100
Behavorial norm, 416
Behaviorism, 543
Benghazi, 17
Bernard, Claude, 24
Big government, 425
Biocultural interaction, 167
Bioethics, 7
Biological heterogeneity, 169
Biological mechanisms of regulation, 104
Biological niches, 28
Biospheric catastrophe, 17
Biotechnology, 263
Bipedalism, 35, 36, 194
Birth control, 51, 113, 218
Birthrate, 4; decline, 320; regulation, 109, world, 323
Birthspacing, 43, 47, 61
Bisa tribe, Zambia, 286
Bitter cassava, 233
Black Death, 118
Blackfoot Indians, 186
Blood type, 73
Boaz, Franz, 184
Bonding, female-male, 193
Bounty mutineers, 111
Brazil, 71
Brazilia, 393
Breakdown of comunity, 426
Breast feeding, 61; in Europe, 62
Bronze Age, 102
Bureaucracy, 24, 26
Burt, Sir Cyril, 105
Bushmen, 113, 228
Butchering, 35
Buto, 201

Calhoun, J. B., 147
Caloric balance, 258

Calvin, John, 496
Camus, Albert, 376
Capital city in nineteenth century, 383
Capitalism, 141, 431, 524
Caribbean, 24
Carrying capacity, 121, 146
Catastrophe, 225; role in evolution, 403
Cattle, 229
Caucasus, 221
Celestial mechanics, 396
Central America, 71
Cereal grasses, 199
Cereals, 231
Chalcolithic, 139
Chaldeans, 396
Charismatic leader, 405
Chellean hand-axe, 88
Children: hunter-gatherer, 113; learning potential, 477
Child-spacing, 61, 208
Chilean governmental model, 427
Chili pepper, 229
Chimpanzee, 34, 36, 38, 39, 68
Choukoutien, 196
Christian church, 494
Christian, John, 147
Christian morality, 353
Chromosomal damage, 78-79
Chromosomes, 69
Chuña, 233
Cisterns, Rome, 221
Cities, 371, 390; abandonment of, 139; adaptations of, 388; advantages of, 377, 391, 393; evils of, 376; growth of, 385, 391; industrial, 378, 379, 381; relationship with country, 374; relationship with nation, 387; Victorian, 382
Civilization, 371
Civil rights movement, 441
Club of Rome, 14
Coadaptation, 83
Coal, 211
Coefficient of inbreeding, 75
Coefficient of relatedness, 95, 155
Cognition, 417
Cognitive capital, 480
Coitus interruptus, 114
Collectivism, 141
Common good, 494, 502, 504
Communal life, 25
Communism, 141
Competence, 259
Competition, interdeme, 76
Computer-assisted manufacturing, 447
Computer science, 418
Conjectural history, 344, 369
Conservation, 400

552

Index

Mother-infant relations, 36
Moral infallibility, 502
Moral theory, 496
Morbidity, 260
Morgan, Louis Henry, 184
Morris, William, 19
Mortality, 211, 260, 323, 329; infant, 14, 137, 208, 308, 408; infant-child, 296
Moslem, 65; Empire, 206; Hakims, 295
Moundville, Ohio, 185
Mumford, Lewis, 372, 381, 388
Murdock, George Peter, 184
Mutation, 84, 130, 131
Mutual defense, 35
Mutual harm, 152
Mutualism, 170

Nagasaki, 2, 85
Napoleon III, 372
National defense, 410
National security, 432
National security state, 432
Natural laws, 354
Natural past, 17
Natural populations, 121
Natural selection, 84, 102, 106, 164, 168, 175
Nature-nurture dichotomy, 104
Naturwissenschaften, 17
Neanderthal man, 115, 197, 290
Near East, 110, 139
Necropolis, 372
Neel, James V., 93, 97, 100
Negentrophy, 220
Nemeskeri, J., 367
Neoconservative entitlement theory, 327
Neo-Epicureanism, 31
Neolithic, 102, 111, 113, 117, 139, 192, 203, 206, 210, 216-17, 219
Neo-Stoic, 31
Nepotism, 151, 152
Neurobiology, 159
New collecting society, 220
New Left, 19
New Lives for Old, 30
New Right, 149
New Yam Festivals, 152
New Zealand, 347-48
Niche specialization, 170
Nietzsche, Friedrich, 523, 526
Nigeria, 114
Nile Valley,
Nitrogen balance, 235
Nomadism, 11, 12, 18, 139
Nonutilitarian products, 202, 204
Noumenal normative rules, 354, 358, 363
Nuclear age, 174

Nuclear energy, 6
Nuclear family, 97, 193, 356
Nuclear war, 1, 85, 409, 429, 449; probability of, 410, 412, 420; weaponry, 214, 448, 451, 470
Nucleotides, 69
Nutrition, 61; deficiencies in, 236; human requirements, 236, 255; stress in, 44

Obligatory urinary nitrogen loss, 235
Odysseus, 87
Ogalala aquifer, 190
Olduvai, 57
Oligarchs, 431
Ontjom, 233
OPEC, 402
Orwell, George, 139, 436, 459
Ostia, Rome, 221
Osteomalacia, 140
Overcrowding, 148
Overcultivation, 230
Overexploitation, resource, 198, 213
Overgrazing, 230
Overpopulation, 4, 118, 137, 146, 164, 191, 193, 197, 205, 212, 216
Overseas Development Organization, 315
Ovulation, 48
Oxford tutorial system, 486

Pacemaker, 311
Paleodemography, 338, 367
Paleoenvironment, 58
Paleolithic, 117, 216
Paleopathology, 147
Paramedical personnel, 293
Parental investment, 43, 46, 49, 51
Parent-child relationships, 330
Parenting behavior, 43
Pareto principle, 507
Parsons, Talcott, 23
People's Temple, 301
Pesticides, 231
Phenotype, 22, 26
Phoenician, 206
Physician, roles for, 309
Picasso, Pablo, 185
Pitcairn Island, 111
Plague, 140
Play and playfulness, 29
Pleistocene, 40, 58, 60, 196, 197, 214
Pliocene, 40, 58
Plio-Pleistocene, 33
Plowshares, 18
Poland, 101
Pollution, 4, 63, 118, 164, 222; air, 5; water, 5
Polygamy, 114